D1000673

Manufacturing and Automation Technology

R. Thomas Wright
Professor Emeritus, Industry and Technology
Ball State University
Muncie, Indiana

Publisher
The Goodheart-Willcox Company, Inc.
Tinley Park, Illinois
www.g-w.com

Manufactured in the United States of America

Library of Congress Catalog Card Number 2004060915

International Standard Book Number 1-59070-484-3

1 2 3 4 5 6 7 8 9 – 06 – 09 08 07 06 05

Library of Congress Cataloging-in-Publication Data

Wright, R. Thomas.
Manufacturing and automation technology / R. Thomas Wright.
p. cm.
ISBN 1-59070-484-3
1. Manufacturing processes—Automation. I. Title.

TS155.63.W75 2005
670.42'7—dc22 2004060915

Safety note. Safety equipment shown in this text may have been removed or opened to clearly illustrate the product and must be in place prior to operation.

Front Cover: These welding robots are fitting automobile parts together on the maunfacturing assembly line.
(Ford Motor Company)

Introduction

Manufacturing and Automation Technology provides complete study of the basic elements of manufacturing and automation and how they affect the world that we live in. It covers the materials, processes, and management techniques used in the industry.

Textbook Organization

This textbook is organized into twelve units of study representing the entire range of manufacturing. These units consist of:

- Introduction to Manufacturing
- Manufacturing Materials
- Primary Manufacturing Processes
- Secondary Manufacturing Processes
- Establishing a Manufacturing Enterprise
- Designing and Engineering Products
- Developing Manufacturing Systems
- Manufacturing the Product
- Marketing the Product
- Performing Financial Activities
- Automating Manufacturing Systems
- Manufacturing and Society

As you use this textbook you will learn that manufacturing is a managed system that draws upon many resources. You will explore a number of materials and material processing techniques common to manufacturing. You will have the opportunity to change the form of materials using material processing techniques. You will also study how people efficiently use resources to manage these activities. You will find out how activities are managed. You may then apply this newly developed knowledge in a realistic setting. You will form a company to design, produce, and market a product. You will be introduced to the concepts of leadership and entrepreneurship while participating in these activities.

Each chapter begins with a list of objectives. You should review these objectives before studying the chapter. New manufacturing-related terms appear in ***bold-itals***. In the margin these terms are defined in a running glossary. At the end of each chapter, a list of these terms is given so you can review. Also given at the end of each chapter are review questions covering the material in the chapter. Several end-of-chapter activities are given so that you can apply the knowledge gained in the chapter just studied.

Special Features

As you use this textbook you will find a number of boxed features. These features were designed to give you additional information about manufacturing and how it relates to the world are around you. The three features include the following:

- **Academic Links.** These provide additional information to highlight how mathematics, science, communications, and social studies are part of the manufacturing world.
- **Career Links.** These provide additional information about careers that are related to manufacturing.
- **Technology Links.** These provide additional information on how manufacturing relates and works with other technologies, such as transportation, communications, agriculture, energy and power, construction, and medical technology.

Safety and Measurement in Manufacturing

Before you begin the study of manufacturing, it is necessary for you to review a few basic concepts from your earlier classes. These concepts can be divided into two groups: safety and measurement.

Safety in Manufacturing

Life is full of hazards. As you move through the day, you encounter situations that could cause harm to you. A car could hit you as you cross the street. You could fall or slip as you climb stairs. However, you can reduce the chance of being hurt by acting responsibly.

Likewise, the manufacturing environment has many hazards. In factories, there are machines and materials being used by people. There are sharp cutters, moving shafts, hot materials, fumes, and many other conditions that could harm workers. The chance of injury can be reduced through thoughtful action. This practice is called safety. It is all the actions people take to reduce the likelihood of personal injury. These safety practices can be grouped into three categories. The categories are safety with people, safety with machines, and safety with materials.

Safety is a state of mind and a series of actions. A person must be concerned about safety. He or she must make safety a high priority. How often have you seen people driving, playing a sport, moving through a building, or working with a tool without regard for safety? We often say that these people "aren't using common sense." They are not directing their actions in responsible ways.

Safety is also a set of actions. There are ways to safely complete tasks. For example, a football coach teaches players to block safely. There are ways to do the task that reduce the chance of injuries. Likewise, there is a series of practices that can make working in manufacturing safer. Often, these safety practices are communicated through rules. These rules, like driving rules, tell how to safely complete a task.

The following sections outline safety rules that will help you review and develop safe work habits. You may want to photocopy them so that you will have a set with you as you work on the manufacturing activities in this course. These are general rules, and they apply to many manufacturing settings and many material-processing activities. More specifically, safety precautions will be presented as your teacher introduces each laboratory activity.

Safety with people—personal safety

1. Concentrate on your work. Watching other people and daydreaming can cause accidents.
2. Dress properly. Avoid loose clothing and open shoes. Remove jewelry and watches.
3. Control your hair. Secure any loose hair that may come into contact with moving machine parts.
4. Protect your eyes with goggles or safety glasses.
5. Protect your hearing by using hearing protectors around loud or high-pitched machines.
6. *Do not* use compressed air to blow chips and dirt from machines or benches.
7. Think before acting. Always think of the consequences before starting an action.
8. *Ask* your teacher questions about any operation of which you are unsure.
9. Seek First Aid for any injuries.
10. *Follow* specific safety practices demonstrated by your teacher.

Safety with people—safety around others

1. Avoid horseplay. What you consider as innocent fun can cause injury to other people.
2. *Do not* talk to anyone using a machine. You may distract him or her and cause injury.
3. *Do not* cause other people harm by carelessly placing materials and tools on benches or machines.

Safety with materials

1. Handle materials properly and with care.
   a. Watch long, moving material. *Do not* hit people and machines with the ends.
   b. Use *extreme* caution when handling sheet metal. The thin, sharp edges can easily cut you and others.
   c. Grip hot materials with pliers or tongs.
   d. Wear gloves when handling hot or sharp materials.
   e. Use *extreme* caution in handling hot liquids and molten metals.
2. Check all material for sharp burred edges and pointed ends. Remove these hazards whenever possible.
3. Place material that gives off odors or fumes in a well-ventilated area, fume hood, or spray booth.
4. Lift material properly. *Do not* overestimate your strength.
5. Dispose of scrap material properly to avoid accidents.

6. Clean up all spills quickly and properly.
7. Dispose of hazardous wastes and rags properly.

Safety with machines—general rules

1. Use only sharp tools and well-maintained machines.
2. Return all tools and machine accessories to their proper storage locations.
3. Use the *right* tool and machine for the *right* job.
4. *Do not* use any tool or machine without permission or proper instruction.
5. Use ALL machine guards.
6. Stop all machines to make measurements and adjustments.
7. *Do not* leave a machine until the cutter has been stopped.
8. Whenever possible, clamp all work.
9. Unplug all machines from the electrical outlet before changing blades or cutters.
10. Remove all chuck keys or wrenches before starting machines.
11. Remove all scraps and tools from the machine before using it.
12. Remove scraps with a push stick.

Safety with machines—casting and molding processes

1. *Do not* try to complete a process that has not been demonstrated to you.
2. Always wear safety glasses.
3. Wear protective clothing, gloves, and face shields when pouring molten metal.
4. *Do not* pour molten material into a mold that is wet or contains water.
5. Carefully secure two-part molds together.
6. Complete casting and molding processes in well-ventilated areas.
7. *Do not* leave hot castings or mold parts where other people could be burned by them.
8. Constantly monitor material and machine temperatures during casting and molding processes.

Safety with machines—separating processes

1. *Do not* try to complete a process that has not been demonstrated to you.
2. Always wear safety glasses.
3. Keep hands away from all moving cutters and blades.
4. Use push sticks to feed all materials into wood cutting machines.

Safety with machines—forming processes

1. *Do not* try to complete a process that has not been demonstrated to you.
2. Always wear safety glasses.
3. Always hold hot materials with pliers or tongs.
4. Place hot parts in a safe place to cool. Keep them away from people and from materials that will burn.
5. Follow correct procedures when lighting torches and furnaces. Use spark lighters, *not* matches.
6. *Never* place hands or foreign objects between mated dies or rolls.

Safety with machines—finishing processes

1. *Do not* try to complete a process that has not been demonstrated to you.
2. Always wear safety glasses.
3. Apply finishes in properly ventilated areas.
4. *Do not* apply finishing materials near an open flame.
5. Always use the proper solvent to thin finishes and clean finishing equipment.

Safety with machines—assembling processes

1. *Do not* try to complete a process that has not been demonstrated to you.
2. Always wear safety glasses.

3. Wear gloves, protective clothing, and goggles for all welding, brazing, and soldering operations.
4. Always light torches with spark lighters.
5. Handle all hot materials with gloves and pliers.
6. Perform welding, brazing, and soldering operations in well-ventilated areas.
7. Use proper tools for all mechanical fastening operations. Be sure screwdrivers and hammers are the proper size.

Safety note: Safety equipment shown in this text may have been removed or opened to clearly illustrate the product and must be in place prior to operation.

Federal and state aspects

Both federal and state laws provide guidelines for the safe operation of tools and equipment. Federal safety regulations are covered in the Occupational Safety and Health Act of 1970. This act (law) gave the Occupational Safety and Health Administration (OSHA), the mission "to assure so far as possible every working man and woman in the nation safe and healthful working conditions and to preserve our human resources."

OSHA's functions include setting standards, inspecting workplaces, leveling citations and penalties for violations, consultation, and conducting training and educational programs.

Employers must comply with OSHA standards by providing and maintaining a safe work environment for workers. This means that workers must be protected from job-related hazards and excessive noise levels. A clean work environment as well as proper lighting and ventilation are also required by OSHA.

Sometimes states develop and operate their own job safety and health programs. These programs must be at least as effective as the federal program. In addition to OSHA standards, many states have additional guidelines that must be met.

Measurement in Manufacturing

All materials have a basic size, shape, or volume. These characteristics can be measured and calculated. This study of manufacturing systems will focus on materials that have a solid structure. These materials, as you will learn, are called engineering materials. In manufacturing, the form of these materials is changed to make them more useful. The materials are reshaped using manufacturing processes. Throughout these processing activities, the size and shape of materials are measured.

Two basic systems of measurement are used worldwide. The one most commonly used in manufacturing is called the SI (International System, or in French, the Systeme Internationale) or metric system. It is based on the meter. This unit is then divided into decimal parts. A centimeter is 1/100 of a meter and the millimeter is 1/1000 meters.

The United States has maintained a separate system called the Conventional system. This system uses the foot as the standard unit. It is divided into twelve equal parts, each of which is called an inch. The inch can be divided into fractional parts. Commonly, we use quarters (1/4 inch), eighths

(1/8 inch), sixteenths (1/16 inch), thirty-seconds (1/32 inch), and sixty-fourths (1/64 inch).

Either of these systems of measurement can be used in the two basic types of measurement. These types are standard and precision.

Standard measurement is widely used in the construction industry, in furniture manufacture, and in sheet metal fabrication. The standard Conventional unit of measurement is the inch and its fractions.

Precision measurement is used for all manufacturing where close tolerances, fits, and dimensional control are required. This type of measurement uses thousandths (1/1000) or ten-thousandths (1/10,000) of an inch.

Directly related to measurement are the names of surfaces and dimensions of a part. These, as reviewed in **Figure In-1**, are:

- **Thickness:** The smallest dimension of a piece of material. It is usually the distance between the two sides or faces.
- **Width:** The second smallest dimension of a piece of material. It is usually the distance between the two edges.
- **Length:** The largest dimension of a piece of material. It is usually the distance between the two ends.
- **Side:** The largest surface of a material.
- **Edge:** Generally, the second largest surface. For wood, it is the second largest surface that does not show end grain.
- **End:** Generally, the smallest surface. For wood, it is the surface that shows end grain.

When giving a part size, convention suggests that the thickness is given first. Then the width is given and, finally, its length. Therefore, if a part is said to be 1/2" × 3" × 10", its thickness is 1/2 inch, its width is 3 inches and its length is 10 inches. Also, if there are no measurement marks such as (") for inches or (') for feet, the measurement is understood to be in inches.

To review your knowledge, select several items such as your textbook, laboratory manual, etc. Measure their width, thickness, and length using SI and Conventional measurements.

Now that you have reviewed safety and measurement, you are ready to start a challenging study of manufacturing. Good luck!

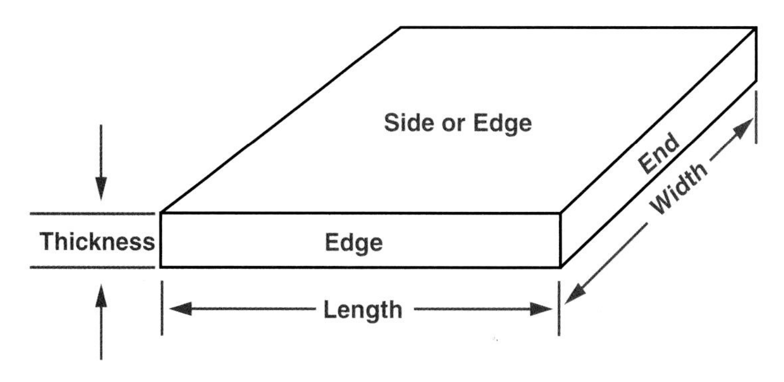

Figure In-1

About the Author

Dr. R. Thomas Wright is one of the leading figures in technology education curriculum development in the United States. He is the author or coauthor of many Goodheart-Willcox technology textbooks. He is the author of ***Manufacturing and Automation Technology, Processes of Manufacturing, Exploring Manufacturing,*** and ***Technology***. He is also the coauthor of ***Exploring Production*** and ***Technology: Design and Applications***.

Dr. Wright has served the profession through many professional offices including President of the International Technology Education Association and President of the Council on Technology Teacher Education. His work has been recognized through the ITEA Academy of Fellows Award and the Award of Distinction, the CTTE Technology Teacher Educator of the Year, Epsilon Pi Tau Laureate Citation and Distinguished Service Citation, the Sagamore of the Wabash Award from the Governor of Indiana, the Bellringer Award from the Indiana Superintendent of Public Instruction, the Ball State University Faculty of the Year Award and the George and Frances Ball Distinguished Professorship, and the EEA-Ship Citation.

Dr. Wright's educational background includes a bachelor's degree from Stout State University, a master of science degree from Ball State University, and a doctoral degree from the University of Maryland. His teaching experience consists of three years as a junior high instructor in California and 37 years as a university instructor at Ball State. In addition, he has also been a visiting professor at Colorado State University, Oregon State University, and Edith Cowan University in Perth, Australia.

Contents at a Glance

Table of Contents

Unit 1 Introduction to Manufacturing

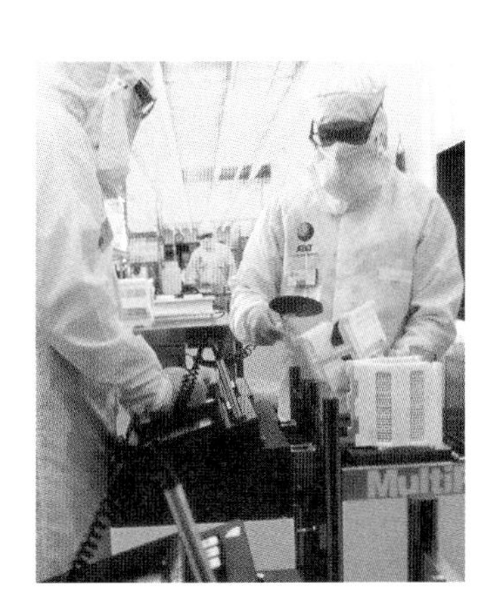

Unit 2 Manufacturing Materials

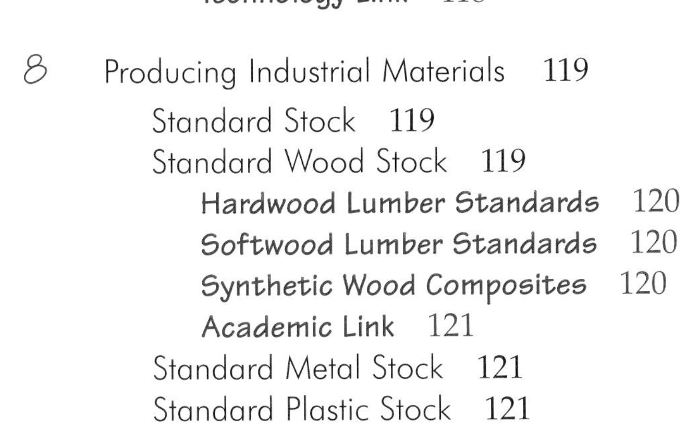

Unit 4 Secondary Manufacturing Processes

ESAB

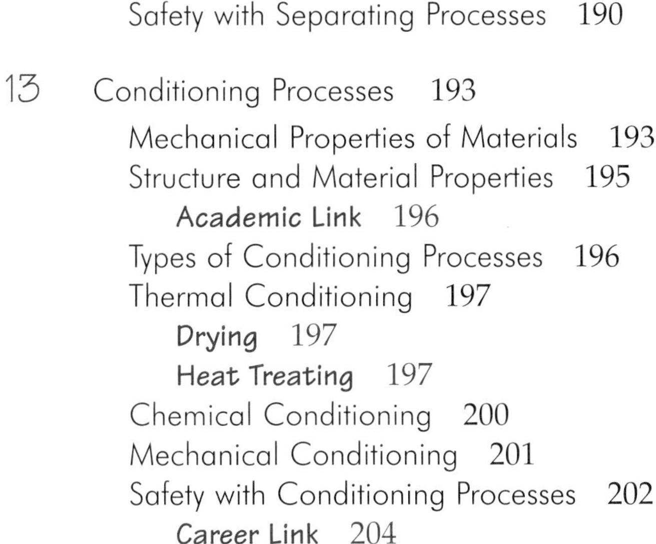

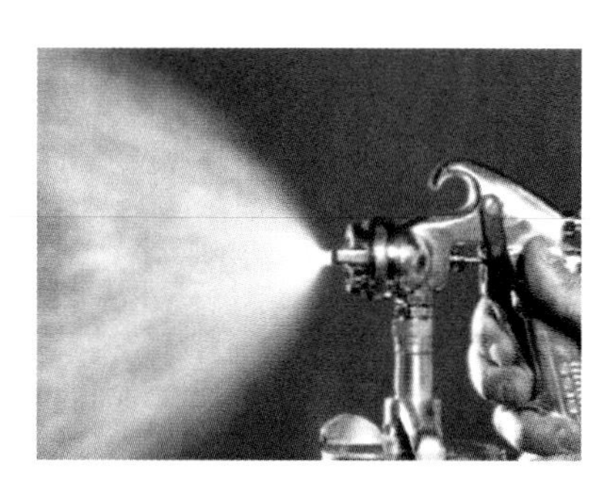

Unit 5 Establishing a Manufacturing Enterprise

Unit 7 Developing Manufacturing Systems

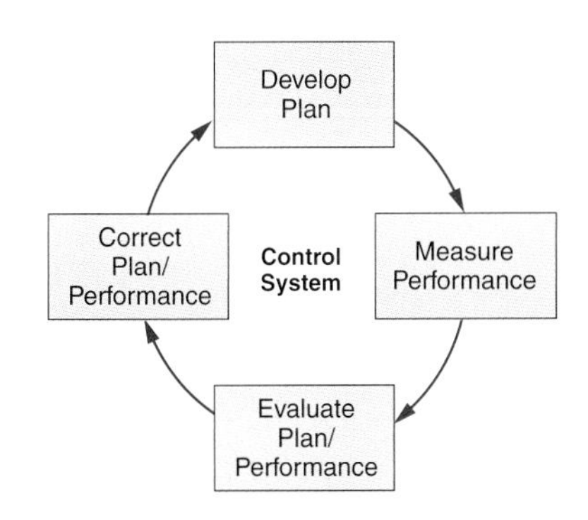

Unit 8 Manufacturing the Product

Unit 9 Marketing the Product

Unit 11 Automating Manufacturing Systems

Unit 12 Manufacturing and Society

Features

Academic Links

Technology Links

Career Links

Unit 1 Introduction to Manufacturing

Chapter 1
Technology and Technological Systems

Objectives

After studying this chapter, you will be able to:

- ✓ List and define the four major groups of knowledge.
- ✓ Explain how technology has an impact on your life every day.
- ✓ Define the five essential elements of a technological system.
- ✓ Give examples of technological systems.
- ✓ List and describe the four major groups of technological systems.

You live in a complex world. This world is a combination of natural and manufactured components. For instance, look at **Figure 1-1.** The hills, water, and sky occur in nature. These are all natural elements. The dam and power plant in the background are manufactured elements. A third element is unseen in the picture. It is human will. People felt that it was appropriate to dam the river to produce electricity.

Figure 1-1. Can you see both natural and manufactured elements in this photograph?

Major Groups of Knowledge

People have developed four different types of knowledge. These can be used to explain and understand the natural and manufactured parts of the world. These groups of knowledge are science, technology, the humanities, and descriptive studies. See **Figure 1-2.**

Science

In studying *science*, you can gain an understanding of the laws and principles that govern the physical universe. See **Figure 1-3.** You may learn about meteorology, biology, mechanics, geology, astronomy, and many other topics.

science. The study of the laws and principles that govern the physical universe.

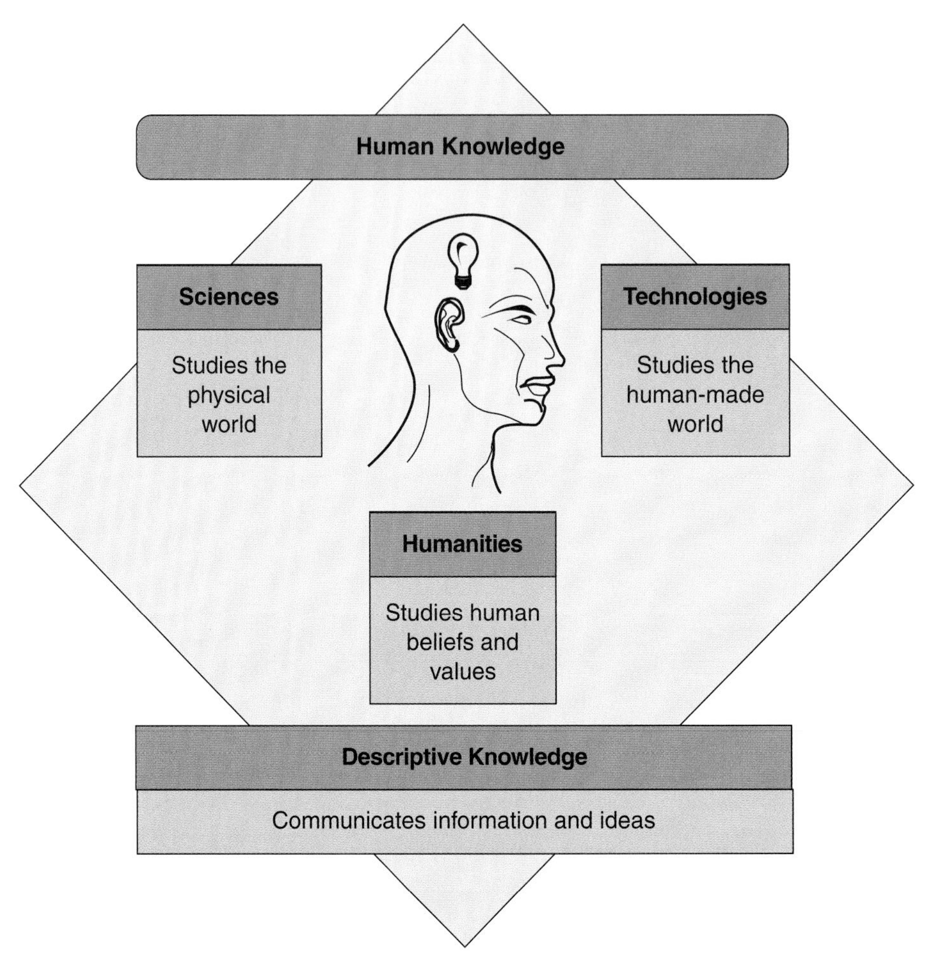

Figure 1-2. To understand the world around us, we must study four major types of knowledge: science, technology, the humanities, and descriptive knowledge.

Figure 1-3. A—Science helps explain how severe windstorms, like the one that damaged this barn, are created. B—Science helps explain how this space shuttle travels into space. (NASA)

For example, refer back to Figure 1-1. You could use science to learn about the natural cycle that causes rain. Through science, you would also gain an understanding of water flow in a river. You could learn about the energy captured in a reservoir. You could gain an understanding of the operation of the dam's turbine and generator.

Technology

There is also knowledge about the things that people make. To develop this knowledge, people study *technology*. Technology is the study of how humans develop and use devices that extend their physical abilities. Technology is information about how people use devices to grow crops, communicate information, treat illnesses, construct housing, use tools, and move goods. It is the study of the development, production, and use of devices that improve the quality of life. The study of technology focuses on how humans design items to change or control the world around them. See **Figure 1-4.**

Figure 1-4. Technology is used to make things that people want and need, such as this house.

technology. **The study of how humans develop and use items to change or control the world around them.**

Technology provided the system used to construct the dam and power plant. It was also used to construct and install turbines and generators in the plant shown in Figure 1-1. It was used to build roads to the facility and power lines that carry the electricity away from the plant.

Humanities

The study of *humanities* relates to areas of philosophy and human concern. As people study the humanities, they develop personal priorities. These help to guide a person in deciding what is and is not desirable. For instance, a study of humanities could help you come to some conclusions about the "proper" use of natural elements such as our rivers. You could take a stand on how much of it should remain "untouched." Should there be free access to the rivers? Should private homes be built on its banks? Should powered boats be allowed? Should fishing be allowed?

humanities. **The study of areas of philosophy that relate to human concerns. Studying the humanities helps people develop personal priorities.**

These kinds of decisions, however, should be made from a knowledge base. People should not decide these issues on emotions. To take an informed stand on an issue, people must understand the scientific and technological principles involved. Issues of proper use of nature and land, rivers, and oceans must balance environmental and economic issues.

Descriptive Knowledge

Underlying all of these studies is the fourth area of human knowledge. This is called *descriptive knowledge*, and includes language and mathematics. These are the "tools" you use to describe what you see and know. Descriptive knowledge is the basis for communication. As you study this book, topics are described through words, drawings, and formulas. Descriptive knowledge allows you to study and understand aspects of science, technology, and the humanities.

descriptive knowledge. **The area of human knowledge that includes language and mathematics. It is the basis for communication.**

Technology

Whether you realize it or not, technology has an impact on your life every day. Technology plays a part in almost all aspects of your life. It impacts the food you eat and the clothes you wear. It affects the home in which you live and the book you are now reading.

What is technology? There seems to be as many definitions for technology as there are people trying to define it. However, some basic elements are included in most of these definitions. These elements, as shown in **Figure 1-5,** suggest that:

- Technology is human knowledge.
- Technology is concerned with tools, systems, and procedures.
- Technology is used to create physical artifacts (structures and products).
- Technology attempts to modify (change) and control the natural and manufactured environment.

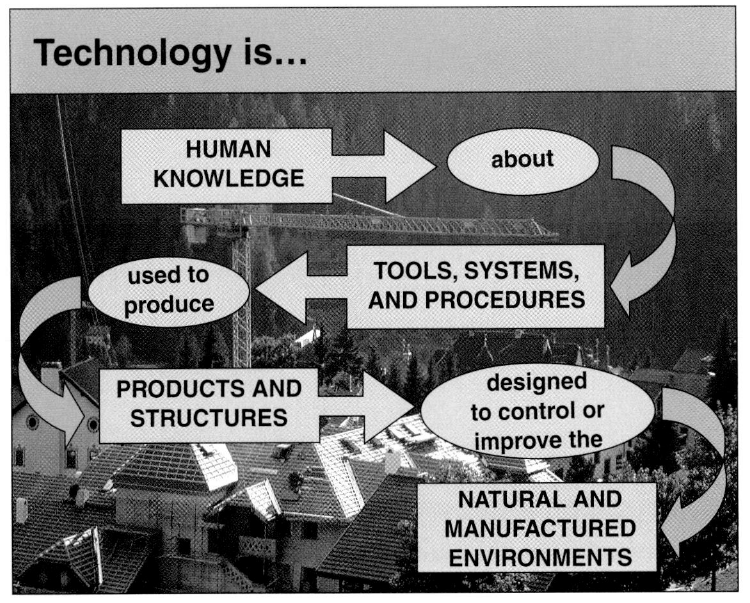

Figure 1-5. These are the common elements in the definition of technology.

Technology is the way people continually extend what they might be able to do, to what they actually can do. This is their potential. It magnifies human ability. Through technological systems, we can build better structures. Tools allow us to shape and form materials into products faster, cheaper, and with better quality. Other technological systems extend our ability to communicate. Devices can convey our ideas and knowledge over long distances. Other devices can store information and retrieve it later. Through technology, we have vehicles that can move people and cargo effectively. Communication and transportation technologies seem to have "shrunk" the world. We can travel long distances in a relatively short time. Information comes to us from all parts of the globe almost instantly. See **Figure 1-6.**

Figure 1-6. Communication and transportation technologies make faraway places seem close to us and easier to access.

Often, technology is thought to be complex tools and machines such as computers, robots, and fiber optic communication devices. In fact, they are often called *"hi-tech"* (high technology) devices. However, these are just the hardware. They are the tools of technology. Technology encompasses (takes in) the complete system. Technology is the knowledge of computer design, material processing, information communication and storage, energy utilization, tools, and a great deal more. It includes knowledge about designing, using, and maintaining products and structures (buildings, roads, dams, etc.). Technology also includes knowledge about appropriate

Figure 1-7. These are examples of the evolving technology of the plow. (Deere and Co.)

actions. It considers the positive and negative impacts of manufactured systems and products. A simple computer-controlled machine will use all of this knowledge and more.

Since technology is always changing, hi-tech is a relative term. This is because a level of technology is related to a point in history. Look at the images in **Figure 1-7.** Each of these represents high technological advancement at a point in agricultural history. The wooden plow replaced the digging stick and represented a major technological advancement. The wooden plow was considered to be a hi-tech device in its day. At the time they were developed, the other three plows were also considered to be hi-tech devices.

The term, hi-tech, is also related to the development of a specific country or region. Not all countries of the world are at the same stage of technological advancement. That is why each of the plows shown may well be a hi-tech device today; depending upon where in the world you find it.

Another example of technological change can be seen in photography. Look again at Figure 1-7. The earlier photographs were only available in black and white. Now color photographs are quite common, even though they were considered to be hi-tech in the past.

It is probably better to think of technological advancements as *evolving technologies*. They will be hi-tech for a period of time but will then be replaced with an even more hi-tech device. Many people can remember when television, the computer, passenger aircraft, plastics, radar, and the dial telephone were first developed. Each of these devices was a major technological advancement. Can you think of new evolving technological devices that have been developed in your own lifetime?

evolving technologies. **Areas of technology that are undergoing continuing change.**

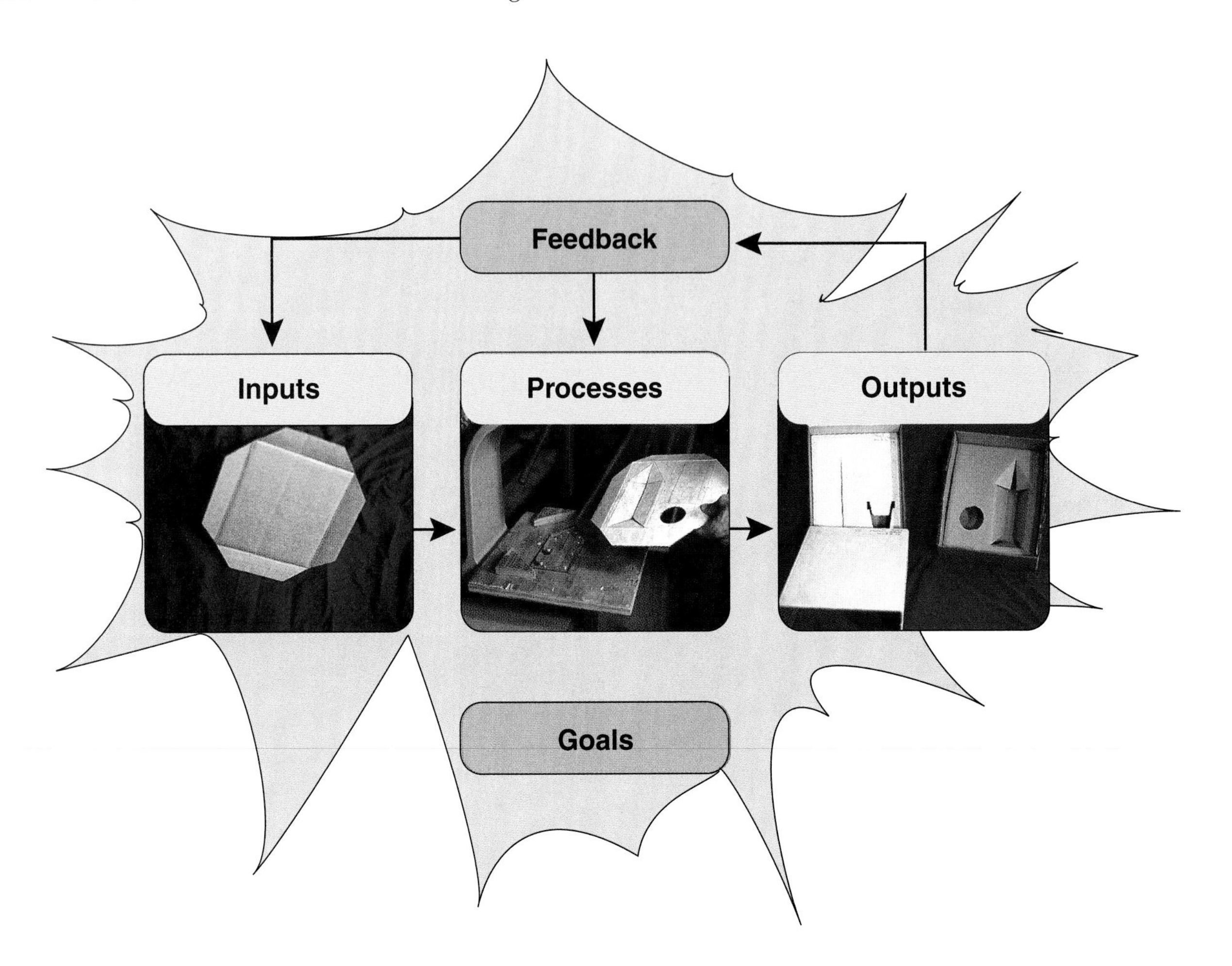

Figure 1-8. These are the elements of technological systems.

Elements of Technological Systems

The world is made up of all types of systems. A system is people *or a group of parts that operate together to achieve a goal*. Nearly everything we see is part of a system. This means that there are parts and these parts are interrelated. An example would be the solar system. The parts include the sun, the planets, and the moons of the planets. Other natural systems can be found in the human body, such as the circulatory system, digestive system, nervous system, etc.

technological systems. **Physical systems consisting of five essential, interrelated elements: inputs, processes, outputs, feedback, and goals.**

Technological systems contain the five essential elements of inputs, processes, outputs, feedback, and goals. See **Figure 1-8.** These elements are interrelated. They must all be present and work with the other elements if the system is to operate properly. A bicycle is a simple example of a system. The *input* is energy (the rider pressing on the pedals). The movement of the pedals and chain to turn the back wheel is the *process*. The *output* is the movement of the bicycle. The rate of speed can be measured and used as *feedback*. All of the parts of this system have a common *goal* of moving the rider along the path or street.

Inputs

All systems start with inputs. *Inputs* are items put into a system to produce outputs and achieve goals. For instance, your digestive system must have food and water (inputs) before it can work. Likewise, technological systems must have inputs. The basic inputs may be grouped into seven categories. These, as shown in **Figure 1-9,** are:

- People
- Natural resources
- Capital (tools and machines)
- Finance (money)
- Knowledge (information)
- Energy
- Time

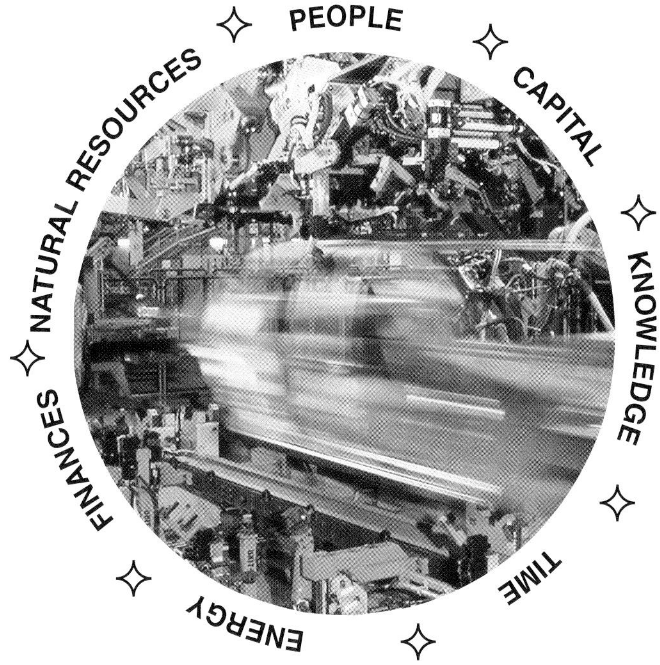

PEOPLE
CAPITAL
KNOWLEDGE
TIME
ENERGY
FINANCES
NATURAL RESOURCES

Figure 1-9. These are seven inputs required by technological systems.

All seven inputs must be present, in their proper balance, for the system to operate effectively. For instance, suppose your class is organized into a system to produce a wooden toy. You have enough students to do the work. You have wood to make into parts. You have power tools to shape the wood. You have money to pay for materials and workers. You have plans (knowledge) for the toy. You have electrical power to run the tools as well as human energy. You would have to have the time to make the toy. When all of these inputs are balanced, the wooden toy can be produced and the system should operate effectively. If one of these inputs happens to fail, then the system will fail.

inputs. **The resources needed to make a product. These inputs may be grouped into seven main classes: natural resources, human resources, capital, knowledge, finances, time, and energy.**

People

Technological systems are developed and operated by people for their benefit. Without people, technology would not exist, nor would there be a reason for it. Technology needs many different types of people. See **Figure 1-10.** There is a need for people with mechanical skills to operate the tools and machines. Engineering, design, and creative skills are needed to design products. Interpersonal skills are needed to manage the human aspects. Logical or critical thinking is needed to schedule activities and to solve problems.

Figure 1-10. People, with their skills and knowledge, are the foundation for technological systems. (Goodyear Tire and Rubber Co.)

Natural resources

All technological systems use materials. These materials may be gases, liquids, or solids. They may be used to support the operation of the system or become part of the system output.

Figure 1-11. All technological systems use tools and equipment. (Jack Klasey)

Gases may be used to weld materials together. Liquids may lubricate machines. Solids may provide the structure for houses or aircraft.

Many technological systems change the form of materials, as shown in **Figure 1-11.** They change copper ore into copper that is used to produce wire and brass products. The leading consumers of copper are wire mills and brass mills, which use the copper to produce copper wire and copper alloys, respectively. End uses of copper include construction materials, electronic products, and transportation equipment. Once refined, copper can be used as a powder in automotive, aerospace, electrical and electronics equipment, in anti-fouling compounds, various chemicals, and medical processes. Compounds of copper include fungicides, wood preservatives, copper plating, pigments, electronic applications, and specialized chemicals. At each step, the material is more useful to people. Its value has been increased.

Capital

capital. **The plant and equipment used to produce products.**

All technological systems use tools and machines. See **Figure 1-12.** These are *capital*. This is what makes them "technological." You and your friend may stand in the hall and talk. You use language to communicate. However, you are not part of a technological system. However, if you talk to your friend with digital cell phones, technology is at work. A technological system has helped you communicate over a distance. Your phone captured your voice. It changed this language-based sound into a digital code called binary code (1s and 0s). The code was transmitted as radio waves through a series of antennas and finally to the receiver in your friend's phone. It was then converted from the binary code back to sound. Your friend heard you talking. All of this was possible through the use of machines.

Figure 1-12. This person is using a machine to help make drawings.

Finance

The old saying, "Money makes the world go around," may not be true. However, it does take money to make technological systems function. It costs money to design, build, and operate these systems. People must be paid for their labor. Materials and machines must be purchased. Owners expect a return on their investments. Taxes must be paid. Finance affects all aspects of a technological system.

Figure 1-13. Early technological systems used animal power. Now, technological systems use complex, powerful machines. (American Petroleum Institute, American Iron and Steel Institute)

Knowledge

All technological systems are based on knowledge. To design and operate technological systems, we must have the knowledge to know how to build products, construct structures, communicate information, convert energy, process food, treat illnesses, and transport goods and people. This knowledge is related to the desired outcome of the system.

Technological systems must also be managed. Goals must be set. Tasks must be set and completed. Outcomes must be evaluated. These actions require managerial knowledge. Managers must know how to combine resources so that they produce efficiently.

Knowledge can come from many different sources. It may be part of what workers bring to the job. It may also be in records, books, magazines, reports, films, videos, or computer memories.

Energy

Energy is needed to do work. Humans use technological systems to extend their abilities beyond their personal strengths. See **Figure 1-13.** Technological systems use many types of energy. Most of the energy comes from burning fossil fuel (petroleum products and coal), falling water (hydroelectric power), and nuclear reactions. Additional energy comes from the wind, the sun (solar energy), chemical action (batteries, fuel cells), and geothermal sources (underground hot water). In some countries, burning wood is a primary source of energy.

energy. The capacity for doing work. It is a necessary input in all technological systems.

Time

The final technological input is time. It includes the human time it takes to produce or operate technological devices and systems. It also includes the time it takes to promote and sell products or services. Time includes machine time used to process materials or information or to convert energy.

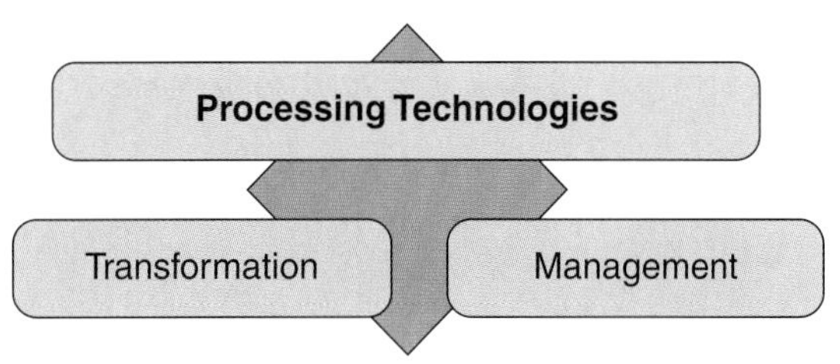

Figure 1-14. These are types of technological processes.

Processes

The second component in a technological system is the process. It is the part of the system that actually does the work. The two major types of processes are transformation and management. See **Figure 1-14.**

Transformation processes deal directly with the inputs. Transformation processes change resources into desired outputs. Each of the major technological systems has specific resources in which they transform resources into outputs. Agriculture and food technology transforms seed into crops and on into edible food. Communication technology uses resources to transform information into media messages. Energy technology changes energy from one form to another form. Manufacturing technology uses resources to transform materials into products. Construction technology uses resources to transform materials and manufactured products into structures. Transportation uses resources to transform energy into power to move people and cargo.

transformation processes. Processes that change resources (inputs) into outputs, such as products.

Management processes are used to ensure the effective use of resources. It includes all the actions used to ensure that transformation acts are efficient and appropriate. Its goals are to see that transformation processes use resources effectively to produce quality outputs. For instance, products and structures must be well made and function properly. Energy must be available and reliable. Information must be accurate and easy to receive. Food must be pure and tasty. People must be transported safely and in comfort. Goods must move quickly with the minimum of damage.

management processes. Processes that are used to assure the efficient and appropriate use of resources.

outputs. The results of a process. Outputs may be desired (such as a product) or undesired (such as pollution, noise, waste, etc.).

Outputs

Technological systems are designed to meet human needs and wants. They produce the *outputs* people want. These are called the *desired* outputs of the system. These outputs may be manufactured products or constructed structures. They may be media messages that deliver information and ideas. They may be goods and people that are in a new location. They may be people that have better health and well-being. The outputs may be designed to meet a basic need or they may help entertain us. See **Figure 1-15.** They may be critical to life or simply add beauty.

Figure 1-15. This parade float is an output of technological activity that entertains us.

Most technological systems have other outputs. Scrap is produced as products are manufactured. Pollution is produced as the system operates. These are examples of *undesired* outputs. Noise, fumes, and liquid chemicals are given off by many technological systems. These outputs are often unavoidable. Carbon monoxide is a product of burning fuels. It cannot be avoided, but must be controlled. If you cut circles out of sheets of plywood, scrap will be produced. Again, it cannot be eliminated, but it can be held to a minimum. Aircraft engines make noise. It cannot be avoided, but it can be reduced.

The designers of technological systems face a common challenge. This is to produce the maximum amount of desired output with the minimum amount of undesired output. This will use resources wisely and minimize the damage to the environment.

Feedback

Systems can use their own information to control themselves. This process is called *feedback*. The output is often measured. If it is not what is expected, corrective action usually results. For instance, consider a thermostat on a furnace. When there is enough warmth in the room, the thermostat sends a signal that shuts off the furnace so that no more heat is produced.

feedback. **Information that a system uses to control itself. By measuring an output, the system gains information needed to take corrective action.**

Many modern manufacturing machines use a feedback system called *adaptive control.* For instance, consider a drill. The drill is being fed into the work. This places force on the motor and causes it to slow down. As the drill slows down, the feedback system senses the reduction in speed. It commands the feed system to reduce the feed rate for the drill. The slower feed rate reduces the force on the motor. It then regains its proper operating speed.

Goals

Technological systems are directed and controlled by two types of goals. These types are system goals and societal goals. The system is established to meet human needs or wants. These are the ***system goals***.

Society has general goals that meet broader objectives. Society wants:

- People employed in challenging jobs.
- People to have a high standard of living.
- Our country to remain economically secure.
- To retain a technological lead in key activities.
- Our country to be strong and respected.

All of these goals are *societal goals*. They address the needs and wants of people as a larger group.

Societal goals. **The meeting of the broad needs and wants of human society.**

Technological systems must be adjusted to meet a balance of system and societal goals. For example, it is unacceptable to us to have workerless manufacturing. Products would be cheaper, but the resulting unemployment would cause many people extreme hardship.

Types of Technological Systems

There are many types of technological systems, as shown in **Figure 1-16.** Each is designed to make lives better. They extend our potential. These systems may be private or industrial. *Private technological systems* are designed to meet an individual person's needs. For instance, you may use a personal technological

Private technological systems. **Designed to meet the demands and needs of an individual.**

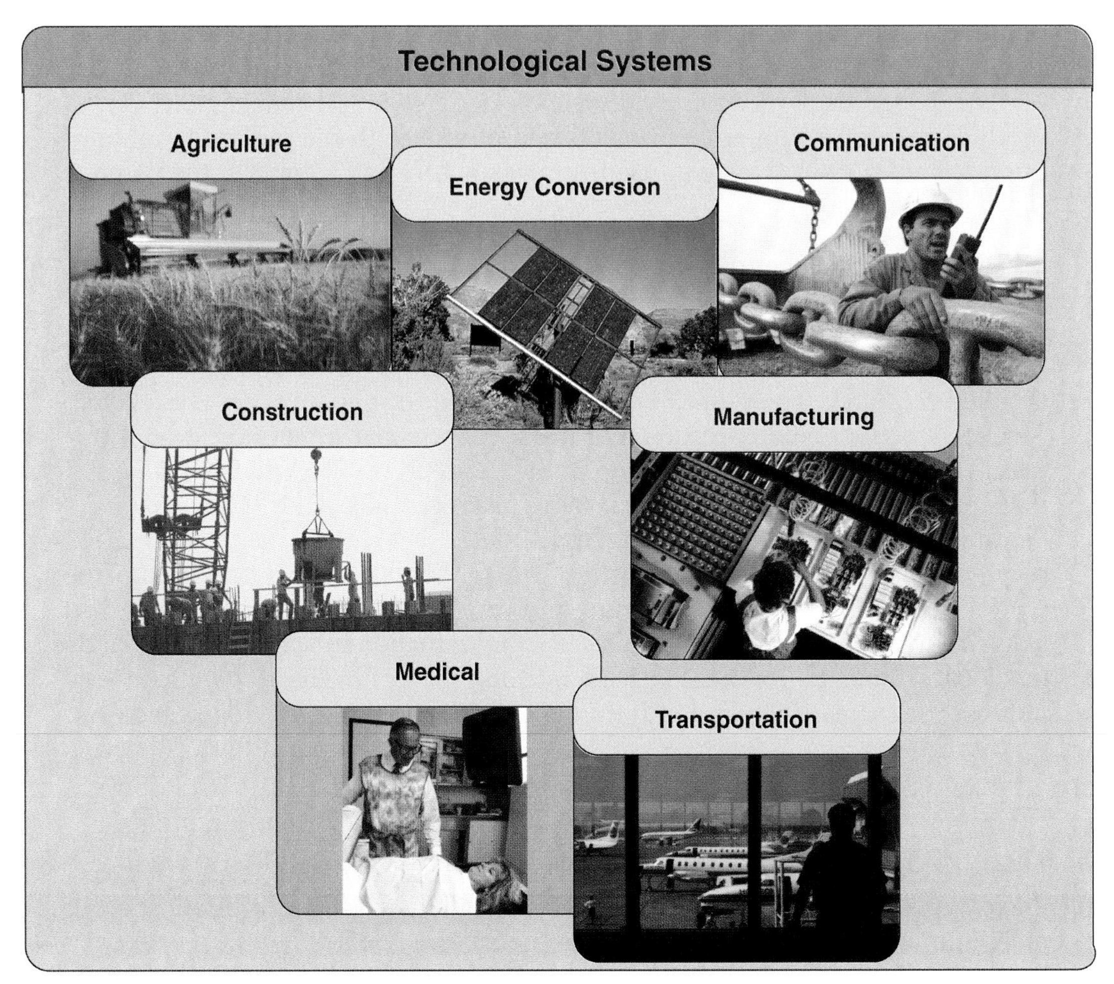

Figure 1-16. These are the seven major types of technological systems.

industrial system. Designed to transform resources into outputs for a large number of people.

system to build a picnic table in your garage. However, most technological systems are *industrial systems*. They are organized systems that transform resources into outputs for large numbers of people.

Both types of technological systems share common elements. They use tools and machines to meet their goals. They transform resources into desired outputs. They may also produce undesired outputs. Technological systems may be divided into at least seven major groups:

- **Agriculture technology.** This includes systems that produce outputs by growing plants and animals. The outputs may be food (meat, vegetables, etc.) or fibers (wool, flax, cotton, etc.).
- **Communication technology.** This includes systems that communicate information and ideas using technical means (equipment). The product is often a media message designed to reach a mass audience.
- **Construction technology.** This includes systems that construct residential, commercial, and industrial buildings and civil structures (roads, pipelines, bridges, dams, etc.). The structures are produced on the site where they are to be used.

- **Energy and power technology.** This includes systems that convert and deliver energy for personal and industrial use. The product of this technology powers machines, heat and light spaces, move vehicles, etc.
- **Manufacturing technology.** This includes systems that transform materials into products in a central location (factory). The products are then transported through a distribution system to the ultimate consumer.
- **Medical technology.** This includes systems that are used to maintain health and treat injuries and illnesses. The "product" of the system is people and animals that have good health and fitness.
- **Transportation technology.** This includes systems designed to move people and goods from one location to another. The systems include a vehicle and support system (terminals, roadways, etc.).

This book will address manufacturing technology as a system. It will give you insight into how manufacturing is used to efficiently produce the products that each of us uses daily.

Summary

Technology is everywhere. It impacts every person. Technology is different from science. Science explains the natural world, while technology explains the manufactured (designed and built) world. Technology is the use of tools, materials, and systems to produce products and structures that modify and control the world. It is knowledge of efficient and appropriate action. All technological systems evolve. They are being improved to better meet human needs and wants.

Technology is a system with inputs, processes, and outputs. It often uses feedback loops to maintain proper operating conditions. The design and operation of technological systems are affected by individual and societal goals.

Technological systems may be small, personal ones or large, industrial ones. They also may be grouped into types. The major types are communication, construction, manufacturing, and transportation. Additionally, extractive and biotechnological systems provide the material resources for the four basic systems.

Key Words

All of the following words have been used in this chapter. Do you know their meanings?

agriculture technology
biotechnology
capital
communication technology
construction technology
descriptive knowledge
energy
evolving technologies
feedback
humanities
industrial system
inputs
management processes
manufacturing technology
medical technology
outputs
private technological systems
science
societal goals
system goals
technological systems
technology
transformation processes

Test Your Knowledge

Please do not write in this text. Place your answers on a separate sheet.

1. List the four major groups of knowledge.
2. *True or False?* A background in both science and technology can help you to arrive at decisions based on knowledge.
3. *True or False?* Technology is always changing.
4. List the five essential elements of a technological system.
5. Is capital an input or an output?
6. _____ processes use resources to create desired outputs.
7. _____ processes are used to increase the effective use of resources.
8. Noise, fumes, and liquid chemicals given off by technological systems are _____.
   a. undesirable outputs.
   b. desirable outputs.
   c. often unavoidable.
   d. Both a and c.
9. Technological systems that transform resources into outputs for large numbers of people are _____ (private or industrial) systems.
10. Technological systems may be divided into at least seven major groups. What are they?

Applying Your Knowledge

Note: Be sure to follow accepted safety practices when working with tools. Your instructor will provide safety instructions.

1. Develop a poster, transparency, or bulletin board that would define technology for someone who is unfamiliar with the word.
2. Distinguish between science and technology. Then give four examples of how each affects your everyday life.
3. Use a managed manufacturing system to produce a product. The tic-tac-toe game shown in **AYK 1-3** on next page is an example of a simple line production project. Follow the steps on next page to select and produce your product.

3. (Continued)
   a. Decide what the class will manufacture. Check catalogs, magazines, and any other literature supplied by your instructor for ideas. Visit local stores for suggestions.
   b. Following a planning session with your instructor, select an appropriate sequence of manufacturing operations.
   c. Determine the tools and materials needed to successfully manufacture the product.
   d. Organize the manufacturing team for the production run. Is every task or job covered?
   e. Review each worker's production line assignment. Are the tools and materials on hand to start the production run?
   f. Following all safety rules, make the production run.

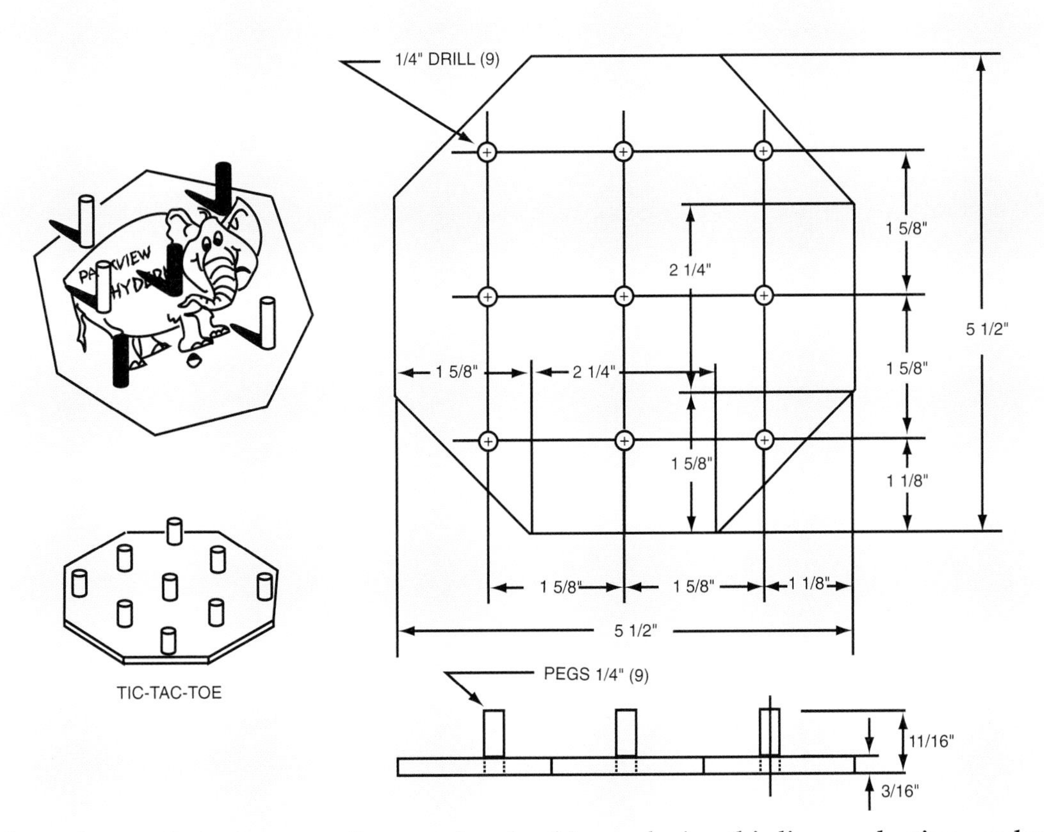

AYK 1-3. Several manufacturing operations are involved in producing this line production product. These plans give details on how to manufacture it.

Career Link

CAD Drafter

A CAD Drafter uses a computer and computer-aided drafting software to prepare technical drawings and plans used by production and construction workers. These plans may be used to build everything from manufactured products, such as toys, toasters, industrial machinery, or spacecraft, to structures, such as houses, office buildings, or oil and gas pipelines.

Their drawings provide visual guidelines, showing the technical details of the products and structures and specifying dimensions, materials to be used, and procedures and processes to be followed. The drawings they produce contain technical details, using drawings, rough sketches, specifications, codes, and calculations previously made by engineers, surveyors, architects, or scientists. They may use their knowledge of standardized building techniques to draw in the details of a structure, or use their knowledge of engineering and manufacturing theory to draw the parts of a machine in order to determine design elements, such as the number and kind of fasteners needed to assemble it.

Chapter 2 Manufacturing Systems

Objectives

After studying this chapter, you will be able to:

✓ Describe how manufacturing has evolved.
✓ List the major components of a manufacturing system.
✓ Identify manufacturing inputs.
✓ Describe manufacturing processes.
✓ Identify manufacturing outputs.

Look around you. Manufactured products are everywhere. The furniture, lighting fixtures, windows, doors, floor coverings, books, and pencils were all manufactured. Even the clothing you are wearing was manufactured. Outside, there are power poles, streetlights, fences, and automobiles, all of which were manufactured. Each product was carefully designed, produced, and marketed. They are the desired outputs of manufacturing systems.

The Evolution of Manufacturing

Manufacturing is as old as human life itself. People have always made things from materials found on earth. Early manufacturing was done for personal use. People made weapons, clothing, and cookware for survival and to make life easier. The processes they used were the manufacturing technology of their time. It represented activity in making and using products to extend human ability. With these products, people could hunt better, stay warmer, and cook food more easily.

Making products for personal use has been called *usufacturing* or "making for use." This earliest manufacturing activity used simple technologies. Crude tools and natural materials were the foundation of this stage of development. Societies based on this type of production were in their handicraft era or age.

usufacturing. Making products for personal use.

Later in history, people started to make products for other people. Individuals developed skills in using certain materials and tools. See **Figure 2-1.** They would concentrate on making a special type of product. This is how some people became carpenters, while others became shoemakers, or weavers, etc. They would trade (barter) their products for things they needed. For instance, a carpenter might trade a stool for a pair of shoes.

Figure 2-1. People in early manufacturing used simple tools and materials.

This was the beginning of *manufacturing* or "making products in a mechanical manner." As specialization grew, manufacturing became concentrated in special buildings called *factories*. Cities grew, and many people left the farms for jobs in the factories. Employees were divided into workers and managers. During this period, manufacturing used machines to produce a large quantity of a specific product. The emphasis was on material processing—people using machines to make products from materials. This resulted in an *industrialized society*.

Today, we are moving into an age where products are first made in the mind. This has been called *mentafacturing* or "making in the human mind." This ability has been made possible with the development of the computer. Emphasis has moved from processing materials to processing information. Computers, like one shown in **Figure 2-2,** can be used to prepare presentations, create drawings, monitor machine operations, inspect products, and maintain production records.

With these developments, society has moved into the information age. Examples of information age products include modern jet aircraft. See **Figure 2-3.** These aircraft are "manufactured" and "flown" on a computer before the actual aircraft are built. With computers, the human mind can conceive, design, build, and test products before the human hand will actually operate machines to process materials.

All these manufacturing methods can be viewed as complete systems. In Chapter 1, you learned that a system has several components. These include inputs, processes, outputs, feedback, and goals.

manufacturing. The process of changing resources into more useful products.

factories. Special buildings where people use machines to manufacture products.

industrialized society. One in which people use machines to manufacture large quantities of products from materials.

mentafacturing. Products are first made in the mind. This has been called "making in the human mind."

Figure 2-2. These business people are using a laptop computer to review information and help plan a presentation.

Figure 2-3. This airplane and service equipment are products of the information age.

For example, communication systems transform information (input) into media messages (output). These messages may be communicated using the printed media, photographs, or electronic means (processes). The messages are designed to inform and persuade people (goal). It is hoped that we will know, move, or act differently because of the communication process.

Transportation is another example. Transportation transforms energy (input) into power to propel vehicles (process). These vehicles move people and cargo safely and efficiently (output). The person or the item will be in a new, desired

Academic Link

The evolution of manufacturing is comparable with the evolution of modern civilization. When studying ***social science*** and how civilizations developed and grew, we can see that production of tools and the products made using these tools also grew.

At one time most humans fended for themselves. They hunted and farmed to feed themselves and their families. They made their own tools from stone and wood. As humans came to live in larger communities, the demand on manufacturing became larger as well.

Because of the increased social demands, demand for materials and processes for making better tools, better housing materials, and better transportation methods increased.

Over the last century in manufacturing, important innovations significantly outnumber the innovations from all previous time combined. A social science perspective would indicate in this 100-year period that human demand was greater for creating more efficient manufacturing processes. What will be the next manufacturing breakthrough? You are not sure? Well, you can be sure that it will have social influences.

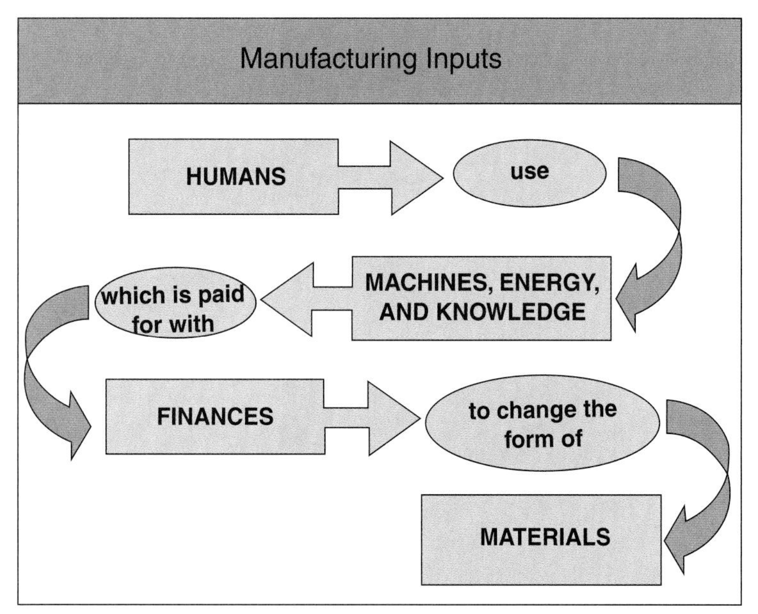

Figure 2-4. Manufacturing uses resources to change the form of materials.

location (goal). The transportation system brings us products and allows us to travel anywhere in the world.

Do you see that every technology has the same components? Now, let's look at these system components as they specifically apply to manufacturing.

Manufacturing as a System

The manufacturing system provides you with nearly all the physical products you use daily. The car in which you ride, the television you watch, the food you eat, and the clothes you wear are all manufactured. Products such as these can make life better. This is the reason humans developed manufacturing in the first place. To understand manufacturing, you should view the three major components of the manufacturing system. These are inputs, processes, and outputs.

Manufacturing Inputs

form utility. **Changing the form of a material to make it more useful.**

transformation technology. **The appropriate use of tools, machines, and systems to convert materials into products.**

material processing technology. **The appropriate use of tools, machines, and systems to convert materials into products.**

managerial technology. **The use of systems and procedures to ensure that transformation actions are efficient and appro-**

All systems have the same inputs. They use human labor, information, capital (machines and equipment), energy, materials, and finances. See **Figure 2-4.** However, each system has an input that is designed to transform (change).

Manufacturing transforms materials into products. See **Figure 2-5.** It changes the form of a material to make it more useful. This action is called *form utility*. For instance, you probably find a plastic CD recording worth more than a pile of plastic pellets. Likewise, most people feel a wood chair is worth more than a pile of boards. The new form has more utility (usefulness) and, therefore, we place a higher value (worth) on it.

Manufacturing Processes

New material forms do not just happen. They are the direct effort of many people working together. They are part of a manufacturing team that applies two basic process technologies. These technologies are used and interact to complete the transformation task. See **Figure 2-6.** The two manufacturing process technologies are transformation and managerial.

The appropriate use of tools, machines, and systems to convert materials into products is called *transformation technology*. This technology is often called *material processing technology*.

The appropriate use of systems and procedures to ensure that transformation actions are efficient and appropriate is called *managerial technology*. Both workers and managers use this technology.

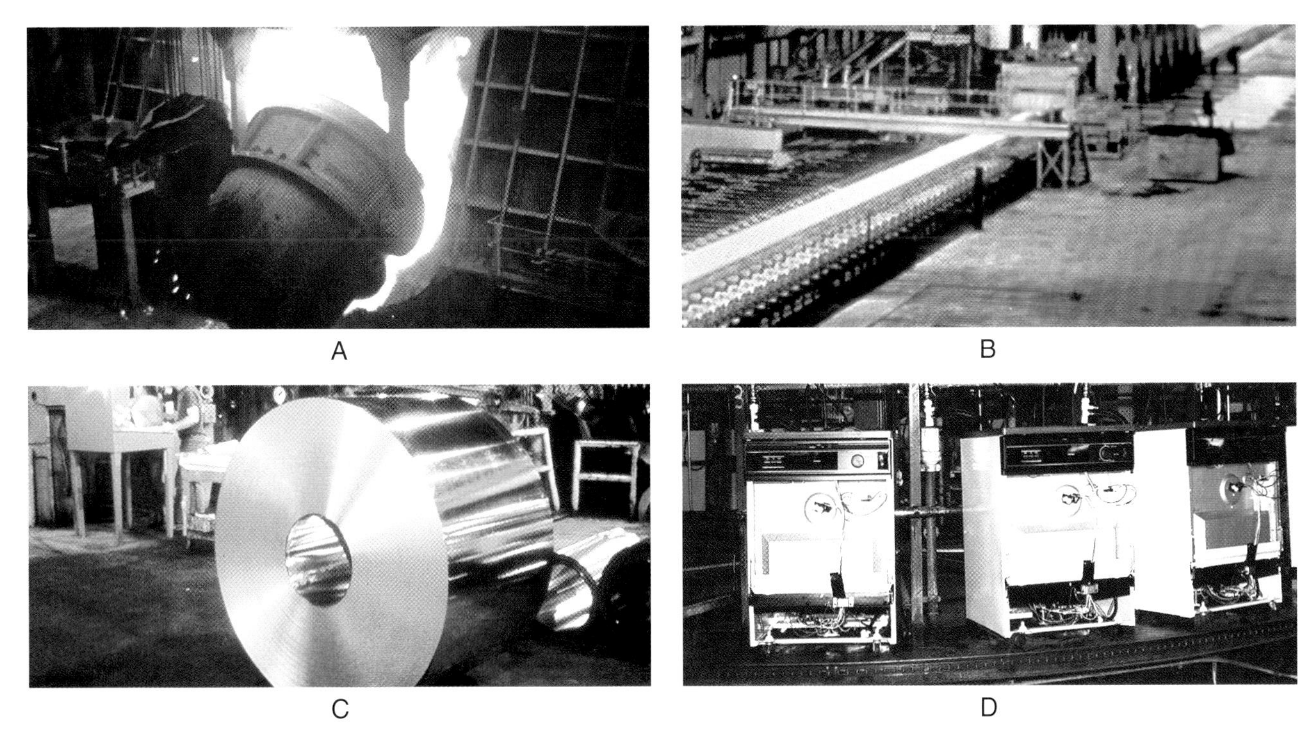

Figure 2-5. Manufacturing changes the form of materials to add to their value. A—Iron ore is changed into pig iron. B—Pig iron is changed into steel. C—Steel is rolled into strips. D—Steel strip is fabricated into a useful product. (US Steel, American Iron and Steel Institute, White Consolidated Industries)

Figure 2-6. These are types of processes in the manufacturing system. The worker on the left assembles a product using a transformation action while, on the right, the worker and manager participate in making a managerial decision.

Manufacturing technologies are the focus of this book. Units 2, 3, and 4 will explain manufacturing materials and processes used in transformation technology. Units 5 through 10 will present management and the managed

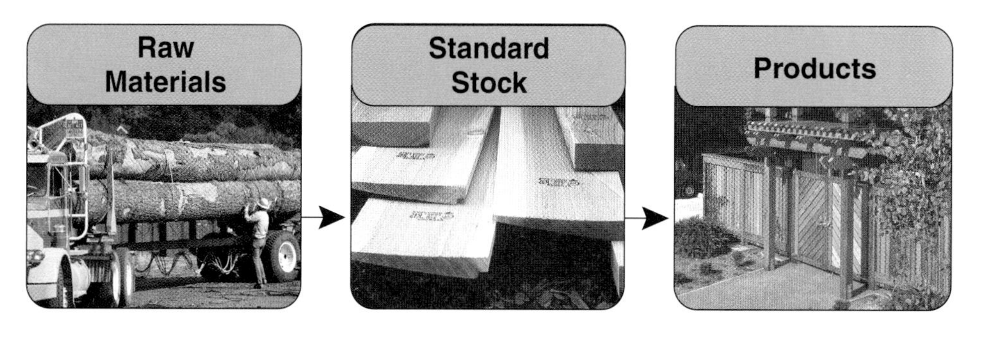

Figure 2-7. These are examples of stages in material transformation. (Georgia-Pacific Corp., California Redwood Assoc.)

enterprise. However, before the technologies are explored in depth, let's look at them in general to see how they interrelate.

Transformation Technology

Transformation technology (material processing) creates form utility. It is people using machines efficiently and appropriately. Material transformation generally takes place in several steps. The steps are shown in **Figure 2-7.**

All materials can be traced to a natural resource. They all are some form of a material found on earth. Most plastics come from natural gas and petroleum. Glass is often made from silica sand and soda ash. Steel is a combination of iron ore, carbon, and other elements. Aluminum is found in bauxite deposits.

Figure 2-8. This photo shows trees being processed into lumber in a sawmill.

The first step in the transformation process involves obtaining the natural resources. This may involve a search for deposits of ores or petroleum. It may require planting, growing, and harvesting trees or other crops. It may also include obtaining materials from the sea. These natural resources are the *raw materials* for the manufacturing process.

raw materials. Natural resources found on or in the earth or seas. Manufacturing starts with raw materials.

primary processing. The first step in transforming raw materials into products. Example: Converting trees into lumber at a sawmill.

standard stock. Material output by primary processing operations, available in standard size units or standard formulations.

Next, natural resources are converted into a usable material. Trees may be transformed into lumber, plywood, particleboard, hardboard, or paper. See **Figure 2-8.** Petroleum may become a fuel, lubricant, or plastic. Metallic ores may be reduced to copper, aluminum, tin, lead, or iron. These activities are considered to be *primary processing*. This first step in transforming materials into products is often performed by basic industries such as steel mills, copper smelters, lumber mills, and petroleum refineries. They provide the basic resources from which other products are made.

Generally, the output is produced in standard sizes. Lumber may be 1 × 12s or 2 × 4s. Plywood, particleboard, and hardboard are made in 4' × 8' sheets. Metal may be produced in standard-sized sheets, bars, or rods. Plastics with specific chemical composition are available. These materials are called *standard stock* or *industrial materials*.

However, standard stock has little value to most of us. What would you do with a truckload of plywood? It only becomes valuable to you when it is made into something useful. This process is called *secondary processing*. These actions continue to change the form of materials. Lumber becomes furniture, and plastic becomes dinnerware. Clay becomes flower pots, steel becomes automobiles, and glass become bottles. See **Figure 2-9.**

Secondary processing shapes and forms materials through casting, forming, and separating actions. Parts are conditioned to give them new properties. Finishes are applied to protect and beautify the product. Then parts are assembled in subassemblies and products.

The result of all these activities is products for us to use. Also, many people find exciting and challenging careers in designing, producing, and marketing the products of manufacturing systems.

Figure 2-9. Glass can be converted into bottles using secondary processing techniques. (Owens-Brockway)

Managerial Technology

Human-made systems must be designed, built, and operated. Goals must be set and courses of action selected. Resources must be obtained and assigned to various activities. The operation of the system must be constantly monitored. Outputs must be evaluated. These actions involve management technology. They involve planning, organizing, actuating (directing), and controlling.

Managerial technology is all the actions that ensure that resources are used efficiently to produce appropriate products. Manufacturing management involves five basic areas of activity. See **Figure 2-10.** These are research and development, production, marketing, financial affairs, and industrial relations.

RESEARCH AND DEVELOPMENT
PRODUCTION
MARKETING
INDUSTRIAL RELATIONS
FINANCIAL AFFAIRS

Figure 2-10. These are the managed areas of activities within a manufacturing enterprise. (Inland Steel Company)

Products and manufacturing processes must be designed and specified. This is the task of *research and development* (R&D).

The specified product must then be built. A production system must be designed. Manufacturing processes must be selected. Machines and tools must be brought into play. Materials are processed into products. The production activity area completes these responsibilities.

The completed product must reach the customer. Its availability and features are presented by advertising. The product is sold to wholesalers and retailers. From them, the final customers buy those products that meet their needs. This movement of products from the manufacturer to customers is called *marketing.*

Research and development, production, and marketing are the three activities directly involved with the product. They design, produce, and distribute the outputs of the manufacturing system. However, other areas are needed. Finances are needed. *Financial affairs* raise money and maintain financial records. Human resources and attitudes must also be developed. *Industrial relations* deal with the human aspects of the system through personnel, labor, and public relations programs.

industrial materials. **Materials ready for secondary processing into manufactured products. Also known as standard stock.**

secondary processing. **Manufacturing methods that change standard stock into finished products.**

research and development. **The area of managerial technology concerned with designing and specifying products.**

marketing. The area of managerial technology concerned with moving a product from the manufacturer to the customer by means of advertising and selling activities.

financial affairs. The area of managerial technology concerned with raising money and maintaining financial records.

industrial relations. The area of managerial technology concerned with the human aspects of the manufacturing enterprise, such as personnel and labor relations.

Without management, efficient production would not happen. Management is essential for all goal-oriented activities, whether they be in industry, sports, or your personal life.

Manufacturing Outputs

Human-made systems are developed for a reason. The manufacturing system has been developed to produce the products we need and want. See **Figure 2-11.** These are the desired outputs of the system.

However, undesired outputs can also result. Factories may create pollution. Fumes can enter the air. Chemicals can mix with groundwater. Also, noise can pollute quiet areas. Some pollution is unavoidable. The challenge is to control these undesirable products of the manufacturing system.

Most outputs of manufacturing systems have a useful life. After this time, the products are often discarded. However, with an increasing concern about pollution and resource depletion, manufactured products should be recycled after they serve their purpose. This allows the worn-out or obsolete products to become a material input for a new manufacturing activity. Large-scale recycling activities are already in place for aluminum, glass, and paper. Others will soon follow. If consumers and manufacturers begin recycling their scrap, waste, and obsolete products, the stress on the environment will be reduced. This will benefit everyone.

Another set of "products" of manufacturing systems is human outputs. Manufacturing creates jobs. Some of these are directly in the factory or in marketing the product. Others support the factory and its workers. Many stores and service businesses disappear when a factory leaves a town or goes out of business. Manufacturing systems also impact people and their lifestyles. Employment means income, physical and mental challenges, and a feeling of worth. See **Figure 2-12.**

New manufacturing systems require people to receive additional education. Recently, a major corporation said that the worker in the future would constantly need to be in training just to stay ahead. Flexible people, who are willing to learn,

Figure 2-11. New products, like these transportation vehicles, are designed to meet human needs and wants. (BEA Systems, Jaguar)

can face a lifelong challenge in manufacturing. Those who cannot or will not adapt to new job requirements most likely will face unemployment. This "change and advance" or "stand still and lose out" attitude can be considered an output of the system.

Figure 2-12. Working in the manufacturing system can be challenging and exciting. (AT&T Archives)

Effects of Developments in Manufacturing

More than ever, people are immersed in a world shaped by human technology. People's responsibilities in a technological age are defined as coming to grips with the problems of living in, and exerting influence upon, the constructed world. The risks and benefits related to technological developments in manufacturing must be weighed so that the world is not influenced negatively. For example, in the future it is discovered there is a development that will cut the cost of manufacturing a product in half. However, the waste produced during the manufacturing process is hazardous and hard to handle and dispose. The amount and level of risks can determine if a development in manufacturing is used or more research and changes are needed to make the development better.

People need to draw relationships to larger world issues and to discuss both the positive and negative aspects of technology. This will help provide an accurate view of how developments in manufacturing play important roles in both meeting human needs and creating human problems.

Assessing Risks and Benefits

The interactions between technology and society sometimes lead to issues that can only be resolved by examining all factors involved. An assessment of the beliefs, values, and alternate solutions associated with each side helps determine benefits. Problem-solving and decision-making strategies are used to aid in determining risks and benefits of developments in manufacturing. Developing alternatives and reaching a rational decision are the goal of these strategies. The four basic principles of making a rational decision are *discovering* the need for a decision, *exploring* the values and goals relevant to the decision opportunity, *developing* alternative courses of action, and *predicting* likely consequences (positive and negative) of alternative courses of action.

There should be an understanding of the relationships between technology and social change, as well as the trade-offs and unexpected side effects that result from technological developments. The effects of the technological innovation should be listed and categorized according to whether the effects were planned or unplanned. The effects should be divided into positive (a benefit to people), negative (harmful to people), or both (have both positive and negative aspects). The effects should be described as local, regional, national, or global. Finally, actions that can be taken to alleviate the negatives should be listed.

Summary

Manufacturing has evolved from producing products for personal use to complex systems that give us a wealth of goods. The manufacturing system changes materials into a more usable form. It moves raw materials into standard stock that is converted into products. Without manufacturing, fewer people would have meaningful jobs. Also, communities, states, and the nation would suffer economic distress.

Key Words

All of the following words have been used in this chapter. Do you know their meanings?

factories
financial affairs
form utility
industrialized society
industrial materials
industrial relations
manufacturing
managerial technology
marketing
material processing technology
mentafacturing
primary processing
raw materials
research and development
secondary processing
standard stock
transformation technology
usufacturing

Test Your Knowledge

Please do not write in this text. Place your answers on a separate sheet.

1. Describe the difference between *usufacture*, *manufacture*, and *mentafacture* in terms of the types of work people do and the machines they use.
2. List the major components of the manufacturing system.
3. Human labor, information, capital, energy, materials, and finances are examples of manufacturing _____.
4. What are the two manufacturing process technologies used by manufacturing systems?
5. *True or False?* Transformation technology creates form utility.
6. *True or False?* All materials can be traced to a natural resource.
7. List the five basic areas of activity involved in manufacturing management.
8. Products and pollution are examples of _____ of a manufacturing system.

Applying Your Knowledge

1. Develop a poster that shows the basic components of a manufacturing system.
   a. Get posterboard or suitable backing from your instructor.
   b. Select a type of manufacturing that appeals to you.
   c. Select a method of presenting your material. Do you wish to draw the parts? Perhaps you would prefer to clip art or photographs from old magazines or advertisements.
   d. Assemble your materials. You may wish to include marking pens for preparing the message.
2. Select a product in your technology education laboratory or from your home to study.
   a. Construct a chart with two columns.
   b. In the first column, list all the material inputs that you think were used to produce it. (Study the product carefully. Not all materials are immediately noticeable!)
   c. In the second column, list as many undesirable outputs as occur to you. These are outputs that may have been the result of the manufacturing system and its processes.

3. Refer to the manufacturing activity (AYK 1-3) described in Chapter 1.
   a. Make a chart similar to the one in **AYK 2-3** below.
   b. List the material inputs for the manufacturing activity in Chapter 1.
   c. Next, list the transformation (changing the material) activities.
   d. Finally list all the outputs of the system.
   e. Assess the risks and benefits of the outputs. Determine if the balance between them is okay or if improvements can and should be done.

Inputs	Process	Outputs
Material:	Transformation:	Desired:
Human labor (types):		
Information:		
Capital (plant and equipment):	Managerial:	Undesired:
Energy (types):		
Finances (product cost):		

AYK 2-3. On a separate sheet of paper, prepare a chart similar to this one on which to record your responses.

Technology Link

Communications

Manufacturing is a unique and broad technology. In basic terms it is the process that changes resources into products. Manufacturing affects many other technologies. Other technologies, such as communications, have an effect on manufacturing as well. The link between manufacturing and communications is a two-way street.

Have you ever thought how books are manufactured? What are the resources used to make this book and how are they manufactured? There is paper, inks, glue, and possibly a few other materials depending on the type of book desired. How then are these resources put together to make a book? Large printing presses are used to put the ink on the paper. Collators and trimmers organize the pages and trim them to the proper size. Binders are used to glue, as well as other means such as staple, spiral wire, or sew, these pages together into book form. Manufacturing processes are also used to make the parts for the computers you use to research information for school. Manufacturing processes are used to make the towers from which radio and television signals are transmitted.

What most people don't realize is the direct effect other technologies have on manufacturing. Have you every wondered how the products you use are developed? Have you wondered why you use the products you do? Many types of communications are used in manufacturing—from the initial design through the marketing of the final product. In the design process, engineers use computer-aided drafting programs, computers, faxes, and phones to develop and deliver product designs. Computers are used to interpret and test product designs and to control manufacturing processes. While it might not commonly be considered a part of the manufacturing process, marketing is a significant ingredient. Whether through television, radio, Internet, or print media, marketing communications can make or break a product.

Career Links

Human Resource Specialists

Human resource specialists provide the link between management and employees. In the past, the specialists have been associated with performing the administrative function of an organization, such as handling employee benefits questions or recruiting, interviewing, and hiring new personnel in accordance with policies and requirements that have been established in conjunction with top management. Today's human resources specialist handles these tasks and, increasingly, consults top executives regarding strategic planning. They have moved from behind-the-scenes staff work to leading the company in suggesting and changing policies. Senior management is recognizing the importance of the human resources department to their bottom line.

Specialists make an effort to improve morale and productivity and limit job turnover. They also help their firms effectively use employee skills, provide training opportunities to enhance those skills, and boost employee satisfaction with their jobs and working conditions. Although some jobs in the human resources field require only limited contact with people outside the office, dealing with people is an essential part of the job.

Because of the diversity of duties and level of responsibility, the educational backgrounds of human resource specialists can vary considerably. In filling entry-level jobs, employers usually seek college graduates. Many employers prefer applicants who have majored in human resources, personnel administration, or industrial and labor relations. Others look for college graduates with a technical or business background or a well-rounded liberal arts education. Many colleges and universities have programs leading to a degree in personnel, human resources, or labor relations. Some offer degree programs in personnel administration or human resources management, training and development, or compensation and benefits.

Unit 2 Manufacturing Materials

Chapter 3 Materials and Manufacturing

Objectives

After studying this chapter, you will be able to:

- ✓ Describe the relationship between materials and manufacturing.
- ✓ Distinguish between organic and inorganic materials.
- ✓ List the three major types of materials.
- ✓ Identify and describe metallic, polymeric, and ceramic materials.
- ✓ Give examples of composite materials.

manufacturing. The process of changing resources into more useful products.

We live in a world of materials. See **Figure 3-1.** Your home, your school, your bike, and the sidewalk on which you ride your bike are all made of materials. Everything on earth is made from one type of material or another.

Throughout history, humans have developed methods of changing many of these materials. We change their shape, their size, or the way they are combined with one another. This process is called *manufacturing*. See **Figure 3-2.**

Materials are used to produce the products that make our lives easier, safer, and more enjoyable. Other materials are used to construct buildings and other structures produced by the construction industry. Still other materials are manufactured into vehicles that transport us over long distances. Materials are used to treat illnesses. See **Figure 3-3.** Natural materials are grown on farms and in forests. Finally, materials are needed for the equipment and media used to communicate information and ideas.

Figure 3-1. Consider the number of manufactured materials that were used to build and furnish this home. Materials were also used in manufacturing the tools and machines used in building the home. (California Redwood Association)

A few materials can be used as they are found on earth. People may use natural gravel and sand to construct a road. Berries may be eaten directly from plants and vines. Trees may be cut into firewood to heat our homes.

However, most materials require processing before being used. Iron ore, coke (processed coal), and limestone can be converted into pig iron. Trees can be sawed into lumber, digested to produce paper, or flaked for particleboard. Bauxite is changed into alumina that is converted into aluminum. Other ores are smelted (refined) to extract

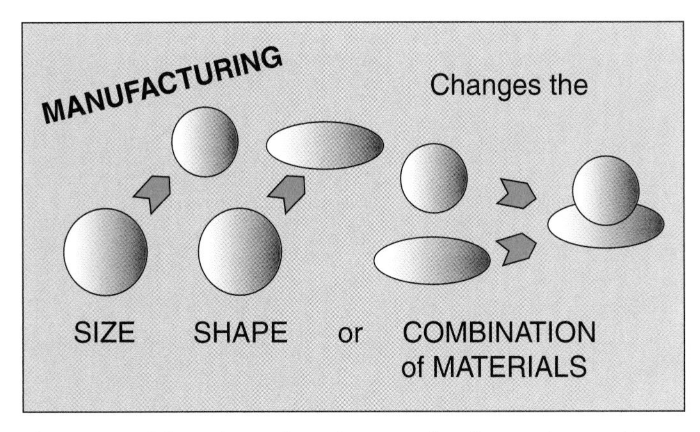

Figure 3-2. Manufacturing changes the form of materials.

copper, gold, silver, and many other metallic materials. Natural gas may be processed to produce plastics. Silica sand may be changed into glass.

Materials used in Manufacturing

Materials come from two basic sources. They were either once living organisms (or part of living organisms) or they never had life. Materials that can be traced back to living things are called *organic materials*. Common examples of these materials are coal, petroleum, wood, and wool. *Inorganic materials* are the nonliving ores and earth elements. Copper, gold, aluminum, glass, cement, and rock are typical inorganic materials.

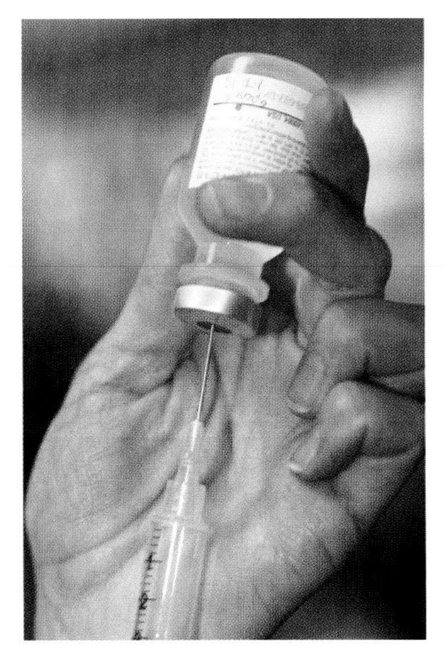

Figure 3-3. Special materials called drugs are used to treat illnesses.

Organic and inorganic materials are found in three physical states: gases, liquids, and solids. See **Figure 3-4.** Solid materials are often called *engineering materials*. These materials have a solid structure. They will hold their shapes without outside support. Gases and liquids are *nonengineering materials*. They must have a container if they are to be confined or held in a particular shape. Gases and liquids that have important industrial applications are fuels, lubricants, and water.

Engineering materials are used to give durable products their structure or framework. These materials, as shown in **Figure 3-5,** can be divided into four basic categories:

- Metallic materials
- Polymeric materials
- Ceramic materials
- Composite materials

organic materials. **Those whose origins can be traced back to living things.**

inorganic materials. **Those that are nonliving; ores and earth elements.**

engineering materials. **Solid materials that will hold their shapes without outside support.**

Metallic Materials

Metals are the most widely used group of engineering materials. Three-fourths of all natural elements are classified as metals. About one-half of these have some industrial application and value. These materials have desirable mechanical and physical properties. They are readily available and can be processed using a wide variety of techniques.

Most commercial metals are obtained from their *ores*. Ores are earth or rock from which metals can be commercially extracted. Metal-bearing rocks that cannot be refined economically are not called ores.

In rare cases, metals appear in veins of pure material. The legendary Mother Lode of Nevada contained such veins of silver. However, most metal is extracted from its ore and concentrated. It is then shipped for final refining into useful industrial materials. During refining, the impurities are separated and removed in the form of slag. See **Figure 3-6.**

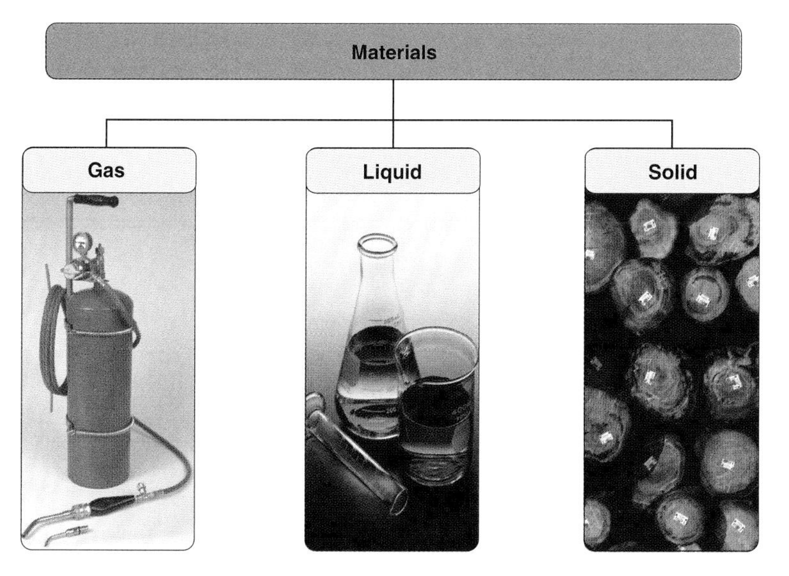

Figure 3-4. These are the kinds of materials as classified by scientists: gases used for welding; liquids such as medicines, chemicals, and paint; and solids such as these logs that will be cut into usable lumber. (Uniweld, Rust-Oleum Corp.)

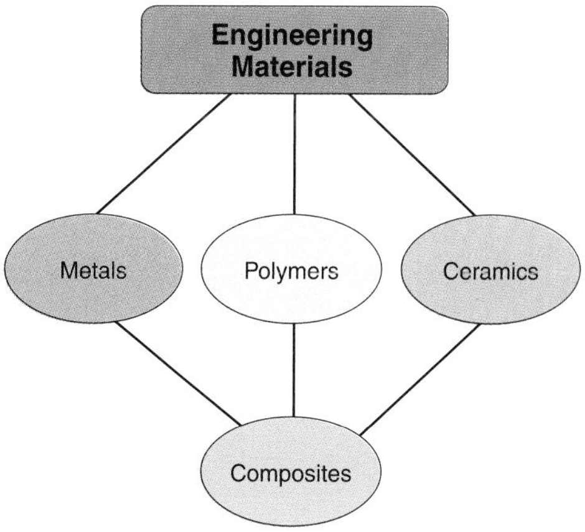

Engineering Materials
Metals
Polymers
Ceramics
Composites

Figure 3-5. The types of engineering materials.

Metallic materials are inorganic, crystalline substances. See **Figure 3-7.** Individual crystals are made up of unit cells. These cells, as will be discussed in Chapter 4, form three-dimensional, crystal lattice structures.

Pure metals have a wide range of physical and mechanical properties. However, they are seldom used in the pure state. Instead, they are generally combined with other metals or inorganic materials to produce an *alloy*. This new material can be engineered to have properties that better match specific needs. Metals are generally grouped into two major categories: ferrous and nonferrous metals.

Ferrous Metals

Iron is the fundamental ingredient for all *ferrous metals*. These materials have a broad range of iron content. Carbon steel and cast iron are composed of over 90 percent (%) iron. Specialty iron alloys (steels) can have nearly 50% alloying elements.

All ferrous metals (irons and steels) are classified as *iron-carbon alloys*. Carbon content in ferrous metals ranges from less than 1% in most steels to 4% of cast iron. Other alloying elements may exceed the percentage of carbon. However, carbon is the most important alloying element in determining the mechanical properties of ferrous metals.

Iron is a very common commercial material. It has been used for many years. Iron appears in a free state only in meteorites. Common iron and steel materials start with iron ore. The first processing step produces pig iron in blast furnaces. This pig iron can be changed into wrought iron that is hammered or rolled into shape. The other form of iron is called cast iron that has about 3% carbon. This material is used for many products including heavy machinery parts and automobile engine blocks. Pig iron, along with scrap steel, can also be the basic ingredient to produce new steel.

nonengineering materials. Gases and liquids that must be confined in a container to hold a particular shape.

ores. Earth or rock from which metal can be commercially extracted.

metallic materials. Inorganic, crystalline substances with a wide range of physical and mechanical properties.

alloy. A mixture of two or more metals or a metal and another inorganic material that yields certain qualities. For example, brass is an alloy of copper and zinc.

ferrous metals. Alloys with varying percentages of iron and carbon.

Figure 3-6. The worker on the left is tapping molten iron from a blast furnace. This class of materials is developed and engineered by chemists to meet specific needs. The view on the right shows molten slag (impurities) being removed from a ladle. (American Iron and Steel Institute, Harsco Corp.)

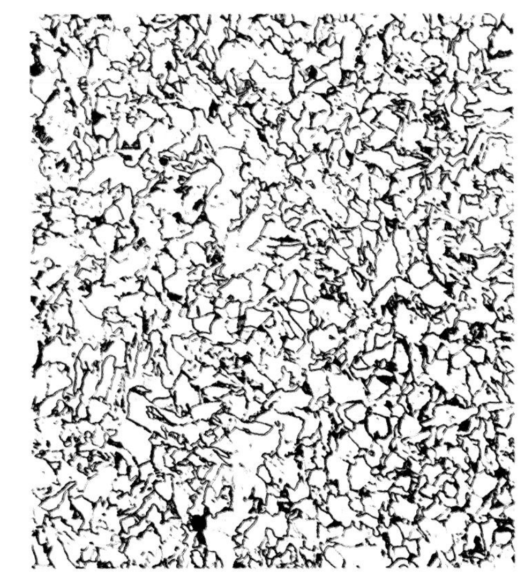

Figure 3-7. This is a piece of mild steel that has been magnified 500 times. Note the crystalline structure.

Steels are grouped into two major types: Carbon steel and high alloy steel. Carbon is the major alloying element in all *carbon steels*. These materials can have up to 1.0% carbon but no more than 1.65% manganese, 0.60% silicon, and 0.60% copper. These steels are classified by a numbering system developed by the American Iron and Steel Institute (AISI). When the first number is 1 in this four-digit system, the steel is a carbon steel. The last two digits indicate the carbon content in points. A point is 1/100 of a percent. For example, AISI 1030 would mean that this carbon steel is 30 × .01 or .30% carbon.

Carbon steels, **Figure 3-8,** are divided into three major classes:

- **Low carbon or mild steels.** These steels have a carbon content up to 0.30% carbon. These materials have low strength and high ductility. They cannot be hardened by heat treating. Cold-working (rolling, drawing, etc.) increases the material's hardness and reduces its ductility.
- **Medium carbon steels.** These have a carbon content between 0.30% to 0.55%. These materials can be case hardened (a thin layer hardened) by heat treating. They are the most widely used steels for structural applications in the automotive, machinery, and agricultural equipment industries.
- **High carbon steels.** These are steels above 0.55% carbon. They are coarse materials that can be hardened by heat treatment. These materials are hard and resist wear. High carbon steels are used for products like springs, saw blades, piano wire, and drills.

carbon steels. **The most common forms of steel, widely used in manufacturing products such as automobiles, machinery, and appliances.**

high alloy steels. **Special steels alloyed with molybdenum, nickel, tungsten, or other elements; tool steel is an example.**

High alloy steels contain significant alloying elements in addition to carbon. Two common high alloy steels are given on the next page:

Figure 3-8. Carbon steels are produced in sheets and coils (left); in bars, rods, and slabs (center); and in structural shapes (right). (United States Steel, American Iron and Steel Institute)

- **High-speed or tool steel.** This is a generic name applied to all steels that hold their sharpness under high temperatures. They generally use molybdenum and tungsten as heat resisting alloys. High-speed steels are widely used for cutting tools and forming dies.
- **Stainless steel.** This is a large family of iron-carbon-chromium alloys that have high-corrosion resistance. This resistance is developed by a layer of chromium oxide that forms on the surface of the metal. Nickel is generally an important additional alloying element in stainless steels.

Figure 3-9. This pot line is used to convert alumina into aluminum. (Alcan Aluminum Co.)

Nonferrous Metals

Metals that do not have iron as their principal ingredient are called *nonferrous metals*. There are about 12 commercially important nonferrous metals. Some of these include aluminum, copper, lead, magnesium, titanium, chromium, silver, gold, and platinum.

nonferrous metals. Metals that do not have iron as their principal ingredient. Copper, aluminum, gold, and silver are all nonferrous metals.

Aluminum

The most commonly used nonferrous metal is ***aluminum.*** The material is the most widely distributed of all metals. It occurs in all common clays. Aluminum is separated from its ore in a two-step process. The ore, which is called bauxite, is chemically processed to form alumina or aluminum oxide. This material is electrolytically treated to produce pure aluminum from the oxide. See **Figure 3-9.** Aluminum is nonmagnetic, oxidizes readily at elevated temperatures, and is very malleable (able to be formed by hammering).

Aluminum is often produced in a commercially pure sheet or ingot. This material is 99.3% pure. Important aluminum alloys contain copper, zinc, tin, and/or lead.

Aluminum conducts heat well and is fairly chemically inert. For these reasons it is widely used in cookware and food preservation (cans, bags, foil, etc.). Aluminum's light weight makes it valuable as a building material, as well as a substitute for iron in engine blocks and in aircraft work.

Copper and copper alloys

Copper is one of the most useful metals and was, most likely, the first one used by humans. The discovery of copper moved us from the Stone Age into the Bronze Age. Copper is a tough, ductile material. It can be drawn or hammered easily into new shapes without breaking or tearing.

Pure or electrolytic copper is 99.9% pure. It is used for electrical conductors and switch contacts, but becomes brittle when heated.

Principal copper alloys are brass, bronze, and nickel silver alloys. *Brass* is a copper-zinc alloy. The alloy can have up to 40% zinc. There are many different brasses that have a broad range of properties. Decorative items, lighting fixtures, and plumbing fixtures are often made of brass.

Bronze is any copper alloy that does not have zinc or nickel as the principle alloying element. Typical bronzes are copper-tin bronzes, and aluminum bronzes. Brazing operations use bronze as a bonding material.

Nickel silvers are copper alloyed with zinc and silver. Copper constitutes about 70% of the alloy while silver makes up 5% to 30%. The balance of the material is zinc. These alloys are used as the base metal for silver-plated dinnerware, springs, electrical connectors, and boat parts.

Other nonferrous metals

Other important nonferrous metals are tin and lead. ***Tin*** is a soft, malleable material that melts at a fairly low temperature. Tin is used as an alloying element and as a coating for steel (such as tinplate for "tin" cans).

There are many materials that are used in manufacturing processes. Some of these materials occur naturally and others are manufactured.

In your science class, the periodic table shows many elements. Many of these elements are used as raw materials for various manufacturing processes. In many instances, a few of these elements are combined to create a material with specific, desired properties. The resulting material is called an alloy.

In the previous discussion about metals, it was stated that nickel, silver, and zinc could be combined to produce alloys used for a number of products. Each of these materials is listed on the periodic table—nickel as Ni, silver as Ag, zinc as Zn. Each of these materials is found in nature in a pure form. However, they are combined to get an alloy that has the desired properties.

Lead is a heavy, soft, malleable metal. Lead is used in storage batteries, a shielding for radiation, and in chemical applications. When combined with tin, lead forms solder that is used in assembling electrical components.

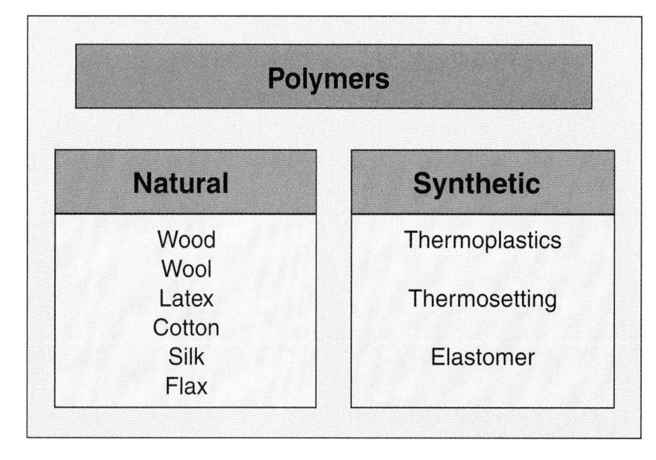

Figure 3-10. These are types of polymers.

Polymeric Materials

Polymeric materials are organic, noncrystalline (without a crystal structure) substances. They are made of many different elements, but carbon and hydrogen always appear in their composition. These elements form long, chainlike molecular structures with a carbon "backbone."

Polymers, as shown in **Figure 3-10,** can be divided into two major groups:

- **Natural polymers.** These are materials that occur in a usable form in nature such as wood, wool, and natural rubber (latex).
- **Synthetic polymers (plastics).** These are materials that are synthetically (human designed and produced) created from natural, organic materials such as petroleum, natural gas, or plant fibers.

polymeric materials. **Organic, noncrystalline substances with long, chainlike molecular structures.**

natural polymers. **Those that occur in a usable form in nature, such as wood, wool, or natural (latex) rubber.**

synthetic polymers. **Polymers designed and produced by humans from natural organic materials.**

hardwoods. **Woods from broad-leafed deciduous trees, such as oak.**

softwoods. **Woods from conifers with needlelike leaves.**

Natural Polymers

There are many natural polymers. One major type is fibers. They are primarily plant fibers and animal hair or hides. Wool and leather are typical animal polymeric materials. Cotton and flax are vegetable fibers that are used to produce fabrics.

Wood is the most common natural polymer that is an engineering material. The term, ***wood,*** is applied to the material derived from trees. Wood, like all vegetable matter, is made up of hollow cellulose fibers. These fibers make up the structure of the material and account for about 70% of the volume.

Lignin is nature's glue that holds the cellulose fiber in place. About 20 to 30% of wood is lignin. The remaining part of wood is composed of extractives (oils, tannins, coloring agents, etc.) and minerals.

Trees that are standing and have a commercial value are called ***timber.*** Boards cut from timber are called ***lumber.*** See **Figure 3-11.** However, lumber is only one product of trees. Logs are sliced into thin layers to produce ***veneer.*** The veneer is then used to produce plywood.

Often, small trees and wood waste are ground into chips. The wood chips may be glued into panels of particleboard or flakeboard. Trees may also be digested into their natural parts—lignin and cellulose fibers. The fibers can be used to produce paper and hardboards. As you can see, wood is a very versatile material that is used in many forms.

All wood, no matter how it is used, may be classified into two groups. These are *hardwoods* and *softwoods*. These groupings have nothing to do with the hardness or softness of the material. Instead, they are based on the type of tree that produced the wood. ***Hardwoods*** are cut from broad-leafed, deciduous trees. ***Softwoods*** are almost always evergreen conifers with needlelike leaves.

Figure 3-11. Trees or timber (left) are cut into logs (center) that are processed into lumber (right). (Weyerhaeuser Co.)

Figure 3-12. This construction worker is using construction lumber and plywood that are both wood products. (Georgia-Pacific Corp.)

Lumber is a major wood product. It is produced in planks (thick, wide, rectangular shapes) and boards (rectangular pieces less that 2" thick). These materials are ***graded*** to indicate proper use and quality.

Most hardwood lumber is used for furniture and architectural moldings (trim for door, windows, etc.). Most softwood lumber is used in building construction, and in door and window manufacture. See **Figure 3-12.**

Synthetic Polymers

Three terms are used to describe the human-made polymers that are in many of the products we use. These terms are polymers, plastics, and synthetic resins. However, *plastics* is becoming the commonly used term to describe this family of materials.

plastics. **Another term for synthetic (manufactured) polymers.**

The first plastic was developed over 100 years ago. This material was celluloid and was designed to replace ivory in billiard balls. About 90 years ago, Bakelite™ was developed by Leo Baekeland. From these early developments has come the world of plastics.

The two basic types of plastics are thermosets and thermoplastics. This classification is based on the structure of the materials that will be discussed more completely in Chapter 4.

All plastics are made up of base organic chains called monomers. There are over 25 families of monomers including ethylene, vinyl chloride, styrene, propylene, acrylic, phenolic, silicon, and epoxy. From these base materials come many different products. You have probably used polyethylene sandwich bags, assembled polypropylene models, and seen acrylic (Plexiglas™) sheets.

Figure 3-13. These aircraft tires are manufactured from elastomers. (Goodyear Tire Co.)

Figure 3-14. This engine is being coated with a paint (synthetic polymer) that was engineered for this application. (Ransburg Corp.)

Thermoplastics are materials that have polymer chains with few cross-links. This structure allows the chains to move past one another. Thermoplastics, therefore, can be heated and reformed repeatedly. Heat will also cause them to lose their shape. Have you ever seen a phonograph record that was left in the sun? It probably warped and could not be played easily. The record was made of a thermoplastic known as vinyl.

Thermosets have cross-linked polymer chains that are set by heat. Once they are molded, these materials resist heat. The plastic handles on a toaster, iron, or kettle are made from thermoset materials-phenolic.

A third type plastic is an *elastomer*. These materials can be stretched to twice their normal length. They return quickly to their original size when the stress is released. Natural and synthetic rubbers are common elastomers. See **Figure 3-13.** Also, rubber and contact cements are elastomer adhesives.

Plastics are the base for what has been called the "material revolution." Wood and metal products are usually designed to meet the characteristics of commonly available materials. However, with plastics, the material is often developed and engineered by chemists to meet specific product needs and characteristics. See **Figure 3-14.**

thermoplastics. Synthetic polymers that can be heated and formed repeatedly.

thermosets. Synthetic polymers that resist heat once they are molded.

elastomers. An adhesive, often called contact cement, that has low strength. It is useful for attaching plastic laminates to panels.

ceramic materials. A range of materials that have a crystalline structure, are inorganic (never living), and can be either metallic or nonmetallic. Ceramic materials are generally stable and are not greatly affected by heat, weather, or chemicals. They have high melting points and are stiff, brittle, and rigid.

Ceramic Materials

Ceramic materials are probably the oldest known materials used by humans. They were the foundation material of the Stone Age. This group contains mostly inorganic (metallic), crystalline materials. Ceramics are a very large family of materials that include:

- **Clay.** These are naturally occurring sediments composed primarily of alumina and silica. These ceramic materials are used to make pottery, brick, pipe, and tiles.

- **Cements.** These are powders that, when mixed with water, set into a solid. The most common cement is Portland cement that is the binder in concrete.
- **Plaster and gypsum.** This is a hydrated calcium sulfate mineral used as building plaster and for wallboard. The drywall board used in many homes is a gypsum product.
- **Refractories.** These are ceramic materials that will withstand high temperatures. They are used to make furnace linings and bricks.
- **Abrasives.** These are hard, sharp ceramic particles. Abrasives are generally an oxide or carbide. They may be used loose, adhered to paper or cloth, or bonded into discs or wheels. Abrasives are used to remove, smooth, or polish materials.
- **Glass.** An amorphous (without set structure) material made from fusing silica (sand). Glass is widely used in packaging containers, in dinnerware, and for windows. See **Figure 3-15.**
- **Porcelain enamel.** A vitrified (fused) ceramic that is extensively used in chemical and electrical applications. Chemistry laboratory equipment and the inside coatings of ovens are often made from porcelain.

Ceramics are the most rigid and least ductile of all materials. They also shatter easily and are very hard. Many ceramics resist chemical action that makes them useful for dinnerware, chemical applications, and sanitary ware (sinks, toilets, etc.).

Composites

composites. **A combination of two or more materials with properties the materials do not have by themselves.**

Composites are materials that are receiving much wider use each year. *Composites* are a combination of two or more materials. However, each material retains its original properties. The new combination is designed so that one material overcomes the weakness of another.

Fiberglass is a common composite. See **Figure 3-16.** The glass fibers (a ceramic) are very strong but highly inflexible. The plastic resin that bonds them

Figure 3-15. These workers are observing sheet glass being produced by the float method. (PPG Industries)

Figure 3-16. This artisan is working on a Fiberglass (composite) body part for an experimental solar powered automobile. (General Motors Corp.)

in place produces a strong material that can withstand flexing and blows. Another common composite is concrete. See **Figure 3-17.** It is a mixture of sand and gravel. These materials are bonded into a solid mass by cement. Wood is a natural polymeric composite. As discussed earlier, the base material is cellulose fibers. These are held in place by lignin.

As you can see from these three examples, all composites contain a structural element—glass fibers, gravel, cellulose, etc. A second material holds the structural materials in a fairly rigid position. The resulting material—a composite—has properties that neither component has all by itself.

Figure 3-17. This foundation for a new home is made from a composite called concrete.

Composites are used for manufacturing hundreds of products including aircraft, automotive and truck bodies, and shower stalls. See **Figure 3-18.**

Purchasing Materials

Engineering materials are purchased by many companies and individuals. These materials are sold in standard sizes or quantities. The typical measurements are units, weight, surface measure, and volume. See **Figure 3-19.**

Figure 3-18. This aircraft propeller is made from a composite. Carbon fiber composites are being widely used in airplane manufacture. (Piaggio Aviation)

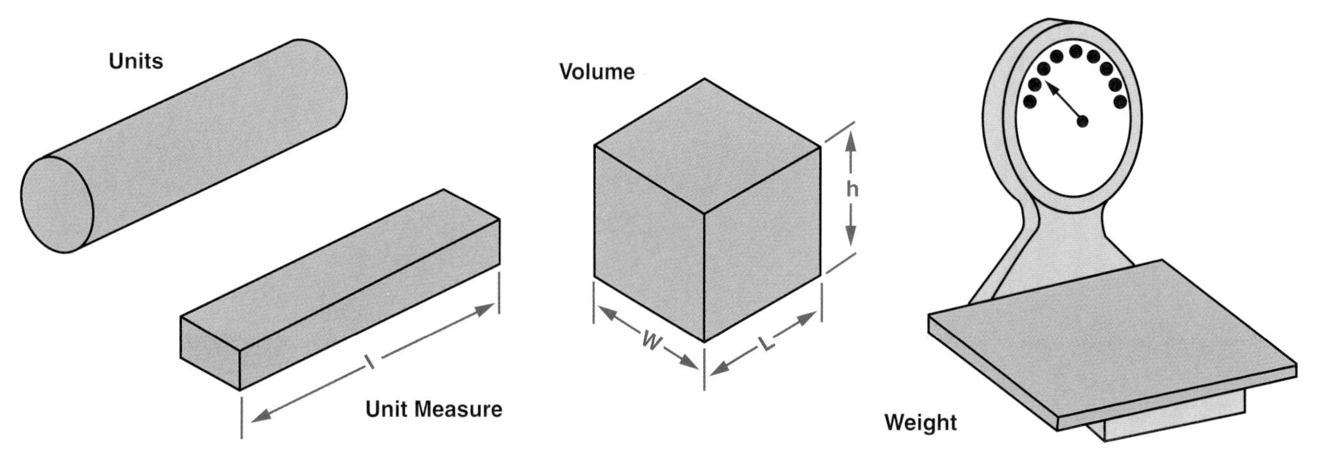

Figure 3-19. Methods of measuring quantities of materials.

Units

Some materials are sold by counting ***units.*** Rolls of plastic are priced by the roll. Screws and bolts are sold by the hundred. Sheets of building materials (plywood, particle board, hardboard, etc.) are sold individually.

Weight

Other materials are sold by their ***weight.*** Ceramic clay and Portland cement are sold by the pound, hundredweight, or ton. Plastic pellets are sold by their weight.

Surface Measure

Surface measure is used in measuring many materials. The two typical measures are linear measure and area. Linear measure is simply the length of the material in a unit. Small quantities of metal rod, strips, and angles are sold by length. Most indoor trim for houses is also sold by the foot. Most retail lumberyards sell wood by the foot or by the piece.

Area

Area is a common way to measure building materials. Plastic laminate (Formica™) is sold by the square foot. Roofing shingles are sold by the square (100 square feet). Area is determined by multiplying the length of the piece by its width.

Volume

The last way to measure materials is by ***volume.*** Two major methods used are cubic yards and board feet. (Lumberyards purchase lumber from mills by the board foot.) Each of these measures requires you to calculate the volume of the material. You must multiply the thickness, width, and length together. Let us look at ways to determine each of these measures. The cubic yard is the measure used for concrete. When you buy concrete, you must decide the volume of material needed. A cubic yard can be thought of as a "container" that is one yard tall, one yard wide, and one yard long. See **Figure 3-20.** The board foot is the measurement for lumber. A board foot is one inch thick and one foot square. See **Figure 3-21.** If the lumber is rough (not surfaced to size), you can use the material's actual size. However, for surfaced lumber, you cannot use the size of the part. Board feet must be figured using the nominal (normal or original) size of the material. For example, the nominal thickness of materials less than one inch thick is one inch. Typical nominal sizes for softwoods are shown in **Figure 3-22.** For softwood finish lumber you use the thickness and width nominal sizes. A 1 × 8 (nominal size) board is actually 3/4" × 7 1/4". However the "1" and the "8" are used in calculating the board measure.

For thicker construction lumber, the same system is used. For example, a 2 × 4 is really 1 1/2" × 3 1/2". Again, for rough lumber you use the actual sizes. Most hardwoods are sold this way.

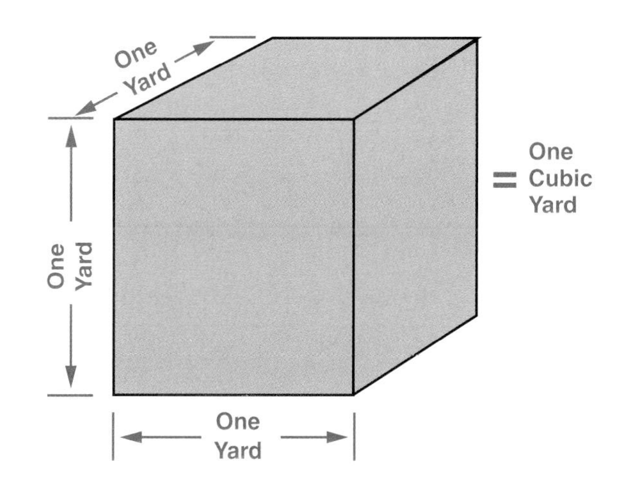

Figure 3-20. A cubic yard is a cube that measures a yard on all sides.

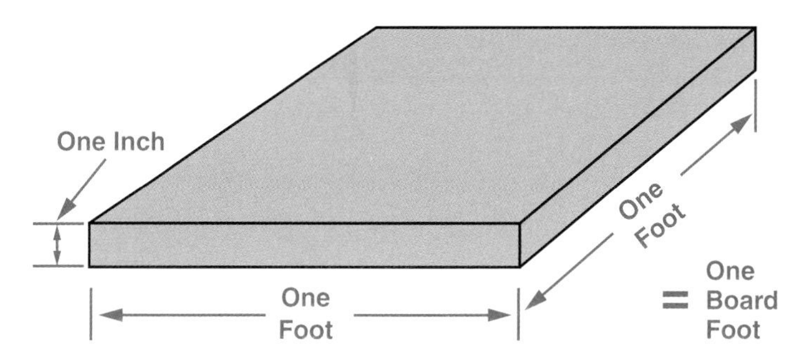

Figure 3-21. A board foot is a measure of volume.

Working with Hazardous Materials

We are surrounded by hazardous materials at school, at work, and at home. Cleaning materials, finishing materials, art and science supplies, lawn and garden chemicals, and gasoline are only a few of the materials that we encounter. Follow these guidelines to prevent accidents:

- Read the Material Safety Data Sheet (MSDS) before using any chemical or material for the first time.
- Follow the instructions in the MSDS on how to use, store, and dispose of the chemical.
- Store flammables in an approved container or in an approved flammables cabinet.
- Wear the appropriate personal protective equipment (i.e., gloves, goggles, respiratory equipment, etc.) when using a hazardous substance.
- Keep emergency phone numbers posted.

You must be able to identify and classify hazardous materials. Be familiar with all hazardous materials in the lab or shop area. Review the MSDSs periodically to refresh your memory and also to make sure you are aware of any new materials.

Softwood Lumber Sizes		
	Nominal Size	Actual Size
Boards	1 × 4	3/4 × 3 1/2
	1 × 6	3/4 × 5 1/2
	1 × 8	3/4 × 7 1/4
	1 × 10	3/4 × 9 1/4
	1 × 12	3/4 × 11 1/4
Dimension Lumbers	2 × 4	1 1/2 × 3 1/2
	2 × 6	1 1/2 × 5 1/2
	2 × 8	1 1/2 × 7 1/4
	2 × 10	1 1/2 × 9 1/4
	2 × 12	1 1/2 × 11 1/4

Figure 3-22. The nominal sizes of lumber are not their actual sizes.

If you encounter a substance and you cannot identify it immediately, assume that it is hazardous. Notify your teacher immediately. Look for nearby containers and then review the MSDSs.

When disposing of hazardous wastes, also follow the requirements specified in the MSDS. Use the proper protective equipment (also specified in the MSDS).

Material Safety Data Sheets

The Occupational Safety and Health Act (OSHA) requires employers to keep a list of all hazardous chemicals and materials used on their premises and maintain a file of Material Safety Data Sheets (MSDS) on the chemicals. They also must train their employees to use the chemicals and respond properly should an accident occur. The MSDS includes the following information:

- Chemical name(s) and symbol.
- Manufacturer's contact information.
- Flammability and reactivity data (vapor pressure and flash point information).
- Health hazards.
- Overexposure limits and effects.
- Spill cleanup procedures.
- Personal protection information.

Additional information is normally included in the MSDS. A sample MSDS is shown on page 484 of the *Appendix*.

Working Safely

The importance of safety in the workplace cannot be overemphasized. Working with hazardous materials and wastes is only one aspect of safety. Other aspects (which are discussed in appropriate locations throughout this text) include working safely with machinery and tools, working safely with electricity, preventing fires, and understanding how to extinguish any fires that may occur.

There are many codes, laws, standards, and regulations related to safety. Organizations such as the *Environmental Protection Agency (EPA), Occupational Safety and Health Administration (OSHA), National Electric Code (NEC),* and the *American Society for Testing Materials (ASTM)* develop safety standards. Often, such standards are adopted into laws. These laws are designed to protect workers from injuries and other health risks.

In addition to safety standards specified by law, most companies have safety standards developed for their specific systems and processes. Many companies have safety training sessions to ensure that employees are familiar with the specific hazards and safety regulations in their workplace.

Summary

Materials are everywhere. These essential building blocks of the world are gases, liquids, or solids. Solid materials are often called engineering materials. These materials can be separated into three groups: metals, polymers, and ceramics. Within these groups are literally thousands of individual materials. Each of these has a particular set of properties that make them useful for specific uses. The structures of these materials will be discussed in Chapter 4, while the properties will be explored in Chapter 5.

Key Words

All of the following words have been used in this chapter. Do you know their meanings?

alloy
carbon steels
ceramic materials
composites
elastomers
engineering materials
ferrous metals
hardwoods
high alloy steels
inorganic materials
manufacturing
metallic materials
nonengineering materials
nonferrous metals
ores
organic materials
plastics
polymeric materials
softwoods
synthetic polymers
thermoplastics
thermosets

Test Your Knowledge

Please do not write in this text. Place your answers on a separate sheet.

1. What is the difference between an organic and an inorganic material?
2. What is the difference between an engineering material and a nonengineering material?
3. List and describe the three major types of materials.
4. Generally, pure metals are combined with other metals or inorganic materials to produce a(n) _____.
5. _____ is the fundamental ingredient for all ferrous metals.
6. List the two major types of steel.
7. The most commonly used nonferrous metal is:
   a. Bronze.
   b. Brass.
   c. Aluminum.
   d. Nickel.
8. What is the difference between a hardwood and a softwood?
9. *True or False?* Plastics is becoming the commonly used term to describe synthetic polymers.
10. *True or False?* Thermosets can be heated and formed repeatedly.
11. Glass, abrasives, clays, and cements are examples of _____ materials.
12. Which of the following is a common composite?
    a. Lead.
    b. Fiberglass.
    c. Thermoplastic.
    d. Construction lumber.

Applying Your Knowledge

1. Identify materials found in a certain product and determine substitute materials.
   a. Select a common product from either your home or the school technology education laboratory.
   b. Study the product. (It should be one made of several different materials.)
   c. Using a chart similar to the one in **AYK 3-1**, identify the materials in the product. (You should be able to identify materials such as metal, ceramic, plastic, or a composite.)
   d. Next, list a substitute for each of the materials used.
2. Collect and label a set of samples that include the following materials:
   a. Natural polymeric composite (wood).
   b. Synthetic polymer (plastic).
   c. Ferrous metal (iron or steel).
   d. Nonferrous metals.
   e. Clay-based ceramic.
   f. Silica-based ceramic (glass).
3. Research the library for information about a new material or a new application of a material.
   a. Go to the periodical section. Search through the contents pages of the magazines for articles on materials.
   b. Prepare a summary of the selected article.
   c. Give a written or oral report to the class.

Product:		
Material	**Type**	**Substitute**

AYK 3-1. On a separate sheet of paper, prepare a chart similar to this one on which to record your responses.

Chapter 4
Structure of Materials

Objectives

After studying this chapter, you will be able to:

- ✓ Choose materials based on knowledge of structures.
- ✓ Distinguish between microstructures and macrostructures.
- ✓ Describe the structure of atoms.
- ✓ Explain how to use the periodic table of elements in finding out about material properties.
- ✓ Describe how the bonding of atoms takes place and list various types of bonds.
- ✓ List three types of structures in which atoms are arranged.
- ✓ Give examples of various types of macrostructures.
- ✓ List and describe four categories of synthetic composites.

You may be asked to select a material for a task or product. How will you make this decision? On what basis will you choose one material over another? Designers and engineers base their choices on a knowledge of the basic structure of a material. See **Figure 4-1.** This structure determines the properties of the material. It determines how a material reacts to stress, temperatures, chemicals, and many other factors.

The basic nature of materials can be viewed at two levels: microstructures and macrostructures. Examples of these are shown in **Figure 4-2.**

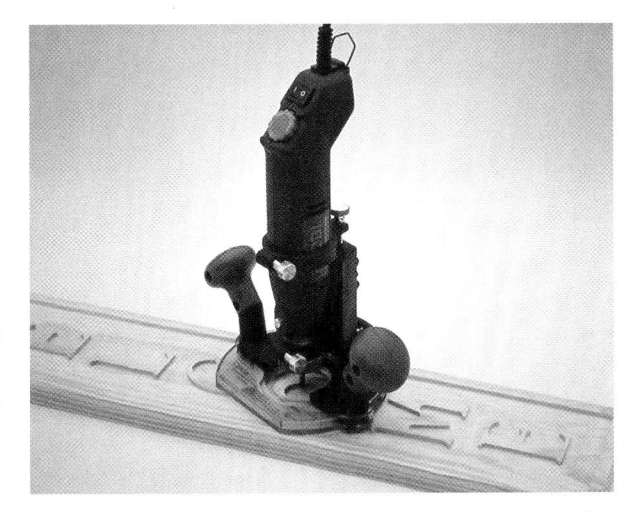

Figure 4-1. The designers considered each material's structure as this router was designed. (Ryobi)

Microstructures

The *microstructure* is the way molecules and crystals are arranged in a material. The microstructure is what makes a material uniform. It determines, for example, whether a material is aluminum, steel, or polyethylene. The microstructure is so small that it must be viewed through a microscope.

microstructure. The way molecules and crystals are arranged in a material.

In studying the microstructure of a material, you will learn about the atomic particles that are present and the way they are held together. This structure, as

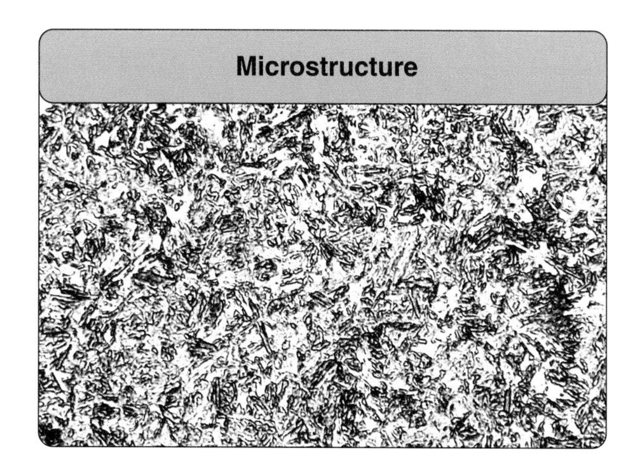

Figure 4-2. Structure of materials: The view on the top is a microstructure of steel magnified 200 times. On the bottom is a macrostructure showing particleboard. Notice the structures of each.
(Bethlehem Steel Corp., Weyerhaeuser Co.)

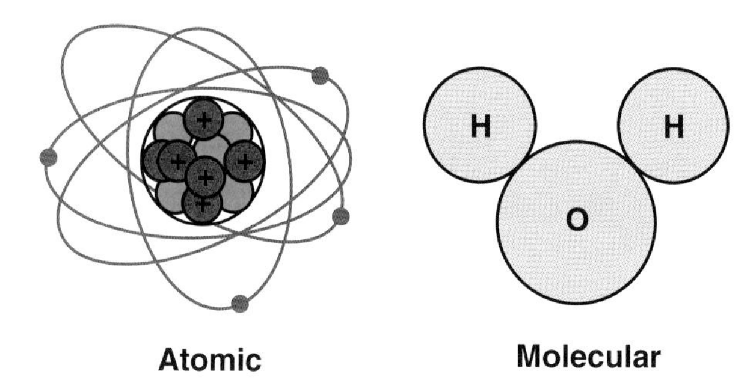

Figure 4-3. Materials are made up of subatomic particles, which form the atomic structure of atoms. Atoms bond with other atoms and form a molecular structure.

shown in **Figure 4-3,** can be studied by exploring the atomic structure and the molecular structure.

Atomic Structure

The *atom* is the basic unit in the study of the microstructures of materials. It was once believed to be the smallest particle that could exist. It was seen as a single unit of mass that could not be further divided.

We now know that the atom is composed of smaller particles. These particles are called *subatomic particles.* There are many types of subatomic matter. Knowing about all types of matter is not essential to understand materials and their characteristics.

atom. **The basic unit in the study of the microstructures of materials.**

subatomic particles. **The smaller particle which make up an atom.**

However, some knowledge of atomic theory is vital. It provides insight into the factors that establish material properties. With this knowledge, for instance, you can understand how oxidation (rusting, tarnishing, corrosion), electrical and thermal conductivity or ductility, and fatigue occur.

Structure of Atoms

Each atom is composed of some basic building blocks. There are electrons, protons, and neutrons.

Bohr's atomic model. **The simplest and oldest model used to describe the atom and its structure.**

A number of models and equations describe the atom and its structure. The simplest and oldest in present use is *Bohr's atomic model.* See **Figure 4-4.** Although it does not fully describe the atom, it provides a basic picture of the physical arrangement of the atom. It suggests that all atoms have a series of electrons that orbit a stationary nucleus in set paths.

Electrons

Electrons are subatomic particles that have a negative electrostatic charge. They orbit around the nucleus (center) of the atom. The positively charged proton in the nucleus attracts the electron. This attraction is offset by the centrifugal force created by the moving electron. (Centrifugal force is the action that forces objects outwardly away from a center of rotation.) As a result, each electron has its own path around the nucleus. Several electrons, however, can orbit at approximately the same distance from the nucleus. They are said to be moving in a band or energy level called a *shell.* See **Figure 4-5.** The shell closest to the nucleus is called the K shell followed outward by the L, M, N, O, P, and Q shells.

Each shell can contain a maximum number of electrons. The number is determined by a simple relationship as shown in **Figure 4-6.** The number of electrons in the outer shell is important. It determines how atoms can bond together to form solid materials. The bonding, in turn, helps determine the material's mechanical properties. These properties will be discussed in Chapter 5.

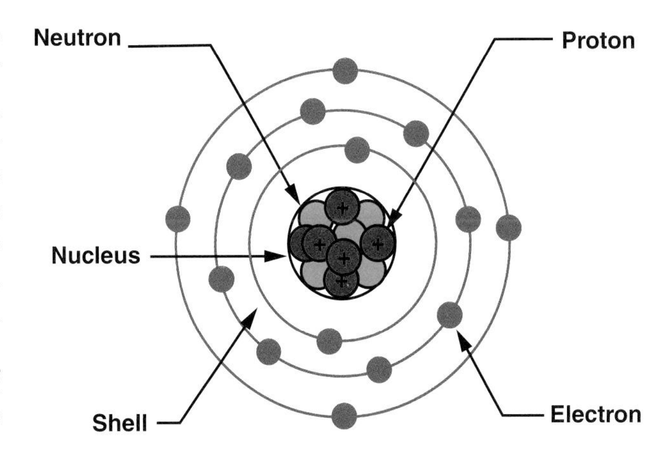

Figure 4-4. This illustration shows the model of an atom.

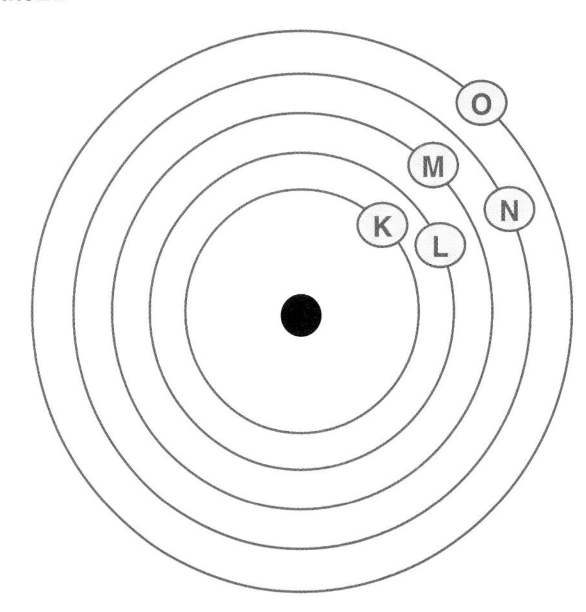

Figure 4-5. Electron shells or energy levels orbit at approximately the same distance from the nucleus.

Nucleus

The *nucleus* is the center of the atom. It is made up of several types of subatomic particles. Two of these are the most important and exist in all atoms except hydrogen. They are protons and neutrons.

Protons

Protons hold a positive electromagnetic charge, while electrons hold a negative charge. Since like charges repel one another and unlike charges attract each other, the positive charge of protons provides the force that holds electrons in orbit. This action is like the gravity of the sun that holds the earth and other planets in their elliptical orbits.

Neutrons

The *neutrons* are neutral particles. The protons and neutrons provide almost all the mass (weight) of an atom. The electrons add very little weight, but their orbits establish the volume or size of the atom.

Shell	Calculation	Maximum Number of Electrons
K	2×1^2	2
L	2×2^2	8
M	2×3^2	18
N	2×4^2	32
O	2×5^2	50

Figure 4-6. Each shell or energy level can contain a maximum number of electrons.

Ions

Hydrogen is the simplest atom in nature. It contains one proton and one electron. The electron and the proton have mutual and equal attraction. The result is an electrostatically stable atom.

ion. An atom with a positive or negative charge.

However, if an atom loses or gains electrons through some action, it becomes an *ion*. An ion is a positively or negatively charged particle. In the case of hydrogen, it becomes a positively charged ion when it loses an electron. The positive charge of the proton in the nucleus is not counterbalanced by a negatively charged electron.

As the hydrogen ion seeks to gain electrostatic balance, it will seek another electron or a negatively charged ion (one that has gained an electron). For instance, if two hydrogen ions find an oxygen ion, a molecule of water (H_2O) will form.

The Periodic Table

periodic table of elements. A listing of the chemical elements arranged by their properties and levels of chemical activity.

Over the years, a table of elements has been developed. This table is called the *periodic table of elements*. See **Figure 4-7.** It is organized around a series of rows and columns.

This table is studied in depth in chemistry classes. In the study of manufacturing, you will be interested in how it deals with material properties. As you study the abbreviated periodic table in **Figure 4-8,** you will be able to make some general observations. These include:

Periodic Table of The Elements

METALS — METALLOIDS AND NONMETALS — TRANSITION METALS

1	2	3	4	5	6	7	8	9	10	11	12	13	14	15	16	17	18
Hydrogen 1.0080 H 1																	Helium 4.003 He 2
Lithium 6.939 Li 3	Beryllium 9.012 Be 4											Boron 10.811 B 5	Carbon 12.01115 C 6	Nitrogen 14.007 N 7	Oxygen 15.999 O 8	Fluorine 18.998 F 9	Neon 20.183 Ne 10
Sodium 22.990 Na 11	Magnesium 24.312 Mg 12											Aluminum 26.981 Al 13	Silicon 28.086 Si 14	Phosphorus 30.974 P 15	Sulfur 32.064 S 16	Chlorine 35.453 Cl 17	Argon 39.948 Ar 18
Potassium 39.102 K 19	Calcium 40.08 Ca 20	Scandium 44.956 Sc 21	Titanium 47.90 Ti 22	Vanadium 50.942 V 23	Chromium 51.996 Cr 24	Manganese 54.938 Mn 25	Iron 55.847 Fe 26	Cobalt 58.933 Co 27	Nickel 58.71 Ni 28	Copper 63.54 Cu 29	Zinc 65.37 Zn 30	Gallium 69.72 Ga 31	Germanium 72.59 Ge 32	Arsenic 74.922 As 33	Selenium 78.96 Se 34	Bromine 79.909 Br 35	Krypton 83.80 Kr 36
Rubidium 85.47 Rb 37	Strontium 87.62 Sr 38	Yttrium 88.905 Y 39	Zirconium 91.22 Zr 40	Niobium 92.906 Nb 41	Molybdenum 95.94 Mo 42	Technetium (99) Tc 43	Ruthenium 101.07 Ru 44	Rhodium 102.91 Rh 45	Palladium 106.4 Pd 46	Silver 107.87 Ag 47	Cadmium 112.40 Cd 48	Indium 114.82 In 49	Tin 118.69 Sn 50	Antimony 121.75 Sb 51	Tellurium 127.60 Te 52	Iodine 126.90 I 53	Xenon 131.30 Xe 54
Cesium 132.90 Cs 55	Barium 137.34 Ba 56	Lanthanum 138.91 La 57	Hafnium 178.49 Hf 72	Tantalum 180.95 Ta 73	Tungsten 183.85 W 74	Rhenium 186.21 Re 75	Osmium 190.2 Os 76	Iridium 192.2 Ir 77	Platinum 195.09 Pt 78	Gold 196.97 Au 79	Mercury 200.59 Hg 80	Thallium 204.37 Tl 81	Lead 207.19 Pb 82	Bismuth 208.98 Bi 83	Polonium (210) Po 84	Astatine (210) At 85	Radon (222) Rn 86
Francium 223 Fr 87	Radium (226) Ra 88	Actinium 227 Ac 89															

RARE EARTH METALS	Cerium 140.12 Ce 58	Praseodymium 140.91 Pr 59	Neodymium 144.24 Nd 60	Promethium (147) Pm 61	Samarium 150.35 Sm 62	Europium 151.96 Eu 63	Gadolinium 157.25 Gd 64	Terbium 158.92 Tb 65	Dysprosium 162.50 Dy 66	Holmium 164.93 Ho 67	Erbium 167.26 Er 68	Thulium 168.93 Tm 69	Ytterbium 173.04 Yb 70	Lutetium 174.97 Lu 71
ACTINIDES	Thorium 232.04 Th 90	Protactinium (231) Pa 91	Uranium 238.03 U 92	Neptunium (237) Np 93	Plutonium (242) Pu 94	Americium (243) Am 95	Curium (247) Cm 96	Berkelium (249) Bk 97	Californium (251) Cf 98	Einsteinium (254) Es 99	Fermium (253) Fm 100	Mendelevium (256) Md 101	Nobelium (254) No 102	Lawrencium (257) Lr 103

Figure 4-7. The periodic table of elements is used in the study of both chemistry and manufacturing.

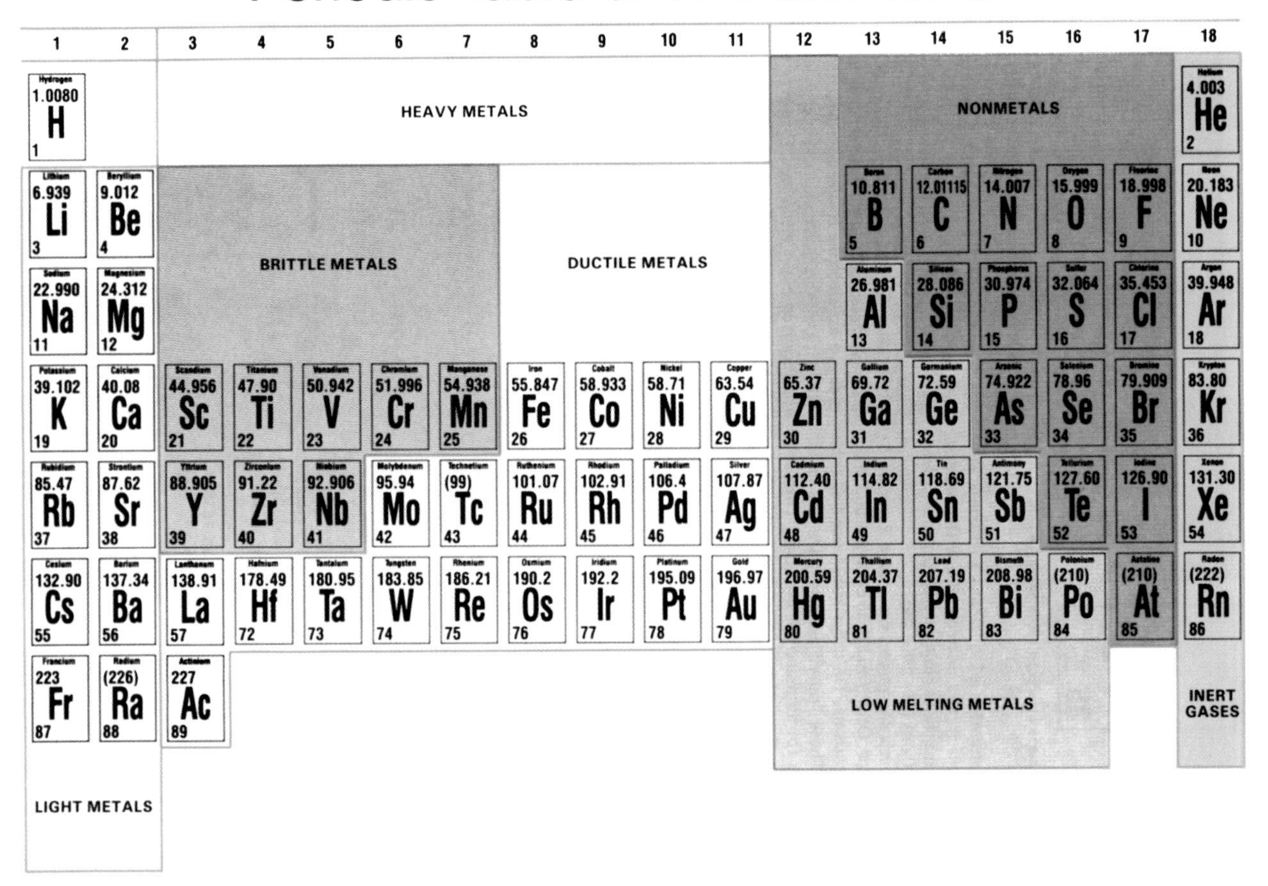

Figure 4-8. This periodic table of elements divides the elements into major groups according to types and properties.

- All the elements in the same vertical row will have similar, but not identical, properties.
- The groups in the left portion of the periodic table easily form positively charged ions.
- Elements in the far right column are inert gases. They will not react with other chemicals or dissolve in metals.
- The other elements on the right side form negatively charged ions.
- Elements on the left and bottom portions of the table are metals. They make up the vast majority of the elements.
- Metals on the left portion of the table are called "light metals."
- Metals on the right portion of the table melt at low temperatures.
- Heavy metals can be divided into brittle and ductile groups by their location on the table.

The Periodic Classification of Elements

The chemical activity of each element is determined by two main factors. These are its atomic number and the grouping of electrons in its outer shell.

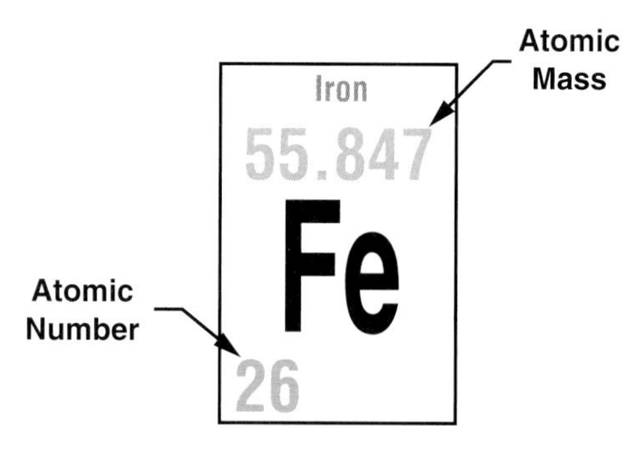

Figure 4-9. This is a periodic table entry for iron.

The grouping of electrons in the outer shell of an element is of special interest to product designers and engineers. It is this grouping that, among other factors, determines:

- The chemical properties of the element.
- The way atoms are bonded together to form solid materials. (This, in turn, determines the mechanical properties of the material.)
- The size of the atom. (This factor is directly related to the material's electrical conductivity.)

Atomic Number and Weight

atomic number. **Represents the number of electrons present if the atom has not been ionized.**

atomic mass number. **Is equal to the sum of the number of protons and neutrons in the nucleus.**

elements. **The "pure" substances that are the basic building blocks for all materials on earth.**

All atoms are designated by two numbers as can be seen in the periodic table in Figure 4-7. The bottom number is the ***atomic number.*** This number represents the number of electrons present if the atom has not been ionized. The atomic number is also equal to the number of protons in the nucleus. For example, iron (Fe), **Figure 4-9,** has an atomic number of 26. Therefore, iron has 26 electrons and 26 protons.

The top number is the ***atomic mass number.*** This number is equal to the sum of the number of protons and neutrons in the nucleus. For example, the atomic mass for iron is 55.85. The whole number, 55, represents the number of nucleons (neutrons and protons) present in the nucleus. Since the atomic number of 26 indicates there are 26 protons, there must be 29 neutrons (55 – 26 = 29).

These numbers are important. They describe a "pure" substance. Only iron has 26 electrons and protons and 29 neutrons. No other substance has this combination. These "pure" materials are the basic building blocks for all materials on earth. They are called ***elements.*** Everything you see around you is composed of fewer than 100 elements that naturally appear on earth.

Elements are stable. Iron does not change into aluminum or manganese or any other element. This stability allows us to:

- Predict the characteristics of pure substances.
- Combine various substances to produce more complex materials.
- Select suitable materials for specific engineering applications.

Academic Link

Mathematic skills have significant uses in manufacturing. Whether we are measuring amounts of materials or figuring production times, mathematics is a necessary part. Without math, how would accountants pay the bills or employees figure out their pay?

As just studied, figuring out the number of neutrons and protons in a pure substance is done with math. In the given example, figuring that iron has 26 protons and 29 neutrons in its nucleus was done with subtraction. Without the ability to subtract the 26 from 55, the number of neutrons could not be determined.

Bonding

Bonding is an attractive force that holds atoms or ions together. All atoms seek to reach a stable structure. This condition exists when there are eight electrons in the outer energy level (except for hydrogen and helium that need only two). Only the inert gases are normally in this state. Therefore, they do not react or bond to other elements. All other elements are active in varying degrees as they seek stability.

bonding. The force that holds atoms or ions together.

compound. Atoms of two or more elements in definite proportions.

There are three ways atoms can reach structural stability:

- Releasing extra electrons, forming positive ions.
- Receiving extra electrons, forming negative ions.
- Sharing electrons with adjacent atoms.

Primary Bonds

The ways atoms reach structural stability are the basis for how atoms bond to form solid materials. They are the foundation for the three primary ways that chemical bonds are developed. See **Figure 4-10.** These methods, which directly impact a material's properties, are ionic bonding, covalent bonding, and metallic bonding.

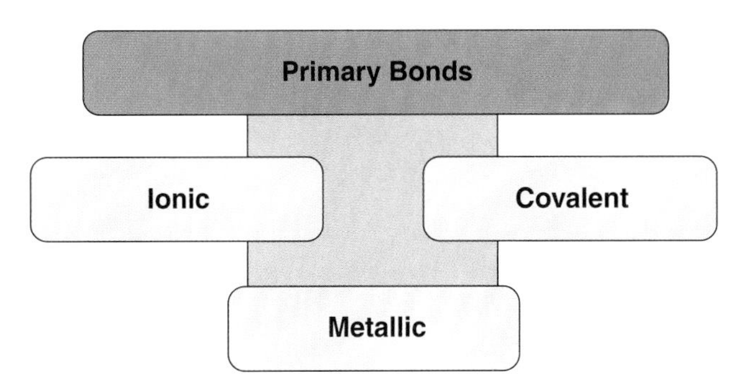

Figure 4-10. These three primary bonds will directly affect a material's properties.

Ionic bonding

As discussed earlier, atoms give up or take on electrons to reach structural stability. This action will produce ***ions.*** For example, the metal, sodium (Na), must lose the one electron in its outer energy level to reach structural stability. The loss of one electron in its M shell leaves the L shell as the outer energy level. This shell has the eight electrons that are required for stability. However, the ion now has ten electrons and eleven protons. It is a positively charged ion, Na^+. See **Figure 4-11.**

Chlorine reaches stability in an opposite action. It has seven electrons in its outer energy shell. It must pick up an electron to become structurally stable. When it gains an electron, its M energy level will have the required eight electrons. However, the ion now has 18 electrons and a negative charge, or Cl^-.

The positively charged sodium ion will attract the negatively charged chlorine ion. This action produces a compound. A *compound* is atoms of two or more different elements in definite proportions. You know this resulting material as table salt, NaCl.

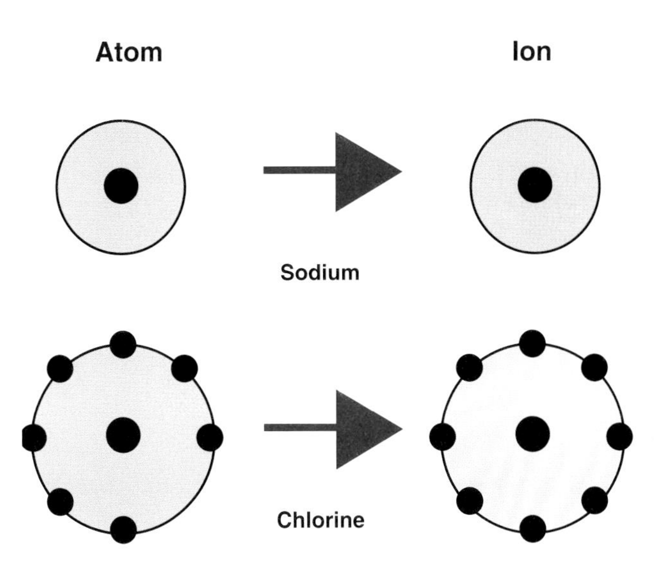

Figure 4-11. This diagram shows the outer energy levels of sodium and chlorine atoms and ions.

However, there will be more than one sodium and one chlorine ion present. Each ion acts the same way. Sodium (positive) ions attract chlorine *ionic bonding.* The bond formed between atoms by the attraction of opposite electrical charges.

monomer. A chemical compound that can unite with other monomers to form a polymer.

(negative) ions and repel other sodium ions. Likewise, the chlorine ions attract sodium ions and repel other chlorine ions. These electrostatic forces cause a very uniform and stable structure. The force is called *ionic bonding*. See **Figure 4-12.**

Ionic bonds are formed between metallic and nonmetallic elements. Metals that have fewer than three electrons in their outer energy levels will become positive ions. The nonmetallic elements will gain electrons from these metals to fill their outer level.

Materials with ionic bonds will dissolve or separate in a solution. They will reform their solid structure when the solvent is removed.

Covalent bonding

The elements toward the center of the periodic table do not ionize easily. Their outer energy levels are half full. Therefore, they do not pick up or lose electrons readily. Instead, these elements share electrons to produce a stable structure. This process is called covalent bonding.

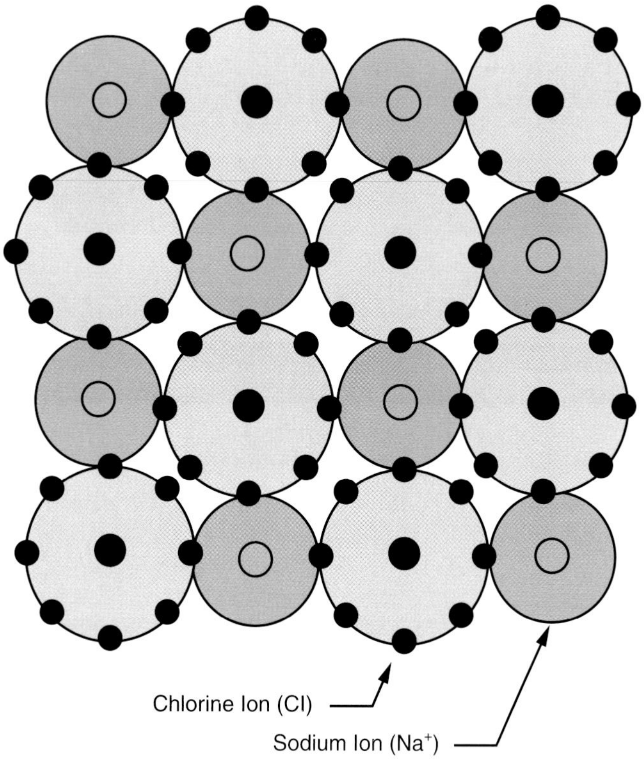

Chlorine Ion (Cl)
Sodium Ion (Na^+)

Figure 4-12. This diagram shows bonding of sodium and chlorine to form table salt.

The simplest example of covalent bonding is hydrogen gas, H_2. Hydrogen has only one energy level. Therefore, it needs two electrons to reach stability. When two hydrogen atoms come close together, the single electron in each atom will act as if it belongs to both atoms. See **Figure 4-13.** Each atom, by sharing its one electron, gives both atoms a stable K energy electron. Most gases form molecules in this manner.

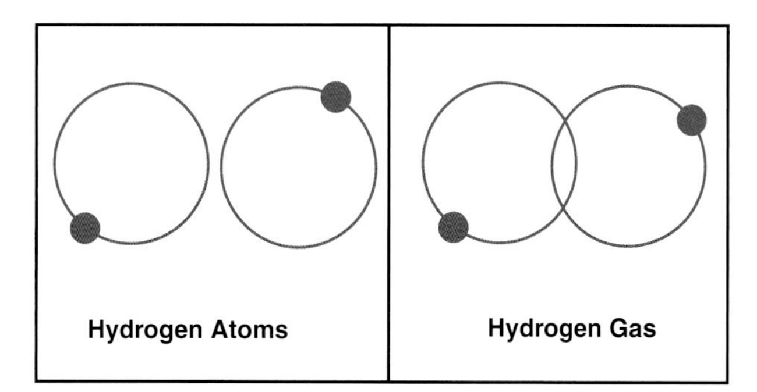

Figure 4-13. Covalent bonds are formed to produce hydrogen gas.

This principle of covalent bonding is also present in solids. An extremely important element that participates in covalent bonding is carbon. This material is the foundation of all organic compounds.

Carbon can combine with other carbon atoms to form a pure solid. A diamond is an example of a pure carbon compound. Carbon also unites with a large number of other elements. These elements unite with carbon's four covalent links. Carbon atoms often develop covalent bonds to form chain-like molecules. The base molecule, like the methane one shown in **Figure 4-14,** is called a *monomer*. Several monomers are united to form a *polymer* in which there is a repeated unit (mer). See **Figure 4-15.** Polymers are often called *plastics*.

Metallic bonding

Metallic elements have a common characteristic. They have one to three electrons in their outer energy level. These electrons are only loosely held by the positive charge in the nucleus of the atom.

These loosely bonded electrons are easily separated from an atom. They form what is called an ***electron cloud*** or ***electron gas.***

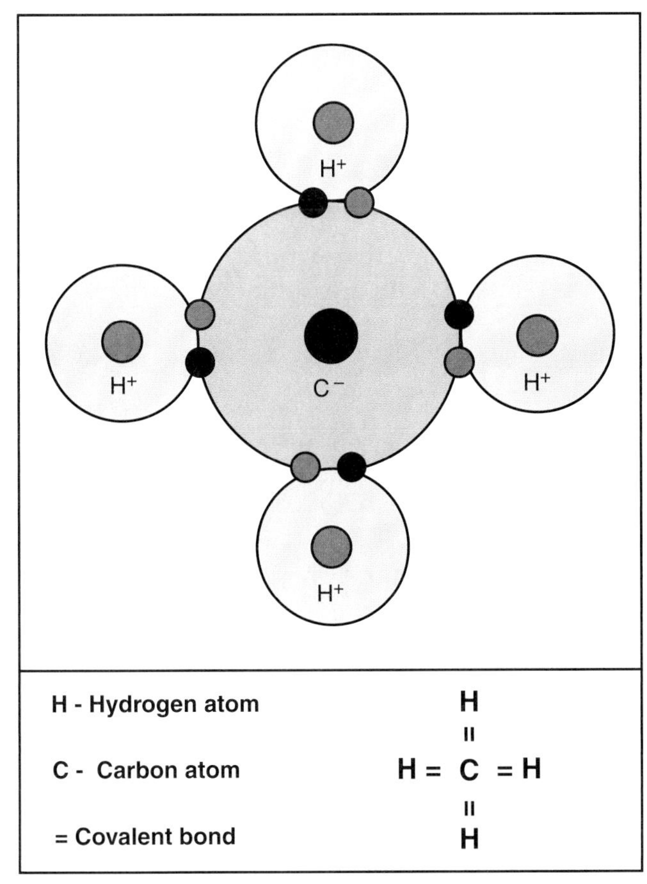

Figure 4-14. Methane is produced by covalent bonds between one carbon and four hydrogen atoms. The top drawing shows the shared electrons. The bottom drawing shows the common way to diagram organic compounds.

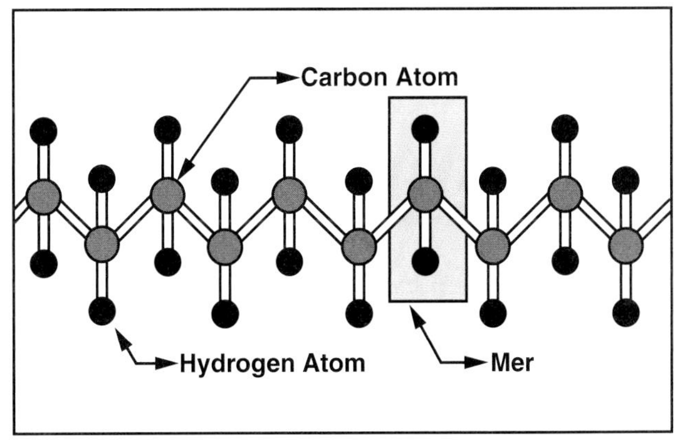

Figure 4-15. This diagram of polyethylene shows a single ethylene monomer and a portion of the polymer chain.

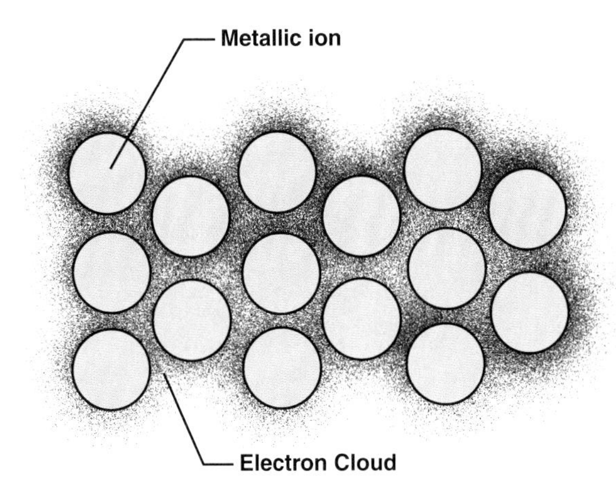

Figure 4-16. Metals are formed with positively charged metallic ions suspended in a negatively charged electron cloud.

Structurally stable metallic ions are formed as the electrons leave the atoms and enter the common electron cloud. These ions have a positive charge. The electron cloud has a negative charge. The electrons in the cloud repel each other causing them to be evenly spaced. Likewise, the metallic ions are spaced evenly due to the repelling positive charges. However, the negative charge of the free electrons and the positive charge of the ions create an attracting force. This combination of repelling and attracting forces, called *metallic bonding*, produces a very rigid structure. It contains positive metallic ions suspended uniformly in a cloud of electrons. See **Figure 4-16.**

polymer. An organic material composed of chain-like molecules of repeating units; plastic.

metallic bonding. A combination of attracting and repelling forces that produces a very rigid structure.

Van der Waals forces. A secondary form of bonding important in determining how a plastic will behave: if the bond is strong, the plastic is thermoset; if weak, a thermoplastic.

Secondary Bonds

There are other bonding forces besides the three primary bonds. The major secondary bond is called *Van der Waals forces*. These forces are present to some extent in all materials. However, they are important in understanding plastics.

The skeleton of a plastic is composed of carbon and other elements. They are covalently bonded into polymer chains. The electrostatic charge along the chain is not uniform. At one point, it may be somewhat positive. At another point, it may be slightly negative. This variation in charge will cause the chains to attract one another, forming the solid material.

If the Van der Waals forces holding the chains are strong, the material is a *thermoset.* Weaker bonds form *thermoplastic* materials.

Arrangement of Atoms

The atoms of all materials are arranged in some manner. This arrangement gives the material its properties. From a microstructural point, atoms are arranged as molecular structures, crystalline structures, or amorphic structures.

Molecular Structures

molecular structures. **A structure made up of polymeric chains held together with covalent bonds.**

crystalline structures. **Those composed of boxlike units, called crystals, arranged in a lattice form.**

amorphous structure. **An internal arrangement of atoms in a material that does not follow any specific pattern.**

Molecular structures of solid materials generally use covalent bonds. The structure is polymeric in nature. The material is made up of a number of *mers.* These basic units are chemically united to form *polymers* (many mers). The result is a very complex, chain-like structure. See **Figure 4-17.**

These structures are common for plastics, wood fibers, and natural rubber. Each of these materials is composed of millions of polymeric chains.

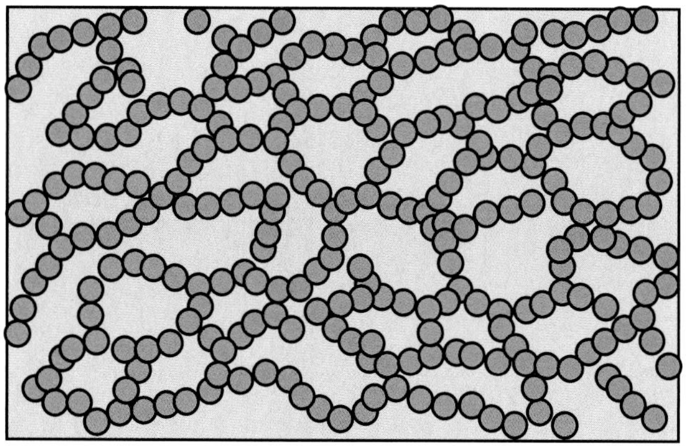

Figure 4-17. The coiled arrangement of mers is typical of polymeric materials.

Crystalline Structures

Metals and most ceramic material have *crystalline structures.* The atoms, like those in molecular structures, are arranged in a regular pattern. They form three-dimensional box-like structures called unit cells or crystals. There are 14 possible crystalline structures. Of these, three are the most common. They include the structures of most metals. These, as shown in **Figure 4-18,** are:

- Face-centered cubic–FCC.
- Body centered cubic–BCC.
- Hexagonal close packed–HCP.

Each crystal has a simplest form. It is called the ***unit cell.*** These cells can combine to form larger structures. Each unit cell always retains its basic three-dimensional structure that is called a ***crystal lattice.*** This lattice is sometimes simply called a crystal.

Crystals unite during the cooling of liquid metals to produce grains of materials. These grains interconnect to form the solid structure. Look back to Figure 4-2. You will see that the grains form in a random pattern. The surface between the grains is called the ***grain boundaries.*** It is along these boundaries that materials separate or fracture. The ease with which this happens determines the machinability and strength of the material.

Amorphous Structures

Amorphous means "without form." This does not mean a material with an *amorphous structure* does not have size and shape. It does mean that the internal structure does not follow any pattern. There is no regular, repeating pattern of the atoms in the material.

All liquids and molten metals are amorphous materials. Glass is the most important amorphous engineering material.

macrostructure. The way in which different substances unite to form complex materials called composites.

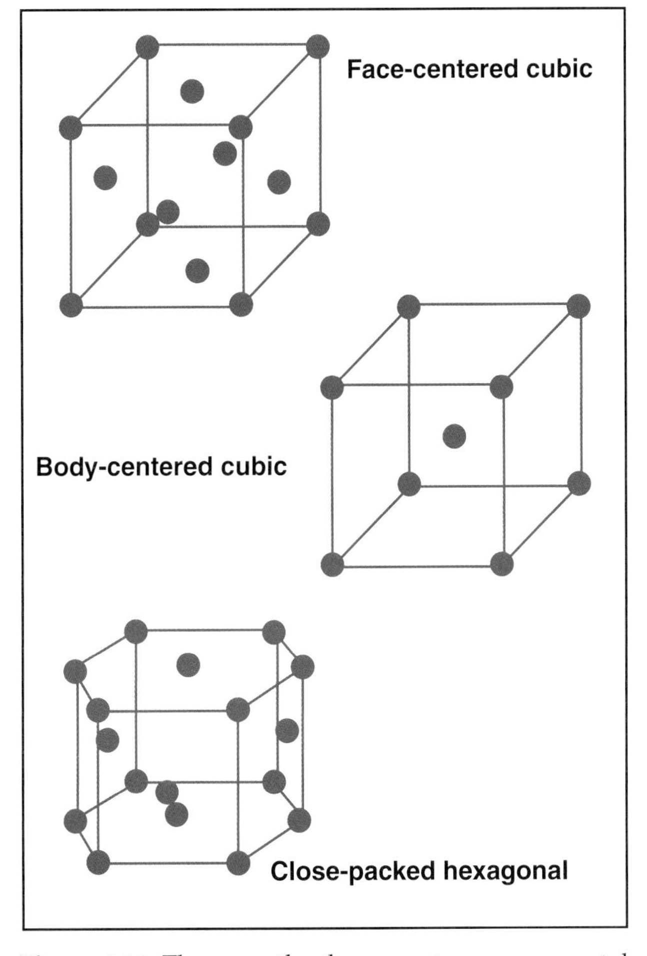

Figure 4-18. These are the three most common crystal structures for metals.

Macrostructure

A material can be viewed from its microstructure—its very small structural pattern. However, that only tells part of the story. A material also has a macrostructure. See **Figure 4-19.** The *macrostructure* is the way different substances unite to form complex materials called composites. These materials, as discussed in Chapter 3, are made up of at least two different substances. Each of these substances retains its own properties. Quite often composites can be separated into their basic components. These basic components will still exhibit their original characteristics.

A macrostructure can be viewed from a large view. For example, plywood is a series of layers of veneer glued together. The wood, as the veneer, has a microstructure. However, the plywood has its own macrostructure. There is a relationship between the layers of veneer and the adhesive that bond them into a single panel.

Concrete is an aggregate (sand and gravel) bonded into a solid by cement. Again, we can view each component's microstructure. However, to understand concrete, we must study its macrostructure.

These materials, as discussed earlier, are called composites. Each material retains its own properties. They are held in place in an external mechanical way. The flakeboard is held by adhesive forces. The same is true of concrete.

Once materials form a composite, the new combination has its own unique properties. These are often the best of the base material's characteristics. Composites are made up of particles, fibers, flakes, layers, and fillers. Their structure is large and can be seen with the unaided eye.

Composites can be natural or manufactured (synthetic). A common natural composite is wood. From wood, people make a number of synthetic composites. These include plywood, flakeboard, particleboard, and hardboard.

Figure 4-19. The flakeboard (oriented-strand board or OSB) used for sheathing for this house has both a microstructure and a macrostructure.

Synthetic composites can be grouped into four categories:

- *Laminate composites.* These are made from layers of material. Typical examples are plywood and six-ply posterboard.
- *Particle composites.* These are made from particles or flakes of material. Typical examples are particleboard, concrete, and flakeboard.
- *Fiber composites.* These are made from fibers of materials. A typical example is hardboard.
- *Dispersed or filler composites.* These are made from a skeleton that is filled with another material. A typical example is a fiberglass panel.

laminate composites. **Materials composed of layers of wood.**

particle composites. **Materials consisting of particles or flakes that are bonded together.**

fiber composites. **Materials consisting of fibers that are bonded together**

dispersed or filler composites. **Materials consisting of a skeleton filled with another material. Fiberglass is such a composite.**

Summary

The properties of materials are determined by their internal structure. This structure may be viewed from the microstructure or macrostructure. The smallest microstructure is the atom. From there, the molecular structure can be studied. This view tells us how atoms combine to form solid materials. This is done through bonding processes. The three bonding methods are ionic, covalent, and metallic.

From an engineering material view, covalent and metallic bonds are most important. Covalent bonding is present in organic materials, while metals are formed through metallic bonds.

The macrostructure approach explains the characteristics of composites. These materials are formed from layers, fibers, particles, and filled structures.

Key Words

All of the following words have been used in this chapter. Do you know their meanings?

amorphous structure
atom
atomic mass number
atomic number
Bohr's atomic model
bonding
compound
crystalline structures
dispersed or filler composites
elements
fiber composites
ion
ionic bonding
laminate composites
macrostructure
metallic bonding
microstructure
molecular structures
monomer
particle composites
periodic table of elements
polymer
subatomic particles
Van der Waals forces

Test Your Knowledge

Please do not write in this text. Place your answers on a separate sheet.

1. What is the difference between a microstructure and a macrostructure?
2. Explain why some knowledge of atomic theory is vital.
3. List the three building blocks of an atom.
4. What is the difference between the atomic number and the mass number?
5. *True or False?* In the periodic table of elements, all the elements in the same vertical row will have similar, but not identical properties.
6. List and describe the three types of bonding discussed in the chapter.
7. *True or False?* Several monomers are united to form a polymer.
8. *True or False?* Metals and most ceramic materials have amorphous structures.
9. Give three examples of macro structures.
10. List four categories of synthetic composites.

Applying Your Knowledge

Note: Be sure to follow accepted safety practices when working with tools. Your instructor will provide safety instructions.

1. Collect samples of three composite materials (such as concrete, fiberglass, plywood, particleboard, or hardboard).
   a. Review information from Figure 4-2 to fix in your mind the meaning of "macrostructure." Also review information from this chapter on the topic.
   b. Study each of the samples collected.
   c. Construct a chart similar to the one shown in **AYK 4-1.**
   d. Describe each material's macrostructure. (For example, are the material parts in sheets, irregular in shape, fine particles

or granules, solid or plastic? Are they fastened mechanically or chemically?)

e. Sketch the macrostructure of each sample.

2. Produce a concrete brick like the one shown in **AYK 4-2.** Describe the basic material needed to produce it and sketch its macrostructure.

Material	Description	Sketch
1.		
2.		
3.		

AYK 4-1. On a separate sheet of paper, prepare a chart similar to this one on which to record your responses.

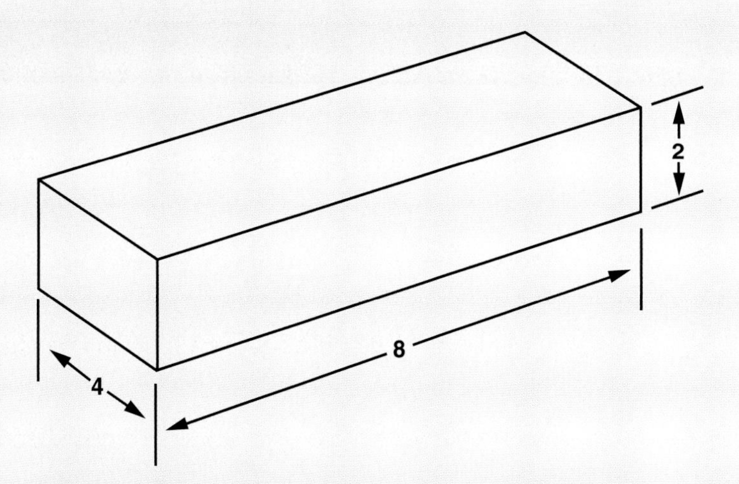

AYK 4-2. Produce a brick like this one and sketch its macrostructure.

Chapter 5
Properties of Materials

Objectives

After studying this chapter, you will be able to:

✓ Select materials based on an understanding of various properties.
✓ Identify the seven major types of material properties.
✓ Name six characteristics that make up the physical properties of a material.
✓ Describe how mechanical properties can affect the way a material will react to an applied force or load.
✓ Cite examples of how chemical properties can affect a material's performance.
✓ List five thermal properties that can affect materials.
✓ Describe electrical and magnetic properties of materials.
✓ Explain how acoustical properties can affect material selection.
✓ Name the three general optical properties.

As you studied earlier, electrons, protons, and neutrons are the base of all matter. The number of these particles in each atom determines what material the matter forms. For instance, aluminum has a different atomic structure than iron has. This atomic structure establishes the material's properties.

These properties are the major factors that distinguish one material from another. We do not count electrons to decide what material to use. Instead, we study its properties. *Properties* are traits or qualities that are specific to a certain material.

properties. **Traits or qualities that are specific to a material.**

Material properties are matched with use or service conditions. A material must perform in the natural environment, which is full of chemicals. Pollutants attack many materials. The oxygen in the air reacts with metals to cause corrosion. See **Figure 5-1.** Water (rain and snow) causes steel to rust, and it causes wood to warp. Plastics are attacked by radiation from the sun. What may seem like a neutral environment to you can be harsh on materials. Materials must also function under varying temperatures. In many areas, temperatures may rise past 100°F in the summer. Then in winter, they may plummet to well below zero. The paint on your car, the doors on your home, and the shoes on your feet must do their job in this wide range of temperatures. However, this range is small compared to internal parts in automobile and jet aircraft engines. Think of the demands temperature changes place on materials.

These two factors are just examples. A material is subjected to countless forces as it is used to complete a task. See **Figure 5-2.**

Figure 5-1. These automobiles are made of materials that must function in a wide range of conditions. (Subaru)

Figure 5-2. The materials used in this lawnmower were selected to meet the forces and other factors present as the owner uses the product. (Toro)

Selecting Materials

Wise material selection, as shown in **Figure 5-3,** is based on an understanding of various properties. These characteristics are determined by the internal structure of the material. You studied these structures in Chapter 4.

All products are carefully designed to meet a human need or want. The designer considers the function of the product. Then various mechanisms and parts are engineered. Selecting the right material is an important function of engineering. From the thousands of materials available, the designer must select the one best suited to each application. The wrong choice can be very costly.

As an example, a major sash and door company manufactured windows with insulated glass. The windowpane consisted of two sheets of glass that formed an envelope. A gas was sealed between the two sheets. The manufacturer purchased the glass unit from another company and installed it in their wood frames. The glass supplier recommended a change in the gas between the glass. The suggestion was followed. However, it was discovered that the gas reacted with sunlight. Over time, the reaction caused the glass to discolor, forming what is called "blue glass." These windows had to be replaced at a high cost to the glass manufacturer. The mistake was costly, but it did not physically harm anyone.

In another example, a poor selection of welding materials caused the weld to fail in World War II cargo ships. The ships broke up in the rough seas. This mistake was costly in terms of human life.

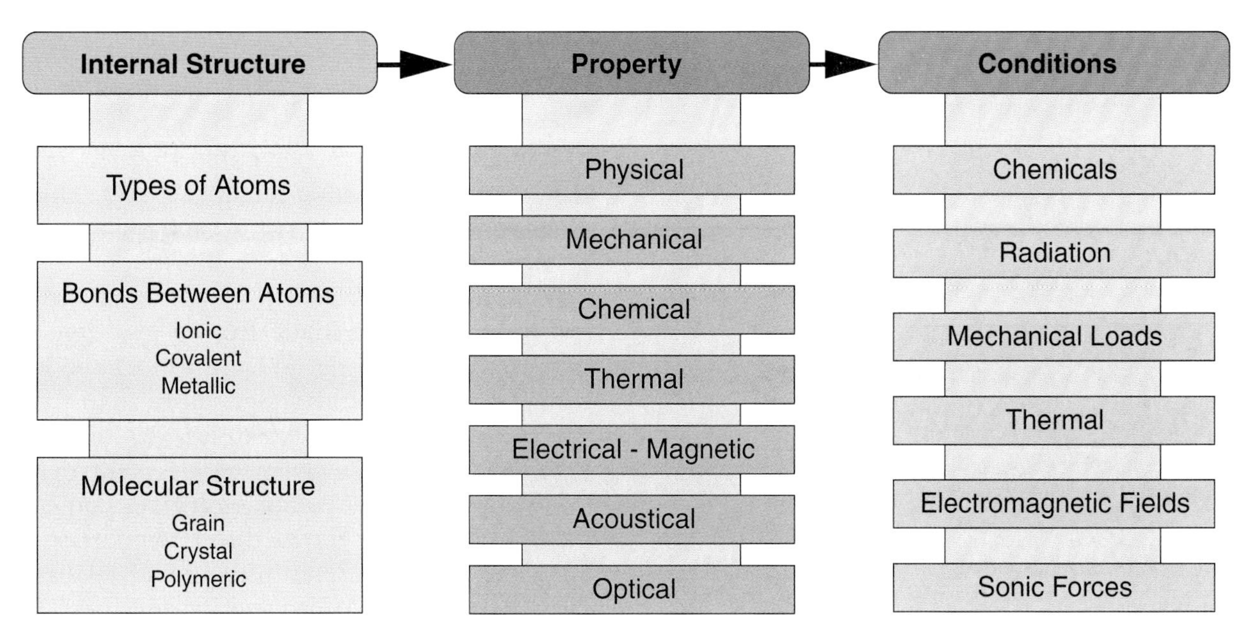

Figure 5-3. The internal structure determines a material's properties that, in turn, determine how well a product performs under use.

The selection of materials cannot be taken lightly. The lives of customers and the reputation of a company may be at stake.

Types of Material Properties

To help engineers select materials, the properties have been grouped into seven major types:

- Physical
- Mechanical
- Chemical
- Thermal
- Electrical and magnetic
- Acoustical
- Optical

Each of these properties is unique. It describes certain characteristics of a material. Also, there are common ways to measure and evaluate these properties. Let's look at each of these properties independently.

Physical Properties

The *physical properties* of a material include several characteristics. See **Figure 5-4.** These include structure, density, moisture content, porosity, size and shape, and surface texture.

physical properties. Characteristics that describe the size, density, or surface texture of a product.

Structure

Structure was discussed in Chapter 4. The structure of a material includes its atomic and molecular structure (microstructure) and its visible structure (macrostructure).

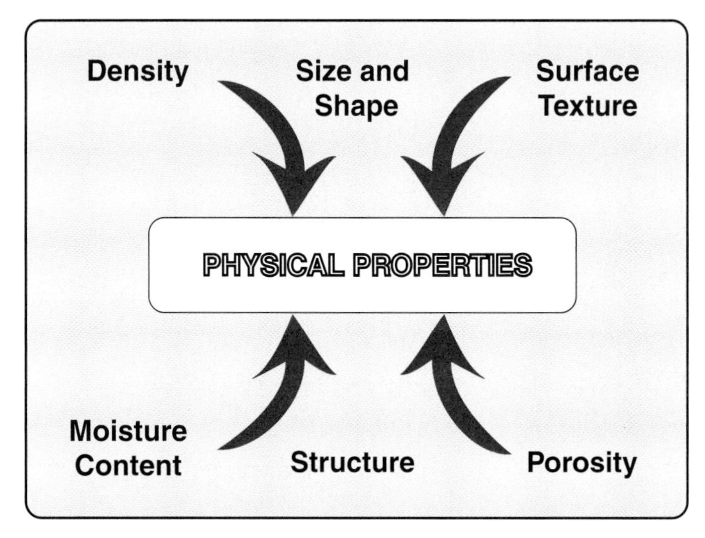

Figure 5-4. These are some typical physical properties.

Density

Density is a measure of the mass of a material. Density is measured in mass per unit volume-pounds per cubic foot (lb/ft^3) or kilograms per cubic meter (kg/m^3). This measure allows you to compare the weights of various materials. Something made from a very dense material will weigh more than it would if it were made from a less dense material. For example, an engine block made from cast iron would weigh more than the same one made from aluminum.

Often, the density of a material is compared to that of water. This comparison gives you an index number that allows direct comparisons between materials. This measure is relative density and is often called ***specific gravity***.

density. **A measure of the weight of a certain volume of material. The volume of the sample objects must be constant. Density is given as pounds per cubic inch (lb/in^3), pounds per cubic foot (lb/ft^3), or grams per liter (g/L).**

moisture content. **The amount of water trapped within a material's structure. It is expressed as a percentage of the dry weight of the material.**

porosity. **This is the relationship of open space to solid space in a material. A porous material will be lighter than a nonporous material. Porous materials contain air within themselves and make good heat insulators.**

surface texture. **The way a material's surface looks or feels: smooth, rough, shiny, dull, etc.**

Moisture Content

Moisture content is a measure of the amount of water trapped within a material's structure. The moisture content is defined as the weight of water as a percentage of the moisture-free weight of the material. It can be calculated using the following formula:

Moisture content (%) = Weight of the water/Oven dry weight × 100

Moisture content is an important measure, especially for wood products. See **Figure 5-5.** It provides the base for calculating the amount of expansion and contraction that can be expected as the wood adjusts to the environment. The moisture content of green (freshly sawed) lumber can range above 225% for western red cedar sapwood to about 35% for Douglas fir heartwood.

Framing lumber (2 × 4s, 2 × 6s, etc.) used for construction purposes is dried to 15 to 19% moisture. Lumber for furniture and interior trim is dried to a 6 to 10% range. As this lumber dries, it shrinks considerably. The volume of a hardwood board will shrink between 13 and 17%. Softwood lumbers will lose from 7 to 12.5% of their volume during drying.

Porosity

Porosity is the ability of a material to absorb air or water. A porous material, such as paper or cotton, will absorb moisture or air. Also, many porous materials will allow air and water to pass through them. For example, a cleaning sponge must be porous so it can absorb liquids. Filters must allow fluids to pass through them while trapping unwanted solid particles.

All materials occupy space. See **Figure 5-6.** Engineering materials have rigid structures and, therefore, have specific *sizes* and *shapes.* These are important properties. In fact, changing these properties is a major goal of manufacturing activities.

Surface Texture

Surface texture refers to the way a surface of a material looks or feels. In some cases, a smooth surface is desired. See **Figure 5-7.** Most furniture and automobiles have smooth surfaces so their finish coatings have a high gloss. In other cases, a textured surface may be desired. For example, stainless steel sinks have a brushed surface so that scratches will be less noticeable.

Figure 5-5. These boards must have low moisture content to reduce warpage and shrinkage.

Mechanical Properties

Mechanical properties govern how a material will react to a force or load. The mechanical properties of a material are determined by three factors. See **Figure 5-8.**

- Type of bonding
- Structural arrangement of atoms or molecules
- Number and types of imperfections in the materials

This last factor is largely determined during the primary processing phase of manufacturing. Materials of the same chemical composition can vary greatly due to improper control of basic refining actions.

There are four major mechanical properties, as shown in **Figure 5-9.** These include strength; elasticity; plasticity, ductility, and malleability; and hardness.

Figure 5-6. Consider the physical properties of the reinforcing bars at this road construction site.

Strength

Strength is the ability of a material to resist forces applied to it. This property determines a material's ability to retain its size and shape and to resist fracturing (breaking). Typically, a part can have four major forces applied to it. These, as shown in **Figure 5-10,** are:

- **Compression forces.** These forces squeeze the material. The resistance to this type of force is called *compression strength.*
- **Tension forces.** These forces try to pull the material apart. The resistance to them is called *tensile strength.*
- **Shear forces.** These are opposing forces that tend to fracture the material along grain lines. The resistance to this force is called *shear strength.*
- **Torsion forces.** These forces twist the material around an axis. The resistance to the force is called *torsion strength.*

Elasticity

Elasticity is the ability of a material to undergo compression or stretching without permanent deformation (shape change). Elastic materials are able to return to their original shapes after an outside force is removed. See **Figure 5-11.**

mechanical properties. Characteristics that govern how the material will react to a force or load, such as compression, tension, shear, and torsion.

strength. The ability of a material to resist an applied force.

compression strength. The ability to resist squeezing forces.

Figure 5-7. This worker is measuring the surface texture of a part. (Rhom and Haas)

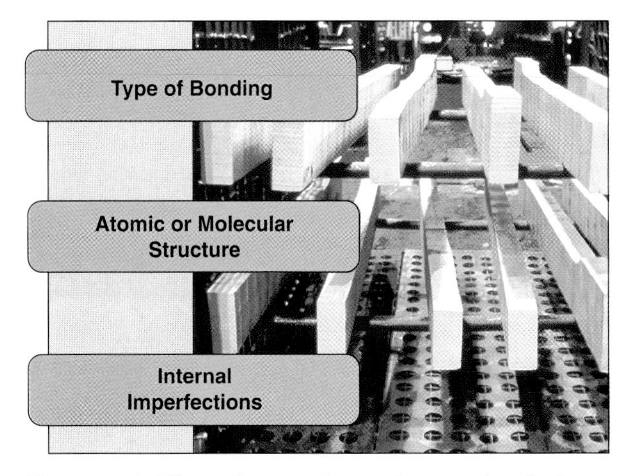

Figure 5-8. These factors determine mechanical properties of materials. (ALCOA)

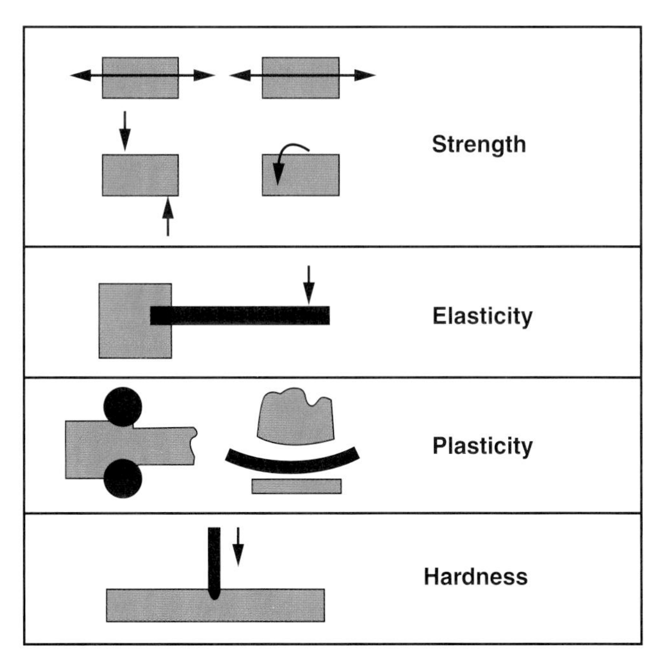

Figure 5-9. These are common mechanical properties of materials.

Plasticity

Another important mechanical property is *plasticity*. It is a material's ability to flow into a new shape under pressure. This characteristic is important to forming processes. Plasticity allows a material to take a new shape when a die or roll applies force.

Two other properties are associated with the plasticity of a material. These are *ductility* and *brittleness*. Ductility is the ability of a material to undergo plastic deformation without rupturing. Materials that are ductile are said to have high malleability, workability, or formability. Brittleness is a measure of the lack of resistance to force. Brittle material will fracture when forces are applied. Glass, cement, and very hard metals are brittle.

tensile strength. The ability to resist forces that would pull the material apart.

shear strength. The ability to resist opposing forces that would fracture the material along grain lines.

torsion strength. The ability to resist twisting forces.

Stress-strain diagram

Strength, elasticity, plasticity, ductility, and brittleness are easily shown in a stress-strain diagram. See **Figure 5-12.** As force is applied to a material, it deforms. It will stretch, compress, or bend. This action builds stress in the material structure. Low levels of force will deform the material only when the force is present. The material will return to its original size and shape when the force is removed. The range of forces, which will temporarily deform a part, is within the material's elastic range. However, at one point, the ***yield point***, the material permanently deforms. The material yields to the force and takes on a new shape. There is a range of pressures above the yield point that will permanently deform the material. At some point, the material can stand no more additional force and will break. This point is called the ***fracture point***.

The range between the yield point and the fracture point is where forming processes operate. Separating processes always operate above the fracture point. Springs, automobile tires, and elastomer adhesives, for instance, operate below the yield point and in the elastic range.

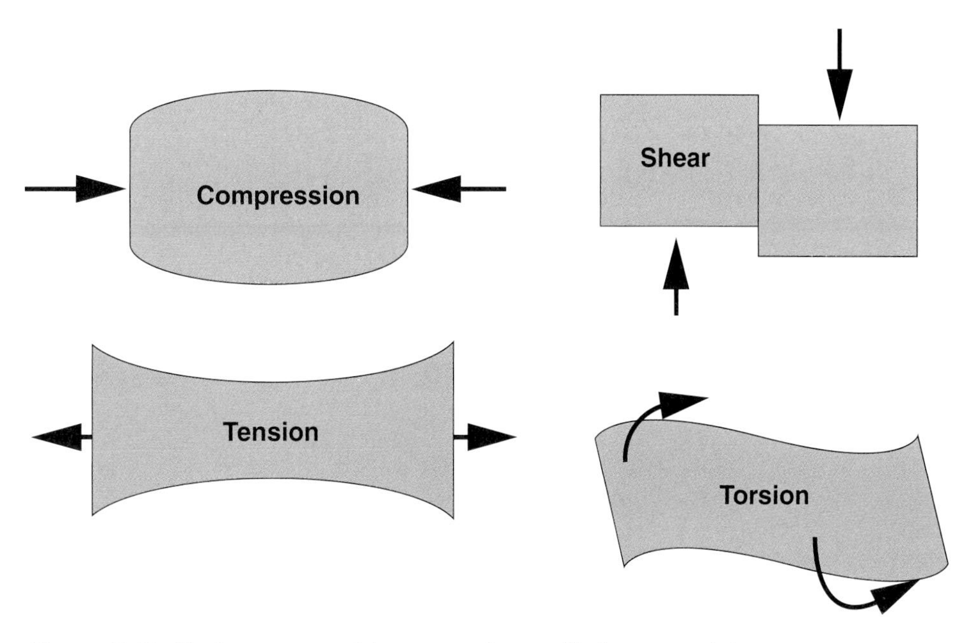

Figure 5-10. Various types of forces can be applied to materials.

Hardness

Hardness is the last mechanical property we will explore. It is a material's ability to resist denting or scratching. See **Figure 5-13.** Hardness and wear resistance are closely related. Hard materials resist wear better than softer materials. However, harder materials are more brittle. Therefore, a balance between hardness and strength must be maintained. Sometimes, hard surfaces are developed on the material (case hardening). The core remains softer and more ductile. This process, which will be discussed in Chapter 13, offers the "best of both worlds." The surface has high wear resistance, and the inside absorbs stress.

There are a number of ways to test mechanical properties. Common to all these techniques is applying force to a specimen. The specimen is a carefully prepared sample of the material.

Often, a universal tester is used. The machine, as shown in **Figure 5-14,** can apply tension, compression, or shear forces. The actual test specimen preparation and test procedures are established by the American Society for Testing and Materials (ASTM).

elasticity. A material's ability to return to its original shape after being deformed by an applied force.

plasticity. A material's ability to flow into a new shape when pressure is applied.

ductility. A property that allows a material to undergo plastic deformation without rupturing.

brittleness. A measure of a material's lack of resistance to force. A brittle material fractures easily.

hardness. The ability of a material to resist scratching or denting.

chemical properties. Characteristics of a material that determine how it will react chemically.

Chemical Properties

All engineering materials will enter into reactions with one or more chemicals. These reactions are controlled by *chemical properties.* For example, the natural elements in the air (oxygen, nitrogen, carbon dioxide, etc.) may react with a material. Often, the action is called corrosion. Iron rusts, copper tarnishes, and aluminum turns dull. In all these cases, a chemical action has taken place. A metallic oxide forms as oxygen joins the pure metal in the material. This action may be unwanted. For instance, we seldom want steel to rust.

Often, materials can be made to resist certain chemicals. For instance, wood can be treated to resist the weather. Plastic and metal food containers can be produced that will not react to the acids in foods. Washing machines can be produced so that surfaces will not be harmed by water, detergents, and bleach.

In many cases, chemical reactions can have advantages. For example, aluminum is anodized to form a protective coating of aluminum oxide. Chemicals are used in a fairly new metal separating process called chemical machining. Glass is etched to make it translucent or to decorate its surfaces. All of these actions are based on the chemical properties of materials. These chemical properties determine

Figure 5-11. The synthetic rubber in these tires has a large elastic range. (Goodyear Tire and Rubber Co.)

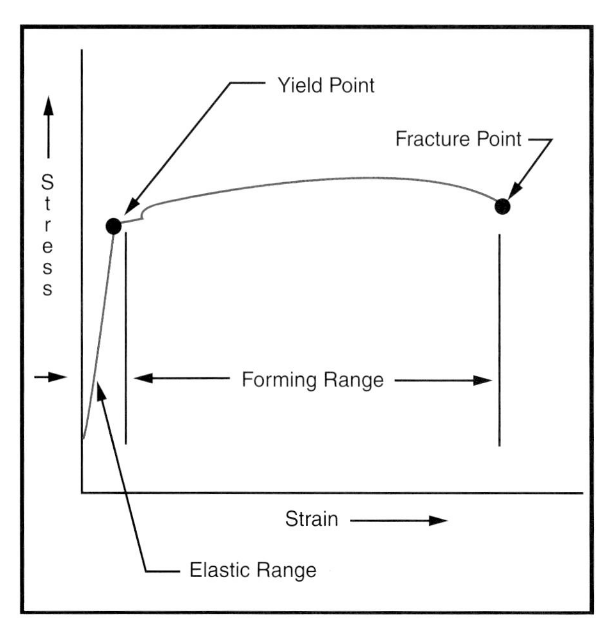

Figure 5-12. This diagram illustrates the concept of stress-strain.

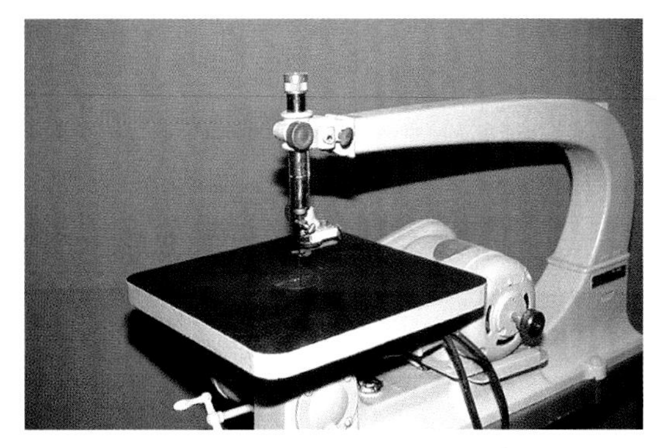

Figure 5-13. The material selected for this saw table resists wear.

the reactions of materials and describe the end results of the reactions.

Many industries are vitally interested in the chemical properties of materials. These industries include the pharmaceutical, plastics, and petroleum industries. See **Figure 5-15.**

Thermal Properties

All materials operate in the presence of heat. You may talk about how cold it is in the winter, but you are really talking about less heat being present. The reactions of materials to varying levels of heat are called their *thermal properties*. These properties include:

thermal properties. **Characteristics of a material that determine how it will react to temperature and to heat and cold at different levels.**

thermal conductivity. **A material's ability to conduct heat.**

- *Thermal conductivity.* This is the ability to conduct heat through the material.

Academic Link

Chemistry is the study of materials, their properties, and interactions with other materials and with energy. In manufacturing, chemists use their abilities to create materials that have specific traits that are desirable.

Chemists also use their skills to test materials. Before products are produced for general use, the manufacturer must be sure that it performs like it should. A chemist may be used to determine that the proper material was used in a particular manufacturing process. For example, chemists are used to develop cooling fluids used to cool machines and products during some manufacturing processes. Without this knowledge, damage to machinery and product waste would be great.

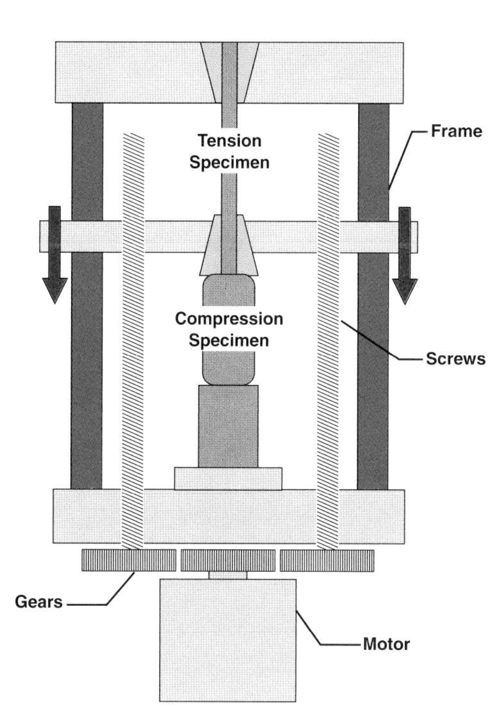

Figure 5-14. This is a schematic of a common universal testing machine.

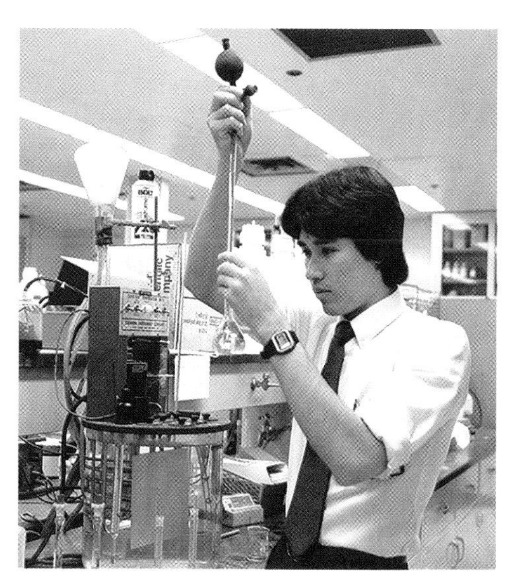

Figure 5-15. This scientist is testing the chemical properties of an industrial material. (Xerox Corp.)

Figure 5-16. This electric-arc steelmaking furnace is lined with firebrick. These ceramic bricks are chemically inactive and good thermal insulators. (American Iron and Steel Institute)

- *Thermal resistance.* This is the ability to resist melting.
- *Thermal expansion.* This is the degree the material expands in the presence of heat (and contracts as heat is removed).
- *Thermal emission.* This is the ability to give off or radiate heat.
- *Thermal shock resistance.* This is the ability to withstand the shock produced by rapid changes in temperature.

Each material has its own assortment of thermal properties. See **Figure 5-16.** Ceramic materials, generally, have high resistance to heat and melt at very high temperatures. This is the reason ceramic materials are used for baking ware. However, many ceramics suffer from mechanical and thermal shock. Dropping a glass pie plate or putting a hot glass dish in cold water can result in cracking or breakage.

Most metals have high melting points. However, they lose their strength and oxidize easily at high temperatures. For instance, a screwdriver is very strong at room temperature. However, if it is heated until it glows red, it will easily bend.

Also, most metals transmit and emit heat readily. Therefore, some metals make good materials for cookware. Metal cookware does, however, require plastic or wooden handles. Metal handles would be too hot to hold. The wood and plastic are thermal insulators, meaning they do not transmit heat readily.

thermal resistance. A material's ability to resist melting.

thermal expansion. The amount that a material expands when heat is applied and contracts when heat is removed.

thermal emission. Ability of a material to give off, or radiate, heat.

thermal shock resistance. The ability of a material to resist shock caused by sudden temperature changes.

electrical properties. The measure of a material's reaction to electrical current.

conductors. Materials that permit a ready flow of electrical current.

semiconductors. Materials that will conduct electricity under certain conditions.

resistors. Materials through which electrical current will not readily flow; insulators.

magnetic properties. The measure of a material's reaction to external electromagnetic forces.

Figure 5-17. This worker is using magnetic inspection to test the quality of a weld and to locate heat-affected zones in a large pipe.
(American Cast Iron Pipe Co.)

Electrical and Magnetic Properties

Electrical and magnetic properties are important in many applications. *Electrical properties* determine a material's ability to carry an electrical current. One major electrical property is ***electrical resistance***. This is the level of resistance the material has to electric current (moving electrons). Sometimes the same property is viewed from the opposite perspective and called ***electrical conductivity***.

Engineering materials can be divided in three groups according to their ability to carry an electric current. The major groups are:

- **Conductors.** Electrical *conductors* are metals with low electrical resistance. This property suggests that the material will carry the current with little heat loss.
- *Semiconductors.* Fall into a range between insulators and conductors.
- **Resistors.** Also referred to as insulators, *resistors* have a high resistance to electrical current.

Closely related to electrical properties are *magnetic properties*. This set of properties governs a material's reaction to a magnetic field. *Magnetic permeability* measures a material's ability to become magnetized. High permeability means the material can be easily magnetized when brought into a magnetic field.

A material's *magnetic conductivity* tells you how well magnetic lines of force are conducted through the material. This property can be used to check products for defects. Magnetic waves (magnetic flux) move in a specific pattern in a material. Internal defects (cracks and impurities) cause the magnetic waves to move differently. Therefore, by plotting the magnetic flux, defects can be located. See **Figure 5-17.**

magnetic permeability. A measurement of a material's ability to become magnetized.

magnetic conductivity. The ability of a material to conduct magnetic lines of force.

Acoustical Properties

Acoustical properties tell you how a material reacts to sound waves. See **Figure 5-18.** The two major acoustical properties are acoustical transmission and acoustical reflection. *Acoustical transmission* describes how well sound travels through materials. The reverse of transmission is insulation. The walls of radio and television studios must have low acoustical transmission. Otherwise, the microphones would pick up the sounds outside the area.

Acoustical reflectivity determines the amount of sound that will bounce off a material. The materials in a theater or a concert hall should have low sound reflectivity. This keep echoes out of the hall and improves the acoustics. Have you ever gone to a concert in a gymnasium? How were the acoustics? Probably not very good because of all the hard (sound-reflecting) materials in the building.

acoustical properties. How a material absorbs, transmits, or reflects sound waves.

acoustical transmission. A measure of a material's ability to transmit sound waves.

acoustical reflectivity. A measure of how well or poorly sound waves bounce off a material.

Figure 5-18. Rooms in homes should have low sound transmission and reflectivity. Wallpapers, carpets, and draperies reduce the noise level in rooms. (Georgia-Pacific Corp.)

Figure 5-19. This large greenhouse is used to raise seedling trees. The covering must let light pass through to stimulate the growth of the trees. (Weyerhaeuser Timber Co.)

Optical Properties

Optical properties determine how a material reacts to light. There are three general optical properties:

- Color
- Light transmission
- Light reflection

Sunlight or light from a lamp may look "white," but it is really made up of different colors. This is called the light spectrum. You can make any color from combinations of the colors of the light spectrum. We see an object in a certain color because that color is reflected by the object, while other colors are absorbed.

Light reflection tells you how much light a material reflects. For instance, a mirror should have a high level of reflection, while window glass should reflect little light. *Light transmission* is almost the opposite of reflection. It determines the amount of light that passes through a material. ***Transparent*** materials allow most of the light to pass through. See **Figure 5-19.** ***Opaque*** materials reflect or absorb most of the light energy. ***Translucent*** materials allow some light to pass through. However, you cannot see objects clearly through translucent materials.

optical properties. **Characteristics that describe how a material reacts to light waves. The two main optical properties are opacity and color.**

light reflection. **A measurement of how well or poorly light waves bounce off a material.**

light transmission. **A measurement of how well or poorly light waves pass through a material.**

Summary

All materials have a series of properties. These properties tell you how a material will act under certain conditions. The seven major types of properties are physical, mechanical, chemical, thermal, electrical-magnetic, acoustical, and optical.

Understanding these properties will allow you to select the correct material for each job. This can reduce material waste, speed processing time, and produce a product that will function properly.

Key Words

All of the following words have been used in this chapter. Do you know their meanings?

acoustical properties
acoustical reflectivity
acoustical transmission
brittleness
chemical properties
compression strength
conductors
density
ductility
elasticity
electrical properties
hardness
light reflection
light transmission
magnetic conductivity
magnetic permeability
magnetic properties
mechanical properties
moisture content
optical properties
physical properties
plasticity
porosity
properties
resistors
semiconductors
shear strength
strength
surface texture
tensile strength
thermal conductivity
thermal emission
thermal expansion
thermal properties
thermal resistance
thermal shock resistance
torsion strength

Test Your Knowledge

Please do not write in this text. Place your answers on a separate sheet.

1. What is the greatest challenge of material selection?
2. Identify and describe the seven major types of properties.
3. Name six major types of physical properties.
4. Which of the following describes a material's ability to undergo compression or stretching without permanent deformation?
   a. Ductility.
   b. Brittleness.
   c. Elasticity.
   d. None of the above.
5. *True or False?* All engineering materials will enter into a reaction with one or more chemicals.
6. Which of the following is a material's ability to give off or radiate heat?
   a. Thermal shock resistance.
   b. Thermal emission.
   c. Thermal conductivity.
   d. Thermal expansion.
7. *True or False?* Electrical conductors have a high resistance to electrical current.
8. Magnetic _____ measures a material's ability to become magnetized, while magnetic _____ of a material tells you how well magnetic lines of force are conducted through the material.
9. What two acoustical properties should be considered in designing a recording booth?
10. If you want a material that will reflect or absorb light energy, you should choose one that is:
    a. Opaque.
    b. Transparent.
    c. Translucent.
    d. None of the above.

Applying Your Knowledge

Note: Be sure to follow accepted safety practices when working with tools. Your instructor will provide safety instructions.

1. Review the plans for the tic-tac-toe board shown in the activity section of Chapter 1. Then:
   a. Draw up a chart like the one in **AYK 5-1.**
   b. Research properties of different materials.
   c. List on the chart five materials you think might be used for making each of the parts (base and pegs).
   d. In the right-hand column, list the properties that made you select each material.
2. Survey the equipment and tools in the technology education laboratory.
   a. Indicate which materials could be worked with the available tools and equipment.
   b. Suggest a set of procedures to be used with one of the materials.
3. Take a 1/4" cross section of a green wood limb. See **AYK 5-3.** Weigh it. Then, following directions from your instructor, place the cross section in a kiln for 24 hours. Weigh it again. Using the formula in the Physical Properties section of this chapter, determine the moisture content of the wood.

Part	Material	Properties
Base:	1.	
	2.	
	3.	
	4.	
	5.	
Pegs:	1.	
	2.	
	3.	
	4.	
	5.	

AYK 5-1. On a separate sheet of paper, prepare a form similar to this one on which to record your responses.

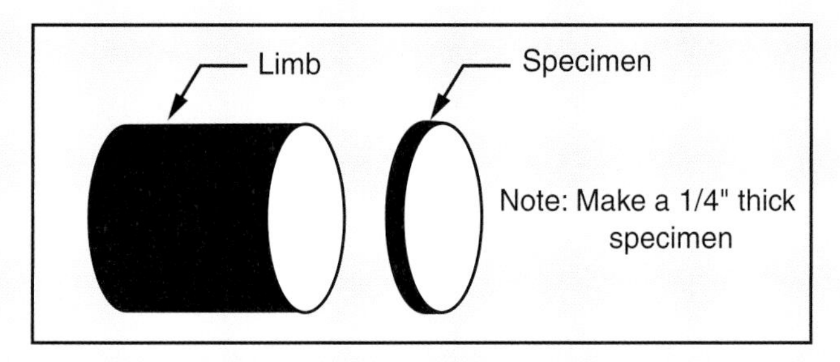

AYK 5-3. Obtain a cross section of a green wood limb.

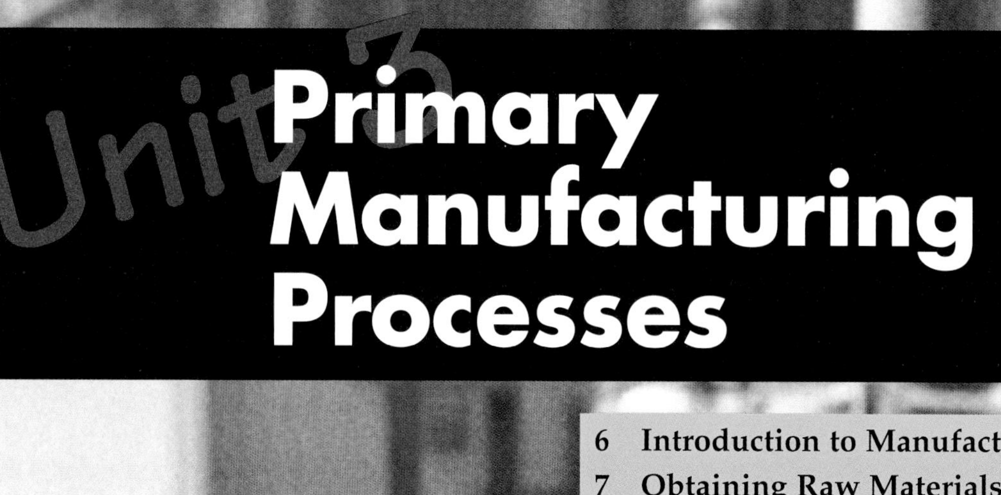

Unit 3 Primary Manufacturing Processes

Chapter 6
Introduction to Manufacturing Processes

Objectives

After studying this chapter, you will be able to:

- ✓ Identify manufacturing processes as primary or secondary.
- ✓ List and discuss major steps in the manufacturing processes.
- ✓ Define primary and secondary processes.
- ✓ Describe six types of secondary processes
- ✓ Name outputs of manufacturing activities.
- ✓ Discuss the impact of manufacturing processes on the environment.

Throughout history, people have changed the size, shape, and looks of materials. They have tried to make materials better fit their needs. In so doing, people have manufactured products.

The actual changing of the form of material is one part of manufacturing. This activity is called *material processing*. Management is the other part of successful manufacturing.

material processing. **Changing the size, shape, and looks of materials to fit human needs. Changing the form of materials takes three steps: obtaining natural resources, producing industrial materials, and making finished products.**

Changing the form of materials takes three major steps. These stages, as shown in **Figure 6-1,** are:

- Obtaining natural resources.
- Producing industrial materials by changing raw materials into them.
- Making finished products from industrial materials.

Keep in mind that we are going to talk mostly about materials. However, this is only one resource in manufacturing. People, energy, time, knowledge, capital, and finances are also important. However, converting materials into products is the task of manufacturing. The other resources (inputs) support this activity.

Obtaining Resources

We cannot build a product without materials. The materials must be located and gathered. The typical processes for collecting materials are listed in **Figure 6-2.**

Figure 6-1. There are three stages of material processing.

Figure 6-2. Resources are gathered in three ways.

These are:

- **Mining.** Digging the material from the earth by means of a hole or tunnel.
- **Drilling.** Pumping material from below the earth's surface through a narrow, round shaft.
- **Harvesting.** Cutting a mature, renewable resource from the land.

Figure 6-3. Primary processing produces industrial materials. (Weyerhaeuser Company)

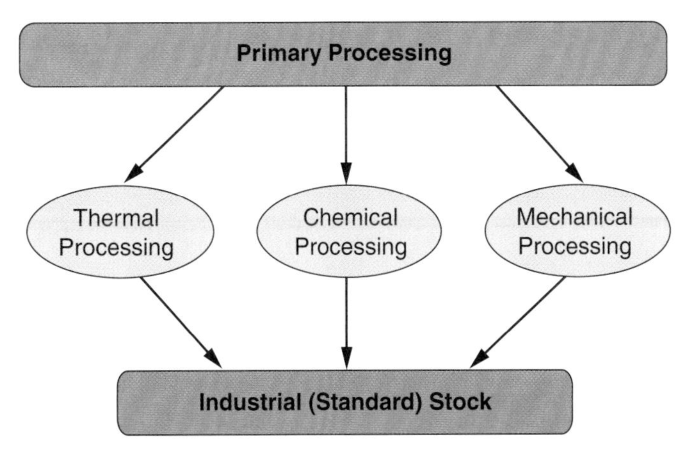

Figure 6-4. Types of primary processing. Sometimes more than one type is used to produce a primary material.

Primary Processing

Once raw materials are obtained, they are moved to a mill. Here the resource is changed into an industrial material. See **Figure 6-3.** Trees are made into lumber, plywood, fiberboard, and particleboard. Iron ore, limestone, and coke are used to produce steel. Natural gas is a feedstock (source) for many plastics. These outputs are called standard stock. They are the products of primary processing.

Primary processing is the first major step in changing the form of materials. Three major processes are used to convert raw materials into industrial materials. These, as shown in **Figure 6-4,** are:

- Thermal (heat) processes
- Mechanical processes
- Chemical processes

These processes are not entirely separate. For example, thermal processes apply heat to crushed (mechanically processed) ores.

Thermal Process

Thermal processes use heat to change the properties of materials. The hot ores enter into a chemical reaction. This reaction separates the metal from impurities. Crushing the ore makes it easier to extract the metal using heat. The two processes work together.

Mechanical Process

Mechanical processes cut or crush resources. Trees are cut into lumber, veneer, and chips. Rocks are crushed into gravel. Wool is sheared from sheep. All are mechanical processes. They use mechanical force to change the resource.

primary processes. Processes that change raw materials into a usable form for further manufacture. Changing logs into lumber is an example.

thermal processes. Processes that use heat to change a material, such as steelmaking changes iron into steel.

mechanical processes. Methods that use mechanical force to change the resource, such as cutting or crushing.

chemical processes. Methods of processing raw materials that use chemical reactions. For example, plastics are formed by chemical reactions.

secondary processes. Processes that change standard stock into finished products.

Chemical Process

Chemical processes use chemical reactions to refine raw materials. Plastics are formed by chemical reactions. Simple compounds are combined to form the complex polymer chains we call plastics. **Figure 6-5** lists some resources that are processed by each type of primary processing.

Primary Processes		
Thermal	**Chemical**	**Mechanical**
Steelmaking Copper smelting Zinc smelting Lead smelting	Aluminum refining Polymer formation Gold refining Papermaking Leather tanning	Lumber manufacture Plywood manufacture Particleboard making Rock crushing

Figure 6-5. Resources processed by each type of primary processing.

Secondary Processing

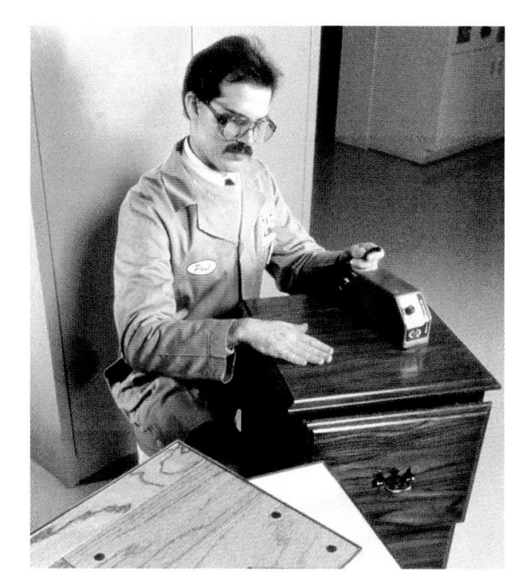

Figure 6-6. Secondary processes produce finished products.
(PPG Industries)

Industrial materials (standard stock) can now be changed into finished products. See **Figure 6-6.** Lumber is made into furniture and houses. Leather is made into shoes and belts. Steel is used to make cars, appliances, and reinforcing bars for concrete. Thousands of products are made from the outputs of primary processing. Secondary processes produce finished products. These activities, as shown in **Figure 6-7,** are:

- **Casting and molding.** Pouring or forcing liquid material into a prepared mold. The material is allowed to become solid. Then it is removed from the mold.
- **Forming.** Using force to cause a material to permanently take a shape. A die, mold, or roll is used to shape the material.
- **Separating.** Changing a material's size and shape by removing excess material. The material is cut or sheared by these processes.
- **Conditioning.** Using heat, mechanical force, or chemical action to change the internal properties of a material.
- **Assembling.** Temporarily or permanently holding two or more parts together.
- **Finishing.** Protecting or improving the appearance of the surface of a material.

Figure 6-7. Secondary processes are casting and molding, forming, separating, conditioning, assembling, and finishing. (Rohm & Haas; Bethlehem Steel; Inland Steel Co.; PPG Industrial; AT&T; DeVilbiss)

Products, Waste, and Scrap

Almost all manufacturing activities create unwanted by-products. Mining creates holes and shafts in the ground. Harvesting timber requires roads through peaceful woods and camps for workers. Steel plants and foundries give off fumes, dust, noise, and wastes. Machining activities create chips, scraps, and shavings. There is no such thing as a scrapless, pollution-free manufacturing activity. People must carefully plan to reduce and control undesirable outputs. By doing this we can make sure that the environment is protected.

The protection of the environment is the responsibility of all citizens. We need clean air, water, and soil. Each of us must do our part to save our precious natural resources. Also, we should expect–*and demand*–that industry help. Companies must consider:

- Their practices in obtaining raw materials.
  - Are they efficient?
  - Is the land returned fully to productive use? See **Figure 6-8.**

Figure 6-8. This is land that was once an open pit mine. (American Electric Power)

Figure 6-9. Scrap metal can be recycled and used in the steelmaking process.

- Pollution caused by manufacturing processes:
  - Can it be avoided and minimized?
  - If not, can it be controlled and cleaned up?
- Scrap and waste:
  - Does the manufacturing process produce unnecessary scrap and waste?
  - Is every effort being made to recycle scrap? See **Figure 6-9.**
  - Is unusable waste material disposed of properly?

Our future will be more secure with industrial and personal resource conservation. You and tomorrow's children can have the "good life" if we use our resources wisely.

Summary

Manufacturing materials are converted from raw materials to finished products in three stages. See **Figure 6-10.** Raw materials are first located. They are then obtained from the earth by harvesting, drilling, or mining. Secondly, the raw materials are converted into standard stock. Thermal, chemical, and mechanical processes are used. Finally, the industrial goods are changed into finished products. The form change is done through casting and molding, separating, forming, conditioning, assembling, and finishing processes.

Throughout the processing, the environment must be protected. Resources must be used wisely. Scrap and waste must be kept to a minimum. Pollution needs to be reduced and controlled.

With environmental concern and efficient manufacturing, we can all live better. We will have useful products and a healthy world.

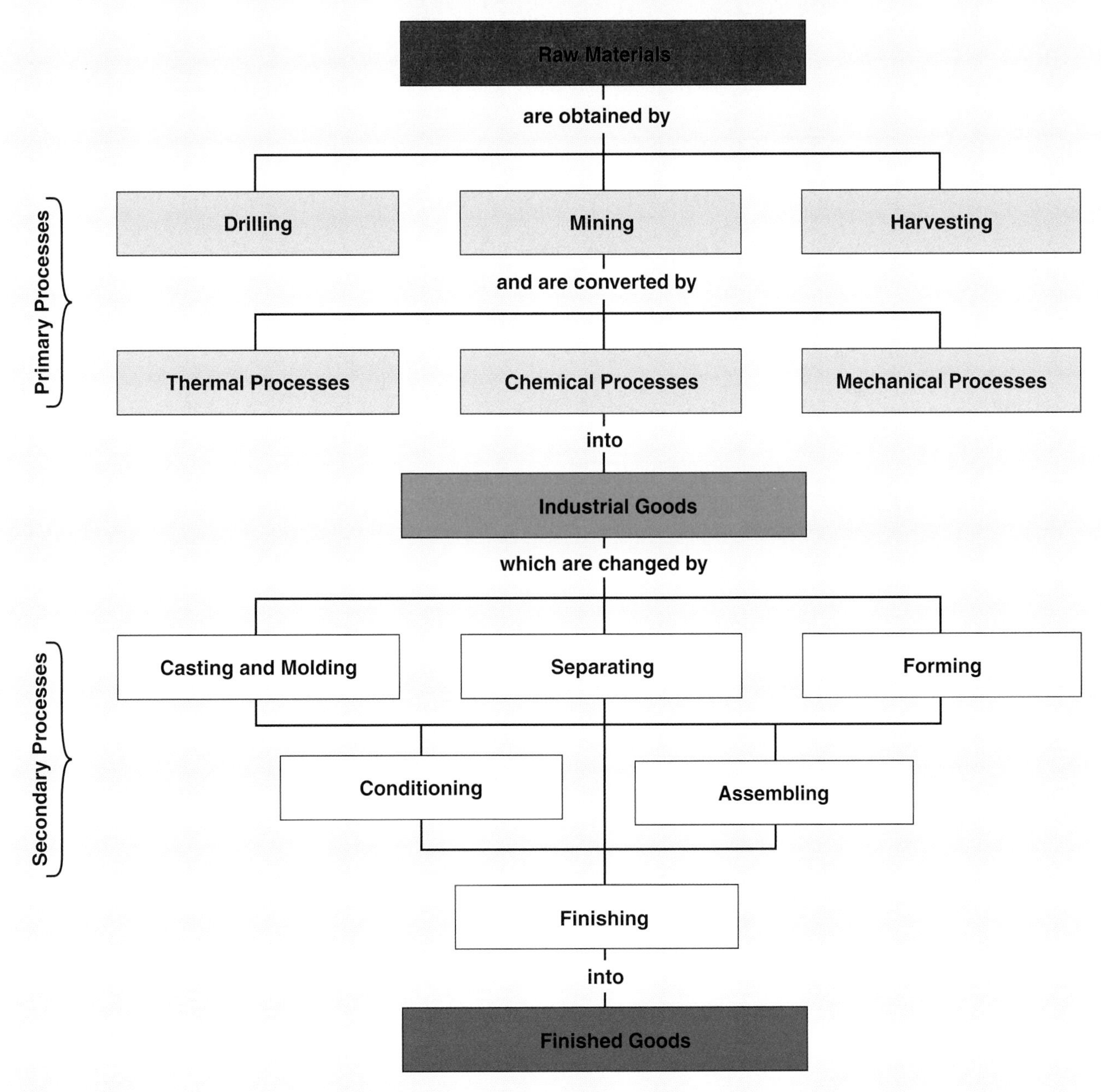

Figure 6-10. A summary diagram of the material processing activities.

Key Words

All of the following words have been used in this chapter. Do you know their meaning?

chemical processes
material processing
mechanical processes
primary processes
secondary processes
thermal processes

Test Your Knowledge

Please do not write in this text. Place your answers on a separate sheet of paper.

1. Changing the form of materials takes (three, four, five) major steps.
2. List and describe three ways of obtaining natural resources.
3. Which of the listed processes are called primary processes?
   a. Chemical processes.
   b Heating processes.
   c. Mechanical processes.
   d. Electrical processes.
   e. Extracting processes.
4. Crushing is a primary process called _____ processing.
5. List and describe the six types of secondary processes.
6. Name the three outputs of manufacturing activity.
7. Protection of the environment is _____ responsibility.

Activities

1. Invite a person from a manufacturing company to discuss the inputs, processes, and outputs of manufacturing.
2. Find out about the Pollution Prevention Act of 1990. Learn how manufacturers or users of a toxic chemical must be responsible to the environment. Begin by writing to the U.S. Environmental Protection Agency, Public Information Center (3404), 401 M Street, SW, Washington, DC 20460 or online at http://www.epa.gov. Request a copy of Guide to Environmental Issues.
3. Go on a field trip to a manufacturing company. Observe and list the resources, sequence of operations, and outputs.

Chapter 7
Obtaining Raw Materials

Objectives

After studying this chapter, you will be able to:

- ✓ Define renewable and exhaustible resources and give examples of each.
- ✓ Describe three different methods for mining raw materials.
- ✓ Define and describe two methods of drilling for oil or gas.
- ✓ Describe three methods of harvesting forests.
- ✓ Explain methods by which raw materials are moved to mills and refineries.

Most products are made from many different materials. All of these materials were once in a "raw" condition. *Raw materials* are natural resources found on or in the earth or seas.

raw materials. **Natural resources found on or in the earth or seas. Manufacturing starts with raw materials.**

All manufacturing starts with raw materials. These materials are of two basic types. See **Figure 7-1.** They are renewable or exhaustible.

Renewable Resources

Renewable resources are biological materials (growing things). Each growing unit (plant or animal) has a life cycle. First, it is planted or born. It then grows through stages to maturity (full size). Finally, it becomes old and dies.

renewable resource. **Biological materials that can be grown to replace the materials we use. Trees and food plants are examples.**

Resources	
Renewable	**Exhaustible**
Trees	Metal ores
Cotton	Petroleum
Wool	Natural gas
Flax	Coal
Animal hides	Clays

Figure 7-1. There is no limit to the supply of renewable resources. Exhaustible resources will be gone one day and cannot be replaced.

Trees are a good example of a renewable resource. See **Figure 7-2.** Nature or people plant them. They grow and after a number of years will reach full size. Then, growth slows and finally stops. Limbs die and fall off. Insects, wind, and decay attack the tree. In time, they die and fall to the forest floor. Then they rot, providing nutrients (food) for other plants.

Figure 7-2. Farm crops and forests are examples of renewable resources.

Figure 7-3. Forests should be managed to ensure that future generations will have trees to harvest.

Managing Resources

Managing a resource means making sure that there is always a supply to use. It means seeing that future generations will also have it to use.

Even though some resources are renewable, managing them is still important. People must plan and work at growing new resources and knowing when and how to harvest them. For instance, a forester should not cut down all the trees in a forest without planting new ones. See **Figure 7-3.**

We depend on many renewable resources for manufactured products. They provide us with wood products (such as furniture), leather, and natural fibers (wool, cotton, silk, and linen) for making cloth.

Exhaustible resources, discussed next, also must be managed. Unlike renewable resources, we cannot replace them. Management means collecting and using resources carefully so they are not wasted.

exhaustible resource. A resource whose supply is finite; it can be used up and no more of the resource will be available.

Exhaustible Resources

Not all natural resources are renewable. Some have a limited supply. There is a fixed amount of them on earth. Once used up, there will be no more. See **Figure 7-4.** These resources are called *exhaustible resources.* They can be used up. Like the dinosaurs, a material can become extinct.

For example, there is only so much petroleum, gold ore, natural gas, or iron ore on the earth. If we use them all up, that's it. Thus, all resources *must* be used wisely.

Figure 7-4. Once the coal being mined here is used up, it cannot be replaced.

Locating Raw Materials

Obtaining raw materials for manufacturing is a three-step process. This includes:

- Locating resources.
- Gathering resources.
- Transporting (moving) resources.

A large part of getting raw materials is finding them. Aerial mapping (using an airplane to take pictures) can help locate trees. Geological (under the ground) searches will find minerals and petroleum. See **Figure 7-5.**

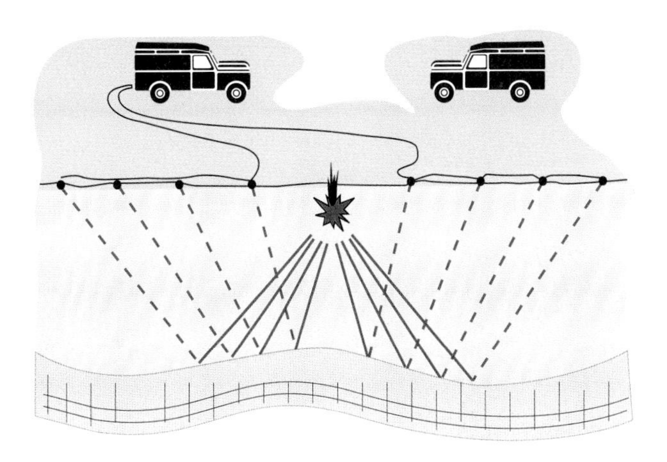

Figure 7-5. Seismic studies send shock waves into the earth to detect promising locations for raw materials. (Shell Oil Co.)

Other resources are easier to find. They are grown commercially (for money). Trees in the south are often grown like a crop. Livestock grown on farms and ranches provides us with leather.

The search for raw materials can be costly and disappointing. For example, our future supplies of petroleum cause us concern. Oil companies spend millions of dollars searching for new pools of oil. Often they come up dry. Several multimillion-dollar "dry holes" have been drilled off the East Coast of the United States. No oil was found.

On a smaller scale, some persons spend a lifetime looking for minerals. The prospectors (gold hunters) of old still live. Today they use better equipment, but are still unlikely to "strike it rich." Working with them are trained geologists (people who study the structure of the earth) from mining companies. Such teams constantly look for gold, silver, uranium, and other metal ores.

Even with all the scientific knowledge and equipment available, finding underground resources is hard, but it is necessary. Our lifestyle is built on materials. Our society must have a continuing flow of raw materials.

Gathering Raw Materials

Once found, raw materials must be gathered. This is done using three major methods. These methods were introduced in Chapter 6, and are shown in **Figure 7-6.**

Figure 7-6. Materials are collected by mining, drilling, and harvesting. (AMAX Corp.; OMI; Boise-Cascade Corp.)

Figure 7-7. An open pit mine in operation. This type of mining is suitable when ores are near the surface.

Mining

Mining involves digging resources out of the earth. If the raw material is close to the surface, it can be mined from the surface. The topsoil is removed and often stored. The mineral is then scooped up by giant power shovels and put in huge trucks. They haul the material to a conveyor or to the surface. See **Figure 7-7.** This is called *open-pit mining* or surface mining.

Often the mineral is in a narrow vein (strip) like many coal deposits. In this case, the topsoil is replaced after mining. The land is returned to productive use. Lakes, pastureland, and farms are developed over the mine pit.

mining. The process of digging resources out of the earth.

open-pit mining. A type of mining used when the resource is close to the surface.

In other cases, the mineral is in a very deep (thick) deposit. As it is removed, a huge pit is produced. The Bingham Canyon copper mine in Utah is over 1/2 mile deep and 2 miles wide. Over 2,000,000,000 (two billion) tons of material has been removed since the mine opened in 1904.

Other ore deposits require digging tunnels to reach the material. This is called *underground mining*. There are three major underground mining methods. As seen in **Figure 7-8,** these are:

- *Shaft mining*. This method is used for deeply buried mineral deposits. A shaft is dug down to the level of the deposit. These shafts can extend several thousand feet into the earth. The main vertical (up-and-down) shaft is used to move people and equipment in and out of the mine. The mineral is also lifted out through the vertical shafts. Other vertical shafts are dug

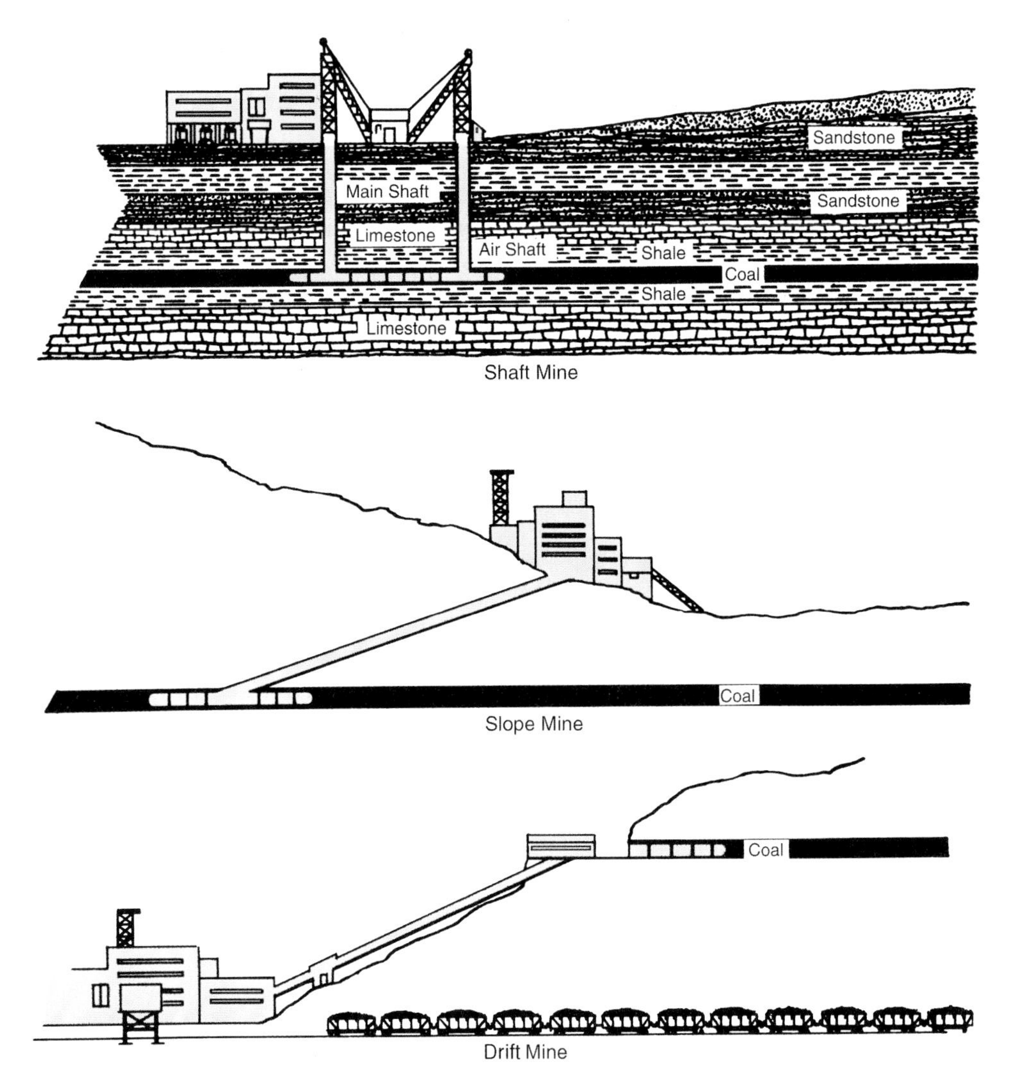

Figure 7-8. Types of underground mines. (National Coal Assoc.)

to bring in fresh air, and remove stale air and gases. The material is mined by digging horizontal (level) tunnels from the vertical shaft.

- *Drift mining*. This method is used when the mineral vein comes to the surface at one point. A tunnel is dug at that point. The tunnel follows the ore vein into the earth. People and materials are moved by railcars that travel along the drift shaft.
- *Slope mining*. This method is used for a shallow mineral deposit. A sloping tunnel is dug down to the deposit. Workers can walk or ride motorized cars down to the deposit. There they dig the mineral out of the vein. The materials are often carried to the surface on a moving platform called a conveyor.

underground mining. A mining method that uses digging tunnels to reach the material. There are three major underground mining methods: shaft mining, drift mining, and slope mining.

shaft mining. Mining method used for deeply buried mineral deposits. A main shaft and an airshaft are dug down to the level of the deposit. The material is mined by digging horizontal tunnels from the vertical shaft.

drift mining. A method used to reach a mineral vein that comes to the surface at some point. A tunnel is dug into the vein, which is serviced by rail cars that move along the drift shaft.

slope mining. A method used for a shallow mineral deposit. A sloping tunnel is dug down to the deposit. The minerals are often carried to the surface on a moving platform called a conveyor.

Drilling

Drilling involves cutting a round hole deep into the earth. See **Figure 7-9.** A drilling rig or derrick is brought to the site. A drilling bit is attached to a drill pipe. The pipe is clamped in a rotary table in the middle of the drilling floor.

drilling. A machining operation that produces a straight, round hole in a workpiece.

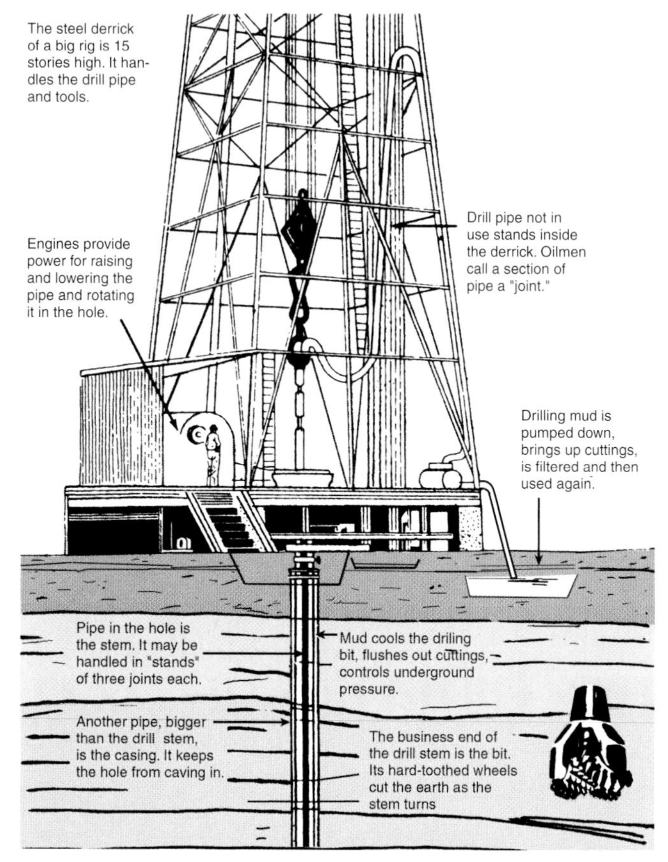

Figure 7-9. A typical oil-drilling rig. The derrick supports a block and tackle for raising and lowering drill pipe. (Shell Oil Co.)

The table turns and the hole is drilled. More drill pipe is added as the hole deepens.

Throughout the drilling, "mud" is pumped down the drill pipe. It comes out through the drill bit. The mud is forced to the surface, carrying with it the cuttings from the bit. Casing (pipe the diameter of the hole) is forced into the hole to keep it from caving in.

Drilling stops when the hole reaches the underground reservoir (pool). The drilling rig is removed. Valves to control the flow and a pump are attached. Drilling is used to reach petroleum, water, natural gas, and other liquids normally found underground.

Drilling may be done either on land or in the water. See **Figure 7-10.** Generally, the drill produces a straight hole. This type of drilling is called *vertical drilling*. Newer techniques allow the drilling of a curved well. This method is called *directional drilling*. It allows us to reach a reservoir that would be hard to get to with vertical drilling. Also, several wells can be drilled from a single platform.

Harvesting

Harvesting is a method used to collect a growing resource. Farmers harvest crops by using a number of different machines, such as potato diggers, grain combines, cotton pickers, and flower pickers. See **Figure 7-11.** Trees are the major "growing" resource that produces engineering materials. Trees are harvested using one of three methods. These, shown in **Figure 7-12,** are:

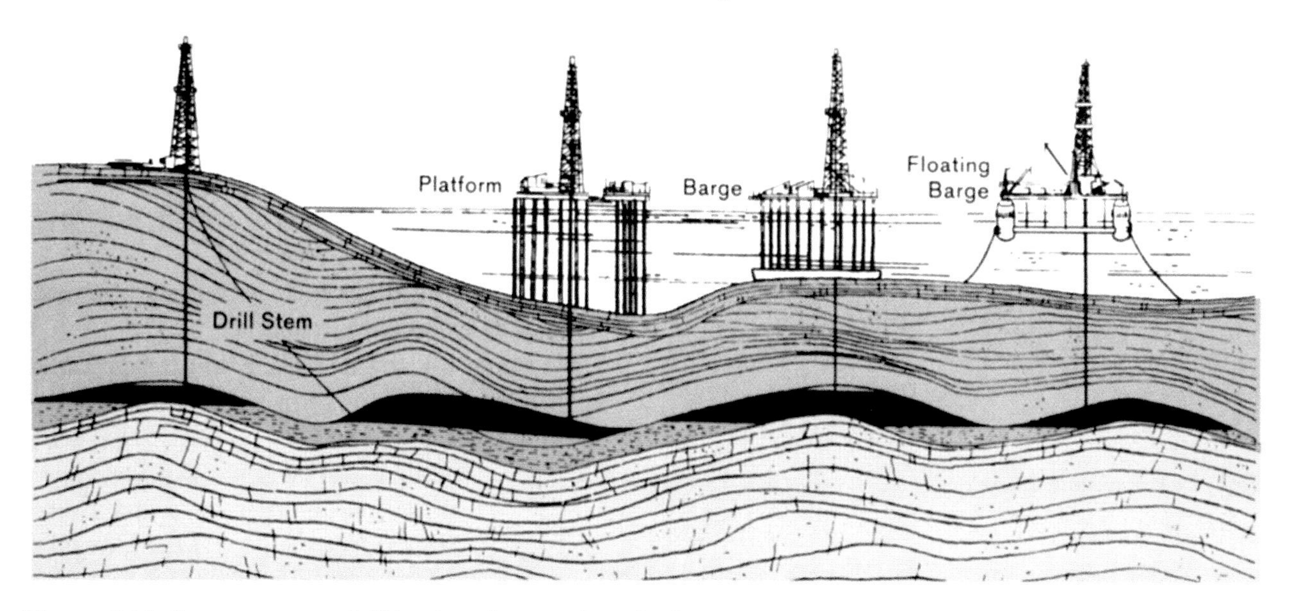

Figure 7-10. Some common drilling locations and methods.

Figure 7-11. Growers harvest renewable resource using many different machines.

Selective Cutting

Clear-Cutting

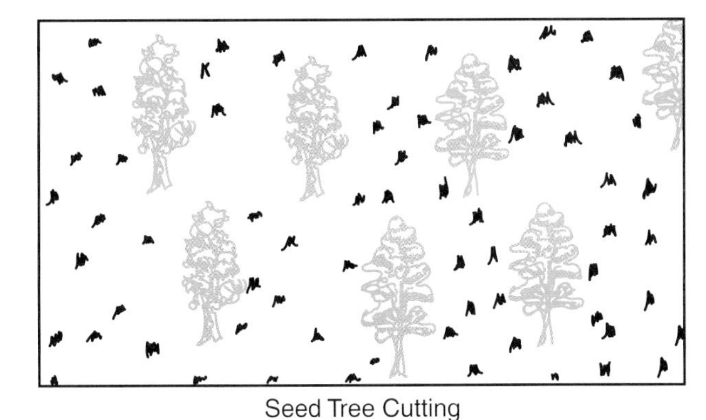

Seed Tree Cutting

Figure 7-12. Types of tree harvesting.

- *Selective cutting.* Mature trees are selected and cut. This method is used for trees that can grow in the shade of others. The stands will have all ages of trees present. Selective cutting is used in western pine areas and in hardwood forests.
- *Clear-cutting* (Block cutting). All trees in a block of about 100 acres are cut. Trees in areas around the block are left alone. They will provide the seed to reforest the area. This technique is used for trees that will not grow in the shade, such as Douglas fir. Clear-cutting is widely used in the western coastal forests.
- *Seed tree cutting.* All trees in an area, except for four or five large ones, are cut. The large trees reseed the area. This technique is used in the southern pine forests.

Harvesting requires several steps. First, either trees or the area to be harvested must be selected. Fallers then fell (cut down) the trees. See **Figure 7-13.** A bucker removes the limbs and top. The tree is cut to standard lengths for the mill.

The lengths, called logs, are moved to a central location for loading on trucks or railcars. This task is called *yarding*. From the yard, the trees are hauled to the mill. They are placed in a pond to help retain their moisture and reduce insect damage.

Newer practices are now using more of the tree. Chippers are placed in the woods. Limbs and tops are chipped up for boiler fuel, hardboard, and paper.

vertical drilling. The most common type of drilling; the result is a hole that runs straight up and down.

directional drilling. A technique that allows wells to be drilled at an angle or along a curve to reach oil or gas.

Transporting Resources

Nearly every type of land and water transportation is used to move raw materials. Pipelines move petroleum and natural gas. Coal can be ground up *harvesting.* A method that collects a growing resource.

selective cutting. A harvesting technique in that mature trees are selected and cut. Younger trees are left standing.

clear-cutting. A method for harvesting trees that cuts down all trees regardless of age or species.

seed tree cutting. All trees in an area, except for four or five large ones, are cut. The large trees reseed the area.

yarding. The method used to move fallen trees to a central point for transport to a sawmill.

slurry. Ground up coal that is mixed with water for pipeline transport.

Academic Link

Forestry is the science or study of forestlands. Through forestry improvements are constantly being made to wood, water sources, wildlife, and recreation. Since trees are the major "growing" resource that produces engineering materials, the study of these resources is of great importance.

In the previous discussion on harvesting, the three methods in which trees are cut were introduced. The study of these methods and their effects on the environment is part of forestry. Many years of studies have been done to see how each of these harvesting methods influence other trees, reseeding of the harvested area, the soil, water sources, and wildlife.

Figure 7-13. A tree is cut or felled by a skilled worker. (Weyerhaeuser Co.)

Figure 7-14. Shown is a truck carrying logs arriving at a sawmill.

and mixed with water, forming a *slurry*, for pipeline transport. Trucks of all kinds and sizes move mineral, farm, and forest products. See **Figure 7-14.** Barges and ships are used on inland waterways and oceans. In short, the most economical method is used to move raw materials from mine, farm, well, or forest.

Summary

Raw materials are the foundation for all manufactured goods. These materials are either a renewable or an exhaustible resource. They are located, gathered, and transported to primary processing mills or refineries.

Commonly, raw materials are gathered through mining, drilling, and harvesting. The gathered resources move over land and on water. There they are transformed (changed) into industrial materials.

Key Words

All the following words have been used in this chapter. Do you know their meaning?

clear-cutting
directional drilling
drift mining
drilling
exhaustible resource
harvesting
mining
open-pit mining
raw materials
renewable resource
seed tree cutting
selective cutting
shaft mining
slope mining
slurry
underground mining
vertical drilling
yarding

Test Your Knowledge

Please do not write in this text. Place your answers on a separate sheet of paper.

1. A material that is still in its natural form is called a(n) _____ material.
2. Explain the difference between a renewable and an exhaustible resource.

Matching questions: For Questions 3 through 6, match the definitions on the left with the correct term on the right.

3. _____ Sloping tunnel is dug to get at shallow deposit.
4. _____ Tunnel follows surface ore vein into earth.
5. _____ Deep shaft is dug straight down to deposit.
6. _____ Deposit is dug from open-pit.

a. Shaft mine.
b. Surface.
c. Drift mine.
d. Slope mine.

7. List 10 renewable resources.
8. List 10 exhaustible resources.
9. _____ searches are used to find underground resources such as petroleum, coal, copper, and other mineral resources.
10. If you want to drill for oil on a certain spot but cannot because a lake is in the way, which of the following methods would you use?
    a. Vertical drilling.
    b. Directional drilling.
11. Describe the difference between the following methods of harvesting trees: selective cutting, clear-cutting, and seed tree cutting.

Activities

1. Visit a drilling, mining, or harvesting operation to see how natural resources are obtained. Then, visit a recycling center to see how manufactured products are recycled for reuse.
2. Working in a team, design a process to separate gravel from a sand-soil-gravel mix.
3. Plant or adopt a tree. Observe its growth throughout the year and the variety of mammals, birds, and insects that utilize it as a resource. Discuss the possible effects on creatures, soil, and air if this tree were removed and not replaced.
4. Start a recycling program at your school or expand an existing program. Recycle all types of waste including paper, cardboard, plastic, glass, and aluminum.

Technology Link

Transportation

Manufacturing is a unique and broad technology. In basic terms it is the process that changes resources into products. Manufacturing affects many other technologies. Other technologies, such as transportation, have an effect on manufacturing as well. The link between manufacturing and transportation is a two-way street.

The next time you are in a car, take a good look around inside. What is the dash made of and how was it manufactured? What about the seat you are sitting on? More important, how are the tires the car is riding on made? From the smallest fasteners to the large door panels, and from the wires in the electrical system to the crankshaft in the engine, all these parts are the result of manufacturing processes. The assembly of all the small parts and small subassemblies is how an automobile is manufactured.

What most people don't realize is the direct effect other technologies have on manufacturing. Transportation has a significant role in the world of manufacturing. There has always been a need for transportation of resources to manufacturing facilities and transportation of final products from the facilities to the marketplace. With the desire of reducing inventory costs and the innovation of just-in-time manufacturing, demands of efficient and timely transportation sources are that much greater. In a world that wants its products "yesterday," transportation of products to the marketplace is that much more important.

Chapter 8
Producing Industrial Materials

Objectives

After studying this chapter, you will be able to:

✓ Explain primary processing.
✓ Define and identify various kinds of standard stock.
✓ Describe four major types of synthetic wood composites.
✓ List forms of plastic standard stock.
✓ Describe steps in making iron and steel stock.
✓ Explain how forest products are converted to standard stock.

When raw materials arrive at the mill, they are ready for a change in form. They will be refined or converted into standard stock. Trees will become lumber or plywood. Natural gas will be changed into plastic. Bauxite will become aluminum sheets. Lime, silica sand, alumina, iron, and gypsum are processed into Portland cement. The first step in manufacturing is called *primary processing*. See **Figure 8-1.** Raw materials are processed into industrial materials. Another name for them is standard stock.

primary processing. **The first step in transforming raw materials into products. Example: Converting trees into lumber at a sawmill.**

Standard Stock

What is standard stock? *Standard stock* is a material that has been changed so it has certain characteristics (qualities). It has a particular grade, size, and shape. See **Figure 8-2.**

Each type of material has its own standards. Let's look at some major types of standard stock.

standard stock. **Material output by primary processing operations, available in standard size units or standard formulations.**

Standard Wood Stock

Forest products include all things made from wood. Even in its natural state, wood is a composite. It is made up of fibers held together by lignin, a natural adhesive. The common natural wood composite is *lumber*.

Fiberboard, particleboard, and laminated boards are synthetic wood composites. Paper is also a synthetic wood composite, but is not an engineering (structural) material.

forest products. **All products that are made from wood.**

lumber. **Pieces of wood (a natural composite) that have been made to a certain size for construction and other uses.**

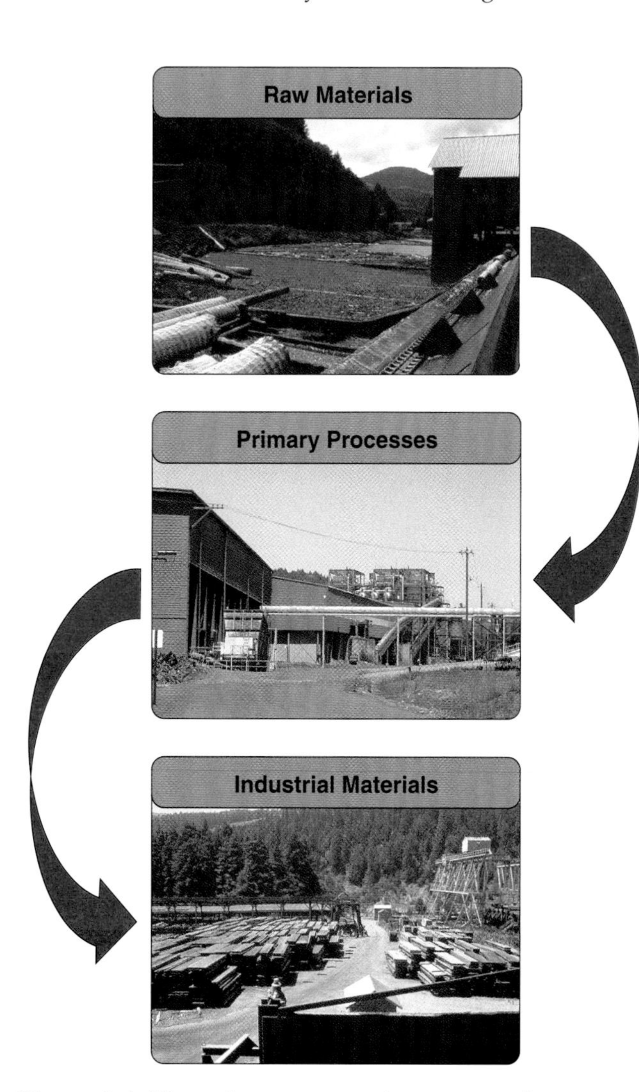

Figure 8-1. The primary processing system changes raw into industrial materials.

Figure 8-2. Standard stock (lumber) is being used to manufacture a mobile home.

Hardwood Lumber Standards

Hardwood is cut to standard thickness—4/4 (four quarters of an inch), 5/4, 6/4, 8/4, and so on. The boards are random (varying) widths and lengths. The diameter and length of the log determine the width and length of the lumber. Other manufacturing steps may produce standard hardwood products. Examples of these are flooring and interior trim for houses.

Softwood Lumber Standards

Softwood lumber is generally produced to standard sizes for all of its dimensions. Boards are cut to set thicknesses, widths, and lengths. For example, you can buy a 2 × 4 × 8, which is actually 1 1/2 inches (") thick by 3 1/2" wide by 8 feet (') long, or a 1 × 8 × 16, which is 3/4" thick by 7 1/4" wide by 16' long. Softwood lumber is sold by its nominal (in name only) size. Remember, a 2 × 4 is not 2" thick and 4" wide, but is 1 1/2" thick by 3 1/2" wide.

Synthetic Wood Composites

Synthetic wood composites are manufactured from thin wood sheets or wood chips and fibers. The typical synthetic wood composites are:

- *Plywood.* A panel composed of a core (middle layer), face layers of veneer (thin sheets of wood), and crossbands. The grain of the core and the face veneers run in the same direction. Except in thin plywood, the crossbands are at right angles to the face veneers. See **Figure 8-3.** Three-ply material has a core that is at right angles to the face veneers. Most plywood is made entirely from a series of layers of veneer. Other plywood has cores made of particleboard or solid lumber (lumber-core). See **Figure 8-4.**
- *Particleboard.* A panel made from chips, shavings, or flakes of wood. The actual name of the material comes from the type of particles used. The most common types are called standard particleboard, waferboard, flakeboard, and oriented strand (placed in a certain direction) board. The particles are held together with synthetic glue.

Academic Link

As you have discovered thus far in this book and will be introduced to more as you continue on, numbers and mathematics is a huge part of the manufacturing world. For example, knowing how to work with fractions is a necessity. Whether converting decimals to fractions, taking measurements, writing product specifications, or hundreds of other instances, you are bound to encounter fractions.

In the previous discussion about standard lumber sizes, it was shown that decimals are very important. With hardwoods if you did not know how to convert and reduce improper fractions (such as 6/4) to common fractions (6/4 = 1 2/4 = 1 1/2), then you would not be able to request the proper size stock. With softwoods standards you now know that a 2 × 4 really measures 1 1/2" × 3 1/2". When figuring measurements, you must know how to add and subtract fractions or your product may be improperly constructed.

plywood. **A synthetic wood composite made of sheets of veneer. The grain of the core and the face veneers run in the same direction. Except in thin plywood, the crossbands are at right angles to the face veneers. Other plywood has cores made of particleboard or solid lumber.**

particleboard. **A synthetic wood composite made from chips, shavings, or flakes held together with a synthetic glue. The most common types are: standard particleboard, waferboard, flakeboard, and oriented strand**

- *Fiberboard.* A panel made from wood fibers. The most common fiberboard is hardboard, commonly referred to as Masonite™. This material is very dense. Fibers are held in place by natural glue (lignin).
- *Laminations.* Heavy timbers produced from a series of layers of veneer or lumber. The grain of all layers runs in the same direction. Synthetic adhesives hold the timber together. Many wood beams in churches, schools, and other buildings are laminations.

Except for laminations, synthetic wood composites are produced in sheets. The most common sheet size is 4' by 8' Materials are made in a number of standard thicknesses. **Figure 8-5** lists the most common ones.

Standard Metal Stock

Like wood, metals also come in several standard sizes and shapes. Basically, the standard determines the cross section and the length of the material. Typical shapes are shown in **Figure 8-6.**

Standard Plastic Stock

Plastic materials are typically produced in pellets and powders, or as sheets and film. Pellets and powders are sold by the pound. They are the inputs for many molding processes.

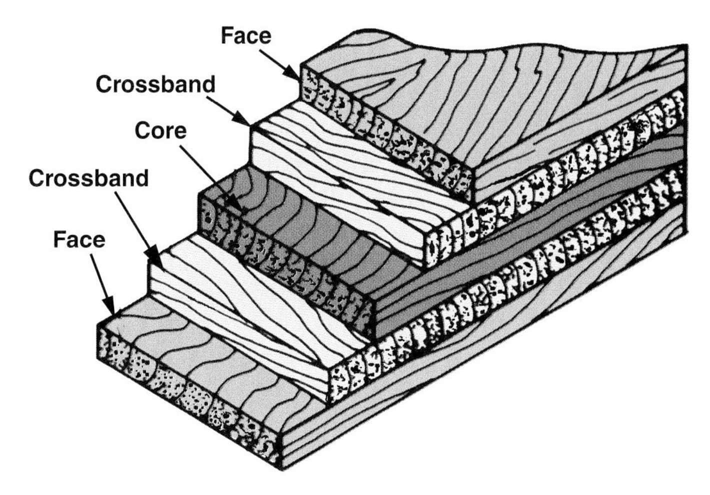

Figure 8-3. A diagram of plywood construction. Note the crossbanding that gives it strength.

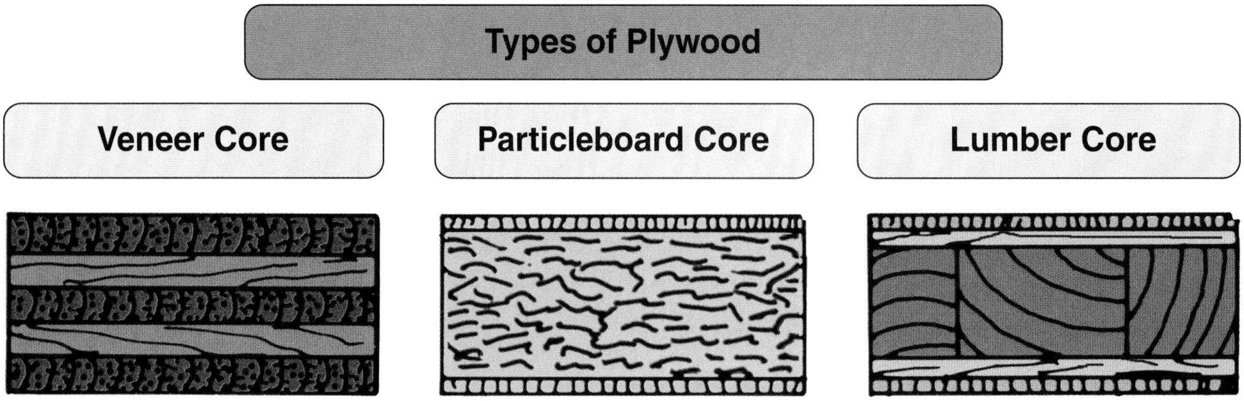

Types of Plywood

Veneer Core

Particleboard Core

Lumber Core

Figure 8-4. Types of plywood are named for their core.

Size	Plywood	Particleboard	Hardboard
1/8			*
1/4	*	*	*
3/8	*	*	
1/2	*	*	
5/8	*	*	
3/4	*	*	
1	*	*	

Figure 8-5. Common thicknesses for sheets of wood composites.

Sheets and films are used for thermoforming processes and in packaging. These materials are sized by thickness and width. The thickness is given in mils. A ***mil*** is 0.001" thick. Width for these materials is given in inches.

Films are sold in single thickness sheets, tubing, and folded forms. See **Figure 8-7.** Folded film is used for shrink packaging, such as is seen on games, puzzles, and compact disc packages. Tubing can be made into bags.

Single-thickness film has many uses. These include vapor barriers (between gypsum board and insulation in homes), heat-sealing wraps (meats and vegetables in grocery stores), and temporary storm windows.

fiberboard. A synthetic composite made from wood fibers. The fibers are held in place by the natural glue in the fibers called lignin.

laminations. Heavy timbers produced from layers of veneer or lumber. The timber is held together with synthetic adhesives.

Producing Standard Stock

Each material has its own primary processing method. Aluminum is refined differently than is copper. Lumber is produced by techniques not used for manufacturing plywood. Each plastic is produced by a special chemical process.

It would be impossible to cover all primary-processing systems. That would take many books. Instead, two common primary processing activities—steelmaking and forest product manufacturing—will be presented. This will give you an idea of the steps a raw material goes through to become standard stock.

Steelmaking

Steelmaking started in North America over 300 years ago. It is now a large industry. The process is a complex one. Complicated chemical actions occur at high temperatures. These processes are called ***thermally activated.***

Steelmaking requires three basic raw materials. These are iron ore, coke, and limestone. Other materials are added as alloys (mixtures of metals) are manufactured. Steelmaking is a four-step process:

- Preparing raw materials
- Making iron

- Making steel
- Producing standard stock

Preparing raw materials

Steelmaking is a process of removing impurities from iron ore. The early processes used high-grade iron ores. These ores then contained large amounts of iron. Now, such deposits are hard to find. Instead, low-grade ores called *taconite*, are often used. They must be preprocessed at the mines. The iron content is increased by removing many unwanted minerals. The remaining material is *sintered* (heated to high temperatures) and made into pellets. The pellets contain about 65% iron combined with oxygen.

Coke also must be produced. It is a clean-burning carbon product made from coal. The coal is loaded into chambers in a coke oven. The ovens, seen in **Figure 8-8,** look like a series of drawers set on edge. Each chamber is about 18" wide, 20' high, and 40' deep. Gases burning between each chamber heat the coal.

As the coal is heated to 2400°F (1316°C), gases, oils, and tar are driven off. These materials are caught and used for many products, such as lipstick and plastics.

After being heated for about 18 hours, the door of the chamber is opened. The coke is pushed into a quench (rapid cooling) car. The cooled coke is a high-carbon fuel.

The final ingredient of steel, limestone, must be mined and purified. It is crushed and screened. At the same time, magnets and the screens remove impurities.

Environmental protection (reducing pollution) is considered during the preprocessing of the iron ore, coke, and limestone. Land and water are reclaimed at the mine sites. Dust is controlled at the taconite mines and limestone crushers. Coal gases are carefully collected at the coke ovens. Workers are careful to collect as many pollutants as possible. Some are useful by-products.

Shape	Appearance	How Sized	Lengths*
Sheet	Less than 1/4" thick	Gage No. (thickness) × width Lengths to 12 ft.	To 12 ft.
Plate	More than 1/4" thick	Thickness and width	
Band		Thickness and width	
Rod		Diameter	
Square		Width	12-20 ft.
Hexagonal		Width across flats	
Octagonal		Width across flats	
Angle		Leg lengths (2) × thickness of legs	
Channel		Width × height × web thickness	To 60 ft.
I-beam		Height × flange thickness × web thickness	

*Special orders (not standard) can be in many other lengths.

Figure 8-6. Standard metal shapes and their names.

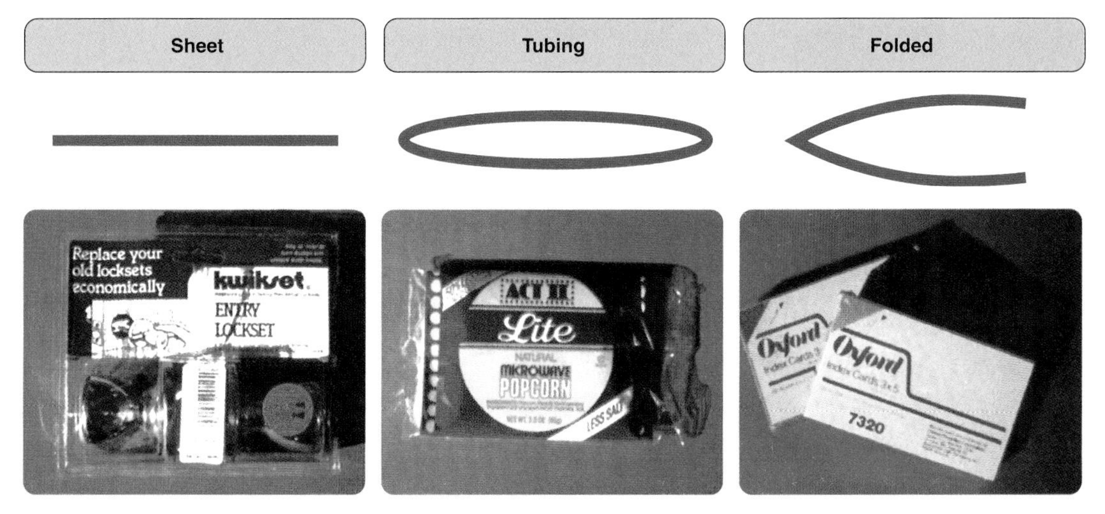

Figure 8-7. Plastic film shapes.

taconite. A low-grade iron ore. Taconite must be pre-processed at the mine.

sintered. Heating a material to a high temperature so that it becomes a solid mass without melting the material.

coke. A clean-burning high-carbon fuel made from coal. Coke is coal with the impurities burnt out of it.

Figure 8-8. This 80-slot coke oven battery can produce 850,000 tons of coke a year. (Bethlehem Steel)

Figure 8-9. This blast furnace produces iron. It towers 270' into the sky. (American Iron and Steel Institute)

Figure 8-10. This 330 ton capacity car carries molten iron away from the blast furnace. (Bethlehem Steel)

Making iron

Iron is the first product of the steelmaking process. Most iron is made in a blast furnace like the one shown in **Figure 8-9.** Layers of coke, limestone, and iron ore are loaded in the top. They move slowly down to the furnace. Super hot gases moving upward ignite the coke.

The burning coke removes oxygen from the iron ore. (Iron ore is really iron oxide, a combined form of iron and oxygen.) The limestone combines with impurities in the molten iron. It produces a substance called slag that is drawn off.

The iron ore, by this time, is a mixture of iron and carbon (from the coke). This material collects at the bottom of the blast furnace. Now and then it is drawn off into railcars. These cars, shown in **Figure 8-10,** move the iron-carbon mixture (called pig iron) to steelmaking furnaces.

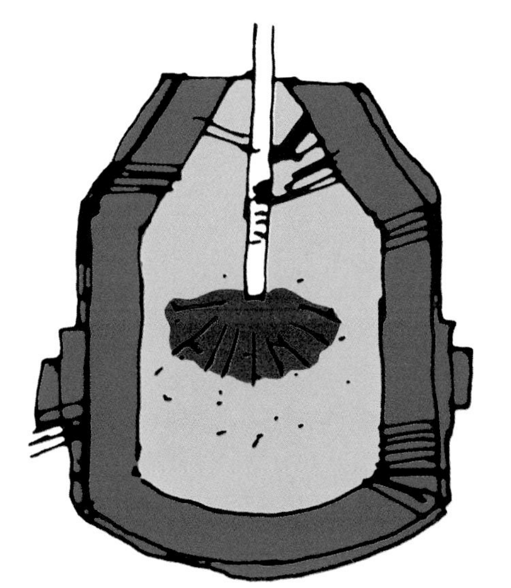

Figure 8-11. The basic oxygen furnace in cross section.

Making steel

Steel is made from pig iron and steel scrap. It can be produced in one of two kinds of furnaces:

- Basic oxygen furnace
- Electric furnace

These newer furnaces are rapidly replacing the older open hearth furnace.

Most steel is produced by the *basic oxygen process.* See **Figure 8-11.** The furnace is charged with steel scrap and hot iron from the blast furnace. See **Figure 8-12.** A water-cooled lance is lowered into the furnace. Pure oxygen is blown through the lance into the metal. The oxygen burns (combines with) the extra carbon and any impurities in the iron. They form gases that are collected and cleaned. Also, lime and other materials are added to absorb more impurities. They form slag that is drawn off. Alloying elements (other metals) are added to the molten steel. The resulting steel is drawn off into a ladle.

Figure 8-13 is a flowchart for the steelmaking process. You can study it to see the steps in producing steel as standard stock. Note the environmental control systems used at each step.

The electric furnace, seen in **Figure 8-14,** is used to recycle steel scrap. These furnaces are generally smaller than basic oxygen furnaces. Many high-alloy steels are produced in them. The output is typically stainless, tool, and specialty steels. Recently, larger electric furnaces have been developed to produce carbon steels.

Figure 8-12. Molten iron from a blast furnace is charged into a basic oxygen furnace. (Bethlehem Steel)

Producing standard steel stock

The molten steel from the steelmaking furnaces must be processed further. It is usually cast into a workable shape first. Then, the steel is rolled into a final shape (standard stock).

basic oxygen process. A steelmaking process that reduces the carbon content of iron by injecting pure oxygen into molten iron.

Older mills first cast an ingot. See **Figure 8-15.** This is a solid casting that will be shaped in later processes. The ingots are first cooled in soaking pits to solidify the steel. Then, they are reheated to a rolling temperature.

More modern mills use a continuous casting process, as shown in **Figure 8-16.** The metal is cast and cut into slabs. See **Figure 8-17.** The hot slabs move directly to the rolling mills. Continuous casting saves large amounts of time and conserves energy.

Workers in control centers, **Figure 8-18,** oversee many rolling processes. These processes change slabs into many shapes. See **Figure 8-19.**

Forest Products Manufacturing

All forest products start with logs. The logs are debarked (bark removed) as they enter the mill. See **Figure 8-20.** In the mill, they move through one of four distinct paths. The logs can be used to make lumber or sheet materials.

Lumber manufacturing

The lumber manufacturing process varies slightly according to the size of the log. Basically, the logs go through seven steps, as seen in **Figure 8-21:**

- Larger logs are squared at a head rig. The outer slabs produce wider boards that are fairly knot-free.
- The square center, called a *cant*, and smaller logs are cut into boards at a gang saw.
- The edges of the boards are sawed parallel by the edger. This process removes wane (sloping edges). Wider boards may also be cut into two or more narrower boards at this step.
- Trim saws cut out defects and square the ends of the boards. They also produce standard length.

cant. The square center of a log that is cut into lumber on a gang saw in a sawmill.

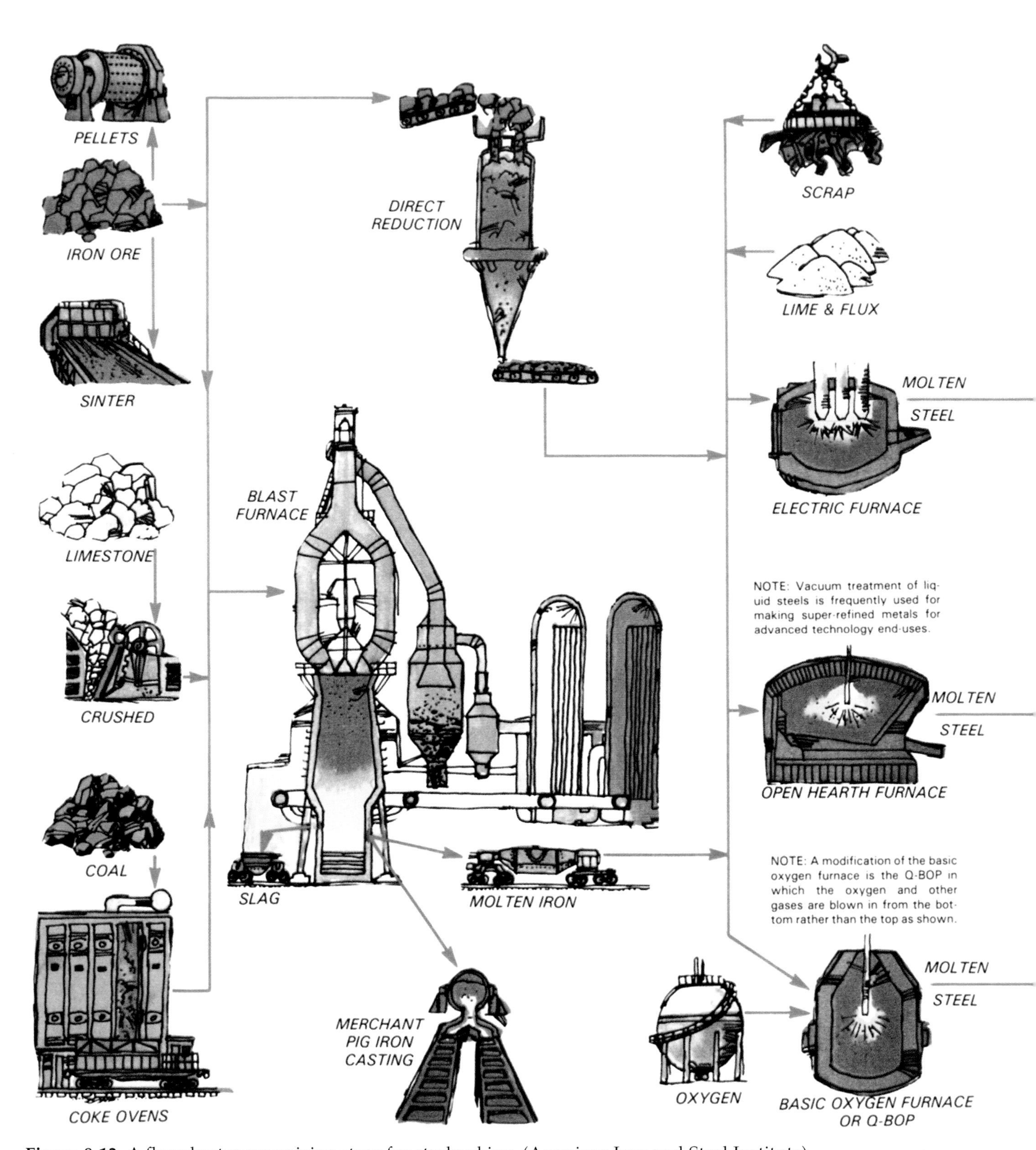

Figure 8-13. A flowchart summarizing steps for steelmaking. (American Iron and Steel Institute)

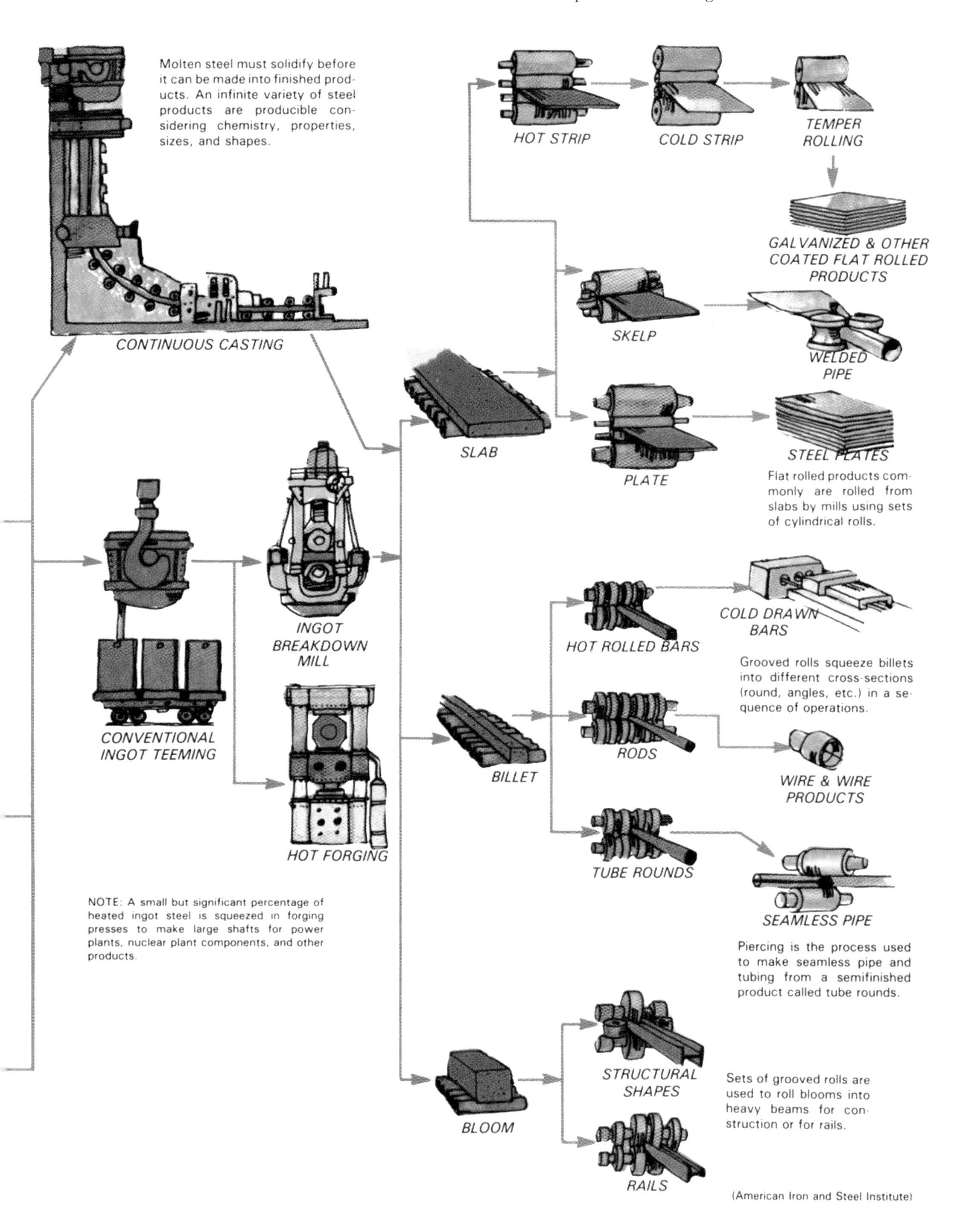

(American Iron and Steel Institute)

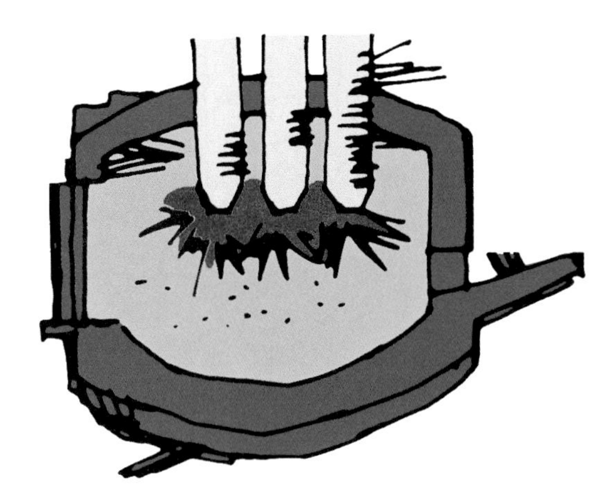

Figure 8-14. The electric furnace uses heat from an electric arc to melt the charge.

Figure 8-15. Molten iron is poured into ingot molds. (Bethlehem Steel)

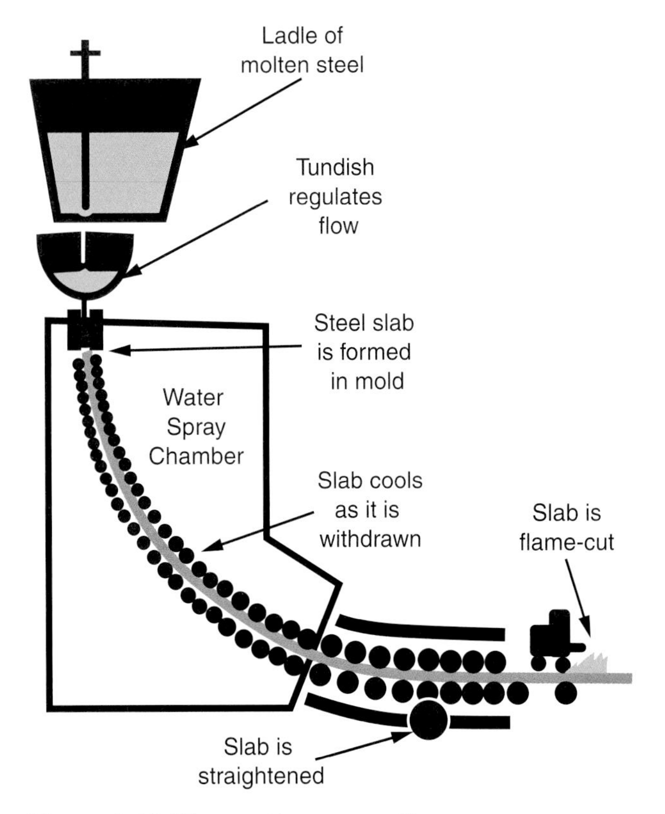

Figure 8-16. The continuous casting process. Molten steel moves from ladle to mold to rollers in an unbroken flow.

Figure 8-17. A steel strand from a continuous caster is flame cut to length. (Bethlehem Steel)

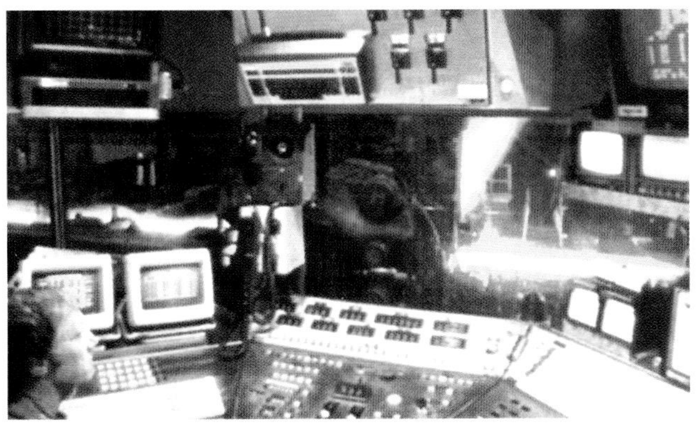

Figure 8-18. Modern electronics help these technicians in an elevated pulpit to control the steel rolling process. (American Iron and Steel Institute)

- The lumber is graded as it moves down a conveyor called a green chain. Different grades are piled in separate stacks.
- The lumber is air and/or kiln (oven) dried.
- The dry lumber is often planed and edged (smoothed) to a standard thickness and width. See **Figure 8-22.**

Figure 8-19. A steel plate is being rolled in a 160 inch wide plate mill. (American Iron and Steel Institute)

Figure 8-20. Logs enter a mill after debarking.

Plywood manufacturing

Most plywood is made from softwood. Douglas fir and southern pine are the main species (types) used. Logs are steamed so they can be cut more easily. They are then placed in a veneer lathe. See **Figure 8-23.** The lathe turns the log against a sharp knife, as shown in **Figure 8-24.** A thin sheet of wood, called *veneer*, is cut from the log. The log is simply "unwound" much like paper from a roll.

The veneer is dried and graded. The better pieces will have knots and defects cut out. Patches are placed in the holes.

The veneer is then laid up into plywood. The inner layers, as shown in **Figure 8-25,** are coated with glue. Veneer strips for outer layers are edge glued to produce a single sheet. They are placed on top of the inner layers. The "veneer sandwich" is placed in a heated press. Heat and pressure squeeze the sheet together and cure the glue.

After a predetermined curing time, the panel is removed from the press. After the sheet has cooled, saws trim it to a standard size and sanders smooth and reduce it to a standard thickness. Finished sheets are inspected, grade marked, and shipped to customers.

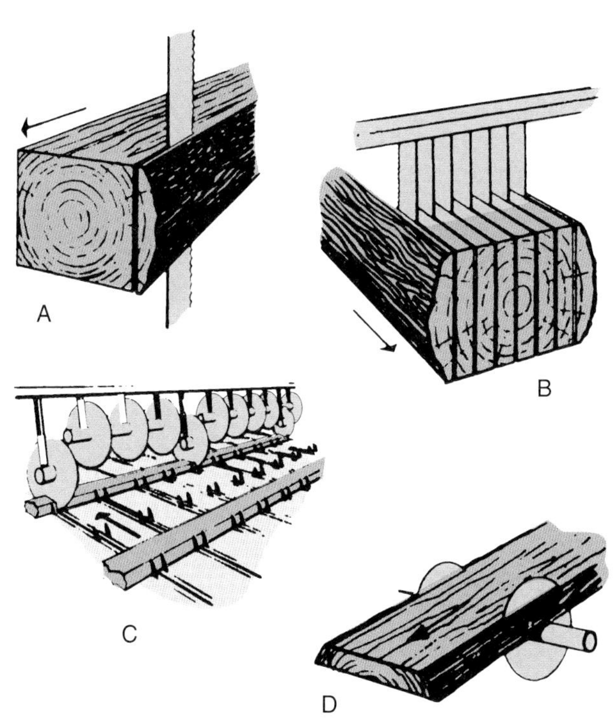

Figure 8-21. Major steps in the lumber manufacturing process. A—Squaring up large logs. B—Sawing a cant into boards. C—Trim saws cut out defects. D—Board moves through the edger.

veneer. **A thin sheet of wood that is cut from a log. The log is unwound much like paper from a roll.**

Particleboard manufacturing

Particleboard is a sheet material made from logs and mill waste. Chips and flakes are mixed with glue. The mixture is formed into a mat. The mat is pressed while heat is applied. The heat cures the glue, producing a rigid sheet. The cured sheet is sanded to thickness and cut to a standard size.

Figure 8-22. Dry lumber is planed and edged to a standard size. It is then sorted and graded. (Weyerhauser Co.)

Figure 8-23. A veneer lathe "unpeels" a log section to make veneer. (American Plywood Association)

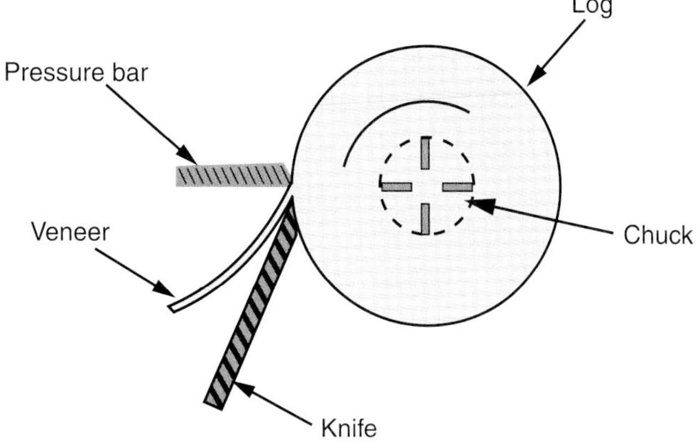

Figure 8-24. A diagram of a veneer lathe. (American Plywood Association)

Hardboard manufacturing

Hardboard also uses mill waste and chips. These materials are broken down into single fibers. The fibers are then formed into mats. The mat is placed in a heated press. Pressure is applied. The natural lignin on the fibers bonds them tightly together. The result is a very hard, dense sheet. The sheet is then cut to a standard width and length. The sheet is so smooth after it leaves the press that it does not need to be sanded.

Figure 8-25. Veneer is coated with an adhesive. The curtain glue spreader is in the center of the picture. (American Plywood Association)

Summary

Raw materials are converted into standard stock. Each material has its own standard sizes, shapes, and compositions. These standards determine the materials that are available to small manufacturers and individuals.

All standard stock is produced by primary processing practices. Each material has its own processing methods. The basic processing activities for steel and forest products were presented. You may want to check books out of a library to read about other materials.

Key Words

All of the following words have been used in this chapter. Do you know their meaning?

basic oxygen process
cant
coke
fiberboard
forest products
laminations
lumber
particleboard
plywood
primary processing
sintered
standard stock
taconite
veneer

Test Your Knowledge

Please do not write in this text. Place your answers on a separate sheet of paper.

1. What is *standard stock*?
2. List and describe the three major types of structural synthetic wood composites.
3. Plastic materials are typically produced in four forms. List them.
4. Describe the major steps in making iron and in making steel.
5. List the steps in manufacturing lumber from logs.
6. How is plywood made?

Activities

1. Write to a producer of forest products or steel. Request information describing their manufacturing processes. Share these materials with your class.
2. Make plywood from several sheets of veneer. Test it for strength and compare it with a piece of solid wood of the same thickness and species.

Unit 4
Secondary
Manufacturing
Processes

Chapter 9
Layout and Measurement

Objectives

After studying this chapter, you will be able to:

✓ State the meaning of the terms *measurement* and *layout*.
✓ Identify surfaces of a part.
✓ Identify special features on a part.
✓ Identify measuring and layout tools.
✓ List principles of measurement for round and flat stock.
✓ Describe how to lay out a part.

measurement. The process of describing a part's size using a standard for comparison. This allows the part to be duplicated from the measurements only.

layout. The suggested arrangement of copy and illustrations for a print advertisement. A layout is submitted by an advertising agency for the company's approval.

Before any part or product can be made it has to be measured and laid out. You need to measure to find the distance between two points. *Measurement* tells you how thick, wide, and long a part is. It tells you the diameter and depth of any holes. You also find the width and depth of grooves and dados by measurement.

The word *layout* means to measure and mark a part so that it can be made. The markings tell you where every feature (hole, notch, etc.) should be on the part. It also shows where a part is to be cut from a larger piece of material. The difference in these terms is shown in **Figure 9-1.**

Before you can lay out or measure parts and features you must know:

- Names of the surfaces on a part.
- Special features or cuts.
- Basic measuring and layout tools.
- Layout practices.

Surfaces of a Part

There are some basic terms that describe a part. Everyone must know these terms so they can communicate with coworkers. These terms are on the top of the next page. See **Figure 9-2.**

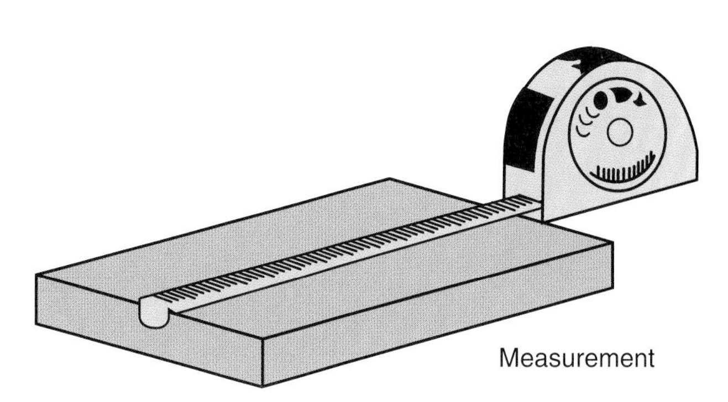

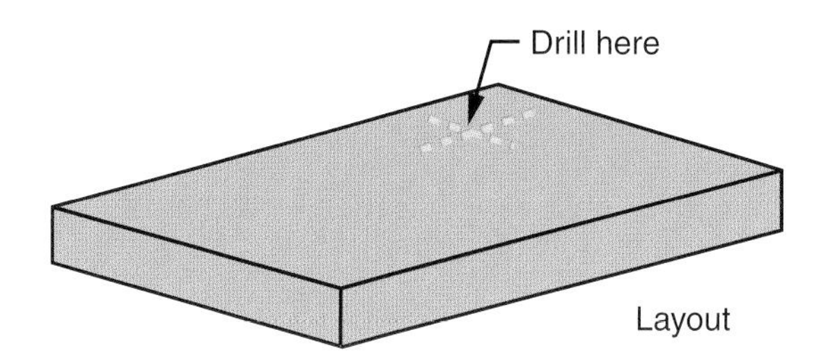

Figure 9-1. These drawings show the difference between measurement and layout.

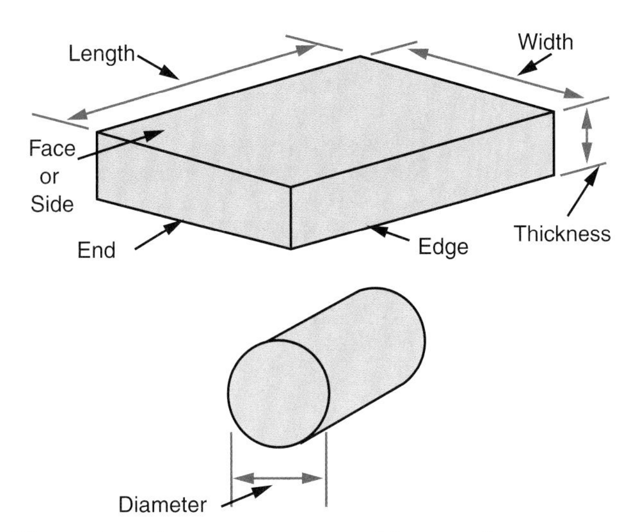

Figure 9-2. Common terms used in most layout and measurement.

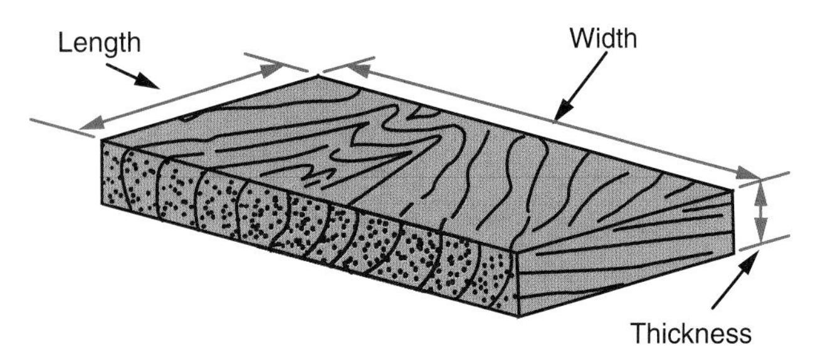

Figure 9-3. Measurement of wood. Length is always with the grain of the wood.

- *Length*. Largest dimension of a part.
- *Width*. The second largest dimension.
- *Thickness*. The smallest dimension of the part.
- *Diameter*. The distance across the end of a round part.
- *Face (or side)*. The largest surface.
- *Edge*. The second largest surface.
- *End*. The smallest surface.

Dimensions are all the size measurements of a part. They are always given in a certain order. For rectangular pieces, the order is "thickness × width × length." A round part's diameter is given first followed by its length.

Wood Measurement

These basic measurement terms are used differently for wood. Length is always measured along the grain of the piece. Width is measured across the grain.

It is possible, as shown in **Figure 9-3,** to have a board wider than it is long. Whenever you see a dimension listed, you can tell grain direction. A piece of wood, 3/4 × 6 × 24, has the grain parallel with the longest dimension. However, a part 1/2 × 12 × 6 is a board with the grain parallel to the 6 inch measurement.

length. The largest dimension of a part.

width. The second largest dimension of a part.

thickness. The smallest dimension of a part.

diameter. The length of a line that passes through the center of a circle.

face. The largest surface of a part.

edge. The second largest surface of a part.

end. The smallest surface of a part.

dimensions. The size measurements for a part.

Special Features

Special features are the cuts that change a part from a square or rectangle. They are holes, cuts for joints, and specially shaped cuts. There are many special features. The most common are pictured in **Figure 9-4.**

Measuring Techniques

Measuring determines sizes of parts and features. Typically, we measure the external (outside) sizes and angles or the internal (inside) sizes and angles.

External measurement basically determines the thickness, width, and length of a part. It also measures the outside diameter of round parts. Special features that change external features, such as angled or shaped edges, must also be measured. See **Figure 9-5.**

Internal measurements determine the sizes of holes, grooves, slots, and other "inside" features. Typical internal measurements are width, inside diameter, angle, and depth. **Figure 9-6** shows some common internal dimensions.

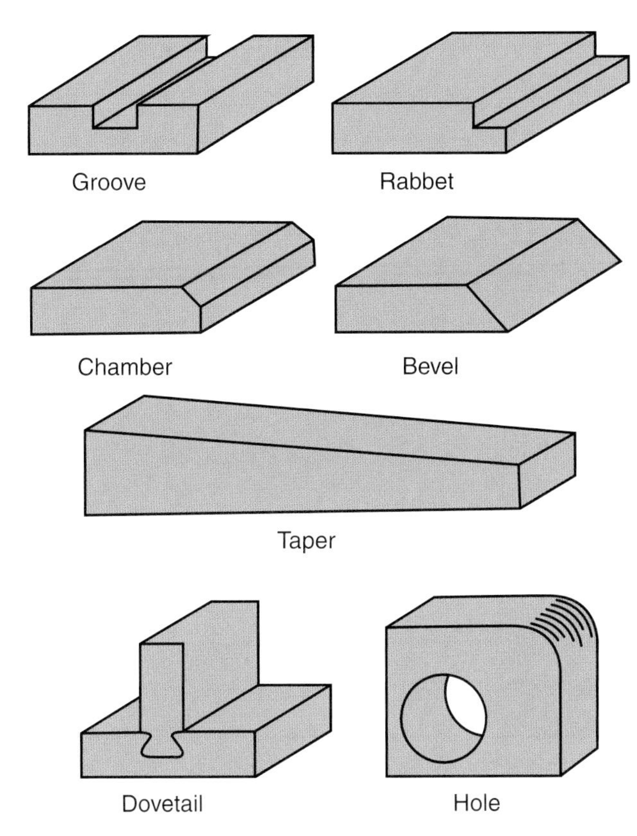

Figure 9-4. Common features on parts change them from a basic square or rectangular shape.

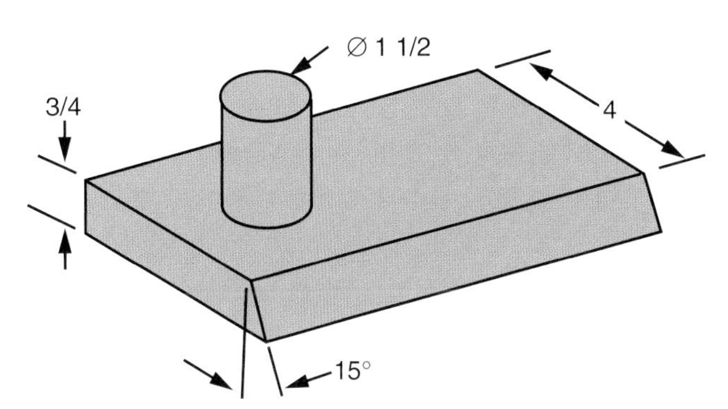

Figure 9-5. Samples of external measurements. They are on the outside surfaces of parts.

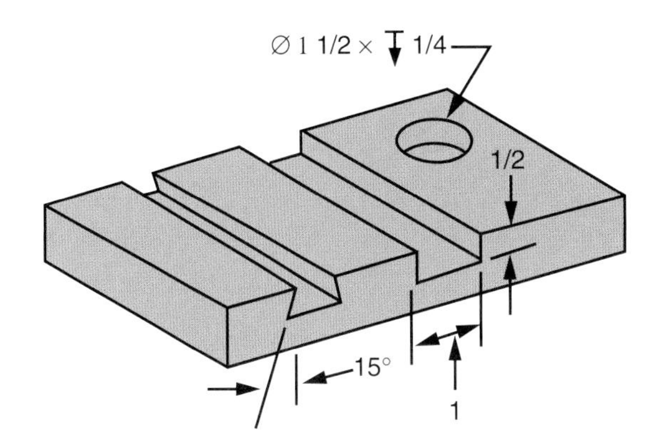

Figure 9-6. Examples of internal measurements. They are on the inside surfaces.

These measurements are generally either standard or precision (very accurate). In *standard measurement*, you would measure to the nearest fraction of an inch. *Precision measurement* is generally given in thousandths of an inch (0.001").

Standard measurement is used for most woodworking, carpentry, and sheet metal work. Precision measurement is commonly used in metal machining, plastic molding, and material-forming practices. The type of measurement to be used depends on the accuracy needed and the stability of the material.

standard measurement. **A type of measurement where the dimensions are held to the nearest fraction of an inch (1/4, 1/8, etc.).**

precision measurement. **A very accurate type of measurement. Readings are given to greater than .001" accuracy.**

SI Metric Measurement

SI metric measurement is used by some U.S. industries and by most countries of the world. In this system, standard measure is taken to the nearest millimeter, which is much smaller than the inch (1 mm = 0.0394"). Precision measurement is accurate to the nearest tenth or hundredth of a millimeter (0.1 mm = 0. 004"; 0.001 mm = 0. 0004").

The more accurate a part must be, the more precise the system of measurement is needed. Engine cylinders or transmission gears must be very accurately manufactured. Therefore, these are held to thousandths (0.001) or ten-thousandths (0.0001) of an inch.

However, some materials do not lend themselves to that type of precision. Wood, for example, is not very stable. It expands and contracts as the humidity of the air changes. It is useless to measure wood in thousandths. It will vary in size from day to day as the weather changes.

tape rule. A common measuring tool that has markings on a flexible tape of steel, cloth, or plastic.

machinist's rule. A ruler that has scales (markings) that will measure down to 1/64".

bench rule. A ruler for measuring small parts, marked in 1/8" increments.

micrometer. A very precise measurement tool used for linear dimensions.

Measuring and Layout Techniques

Layout and measuring is always done with the aid of tools and gages. The basic measuring tools may be grouped by what they measure:

- Linear (length) distances
- Diameters
- Angles

Linear Distance Measuring Tools

Certain tools measure distances between two points in a straight line. They will determine the basic size measurements: thickness, width, and length. The most common measuring tool is the *tape rule.* It provides an easy way to measure long parts. Shorter parts may be measured with *machinist's rules* and *bench rules.* See **Figure 9-7.** Machinist's rules usually have scales (markings) that will measure down to 1/64 in.

A micrometer, **Figure 9-8,** makes precision measurements. The most common type is an outside micrometer. They come in various sizes. The 0-1 inch micrometer, which measures in thousandths of an inch, is typical. Others measure larger sizes, but within a one inch range. For example, a 1-2 inch micrometer is available. So is a 3-4 inch.

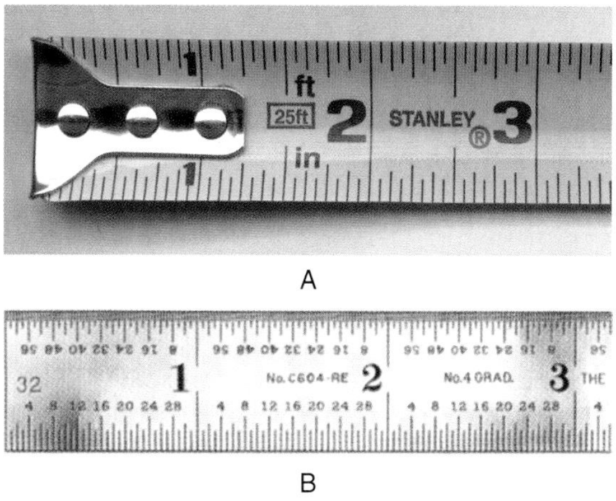

Calipers and *dividers,* can also be used to measure linear distance, The caliper, **Figure 9-9,** is set against the two parallel surfaces. It is then removed from the part. The distance between the caliper legs is measured with a rule. Dividers are similar, but they measure between two individual points.

Depth is also a linear measurement. It may be the depth of a hole, groove, or other feature. Measurements can be made with a rule if it will fit into the feature. Also, special *depth gages,* as shown in **Figure 9-10,** are available.

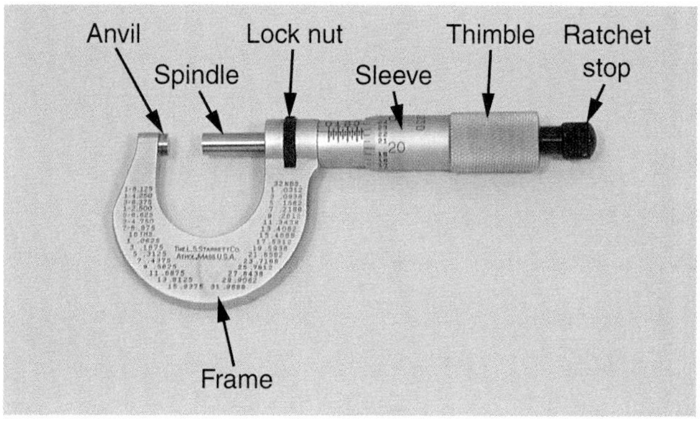

A

B

Figure 9-7. Rules such as these measure distance along a straight line. A—Tape rule. B—Machinist's rule. (The L. S. Starrett Co.)

Anvil
Spindle
Lock nut
Sleeve
Thimble
Ratchet stop
Frame

Figure 9-8. Basic parts of a micrometer caliper.

Diameter Measuring Tools

Rules can be used to measure diameter. However, accurate measurements are hard to make. Typically, outside diameters are measured with a micrometer, **Figure 9-11,** or calipers. Micrometers provide precision measurements while calipers are accurate up to 1/64 inch.

Inside diameters can be measured with inside micrometers or inside calipers. Again, micrometers are more accurate than calipers. See **Figure 9-12.**

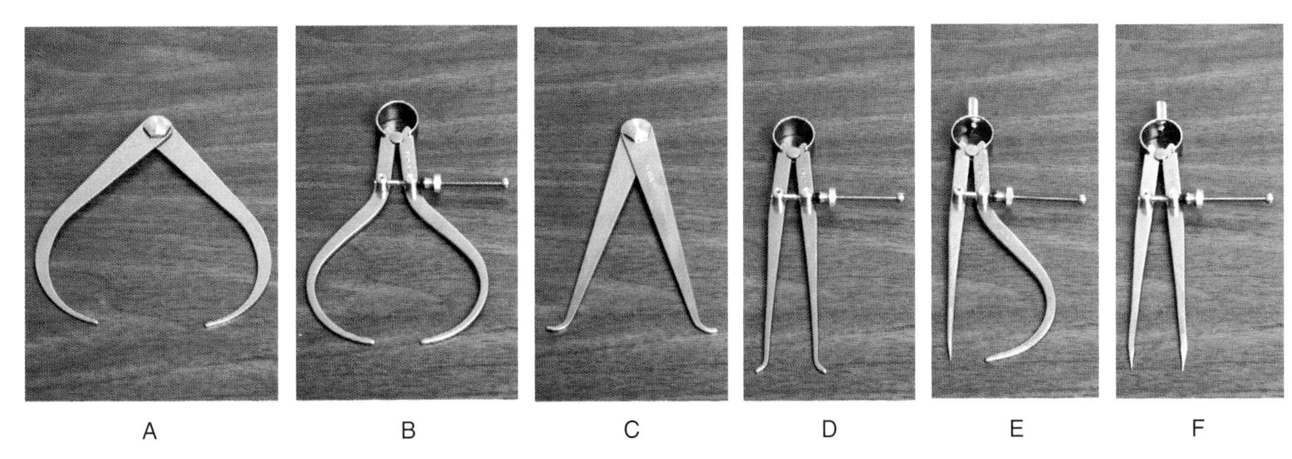

Figure 9-9. Calipers and divider for transferring measurements. A—Firm joint outside caliper. B—Bow string outside caliper. C—Firm joint inside caliper. D—Bow spring inside caliper. E—Hermaphrodite caliper. F—Divider.

Angle Measuring Tools

Angles can be any value either greater or smaller than 90°. The common 90° or right angle is checked with a *square*. There are many kinds of squares, as shown in **Figure 9-13.** These include rafter squares, try squares, and combination squares. Combination squares will check both 90° and 45° angles.

Angles other than 90° or 45° can be checked with a *protractor*. See **Figure 9-14.** This may be a separate tool or part of a combination set. The *combination set* has a 45° and 90° head, a protractor head, and a center-finding head. All the heads will fit on a single rule, but are used separately. See **Figure 9-15.**

Figure 9-10. A direct reading depth gage is used to check depth of a tire mold. (Goodyear Tire and Rubber Co.)

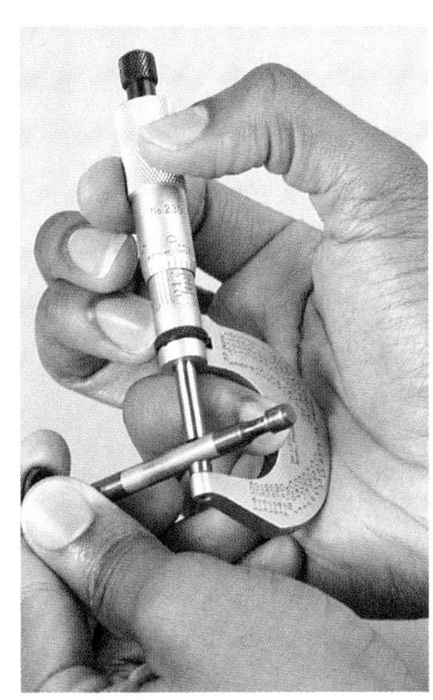

Figure 9-11. Measuring a piece using a micrometer.

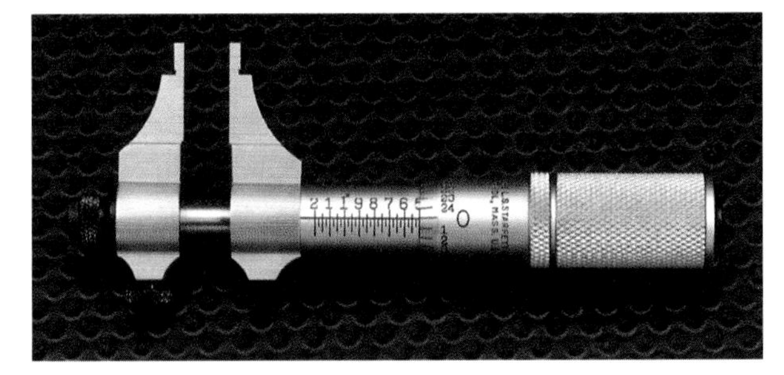

Figure 9-12. Inside micrometers allow quick and accurate measurements of internal features. (The L.S. Starrett Co.)

Layout Practices

One principal use of measuring tools, as shown in **Figure 9-16,** is layout. The shape of the part itself must be laid out (drawn) on the standard stock. Proper layout is essential to making good parts. Layout should follow some basic steps.

First, the part size is measured off on the stock. Lines for length and width are scribed or drawn on the piece of material. Look at **Figure 9-17.** It shows a way to use the square to do this. The part blank is then cut out of the standard stock.

caliper. A tool that is used to measure two parallel surfaces. The distance between the caliper legs is measured with a rule. There are inside and outside calipers.

divider. A tool used to measure the distance between two points. The points of the divider are placed on the two points to be measured. The divider is placed next to a rule and the distance is read off.

depth gage. A tool used to measure the depth of a hole, groove, or other feature.

square. A measuring tool used to check right (90°) angles.

protractor. A tool used to check angles other than 90° or 45°.

combination set. A combination square and additional measuring tools that attach to the square. The combination square has a 90° edge and a 45° edge. The protractor head can measure angles accurately. The center-finding head is used to find the center point of a circular face.

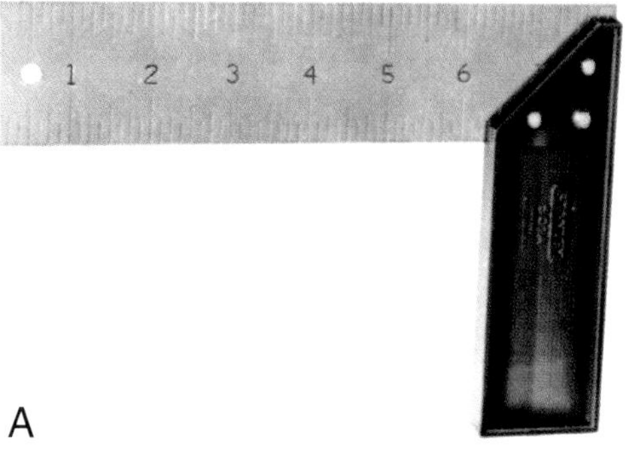

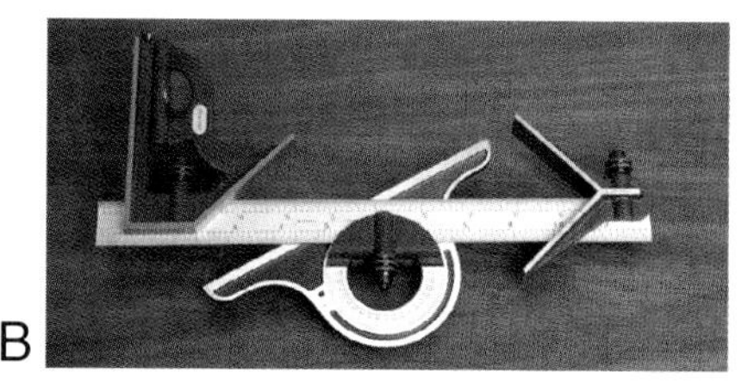

Figure 9-13. Types of common squares. A—Try square. B—Combination set. C—Steel square. D—Machinist's square. (Stanley Tools; The L. S. Starrett Co.)

Figure 9-14. A protractor can check many angles. (The L.S. Starrett Co.)

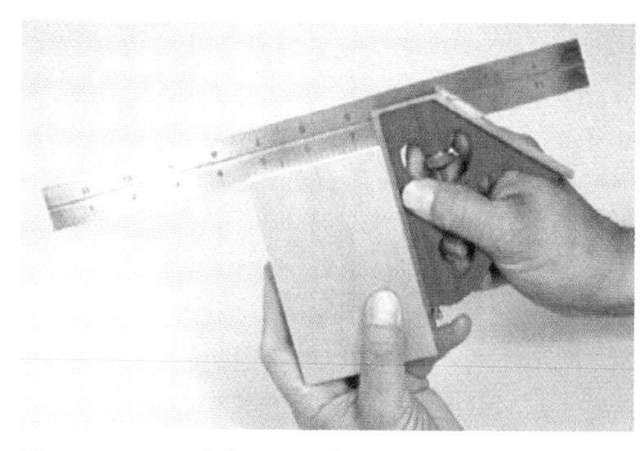

Figure 9-15. The combination square can be used to check squareness and measure depth.

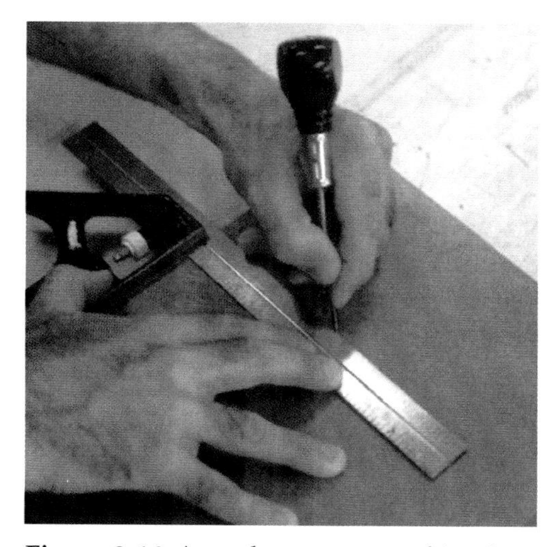

Figure 9-16. A worker uses a combination square to lay out a piece of wood.

Second, the features of the part should be located on the blank. The centerlines for holes and arcs are drawn first. Rules and squares are used to make these lines square and parallel.

Holes and arcs are then located. Dividers or a compass are used to lay out the circumference of (distance around) circles.

Centers of holes should be marked. A prick punch or center punch is used for this task. Tangent lines (lines that connect circles) are then drawn. Finally, straight cuts are laid out. **Figure 9-18** shows a typical layout process.

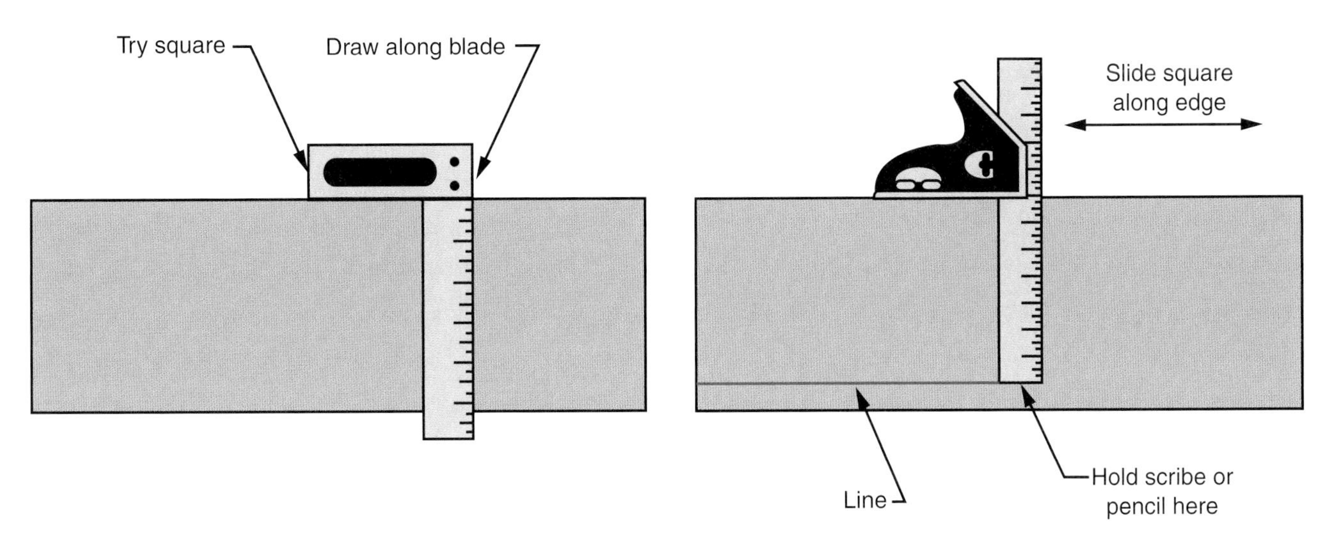

Figure 9-17. Laying out the length and width of a part. Top—Marking the length. Bottom—Using a combination square to scribe a parallel line for width.

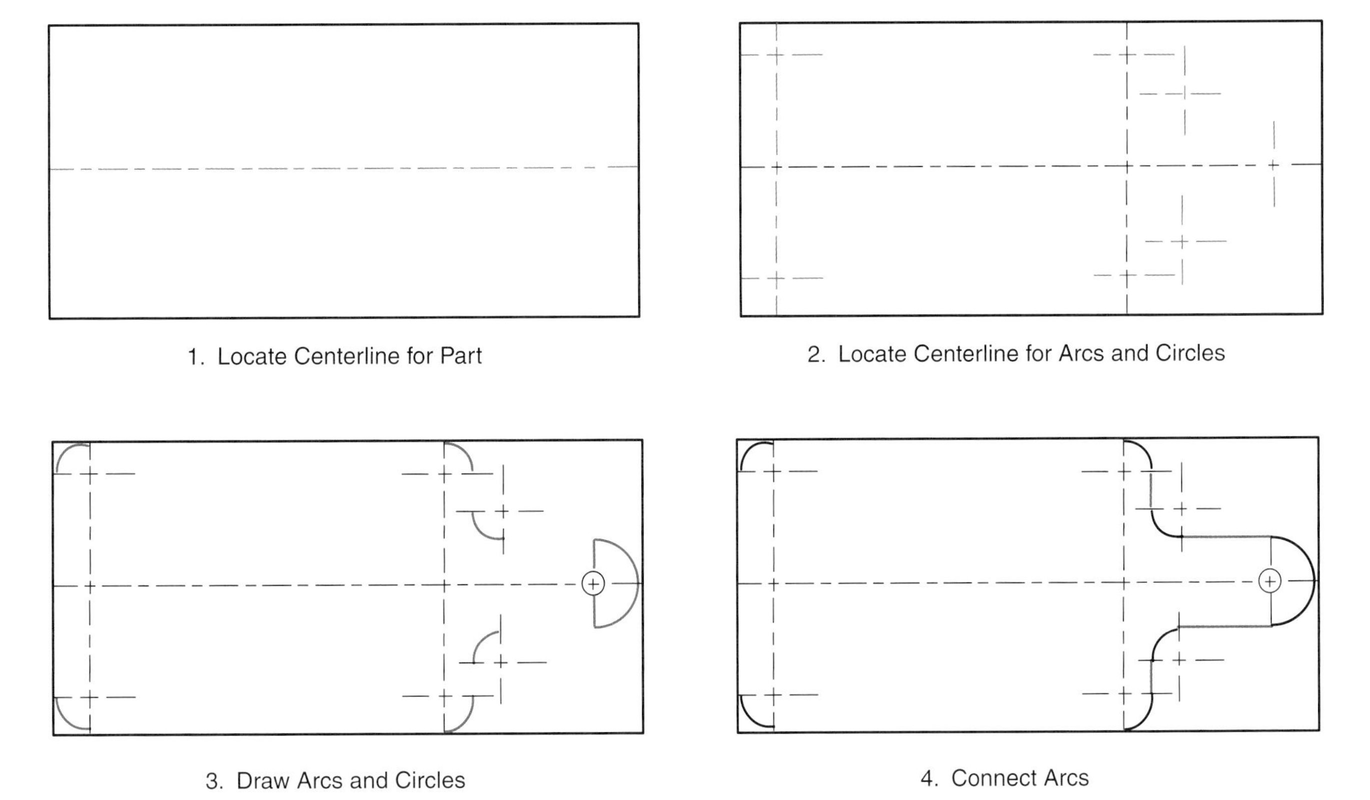

Figure 9-18. Procedure for laying out product features.

Summary

Layout and measurement are important in all manufacturing processes. Measurement determines sizes of a manufactured part. Layout determines the location of features and the size of parts to be produced. We lay out something to be made. We measure the part after manufacturing operations are completed.

The principal surfaces of a part are its face, edge, and end. They are measured as a width and thickness (or a diameter) and a length.

Measurement may be very accurate (precision) or simply in fractions of an inch. These measurements may be for internal or external surfaces and features. All measurements are done with tools that measure linear distances, diameters, and/or angles.

General Safety

1. Always think "SAFETY FIRST" before performing any operation.
2. Dress properly. Avoid wearing loose clothing and open-toed shoes. Remove jewelry and watches.
3. Control your hair. Secure hair that may come into contact with machine parts.
4. Wear safety glasses, goggles, or a face shield in danger zones or other areas where they are required.
5. Wear ear protection around machines or equipment that produce a loud or high-pitched sound.
6. Do not use tools, machines, or equipment unless the instructor has demonstrated their use to you.
7. Stay alert! Always keep your mind on what you are doing.
8. Before using any tools, machines, or equipment, notify your instructor of any unsafe conditions.
9. Keep the floor clear of scrap materials.
10. Report even the slightest injury to your instructor. Small cuts or other minor injuries may become serious if left untreated.
11. Respect the rights of other people and their property.
12. Avoid horseplay. Do not distract the person performing an operation.
13. Concentrate on your work. Daydreaming and watching other people can cause accidents.
14. Know your own strength. Do not try to lift or move heavy materials or machines.
15. Follow all specific safety rules for tools and machines.
16. Do not use compressed air to blow chips or dust from machines, workbenches, or other surfaces. Use a brush.
17. If you have any questions about an operation, always ask them first before beginning.

Key Words

All of the following words have been used in this chapter. Do you know their meaning?

bench rule
calipers
combination set
depth gage
diameter
dimensions
dividers
edge
end
face
layout
length
machinist's rule
measurement
micrometer
precision measurement
protractor
square
standard measurement
tape rule
thickness
width

Test Your Knowledge

Please do not write in this text. Place your answers on a separate sheet of paper.

1. What is meant by precision measurement?
2. Which measuring tool would you use to measure the following features?
   a. Diameter of a hole in thousandths.
   b. Length of a board.
   c. Diameter of a wood dowel.
   d. Diameter of a steel shaft in thousandths.
   e. Length of a metal part in 1/64ths.
3. The proper order for listing the dimensions of a part is: _____.
   a. For rectangular parts: thickness × width × length; for round parts: diameter × length.
   b. Always length × width × thickness.
   c. Always list in descending order of size.
4. _____ _____ are the cuts that change a part from a square or a rectangle.
5. In laying out and measuring a part, which would you do first?
   a. Locate features on the piece of stock.
   b. Saw off the piece of stock to right length.
   c. Lay out the length and width of the part on the stock.
6. _____ of holes should be marked before they are drilled. A(n) _____ _____ is used for this task.

Activities

1. Working with a partner, design a device to measure the angle of cuts made on a saw. The device should be able to measure 0° to 45° in 1" increments.
2. Measure the lengths, diameters, and angles of various objects in the room using linear distance, diameter, and angle measuring tools. Lay out the features of a door. Use tools that measure linear distances, diameter, and/or angles.

Tool and Die Maker

Tool and die makers are among the most highly skilled production workers in the economy. These workers produce tools, dies, jigs, and fixtures that enable machines to manufacture a variety of products—from shoes, clothing, and furniture to heavy equipment and parts for cars, farm equipment, and aircraft.

Toolmakers craft precision tools that are used to cut, shape, and form metal and other materials. Die makers construct metal forms (dies) that are used to shape metal in stamping and forging operations. They also make molds for diecasting and for molding plastics, ceramics, and composite materials. In addition to developing, designing and producing new tools and dies, these workers may also repair worn or damaged tools, dies, gauges, jigs, and fixtures.

Tool and die makers may use many machine tools and precision measuring instruments. They must also be familiar with the machining properties, such as hardness and heat tolerance, of a wide variety of common engineering materials. As a result, tool and die makers are knowledgeable in machining operations, mathematics, and print reading. In fact, tool and die makers are often considered highly specialized machinists.

From drawings and prints, tool and die makers plan the sequence of operations necessary to manufacture the tool or die. They then measure and mark the workpieces that will be cut to form parts of the final product. Then, tool and die makers cut, drill, or bore the part as required, checking to ensure that the final product meets specifications. Finally, they assemble the parts and perform finishing jobs such as filing, grinding, and polishing surfaces.

Chapter 10
Casting and Molding Processes

Objectives

After studying this chapter, you will be able to:

- ✓ Explain how casting and molding have evolved.
- ✓ List the steps involved in casting and molding.
- ✓ Describe how each step in the casting and molding process is completed.
- ✓ Give examples of items that result from casting or molding processes.

Casting and molding, as shown in **Figure 10-1,** are a group of manufacturing processes in which an industrial material is made into a liquid. Then the material is introduced (poured or forced) into a prepared mold of proper design. The material is allowed to solidify (become hard) before being extracted (removed) from the mold. The finished item is called a casting or a molded part. Casting is the term generally used when working with metal and ceramic materials. The term molding is used when working with plastics.

Casting and molding are the most direct routes from raw material to finished product. See **Figure 10-2.** This accounts for the development of casting and molding early in history.

How Casting and Molding Evolved

Casting and molding date back almost 6000 years. Legend suggests that the first metal extracted (separated from the earth) was copper. It is believed that a person banked a cooking fire for the night with rusty-looking rock. As the wood changed into charcoal, the fire became very hot. This heat probably melted copper from the ore (metal bearing rock). When the ashes were scattered the next morning, the first pure metal was discovered. Uses were soon found for this metal. This occurrence propelled humans out of the Stone Age and into the Bronze Age.

The first use of this new material was for forged (hammered) tools and weapons. These were shaped using crude methods as shown in **Figure 10-3.** Later, metalworkers learned that copper could be formed more easily if it was first heated. During the hot forging process, some of the metal may have become too hot and melted. The melted metal became the standard stock for casting activity.

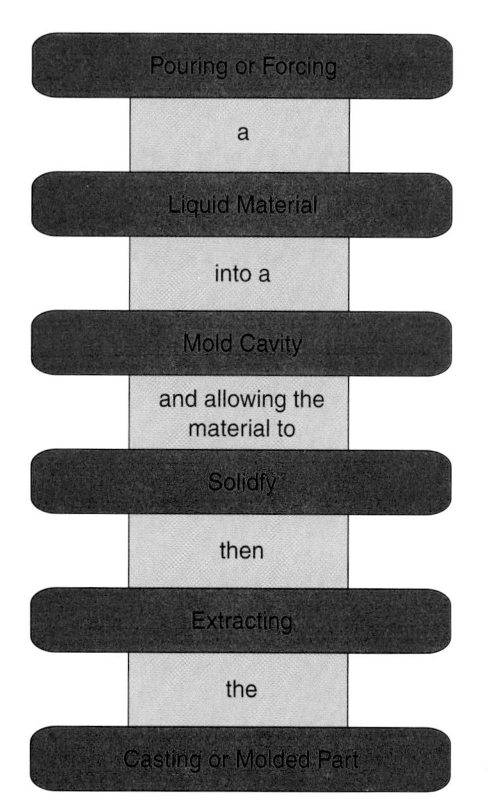

Figure 10-1. Casting and molding produce a simple or complex shape in a single step. (American Cast Iron Pipe Co.)

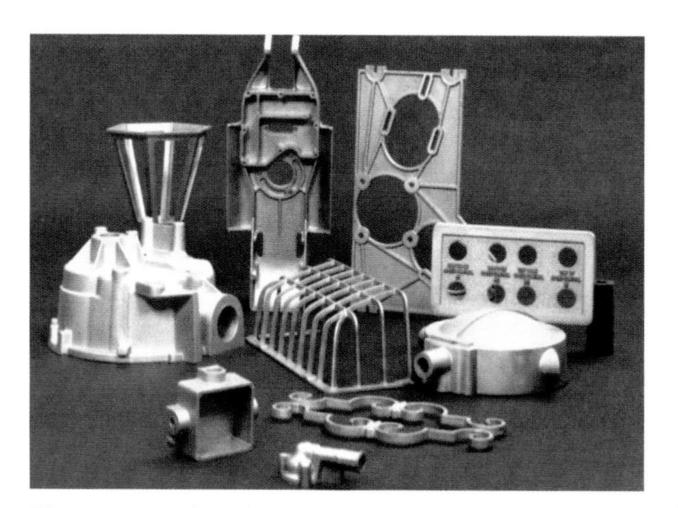

Figure 10-2. This photo shows some typical cast parts.

Figure 10-3. Metal was first used for making hand tools and weapons. (Miller Electric Manufacturing Co.)

The art of casting was most likely developed by accident. However, no matter how it was born, casting became the most direct route from raw material to finished part. See **Figure 10-4.** Casting began to replace the forge and the hammer.

Liquid metal was first poured into molds of sand. Later, molds cut in stone, formed in soft limestone, or pressed out of sunbaked clay were used.

Many developments in casting came from the Orient. The art of casting matured there before it developed in Europe. Before 1000 A.D., the Chinese developed ways of casting iron. Later, the technique of casting crucible steel was invented in India.

Academic Link

The social effects of the development of manufacturing processes are sometimes forgotten. Since its discovery, metal casting has played a critical role in the development and advancement of human cultures and civilization. For example, early social impacts of metal casting were the production of tools for farming and construction, weapons for protecting property, and church bells for communication.

After 5000 years of technological advances, metal casting plays a greater part in our everyday lives and is more essential than it has ever been. In more modern times, many automobiles (old and new) have parts that are cast metal. Just think about the social impact that the automobile has had over the past century. In 1900 in this country, virtually everyone traveled by feet, by horses, or steam locomotive. Years ago, travel was not as easy. It may have taken a person a number of days to travel a couple of hundred miles. A hundred years later, most people have a car or access to motorized transportation. Now we might not think twice about getting in a car and traveling 100 miles.

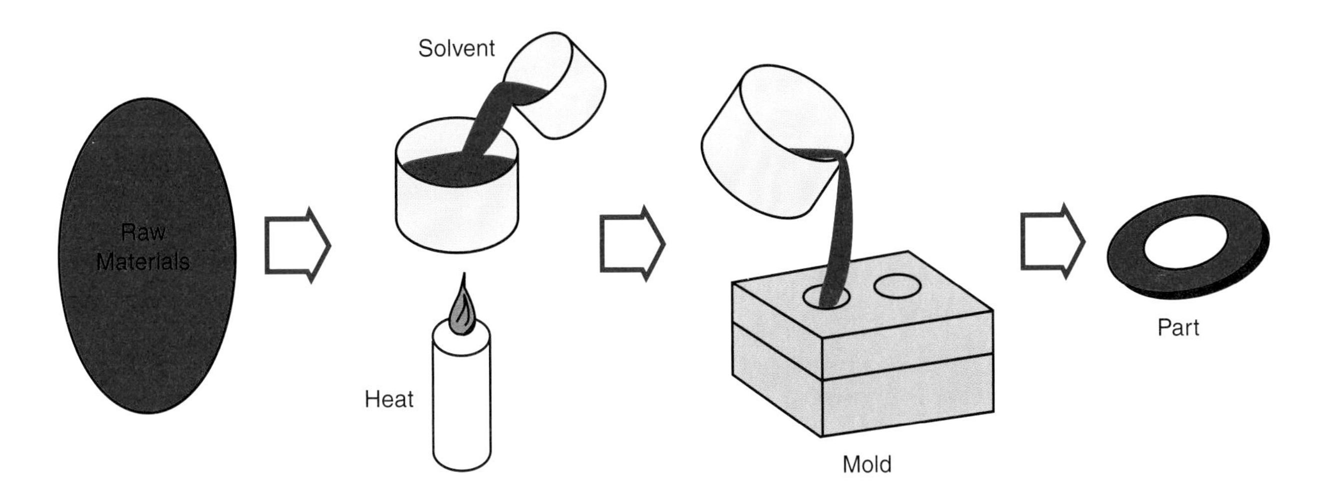

Figure 10-4. Casting is the most direct route from raw materials to the finished part.

Oriental developments in casting technology were brought westward to the Middle East and Europe. Cast bells and cannons became important European products. The first cast iron gun was produced in England in about 1500 A.D.

The first casting in America was done from the famous "Saugus Pot." It was located at the Saugus Iron Works near Boston, Massachusetts. From this lowly start, the modern foundry (metal casting) industry has grown.

Steps in Casting and Molding

Casting and molding processes belong to a single family. Like the members of a human family, there are family similarities and individual differences.

Each casting technique is unique and is used to manufacture products with special characteristics. However, whether the material is clay, metal, plastic, wax, or glass, all casting and molding processes follow the same five steps:

1. A mold of proper design is produced.
2. The material to be cast or molded is prepared (made liquid).
3. The material is poured or forced into the mold cavity.
4. The material is allowed or caused to solidify (harden).
5. The finished casting or molded part is extracted from the mold.

Expendable
Greensand Dry Sand Shell Investment Plaster

Permanent
Die Centrifugal Pressure Plaster Investment

Figure 10-5. There are several types of molds that can be grouped under two types.

Molds

All casting processes need a container to hold the molten material until it becomes hard. This container is called a *mold*. The mold is constructed with care and accuracy. The *cavity* (shaped hole) inside the mold is formed to produce a part of proper size and shape.

Molds for casting and molding fall into two major groups. These, as shown in **Figure 10-5,** are:

- Expendable (one-shot) molds.
- Permanent molds.

Expendable Molds

Many molds for casting processes are used only once. Removing the finished casting destroys them. These molds, which are called *expendable molds* or one-shot molds, must be easy and inexpensive to produce. The most common expendable mold materials are sand and plaster.

Producing expendable molds follows a two-step process, as shown in **Figure 10-6.** First, a *pattern* of proper size is made. A pattern is a shape made from wood, metal, or epoxy. This pattern must have *draft* (angled sides) so it can be removed from the mold.

The pattern gives the mold its shape. The mold is generally made in two or more parts. See **Figure 10-7.** This allows the mold to be separated to remove the pattern. The mold material is poured or packed around the pattern. Then the pattern is removed and the mold is reassembled. It is now ready to receive the molten material.

mold. **A container to hold a liquid material until it solidifies. A correctly shaped cavity is built into the mold. The two types of molds are expendable and permanent.**

cavity. **The void in a casting or forming mold where the part will take shape.**

expendable molds. **A mold that is destroyed after one use.**

Greensand casting molds

Metal casting techniques account for nearly all expendable molds used. The most common of these techniques is *greensand casting*.

The first step is to make a pattern in the exact shape of the finished product. The pattern will be larger than the desired product. This compensates for the shrinkage that occurs when molten materials cool. The amount the pattern is enlarged depends on the metal being cast. This added size is called the ***shrink allowance***. The mold is made in a rectangular metal box called a ***flask***. The flask has two parts: the ***cope*** (top) and ***drag*** (bottom). Each half has four sides, but no top or bottom. Creating these molds follows the steps shown in **Figure 10-8.**

1. The pattern is placed inside the drag and rests on a bottom board. It is covered with a thin layer of ***parting compound.*** This fine powder allows for easy removal of the pattern from the mold.
2. Sand is ***riddled*** (sifted through a screen) over the pattern and pressed around it. The drag is then filled with sand and ***rammed*** (tamped down).
3. The excess sand is struck off to provide a smooth top for the mold.
4. The drag is turned over and placed on a board. The cope is then put in place. If a two-part (split) pattern is used, the second half is placed in the cope.
5. Again, parting compound is applied, and the sand is riddled, packed, rammed, and struck off.
6. Making a greensand mold involves pouring the metal through an opening on one side of the mold called the ***sprue.*** Channels or ***gates*** are made between the sprue and the mold cavity. The molten metal will fill the cavity and flow out a gate on the other side. This second gate allows excess metal and gases to escape to another opening called the ***riser.*** The sprue is made with a sprue pin and the riser is made with a riser pin.
7. Next, the cope and drag are separated. A gate is made from the mold edge to the sprue and another is made from the mold edge to the riser.
8. The pattern is carefully removed. This leaves a cavity for the molten metal that is the proper size and shape.
9. Cores are placed in the cavity to produce any desired holes in the finished casting.
10. The cope and the drag are carefully reassembled.
11. Molten metal is poured into the mold.
12. After the casting cools, the top half is removed exposing the casting.
13. The casting is removed and the sprue, riser, and gates are broken off and it is ready for finishing.

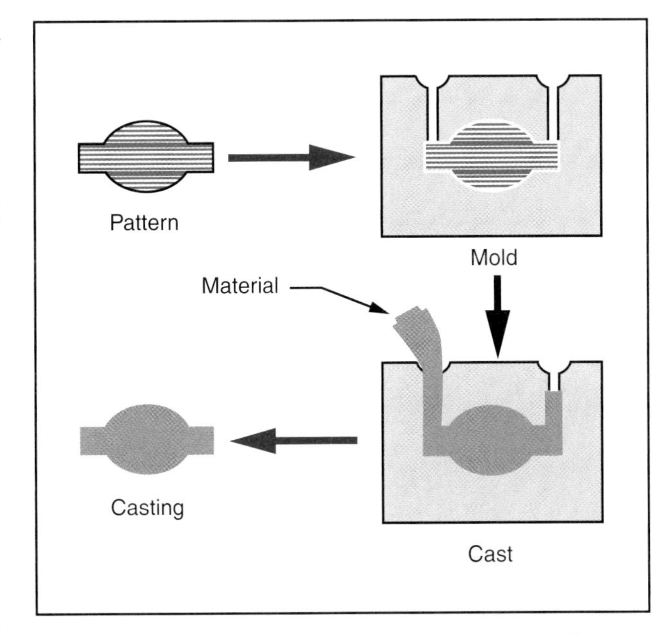

Figure 10-6. Expendable casting processes involve creating a pattern, making a mold, and producing a casting.

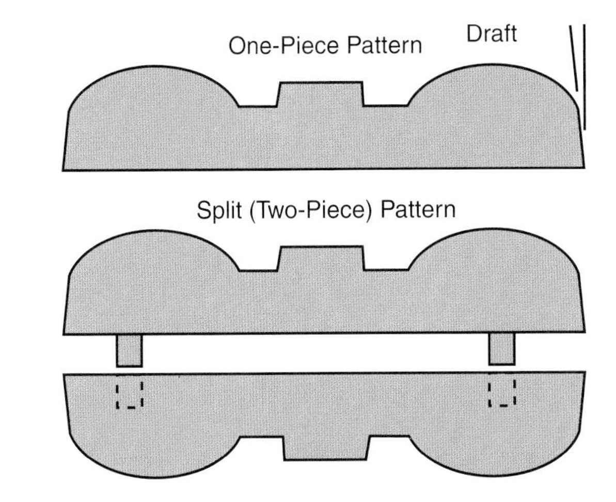

Figure 10-7. There are two basic types of patterns.

pattern. **A pattern is a device that is the exact shape of the finished part. Patterns are used to make expendable molds.**

draft. **Angled sides on a pattern that allow it to be removed from a mold.**

Shell molds

Shell molds were developed during World War II. The technique involves using a hot metal pattern to produce each half of the mold. The pattern is covered with a sand and resin mix. The heat from the pattern melts the resin. This

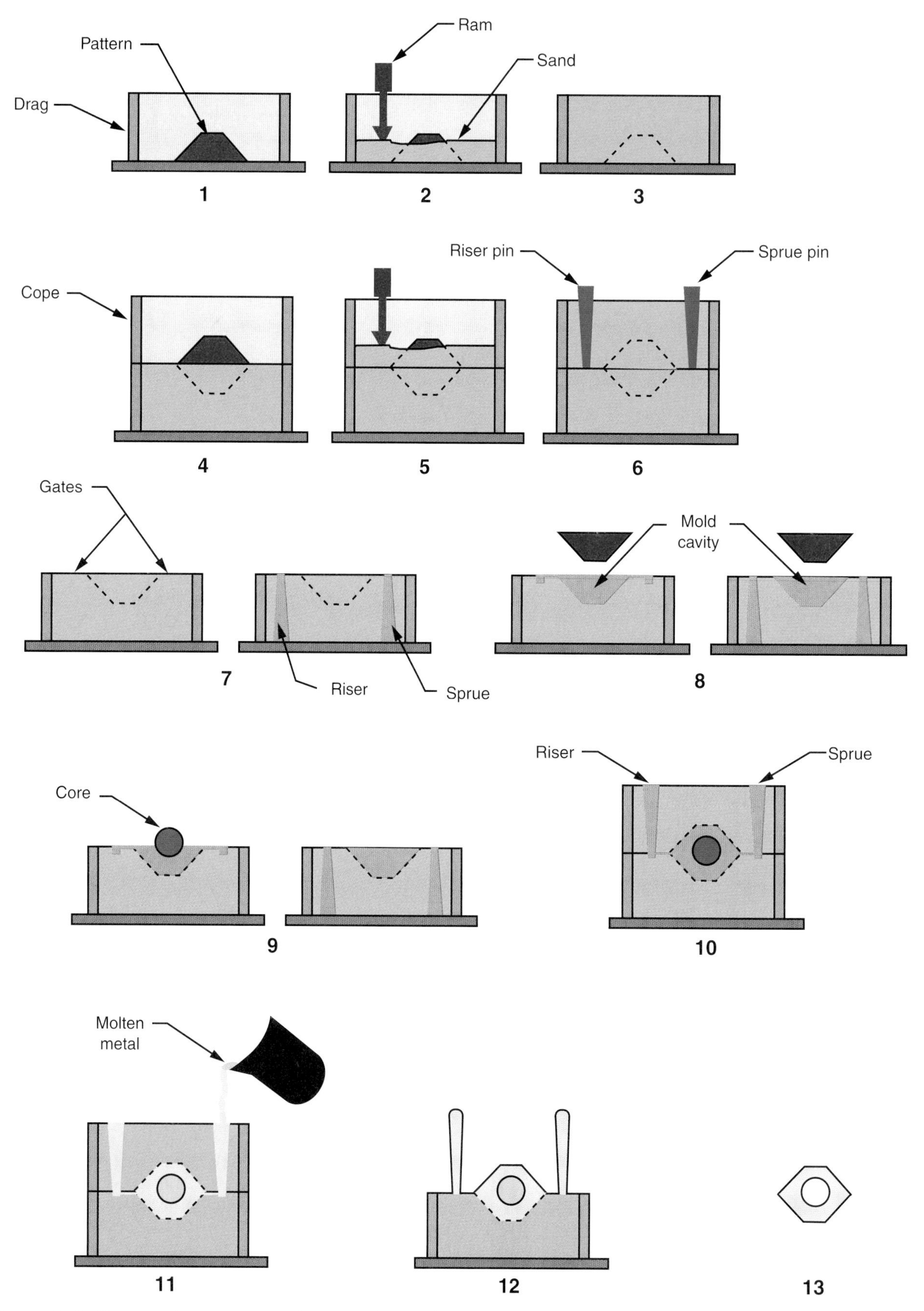

Figure 10-8. The step-by-step procedure used in making a greensand mold. Each step is explained in the text of this chapter.

Figure 10-9. This is a shell mold that has sand-resin core inserts in place. (Brush-Wellman Co.)

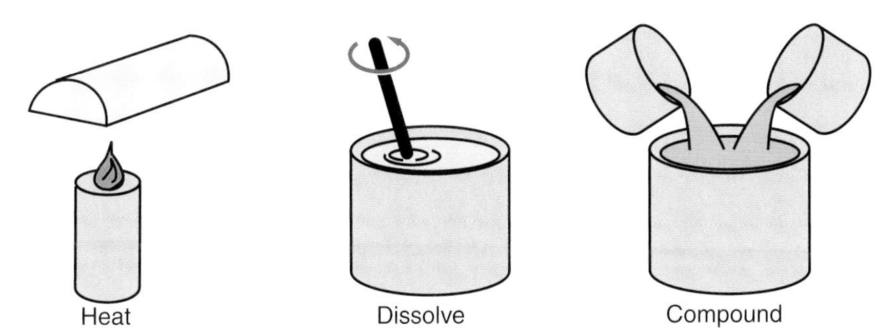

Figure 10-10. Material may be prepared for casting in three basic ways.

action forms a sand crust (shell) over it. When the shell has reached a proper thickness, it is carefully removed from the pattern and cured in an oven. See **Figure 10-9.** The two cured mold halves are clamped together.

This makes them ready to receive the molten metal. Shell molds produce a more accurate part than green sand molding. Also, the part has a smoother surface.

greensand casting molds. A casting process that uses moist sand packed around a pattern.

shell molds. Molds formed as thin resin and sand shells on heated metal patterns.

Plaster investment molds

Plaster investment molds are one of the oldest types of molds. They were developed in China and Italy. This technique became an important industrial process during World War II. Plaster investment molding starts with a wax pattern. Plaster is poured around the pattern and allowed to harden. When the mold is heated, the wax melts. The liquid wax is poured out leaving a very smooth, accurate mold cavity.

plaster investment molds. Expendable molds formed by pouring plaster around a wax pattern. Once the plaster hardens, heat is applied to melt and remove the wax pattern.

Permanent Molds

Some casting processes use a mold that is designed to produce a large number of castings before it must be discarded or repaired. These molds are called *permanent molds*. They are used to cast nonferrous (not iron) metals, plastics, and ceramics.

Permanent molds have cavities machined or otherwise formed into them. The cavities must be produced with the same concern for draft and material shrinkage. In addition, the molds must have a way to ensure that the two halves line up with each casting.

Permanent molds are more expensive to make than expendable molds. However, their ability to withstand the pressures created by forcing materials into them is an advantage. Also, permanent molds generally provide better dimensional control of the part and a smoother surface finish.

permanent molds. Molds that can be used again and again to cast or mold a part. They will produce many parts before they wear out. Permanent molds are more expensive than expendable molds. There are two types of permanent molds: gravity and pressure.

Preparing the Material

Material to be cast must become a liquid or semiliquid. Several methods can be used to change the material to this physical state. See **Figure 10-10.** These methods are melting, dissolving, and compounding.

melted. Heated until a liquid or semi-liquid state is attained.

dissolved. Broken up into tiny particles and mixed with a liquid.

slip. Fine particles of clay suspended in liquid for ease in molding ceramic products.

compounded. Mixed together; used to describe blending of liquid or finely divided solid materials in preparation for molding or casting.

Most metals and plastics are *melted* or softened. Some processes use a separate furnace to melt metals.

Other materials, such as clay, can be *dissolved* or suspended in a liquid. The suspended material is called *slip*. The particles in the liquid remain solid, but the liquid carries them into the mold. Bathroom fixtures, ceramic decorative items, and some dinnerware pieces are produced from this type of material.

Still other materials, such as some plastics, are *compounded* (mixed) or manufactured in a liquid form. They can be poured or forced into a mold in the natural state. During preparation, additives may be mixed with the material to be cast. Coloring agents may be added to plastics and clays. A catalyst (material that starts or speeds up a chemical action) may be added to plastics to cause them to harden. Other materials may be added to the original material to change properties. For example, plastic can be made harder, more flexible, stronger, or more durable.

Introducing the Material into the Mold

There are two basic ways of introducing the material into the mold. The material, as shown in **Figure 10-11,** may be either poured or forced into the cavity.

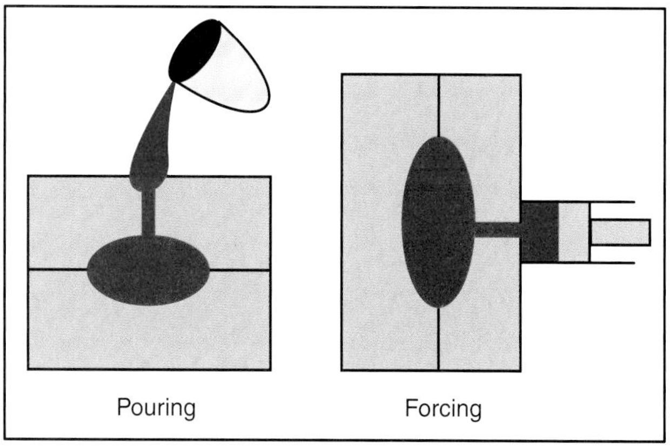

Pouring

Forcing

Figure 10-11. Materials can be introduced into molds by pouring (gravity) or forcing.

Gravity causes the mold to fill with liquid material during pouring. See **Figure 10-12.** Most expendable molds are filled this way since they cannot withstand much pressure. Poured molds include greensand and shell molds for metals; slip casting molds for ceramics; and nonpressure permanent molds for metals, plastics, and wax.

Some materials are forced into the mold cavity. The two major techniques using force are die casting of nonferrous metals and injection molding of plastics.

In *die casting*, the metal is melted in a "pot." Then a ram (plunger) forces the metal into a permanent, water-cooled steel mold. See **Figure 10-13.** Here, the material solidifies. The mold opens and the finished part is ejected. **Figure 10-14** shows a typical die casting operation. As with all casting operations, the die cast material goes from raw (industrial) material to a shaped part in one step. The castings require only a minimum amount of machining before they are ready to use. Die cast products include toys and some automotive parts.

Injection molding is very much like die casting. The biggest difference lies in the material being processed. Die casting casts nonferrous metals, while injection molding shapes plastics. Again, the material is heated in the machine. Then a screw or ram forces it into a cavity. The material solidifies and is then ejected. Plastic and many synthetic (human-made) materials can be

Figure 10-12. This worker is pouring precious metals into a mold. (FMC Corp.)

produced by injection molding. See **Figure 10-15.** These molded parts include models, toys, game pieces, and other items.

Like die casting, *centrifugal casting* forces molten material into the mold. A spinning action generates the force. This is the same force that "throws" a person against the sides of a spinning carnival ride. In centrifugal casting, the molten metal is placed in the mold. The mold is then spun. This action causes the metal to be thrown to the outside of the revolving mold. A major product of centrifugal casting, as shown in **Figure 10-16,** is cast iron pipe.

In *slip casting*, liquid slip (clay suspended in water) is poured into a plaster of paris mold. The plaster mold wall absorbs water from the slip causing it to harden from the outside inward. The clay is allowed to build wall thickness by the drying action. If the product is to be hollow, the liquid material in the middle is poured out leaving the clay shell. Solid casting allows the clay to totally harden. After a drying time in the mold, the soft clay casting is removed. It is then allowed to dry further before it is processed into a finished product. Ceramic figurines, cream pitchers, and bathroom fixtures are made by this process.

die casting. **A forming process in which melted metal is forced into a steel mold, or die. The mold is water-cooled and ejects the solidified part.**

injection molding. **A process similar to die casting, but usually used for plastics rather than metals.**

centrifugal casting. **A forming process in which molten metal is placed in a spinning mold to form a hollow product.**

slip casting. **A process in which a solution of clay (particles suspended in water) is poured into a mold. The water evaporates, leaving the clay behind. Either solid or hollow objects can be slip cast.**

Figure 10-13. Injection molding machines use pressure to fill molds with plastics. (Stokes)

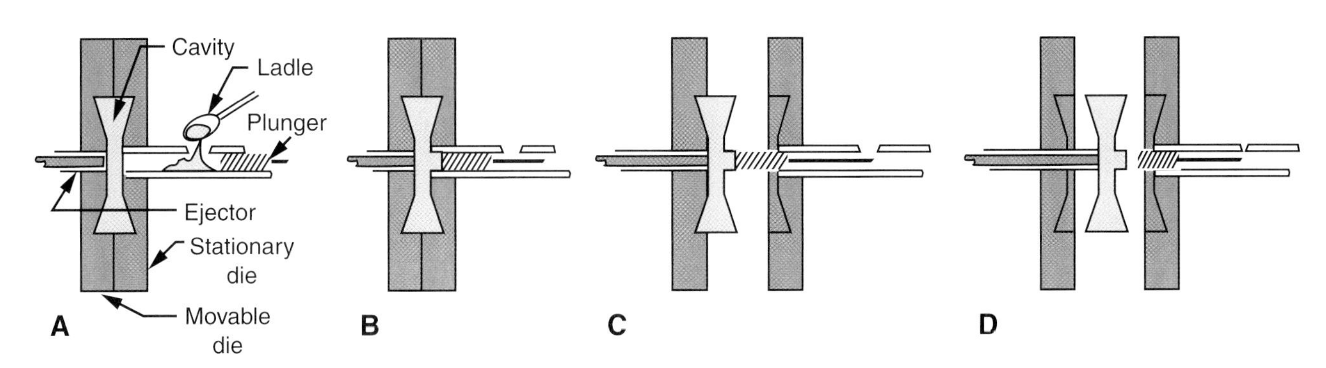

Figure 10-14. Steps in cold-chamber die casting. A—Metal is placed into the chamber. B—A plunger forces the molten metal into the die. C—The metal solidifies. Then the die opens. D—The finished part is ejected.

Figure 10-15. This automobile bumper was molded using the injection molders in the background. (Cincinnati Milacron)

Slush casting can be used to cast hollow metal objects. The molten metal is poured into the mold as in other metal casting processes. However, before the metal hardens, the process is interrupted. As the mold absorbs heat, a solid shell of metal is formed on the mold walls. When the wall thickness is right, the excess metal is poured out of the center of the casting.

Plastic objects can also be shaped using the slush casting technique. However, in this technique, powdered plastic is placed in a heated mold. The heat causes the plastic to fuse, forming a thin shell next to the mold. The longer the plastic is allowed to remain in the mold, the thicker the shell will become. As with metal, the excess plastic is poured out when the wall is thick enough.

slush casting. A process in which a thin-walled hollow casting is formed from molten metal or powdered plastic.

solidifying. The hardening of a liquid or solid material in a mold.

Solidifying Methods

Once the material has been placed in the mold, it must harden for the product to be usable. This is called *solidifying.* Solidifying can be done using cooling, drying, or chemical action.

Cooling

One way to harden a material is to remove heat. *Cooling* water creates ice. Likewise, removing heat from molten metal or plastic will return the material to its normal solid state.

In the case of melted plastics, melted metals, or other materials, air or water may be used to cool the molten materials. Water may be circulated through channels in a permanent die casting or injection mold. The fluid carries away

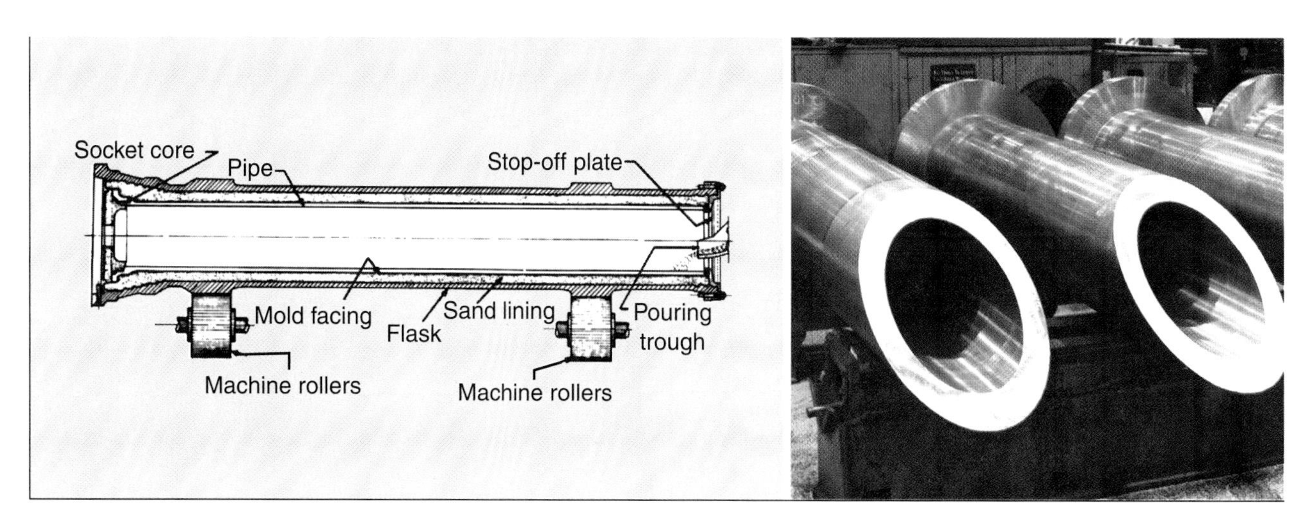

Figure 10-16. Cast iron and steel pipe, like that shown on the right, can be cast using the centrifugal casting technique diagrammed on the left. (American Cast Iron Pipe Co.)

the heat from the mold and part. The system works like the cooling system of an automobile that removes the combustion heat from an engine. The air in the room may be used to carry away the heat in greensand casting and shell molding operations. This process is slower, but cheaper, than fluid cooling of permanent molds.

Drying

A second method of solidifying materials is *drying*. This method removes solvent from a dissolved or suspended material. Most cast ceramic materials are solidified in this manner. Of all the ceramic casting process that use drying, slip casting is most common. See **Figure 10-17.**

Figure 10-17. Top. The major steps used in the slip casting process. Bottom. Photos showing slip casting of a toilet base. (Kohler)

Chemical Action

The third way to harden a material is by *chemical action.* Liquid materials, like plaster of paris, concrete, and casting acrylic (plastic), become hard because of internal action. Their chemical structures are altered, which changes the liquid material into a solid. Adding water may be enough to start the action. Plaster of paris and concrete set (become hard) in this manner. Other materials, like liquid acrylics, harden after a *catalyst* is added. This agent starts the chemical action needed to solidify the material.

Once the chemical action starts, it cannot be stopped. This is why you will see people working concrete rapidly. They have a limited time to smooth and finish the surface before the material becomes too hard to work.

Extracting the Casting or Molded Item

extracting. **The process of removing a finished part from a mold.**

The *extracting* step involves removing the casting or molded item from the mold. The technique used is determined by the type of mold used.

Expendable molds are broken apart to remove the product. Sand molds are vibrated to cause them to collapse. Plaster and shell molds are fractured to remove the casting or molded part.

Permanent molds open and close easily. Often, the casting or molding machine automatically opens the mold. Then ejector pins push out the finished part. The mold automatically closes, and the casting operation can be repeated.

Summary

You have learned that casting and molding processes provide the most direct route from raw material to shaped part. All these processes use the same five basic steps. First, a permanent or expendable mold is produced. Then, the material is prepared to be introduced into the mold. It may be heated to its melting point, dissolved or suspended in a liquid, or compounded. The molten material is introduced into the mold by gravity or with force. The material is then caused to solidify through cooling, drying, or chemical action. Finally, the finished casting or molded part is removed from the mold. Expendable molds are destroyed to remove the product. Permanent molds are opened, and the part is removed.

Safety with Casting and Molding Processes

- Do not attempt a process that has not been demonstrated to you by the instructor.
- Always wear safety glasses, goggles, or a face shield.
- Wear protective clothing, gloves, and a face shield when pouring molten material.
- Do not pour molten material into a mold that is wet or contains water.
- Carefully secure two-part molds together.
- Perform casting and molding procedures in a well-ventilated area.
- Do not leave hot castings or molded parts where they could burn other people.
- Constantly monitor the material and equipment temperatures during casting or molding processes.

Key Words

All of the following words have been used in this chapter. Do you know their meanings?

cavity
centrifugal casting
compounded
die casting
dissolved
draft
expendable molds
extracting
greensand casting
injection molding
melted
mold
pattern
permanent molds
plaster investment molds
shell molds
slip
slip casting
slush casting
solidifying

Test Your Knowledge

Please do not write in this text. Place your answers on a separate sheet.

1. Distinguish between casting and molding.
2. List five steps that are common to all casting and molding processes.
3. *True or False?* Permanent molds, called one-shot molds, are easy and inexpensive to produce.
4. List the three methods of changing materials to a liquid or semiliquid state.
5. _____ _____ casts nonferrous metals, while _____ _____ shapes plastics.
6. Name three methods of solidifying materials.
7. During the extracting step, the casting or molded item:
   a. is added to the mold.
   b. is removed from the mold.
   c. becomes solid in the mold.
   d. None of the above.

Applying Your Knowledge

Note: Be sure to follow accepted safety practices when working with tools. Your instructor will provide safety instructions.

1. Manufacture a part or product using a common casting process.
   a. Before making a selection of a product, be sure that your technology education laboratory has the necessary tools and equipment. Suggestions for your consideration:
      - A candle made in a greensand mold.
      - A plaster wall decoration using a permanent mold.
      - Plastic screwdriver handle. See **AYK 10-1.**

   b. Review the casting processes you have studied. Then, prepare a set of procedures for making the product you have chosen.
   c. Review all casting procedures with your instructor.
2. Look around the room. List as many products or parts that you can see that were produced using a casting process. Next to each one, list the process you think was used and determine whether a permanent or expendable mold was used. Present your data on a chart like the one shown in **AYK 10-2.**

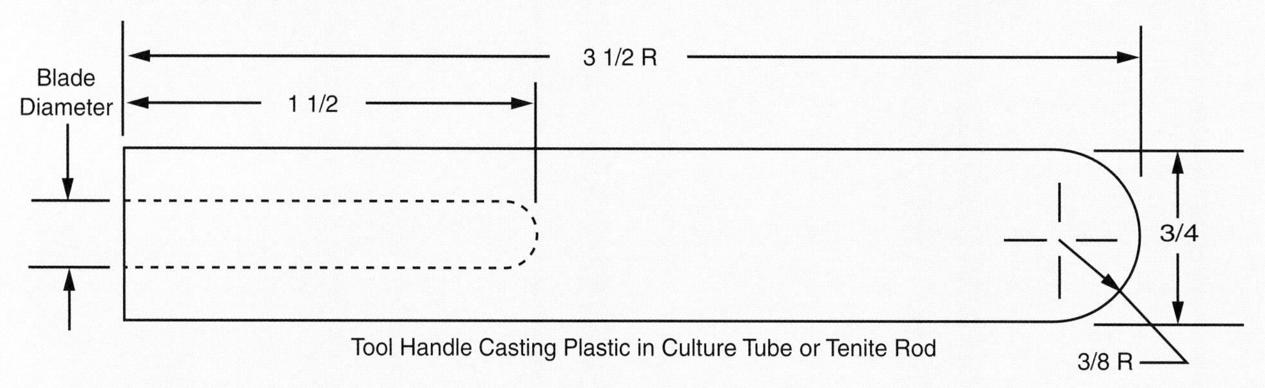

AYK 10-1. Use these plans to cast a screwdriver handle.

Product or Part	Process	Mold Used

AYK 10-2. On a separate sheet of paper, prepare a chart similar to this one on which to record your answers.

Chapter 11 Forming Processes

Objectives

After studying this chapter, you will be able to:

✓ Explain how forming has developed.
✓ List the three essential elements that must be present in order for forming processes to take place.
✓ Identify various types of shaping devices used in forming processes.
✓ Explain why the temperature of a material is an important consideration in all forming operations.
✓ Name and describe three types of forming forces.

Forming changes the size and shape, but not the volume of a workpiece. Applying force to the material with a shaping device makes this change. This force must be strong enough to make a material take on a new, permanent shape. See **Figure 11-1.**

Figure 11-1. This hot metal part is being removed from a forging die. (Budd Co.)

Development of Forming

Hammering by hand was the first forming method used to shape metals. Since the Bronze Age, about 4000 years ago, humans have melted metals, poured them into molds, and hammered the materials into new shapes. This hammering action is a forming process. Like all forming techniques, it changes the size and shape of the base material. However, it does not involve removing material through cutting or shearing. Rather, the material is simply reshaped with no weight loss.

One of the earliest uses of forming processes was in producing hunting weapons. The development of metal-pointed spears and knives made hunting and food gathering easier. Cutting trees and brush became simpler with the metal ax. Metal knives gave much better results than the stone-edged tools commonly used for cutting food.

rolling. A forming process that uses a rotating applied force to change the thickness of a piece of steel or other material. Also, a finishing method in which a coating is applied by a roller.

drawing. A forming process that involves stretching metal into the desired shape.

yield point. Beyond this point, additional stress will permanently deform a material.

fracture point. Point a material breaks into two or more parts after being stressed.

Many early military and political conquests were won using weapons formed from solid metal. Solomon, and later, Alexander the Great, depended on forged weapons to equip their armies. Vast territories were won or lost on the strength of forged metal.

In the New World, formed products were important to early settlers. The village blacksmith was the first colonial forming expert. Using a hammering force on the anvil, the blacksmith produced forged items such as wagon tires, chain, and horseshoes.

Rolling became an important industrial forming process. *Rolling* causes a material to flow into a new shape as a result of an applied force. This process is similar to a cook using a rolling pin to roll out piecrust dough to bring it to the proper thickness.

Paul Revere practiced a third type of forming. He produced his famous silver bowls through a *drawing* process. He carefully stretched and drew silver and pewter sheets into useful shapes with the hand tools of the day. Today, large presses perform the same tasks in a fraction of the time.

Essentials of Forming

Forming uses pressure to shape and size materials. All engineering materials being formed go through two major stages as stress (pressure or force) is applied. These stages are the elastic and plastic stages. See **Figure 11-2.**

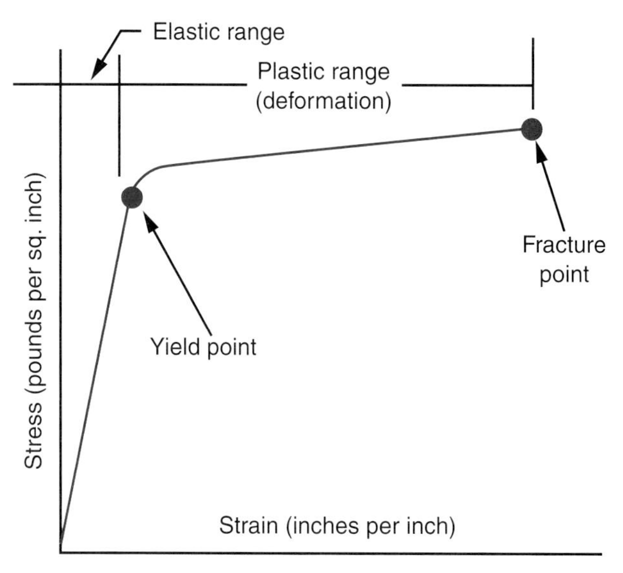

Figure 11-2. This is a simple stress-strain curve showing material forming stages. The part of the curve that is nearly vertical is the elastic range. Up to the yield point, the material will return to its original shape if the stress is removed. Past the yield point, permanent changes in shape occur.

Elastic Stage

Stress will stretch the material in the *elastic stage*. However, when the stress is removed, the material will return to its original size and shape. At the end of the elastic range is the *yield point*. Beyond this point, additional stress will permanently deform (change shape of) the material.

Plastic Stage

This stage starts at the yield point. During the *plastic stage* the material starts to give in (yield) to the stress. It will be permanently stretched into a new shape. Added stress will cause more change in the shape. The plastic stage continues up to the *fracture point*. At this point the material breaks into two or more parts. Forming is always done between the yield point and the fracture point. Pressure is applied to cause permanent reshaping (deformation) without breaking the material.

Forming Practices

All forming practices have three major things in common. These, as shown in **Figure 11-3,** include:

- A shaping device
- An appropriate material temperature
- A sufficient forming force

Figure 11-3. These are the essential elements of all forming processes.

Shaping Devices

The shaping device for a forming process does the same job that the mold does for casting and molding operations. The *shaping device* determines the final shape of the part or product being produced. Two major types of shaping devices are used in forming processes. These are dies and rolls.

shaping device. A machine or tool used to determine the final shape of a part or product.

dies. Shaping devices which material is squeezed between or formed over to achieve a new shape. There are three types of dies: open dies, die sets, and shaped dies. Dies can also have blades designed to cut special shapes such as curves, circles, or whole outlines.

Dies

Dies are common shaping devices used in many forming operations. These devices are blocks of hard steel called *tool steel.* They have a shape cut into or onto them. They can be used to form any material that is softer than the die. Commonly, dies are used to form metals and plastics.

Dies can be grouped into three major categories: open, closed, and one-piece shaped dies. See **Figure 11-4.**

Open dies

The simplest type is called an *open die*. It consists of two flat die halves. The part forms as a result of force applied between the die parts.

open die. The simplest type of die. An open die is two flat, hard plates. One half of the die does not move. The other half moves to hit (hammer) or put pressure (squeeze) on the material between the dies.

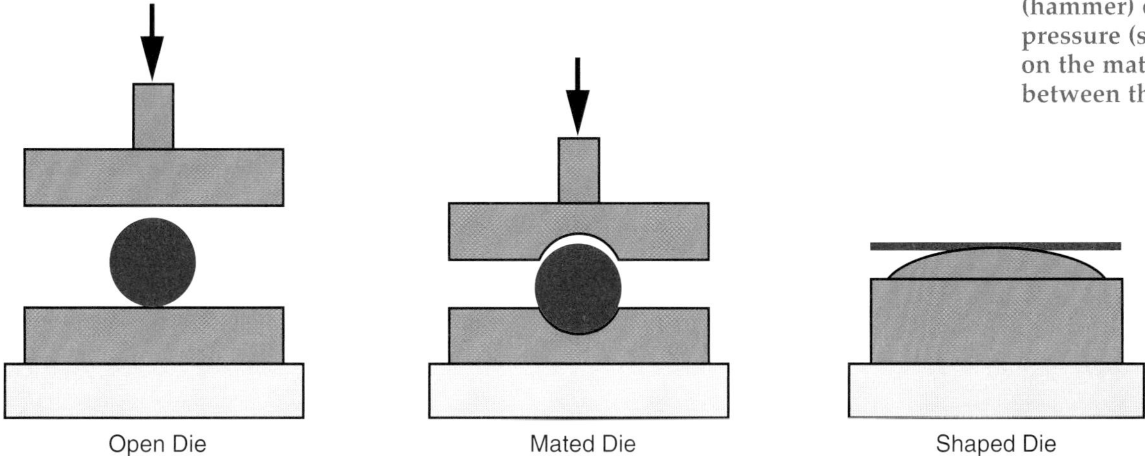

Figure 11-4. These are types of forming dies.

Figure 11-5. This huge, red-hot metal workpiece is being formed using a forge.

The village blacksmith's anvil and hammer were crude examples of this type. The anvil was a stationary (unmoving) die. The hammer was a moving die that applied force to the work. The hammer striking the work caused the hot metal to take a new shape.

A once common type of open die forming is *smith forging.* See **Figure 11-5.** In this process, a flat die is fixed to the solid base of a press or hammer. The other die part is attached to a movable plunger. The operator places the heated metal workpiece between the die parts. The hammer or press is then activated. The heavy, upper die part drops or is driven downward and strikes the hot metal. This impact causes the metal to flow into a new shape. The metal flow is called *plastic deformation.*

In smith forging, the size and shape of the formed product is controlled, in large part, by the operator. A high degree of operator skill is needed. Because so much depends on the skill of the operator, it is difficult to produce consistently accurate parts. Today, few finished shapes are produced using smith forging.

mated dies. **A set of dies with a raised shape on one die and a matching cavity in the other. Material is squeezed between the two to take the desired shape.**

Mated dies

Most dies have machined or engraved cavities in their faces. These dies, shown in **Figure 11-6,** are called *mated dies* or closed dies. When the two halves of the die set close, the material between them takes the shape of the die cavities.

Several important forming techniques use closed dies to form metals, ceramics, and plastics. One of these is *forging,* which is primarily used to form metals. In this process, the metal is heated to a temperature close to its melting point. It is then placed between the two open die halves. The dies are closed with such force that the hot metal flows into the die cavities.

Figure 11-6. An example of a two-piece die with cavities machined in them.

Forging processes may use either a hammer or a press to close the die set. A hammer is used for *drop forging.* The dies are quickly driven or dropped to give a sharp blow. *Press forging* gives a steady, squeezing force. This is provided by a hydraulic ram closing the dies on the material.

Forged parts have high strength because the grain structure produced flows with the part shape. See **Figure 11-7.** Forging techniques are used to produce items such as aircraft and automotive engine parts, tools, and flatware (knives, forks, and spoons).

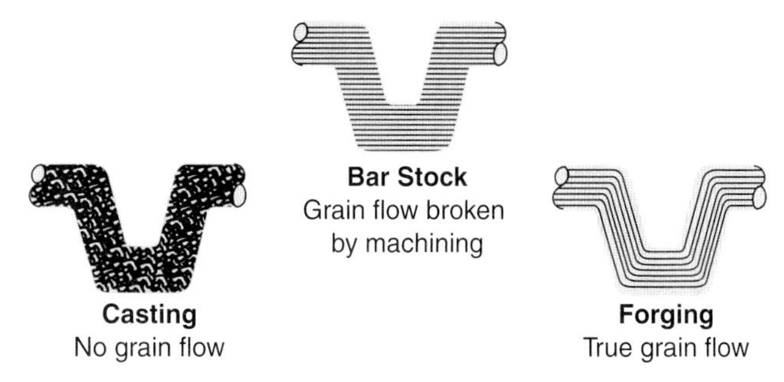

Figure 11-7. These drawings compare the grain structure of a casting, bar stock, and a forging. (Forging Industry Assoc.)

Shaped dies

One-piece *shaped dies* can be used to form ceramic, metal, and plastic materials. There are many different processes.

One is *thermoforming.* This process is used to form plastic sheets into packaging trays and covers, freezer boxes, automotive parts, and many other products.

In thermoforming, as shown in **Figure 11-8,** sheet plastic is first heated. The hot material is placed over the one-piece shaped die that is called a mold. The air in the mold is evacuated with a vacuum pump. The atmospheric pressure above the hot plastic forces it into the mold where it cools. The cool plastic will retain its new shape.

The thermoforming mold may have a concave cavity into which the plastic is drawn, or the mold may be a convex shape over which the hot material is formed. See **Figure 11-9.**

A second process that uses a shaped die is *metal spinning*. See **Figure 11-10.** The die is called a *chuck*. It is mounted on a headstock of a special lathe. A soft metal disc is clamped between the chuck and a tailstock pad. The lathe is turned on, and a tool applies pressure to the disc. The pressure causes the metal to flow (plastic deformation) around the chuck. The process work hardens the material so that it easily retains the new shape. Many conical parts are formed with metal spinning. These items include reflectors for lighting fixtures and ends for tanks.

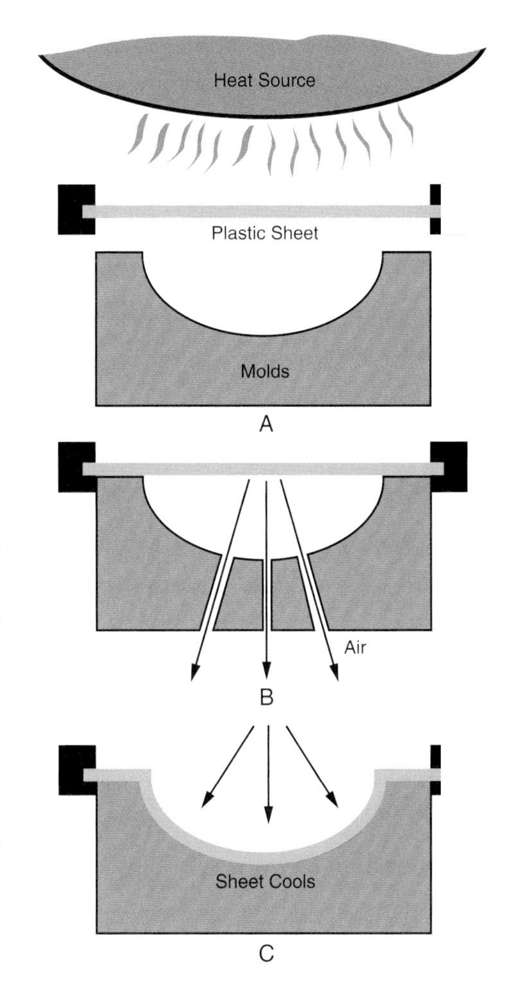

Figure 11-8. This drawing shows the simplest thermoforming technique.

Rolls

The second major shaping device is a roll. There are two major types. These are smooth rolls and shaped rolls.

Smooth rolls are used to press materials into a desired thickness during primary processing. They are used to change aluminum and steel ingots into sheets and plates.

During secondary processing, smooth rolls form curves out of sheet metal, tubing, and bar stock. See **Figure 11-11.** During this process, the material passes between three rotating rolls. The bottom two rolls are stationary, and support the material. The upper roll can be fed downward. This movement causes the material to flex and curve upward forming an arc. The farther downward the upper roll is fed, the greater the resulting curve is. This process is used to produce round tanks, plates for rocket bodies, and large pipes.

Some roll forming techniques use *shaped rolls*. These rolls have a series of grooves machined in them. As the material passes through the machine, each set of rolls produces a different part of the new shape.

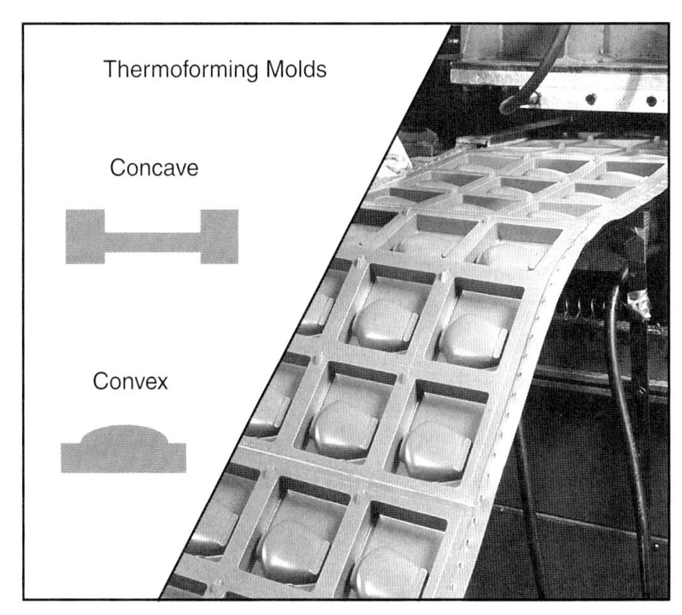

Figure 11-9. The types of thermoforming molds are listed on the left. A typical thermoformed sheet for electric razor packages is shown on the right.

Material Temperatures

The temperature of a material is an important consideration in all forming operations. The material must flow into a new shape *drop forging.* **Shaping of material between two dies using the force of a falling (drop) hammer.**

press forging. **Shaping of material between two dies using the force of a hydraulic ram.**

shaped die. **A one-piece die over which material is formed.**

thermoforming. **A forming process that holds and heats sheet material in a frame. The hot material is lowered over a mold. The air in the mold is drawn out. Atmospheric pressure forces the plastic into the cavity.**

metal spinning. **A forming method in which a rotating disc of soft metal is deformed by pressure of a tool.**

smooth rolls. **Rotating forming devices used to squeeze material to the desired thickness.**

without becoming a liquid. Generally, the hotter a material is, the easier it is to form.

Forming processes can be divided into two categories, according to the working temperature. These categories are hot forming and cold forming.

Metals go through several stages as they are heated. At each stage, a material has specific internal crystal structures. At one of these points, the metal can be formed without developing internal stress within the crystal structure. This point is called the *recrystallization point.* All forming techniques that shape materials above this point are called *hot forming* processes.

Among the hot forming processes are metal forging, extrusion, roll forming, and drawing. Plastics are worked hot in thermoforming, blow molding, and extrusion. Ceramics are often blow molded and hot pressed into shape. See **Figure 11-12.**

Forming done below the recrystallization temperature is called *cold forming.* The material may be at room temperature or it may be heated. Forming done above room temperature and below the recrystallization point is often called *warm forming.*

The choice between cold and hot forming is made after considering several factors. These are included in the chart in **Figure 11-13.**

Types of Forming Forces

Forming is done by using a combination of three basic forces. These, as shown in **Figure 11-14,** are:

- Squeezing forces
- Drawing forces
- Bending forces

Squeezing forces

Squeezing forces compress a material to cause plastic deformation. As the material resists this force, it flows into a new shape. An example of a forming process that uses compression is extrusion. This technique produces long lengths of material with a common cross-sectional shape. See **Figure 11-15.**

A

B

C

Figure 11-10. The major steps in metal spinning are shown above. A—The aluminum disc is placed in the lathe. B—The rotating disc is formed over the chuck. C—The finished part is removed. (Crouse-Hinds Co.)

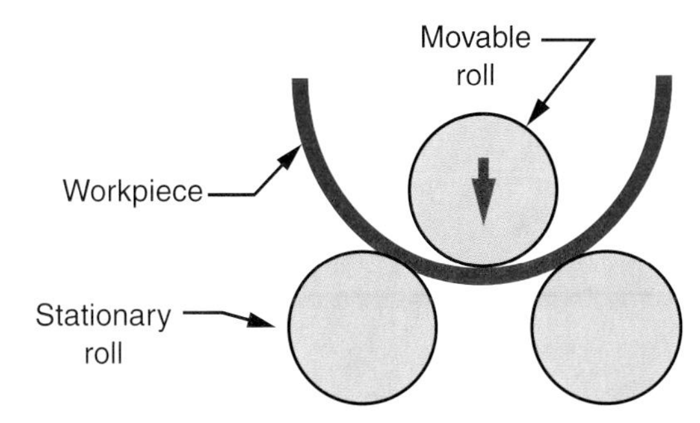

Figure 11-11. The principle of roll forming.

Cold Forming	Hot Forming
• Produces a smooth, oxide-free surface. • Requires heavier forces. • Improves strength properties. • Creates stress. • Increases hardness.	• Produces surface scale. • Requires lower pressures. • Does not change properties.

Figure 11-13. These are some of the characteristics of cold forming and hot forming. The choice between cold and hot forming is based on these characteristics.

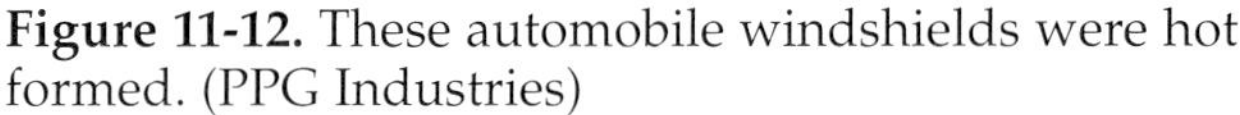

Figure 11-12. These automobile windshields were hot formed. (PPG Industries)

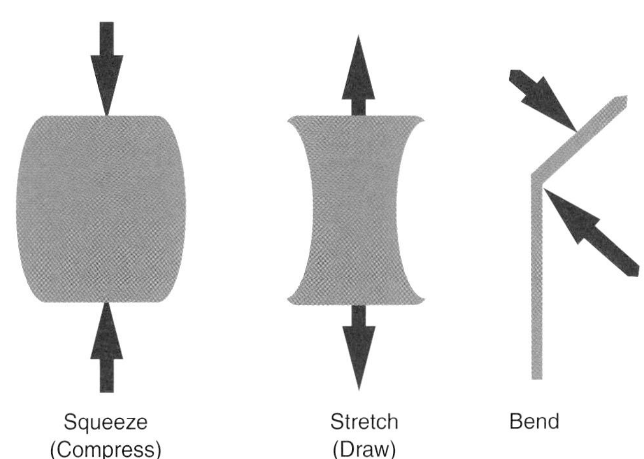

Figure 11-14. Types of forming forces are shown above.

In this process, a one-piece shaped die is produced. The die has a hole that is the same shape as the cross section of the extrusion. A billet (bar) of material is placed in a chamber behind the die. A ram or plunger is forced into the end on this chamber. As the plunger moves forward, the extrusion material is forced through the die opening. The material's cross-sectional shape will be exactly the same as the die opening. Squeezing toothpaste from a tube is a simple process that is very much like extrusion. Building bricks, concrete blocks, plastic molding, and aluminum sash (window) and door parts are extruded.

Drawing forces

Drawing forces or stretching forces produce a wide variety of metal and plastic products. The aluminum beverage cans, shown in **Figure 11-16,** were produced using a *deep drawing* process. In this process, a soft sheet of aluminum was clamped over the lower die half of a mated die set. The upper die half was pressed into the lower die. This action caused the metal to flow into the lower die cavity creating the finished shape.

Drawing produces items like flashlight battery cases and automotive body parts. Thermoforming, which was discussed earlier in this chapter, also uses drawing forces.

shaped rolls. Rotating forming devices with grooves machined into them

hot forming. Forming of material that is heated above the recrystallization point.

cold forming. Forming of material at temperatures below the recrystallization point.

warm forming. Describes forming heated materials that are not above the point of recrystallization.

squeezing forces. These forces compress a material to change its shape or thickness.

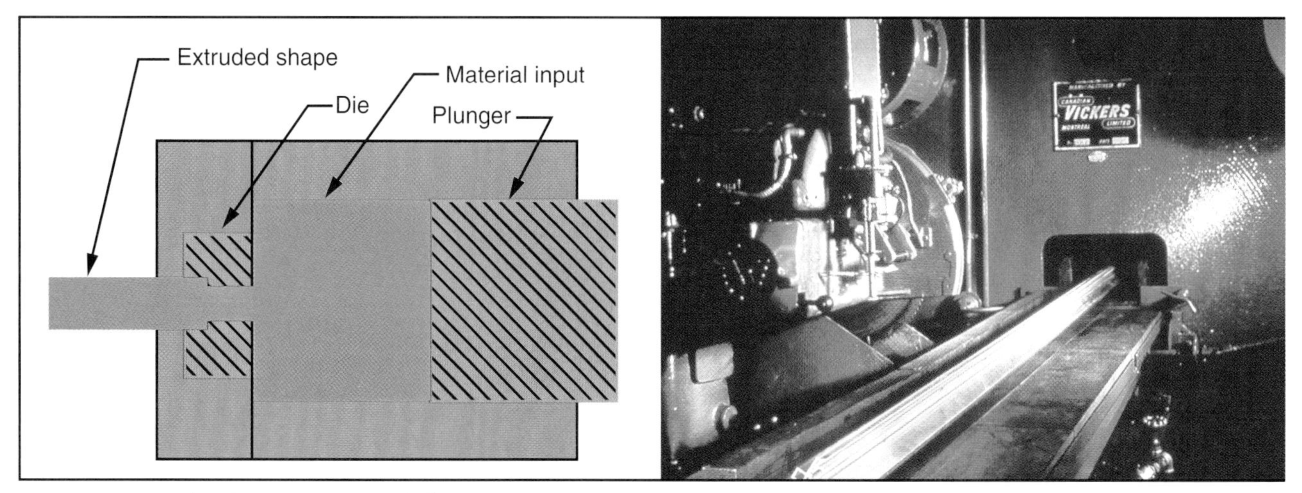

Figure 11-15. The drawing on the left shows forward extrusion. The photo on the right shows a complex architectural aluminum shape being extruded through a shaped hole in the die. (Alcan Aluminum Co.)

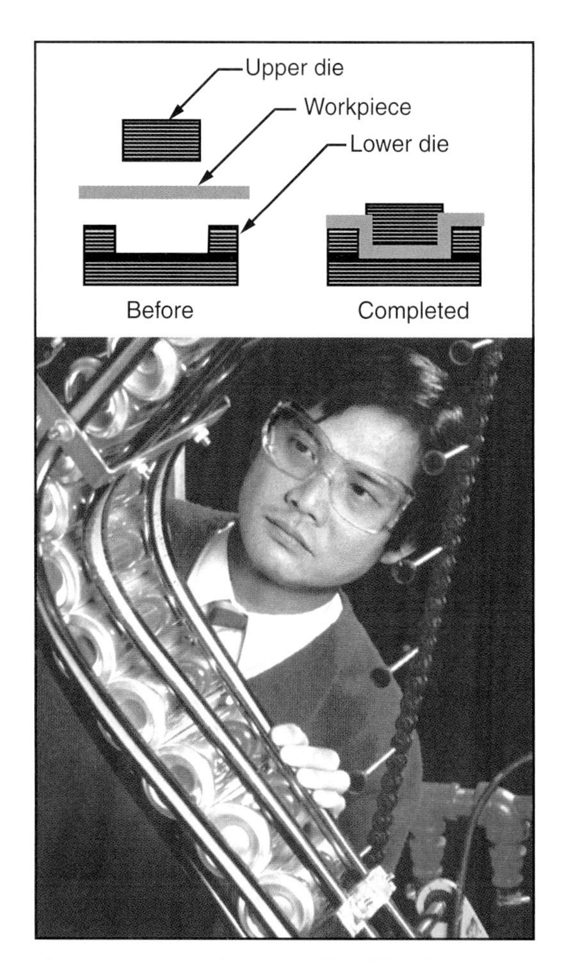

Figure 11-16. A drawing die, like the one shown in the sketch, is used to form the beverage cans in the photograph. (Reynolds Metals Co.)

Bending forces

Bending forces are used to form sheet materials, bars, and tubing into new shapes. See **Figure 11-17.** Bending is done by gripping one end or edge of a material. The free end or edge is then folded or curved around an axis. Bending forces produce straight bends along a sheet or curves in bars, rods, or sheet materials.

Sources of Forming Pressures

Forming forces must be generated by some outside means. Machine tools are the most commonly used source of these forces. **Figure 11-18** illustrates the four basic machines that are used in forming operations. These are:

- Hammers
- Presses
- Rolling mills
- Draw benches

Hammers

Hammers and presses are closely related. This can be seen by studying the diagrams in **Figure 11-19.** The main difference between hammers and presses lies in the type of force they deliver.

A *hammer* delivers a sharp blow. Many forging processes use this type of force. The quick, hard blow is delivered by dropping or propelling the movable upper head and die onto the bottom die. Some operations use a drop hammer that simply releases the upper die and plunger. The weight of the falling die assembly delivers the impact. Other hammers use either steam or hydraulic pressure to drive the upper assembly downward, which delivers a rapid blow.

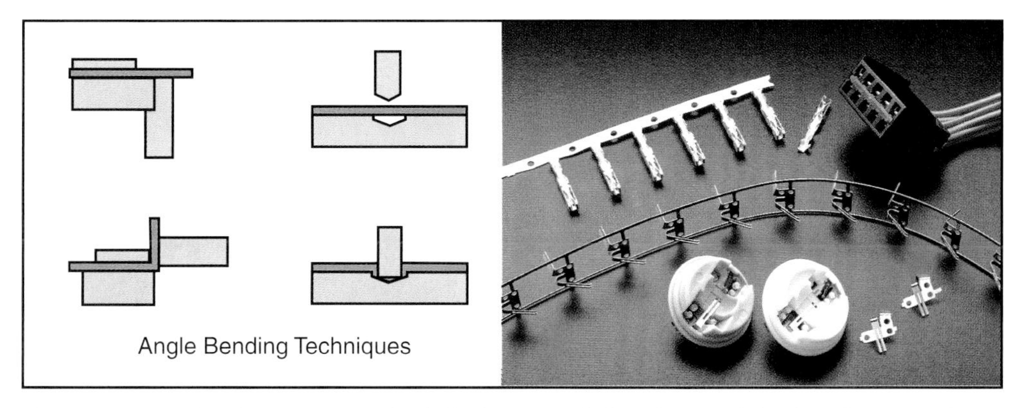

Figure 11-17. The drawings on the left show angle bending, while the photo on the right shows parts that were shaped by bending and shearing. (Brush-Wellman)

extrusion forging. A material is forced through a shaped hole in a one-piece die. The material flows through the die taking on the shape of the opening. Extrusion can produce shapes in metals, plastics, and ceramics.

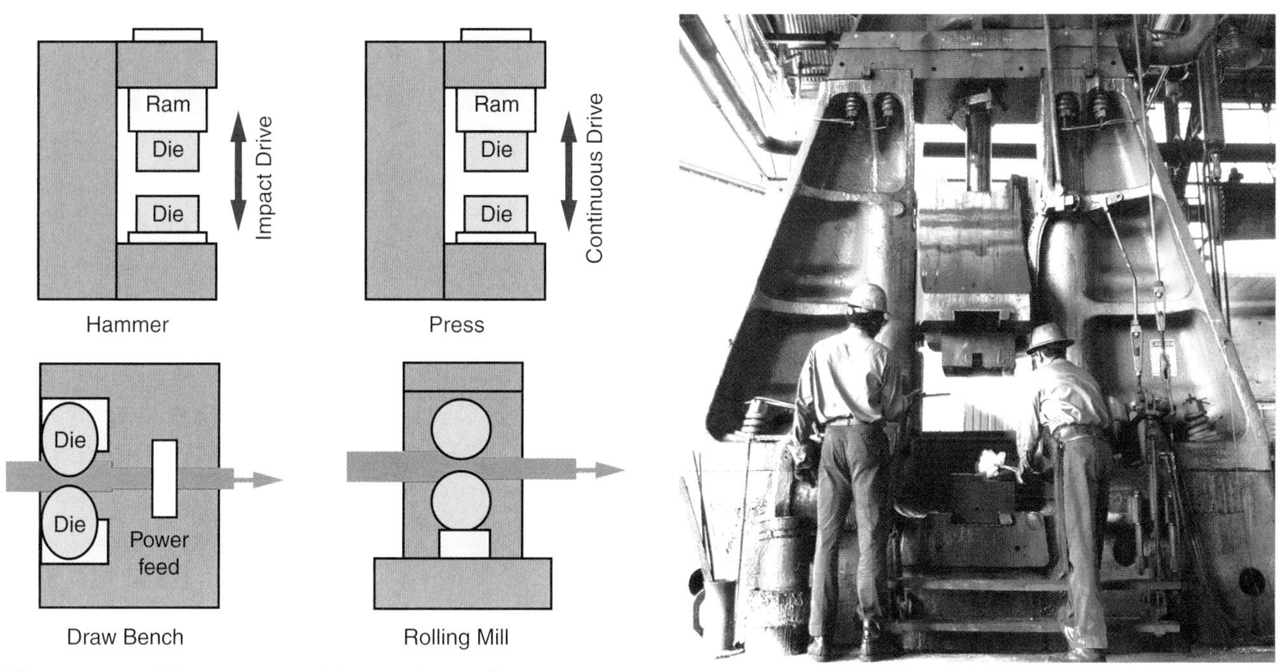

Figure 11-18. These are machine tools used to generate forming forces.

Figure 11-19. This hammer is forging a steel part. (Fansteel Inc.)

Presses

A *press*, like a hammer, has a movable head to which a die is attached. Unlike the hammer, the press moves the head in the steady squeezing action needed for drawing, bending, shearing, and press forging.

Presses can be used for both forming (contour producing) and shearing (shape producing, cutting) activities. Often, these two processes are combined into one die to produce a metal part called a *stamping*. See **Figure 11-20.**

bending forces. Used to form sheet metal, tubing, and bar stock into new shapes.

drawing forces. Forces that pull and stretch material into the desired shape.

Rolling machines

A *rolling machine* or roll is the third type of forming machine tool. It has two or more smooth or shaped rolls. They revolve in opposite directions producing the required forming forces.

The force generated between the rolls draws the material into the machine and changes its thickness or shape. The type of material being formed, its hardness, and its temperature will determine the amount of change that can be developed in one pass through a rolling machine.

Figure 11-20. These workers are removing a metal stamping from a forming press. (Budd Co.)

Figure 11-21. A draw bench (wire drawing machine), produces wire through a series of forming dies. (The Wire Association)

Figure 11-22. This powerful stretch forming press is being used to form a large beam.

hammer. A machine that delivers a sharp blow with a movable upper head to forge metal.

press. A machine that shapes or cuts material by applying steady pressure.

rolling machine. A device that changes the thickness of a material by squeezing it between rotating rolls.

draw bench. A forming machine with a shaped one-piece die and a drive mechanism. Draw benches are used to stretch a bar or sheet over a die, and to pull a wire or bar through a hole in a die.

Draw bench

A *draw bench* is a forming machine with a one-piece shaped die. A typical drawing machine produces wire. See **Figure 11-21.** In the *wire drawing* process, a metal rod is pulled through a number of dies. Each die has a smaller hole than the one before it. Therefore, the rod is drawn into finer and finer wire until the desired diameter is reached.

Draw benches can also be used for *stretch forming*. See **Figure 11-22.** In this process, the material is gripped at its ends. The grippers move outward to stretch and pull the material tight. Then a die under the material is forced upward. The material under stress takes the shape of the upwardly moving die. This process is used to form large sheet metal shapes for the aircraft industry.

Other Forming Forces

Machine tools are not the only source of forming forces. Air pressure and high-energy sources can also be used.

A typical process using air pressure is *blow molding*. This process produces the plastic and glass jars and bottles you see every day. **Figure 11-23** shows a glass bottle blow molding machine in operation. In this process, a glob of hot glass is formed. It is dropped into an open steel mold. The mold closes, pinching

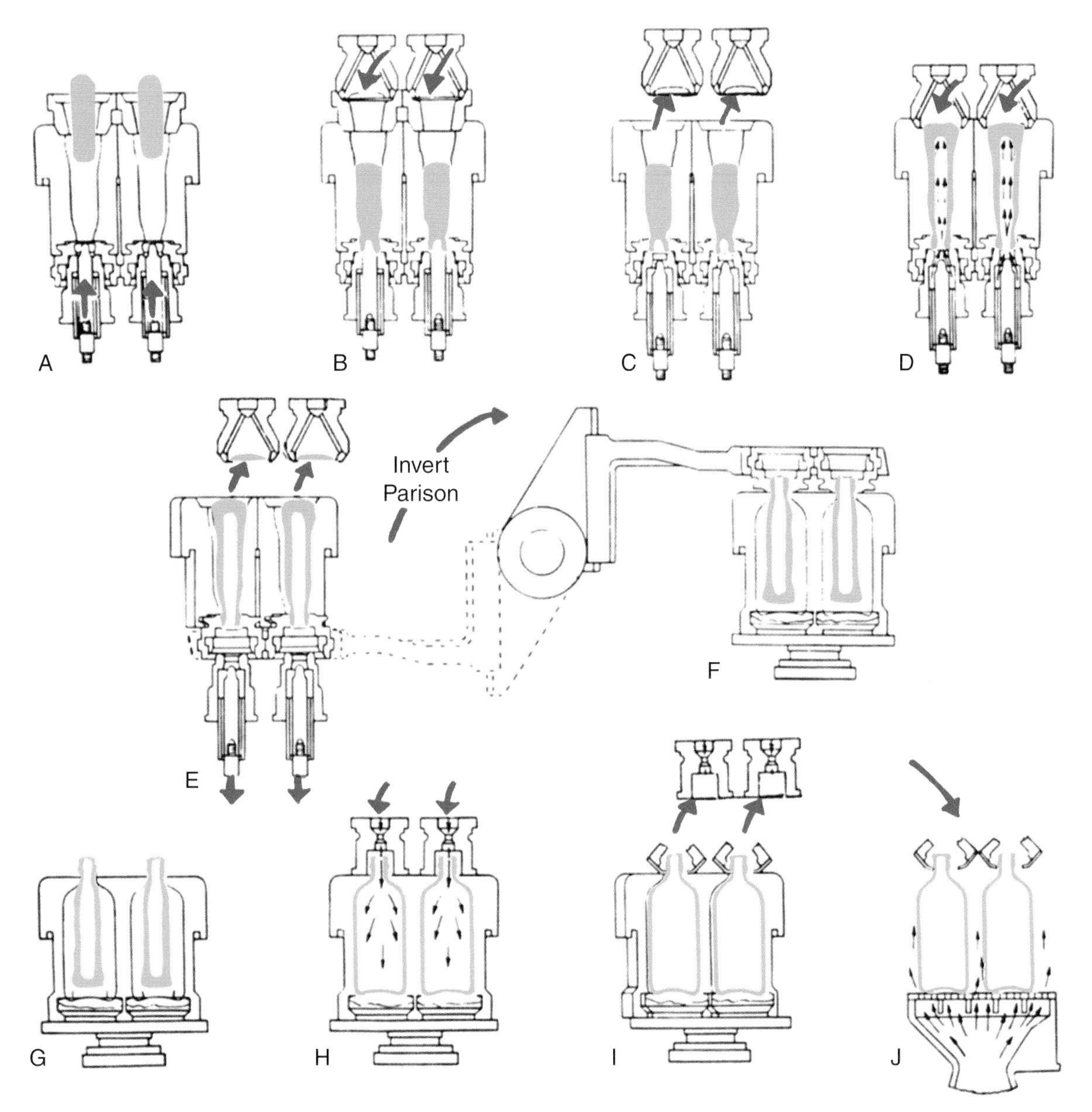

Figure 11-23. Diagram of a two-step blow molding process for producing glass bottles. A—Gob of heated glass is delivered into closed blank molds. B—Subtle blows, pushing glass downward onto plunger. C—Gob reheats. D—Air from plungers shapes parison (gob). E— Open; parison allowed to reheat. F—Parison released into blow molds. G—Parison reheats and gets longer. H—Bottles formed to blow mold shape. I—Molds open. Bottles taken out. J—Bottles released and swept off onto conveyor. (Owens Illinois)

off the gob forming the neck of the container. Air is then blown into the gob causing the glass to expand outward against the mold walls. After the glass cools enough to hold its shape, the mold opens. The finished container is ejected.

blow molding. A process that uses air pressure to form hollow glass or plastic objects.

High-energy rate (HER) actions provide another source of forming pressures. The two major HER forming techniques are explosive forming and electromagnetic forming.

In *explosive forming*, a sheet of material is clamped in a steel container. See **Figure 11-24.** The lower half of the container is the die with a shaped cavity in

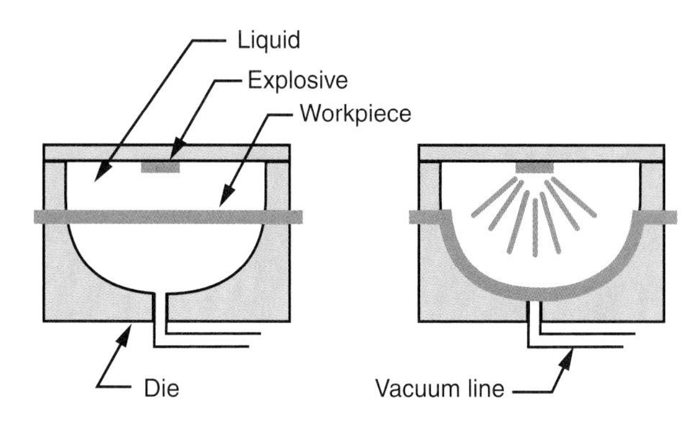

Figure 11-24. This simple illustration shows explosive forming.

it. The upper half is the explosion chamber that is filled with air or a liquid. An explosive charge is detonated in the chamber. The shock waves from the explosion drive the material downward where it takes the shape of the die cavity.

Electromagnetic forming uses the magnetic field formed around a coil to form materials. The material to be formed is placed inside the coil. Its high electrical charge is rapidly discharged through the coil. This causes a very strong magnetic field to quickly develop around the coil. The magnetic forces induce current in the workpiece and cause a permanent change in its shape. This process is used to shrink steel tubing around steel parts. The ends of automotive struts are attached to the strut body using this process.

explosive forming. **A forming process that uses explosive charges to force material against a die.**

electromagnetic forming. **A process used to change the shape of a material through use of a very strong magnetic field.**

Summary

Forming is a widely used family of processes that provide contour to materials. These processes use shaping devices to change the shape of materials by applying force above the material's yield point and below its fracture point.

Forming operations squeeze or compress, draw or stretch, or bend materials. In all cases, however, the material takes on a permanent shape without changing in weight or volume.

Safety with Forming Processes

- Do not attempt a process that has not been demonstrated to you by the instructor.
- Always wear safety glasses, goggles, or a face shield.
- Always hold hot materials with a pair of pliers or tongs
- Place hot parts in a safe place to cool, keeping them away from people and materials.
- Follow correct procedures when lighting torches and furnaces.
- Never place your hands or foreign objects between mated dies or rolls.
- Always use a spark lighter to light a propane torch or gas furnace.

Key Words

All of the following words have been used in this chapter. Do you know their meanings?

bending forces
blow molding
cold forming
dies
draw bench
drawing
drawing forces
drop forging
electromagnetic forming
explosive forming
fracture point
hammer
hot forming
mated dies
metal spinning
open die
press
press forging
rolling
rolling machine
shaped die
shaped rolls
shaping device
smooth rolls
squeezing forces
thermoforming
warm forming
yield point

Test Your Knowledge

Please do not write in this text. Place your answers on a separate sheet.

1. The first forming method used to shape metals was:
   a. Melting.
   b. Rolling.
   c. Hammering.
   d. None of the above.
2. List the three essential elements that must be present in order for forming processes to take place.
3. Name the two major types of shaping devices used in forming processes.
4. Explain why few finished shapes are produced using open dies.
5. Generally, the _____ a material is, the easier it is to form.
6. Extrusion is an example of a:
   a. Squeezing force.
   b. Drawing force.
   c. Bending force.
   d. None of the above.
7. *True or False?* Unlike the hammer, the press moves the head in the steady squeezing action needed for drawing, bending, shearing, and press forging.
8. List the two high-energy rate (HER) actions that provide forming pressures and describe each process.

Applying Your Knowledge

Be sure to follow accepted safety practices when working with tools. Your instructor will provide safety instructions.

1. Manufacture a part or product using a common forming process.
   a. Select an appropriate product. This might be a tack puller or screwdriver blade. See **AYK 11-1.** It will fit the handle you may have made in the casting activity for Chapter 10.
   b. An alternative might be to produce a set of injection molded checkers for the checkerboard described in the separating activity for Chapter 12.
   c. Whatever your project, make sure you have a set of procedures worked out before you begin. Check the procedures with your instructor. Be sure to include all the operations needed to complete the product. Try to determine which operations should be done first. Select the most appropriate machines or tools in your laboratory to accomplish all tasks.
2. Talk with someone who has made a piecrust. What forming process did the person use to achieve the proper crust thickness?
3. Obtain a flat piece of metal. Use a hammer to cause it to bend into a right angle curve. Carefully examine the metal and describe the effects of this bend forming action.

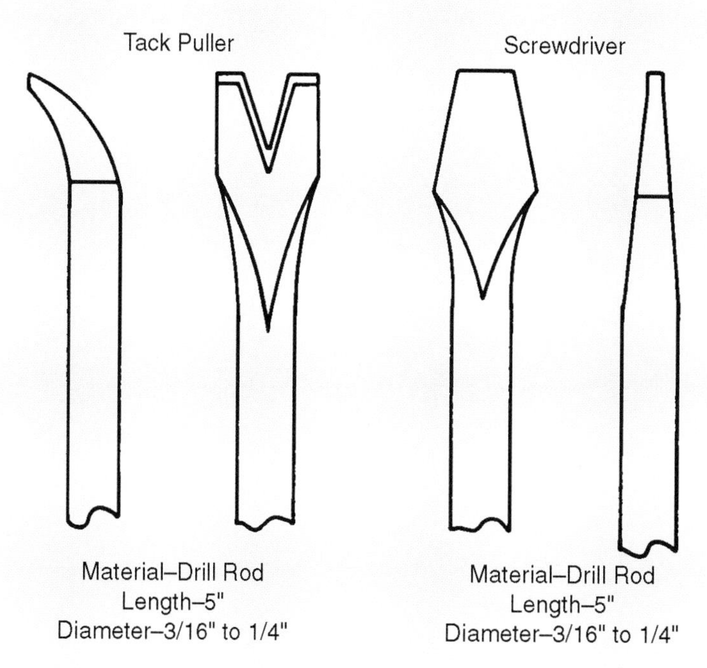

AYK 11-1. Manufacture a tack puller or screwdriver blade using a common forming process described in this chapter.

Chapter 12
Separating Processes

Objectives

After studying this chapter, you will be able to:

✓ Distinguish between the processes of machining and shearing.
✓ Describe how the various separating processes developed.
✓ List the three essential elements of separating processes.
✓ Name six basic machines that have been developed to separate materials.
✓ Explain how separating machines operate.

As discussed in previous chapters, materials may be sized and shaped using casting and forming techniques. Another family of processes to accomplish this task is separating. Separating processes remove excess material, creating the desired size and shape of a workpiece or part.

The two major ways to separate materials are machining and shearing. See **Figure 12-1.** *Machining* changes the size and shape of materials by removing the excess material in the form of chips or small particles. *Shearing* uses opposed edges to fracture the excess material away from the workpiece.

Shearing

Machining

Figure 12-1. The two types of separating processes are machining and shearing.

How Separating Developed

Early humans used crude separating practices as they cut (sheared) meat and other foods with stone tools. Soon they learned to separate wood and stone materials to create useful items. Using crude axes and knives, they produced tools and shelters, harvested plants, hunted game, and caught fish.

Later, they developed the ability to use tools to drill holes in various materials. The bow drill, shown in **Figure 12-2,** was possibly the first successful design for a hole-drilling tool. Other early separating devices include the saw, chisel, and knife.

As humans moved into the Iron Age, machines were needed to work metals. One of the earliest power-driven metal-cutting machines was the lathe. See **Figure 12-3.** This machine allowed the workers to produce accurate round parts

shearing. A process that uses force applied to opposed edges to fracture excess material away from the workpiece.

Figure 12-2. Separating materials became easier when our ancestors invented the bow drill to produce holes.

Figure 12-3. This modern lathe uses the same machining principles as historical lathes did.

Figure 12-4. This is a modern milling machine that operates much like historical ones did.

such as axles and shafts. The lathe was used in Europe before the Middle Ages (about 500 to 1400 AD). Early clockmakers used the lathe quite often in their precision metalworking activities. Later, lathes were developed for heavier metal machining operations and played an important part in the industrial revolution.

Still other machines followed the development of the lathe. One of these machine tools, called a ***milling machine,*** was developed in the 1800s. See **Figure 12-4.** This machine allowed the artisan to cut notches and to smooth the surfaces of flat stock. Eli Whitney used this machine tool in his early development of interchangeable parts for muskets. Other separating machines developed for the industrial revolution included:

- Power driven saws
- Metalworking shapers and planers
- Grinding machines
- Shearing machines

America has developed one of the greatest production capacities the world has ever experienced. Separating and shearing machines are the foundation of this production system. These machines shape and size materials into interchangeable parts. They are used to build the jigs, fixtures, templates, and patterns that support other manufacturing processes. Separating machines are also used to make other machines that produce the parts for everyday products.

Essential Elements of Separating

All separating activities share some common characteristics. These essential elements of separating, as shown in **Figure 12-5,** are:

- A tool or other *cutting element* is always used.
- There is ***motion*** between the work and the cutting element.
- The work or the tool is given ***support*** (clamped or held in one position).

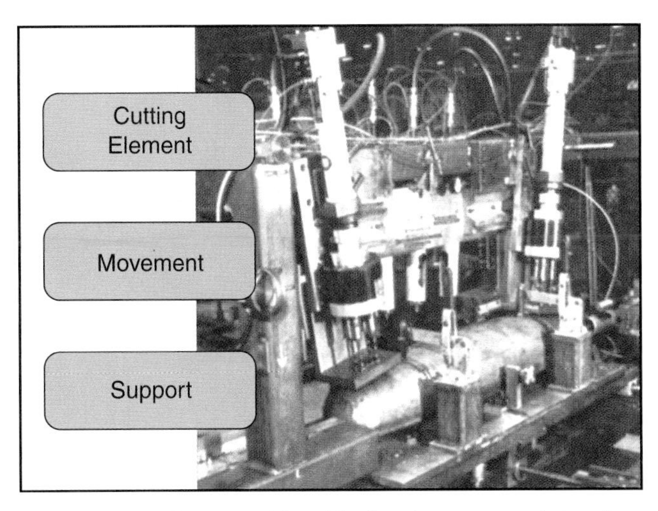

Figure 12-5. Can you identify the three essentials of separating in this specially built drilling machine?

Tools or Cutting Elements

Separating processes can be grouped according to the cutting elements used. The major differences among separating processes are shown in **Figure 12-6.** Shearing and machining by chip removal use a cutting tool that is called either a cutter or a blade. ***Cutting elements*** are made from materials that are harder than the material to be cut. Modern cutting tools are usually made from tool steel or ceramic materials (carbides).

Other separating processes use burning gases, light and sound waves, or electrical sparks to separate materials. High pressure liquids can also be used in separating processes.

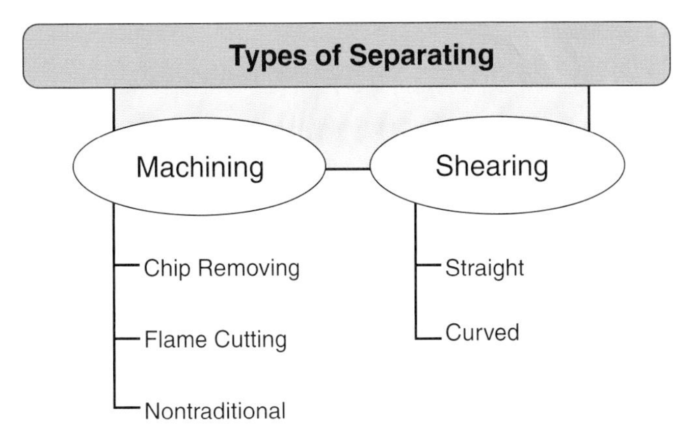

Figure 12-6. These are the two major separating processes with their subdivisions.

Chip Removing Tools

Traditional (chip removal) machining and shearing is based on the hardness properties of materials. Any material will cut a softer material. For instance, a diamond will cut glass. Carbides will cut tool steel and tool steel will cut mild steel. Carbon steel will cut wood and nonferrous metals. Cardboard (a wood product) will cut butter.

When the material is properly shaped, the resulting cutting element is called a ***tool***. All tools may be divided into the two classes of single point tools and multiple point tools.

Single Point Tools

Single point tools, like the one shown in **Figure 12-7,** are the simplest of all tools. They have one cutting edge that forms the chip. Most lathe, metal shaper, and metal planer tools are single point devices. Woodworking planes, chisels, scrapers, and gouges are common single point hand tools. See **Figure 12-8.**

Multiple Point Tools

Multiple point tools are two or more single point tools arranged to form a cutting device. See **Figure 12-9.** Each point of the tool acts like a single point

cutting elements. Tools or blades that are made of material harder than the material to be cut.

single point tools. Simple tools with only one cutting edge.

multiple point tools. An arrangement of two or more single-point tools to form a cutting device.

Figure 12-7. A single-point tool is the simplest type. It has one cutting edge. (Inland Steel Co.)

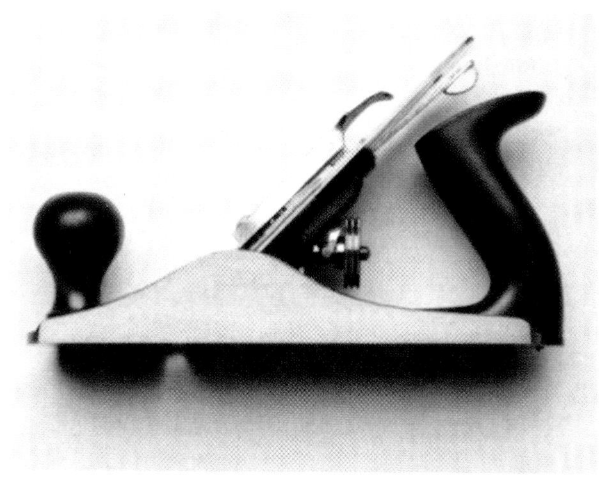

Figure 12-8. A jack plane is a typical single-point hand tool.

Figure 12-9. Most cutting tools are multiple point devices. (Inland Steel Co.)

tool. Arranging several single point cutting edges into a multiple point tool speeds the cutting process.

Several types of equipment use multiple point tools. These include circular, band, and hacksaws; woodworking shapers, planers, and jointers; drill presses; sanding and grinding machines, and routers.

The cutting points on multiple point tools are arranged in two major patterns. Some are randomly arranged, while others have a set pattern.

With abrasive cutting devices, such as abrasive papers and grinding wheels, the single point cutting edges (abrasive grains) are randomly arranged. They have no set pattern or spacing.

Cutters, such as saw blades and milling cutters, have a specific pattern to the cutting points. They are carefully spaced around the cutter in a pattern to generate the cutting action. See **Figure 12-10.**

Cutting Tool Design

Cutting tools come in many shapes and types. The material to be cut and the machine that will do the cutting help determine the tool to be used. However, each cutting surface of a tool must be properly designed. The tool designer must consider at least four factors:

- Each tooth must be the correct shape to develop a chip and to produce a desired contour to the material. See **Figure 12-11.**

Figure 12-10. Multiple point cutting tools may have uniform or random spaced teeth.

Figure 12-11. These are some of the common shapes that are available in multiple point wood shaper cutters.

- Sharp points are avoided whenever possible. They wear rapidly and are easily broken or chipped. To test this, try to cut a piece of wood first with a knifepoint and then with the edge of the blade. You will obtain a cleaner cut with the edge of the knife.
- *Relief angles* must be provided behind all parts of the cutting edge that generates the chip. See **Figure 12-12.** These angles prevent the body of the cutter from rubbing against the material and creating excess heat. Also, without relief angles, the cutting edge cannot easily penetrate the material. Using your knife again, lay the blade flat on the wood. You cannot get it to cut. Also, if you rub it back and forth, the blade will become hot. Now tilt the blade up at a 10° to 15° angle. The blade will cut the material.
- *Rake* is required on all cutting edges. This angle, as shown in **Figure 12-13,** is sloped away from the cutting edge. Rake reduces the force it takes to produce a chip and lifts the chip away from the material.

relief angle. The clearance behind the cutting parts of the tool. Relief angles also keep most of the tool from rubbing on the workpiece.

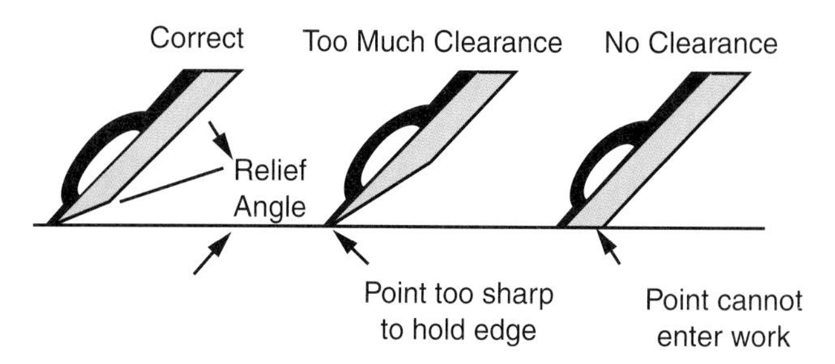

Figure 12-12. Cutting tools must have relief angles if they are to cut properly.

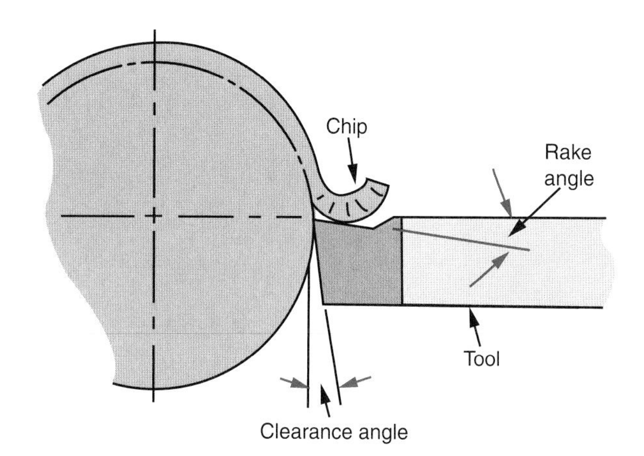

Figure 12-13. This diagram shows the typical rake and clearance angles that must be developed when producing a single point metal lathe tool.

Figure 12-14. This worker is monitoring a multiple head, computer-controlled flame cutting machine. (Federal Industries)

Nonchip Cutting Elements

Nontraditional machining and flame cutting do not use a tool in the normal sense of the word. These processes rely upon electric arcs, sound waves, light beams, chemical reactions, and burning gases to separate excess material from the workpiece. Many times, the "chip" is a fine particle of material or a glob of molten metal.

rake. **An angle that slopes away from the cutting edge of a tool. It reduces the force needed to produce a chip.**

flame cutting. **A process in which a mixtures of gases is burned to melt a path between the workpiece and the excess material.**

electrical discharge machining (EDM). **A process that uses a spark (electrical discharge) to erode a small chip from the workpiece. EDM is widely used to make cavities for forging and stamping dies.**

Flame cutting is probably the oldest of the nonchip separating processes. With this process, a mixture of gas (usually oxygen and acetylene) is burned to melt a path between the part and the excess material. See **Figure 12-14.** The flame preheats the metal. Then a stream of oxygen causes the base metal to oxidize rapidly. This chemical reaction turns the metal into its molten state. The liquid metal falls away from the cooler metal on each side of the cutting path.

The oldest of the nontraditional machining processes is *electrical discharge machining (EDM)*. This process, which is shown in **Figure 12-15,** uses a shaped carbon electrode or tool that is mounted on a movable head. The workpiece is placed in a dielectric liquid (nonconductor of electricity). The tool is slowly lowered until an electric arc jumps between the workpiece and the tool. The spark becomes the cutting element and dislodges fine particles of the workpiece. As this erosion continues, the tool is fed downward, creating a shaped cavity in the part.

EDM is widely used to produce accurate cavities in forging and stamping dies and in very hard metallic materials. Recently, processes similar to EDM have been developed. All of them use an electrical spark between a tool and the work as a cutting element. The most common of these processes are:

- *Electrical discharge grinding (EDG)*. This process removes material using an electrical spark between a rotating disc and the work as a cutting tool.

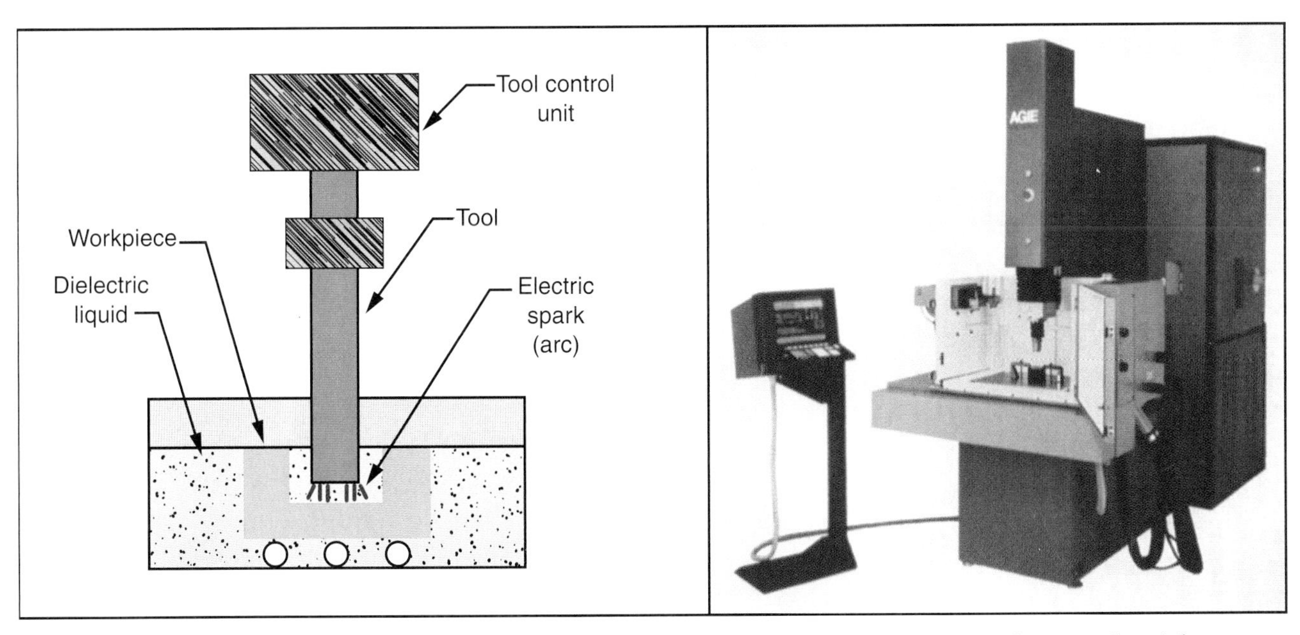

Figure 12-15. The principle shown in the drawing on the left shows how the EDM machine on the right operates. (AGIE USA Ltd.)

- *Electrical discharge sawing (EDS).* In this process, an electrical spark between a moving knife edge band and the work serves as a cutting tool.
- *Electrical discharge wire cutting (EDWC).* The cutting tool for this process is an electrical spark between a taut, traveling wire and the work, **Figure 12-16.**

Figure 12-16. An electro-discharge wire-cutting machine in operation. (Agie Losome)

Recently, additional processes have been developed using nontraditional cutting elements. The simplest of these is *chemical machining.* This process uses the same chemical action that causes metal to corrode (rust or tarnish). In chemical machining a photographic negative of the area to be machined is produced. The metal is cleaned and coated with a ***photoresist***, which is a material that will react to strong light. The negative rests on the coated metal while a powerful light exposes the resist through the negative. The resist is developed with chemicals much like a photographic print is developed. This leaves the area that is to be machined uncoated. Then chemicals, which attack and etch (eat away) the uncoated area, are applied to the workpiece.

Another new, nontraditional machining process uses beams of focused light as a cutting element. This process, called *laser beam machining (LBM),* directs a concentrated beam of monochromatic (one color) light on the work. The light produces intense heat that melts a path in the material. See **Figure 12-17.** Laser cutting can be used on a wide variety of materials ranging from paper and wood to metal and ceramic.

electrical discharge grinding (EDG). **A process that uses a spark as a cutting tool between a rotating disc and the workpiece.**

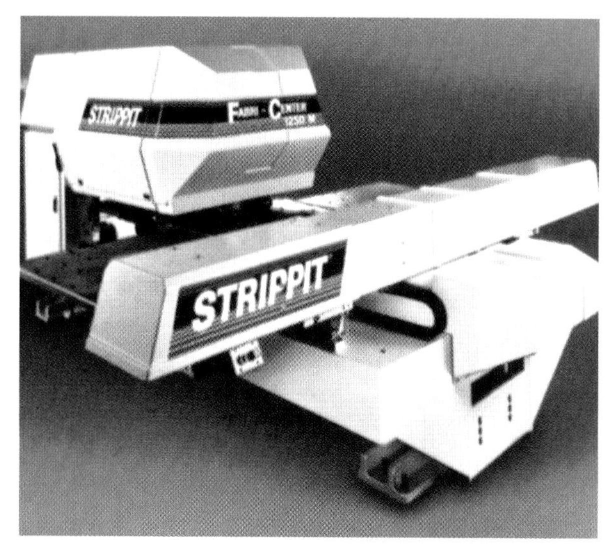

Figure 12-17. A general, purpose laser-cutting machine (Strippit, Inc.)

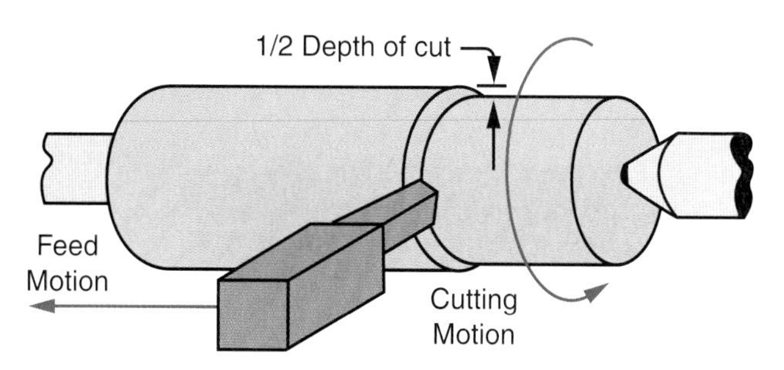

Figure 12-18. This diagram shows the cutting and feed motions on a lathe.

Motion

All separating operations use motion between the tool and the cutting element. There are generally two types of motion. These are cutting motion and feed motion. See **Figure 12-18.**

Cutting Motion

Cutting motion is movement between the work and the cutting element that causes material to be removed. Three basic types of movement, **Figure 12-19,** can generate cutting motion. The tool or the work can *rotate* or revolve to create the motion. The work or the tool may *reciprocate* (move back and forth). A *linear* or straight-line motion may also be used to produce the chips.

Examples of specific machines that use each of these motions will be given later in this chapter. We will compare the basic separating machines in terms of the type of tools used, motions they generate, and the methods used to support the tools and the work.

Cutting speed is the rate of travel of the tool or workpiece during a cutting motion. In either case, the cutting speed is measured in feet per minute or meters per minute.

Feed Motion

The movement that brings new material in contact with the cutting element is called *feed motion.* Feed motion is generally linear movement generated in each cycle of the cutting action. It is produced by either moving the workpiece into the cutting element or moving the cutting element into the workpiece.

Feed is generally very slow when compared to the cutting motion. It is expressed in either inches or millimeters per revolution (for rotating movement), in inches or millimeters per stroke (for reciprocating movement), or in inches per foot (for linear movement).

To understand the difference between feed and cutting motions, carefully study the diagram of a band saw shown in **Figure 12-20.** The blade moves downward to produce a linear cutting action. The chip is produced as a tooth passes through the workpiece. Think about this experiment. Suppose a band saw is not running. If you place a board against the blade and start the machine, the first tooth will produce a chip and remove some material as the blade moves downward. This is the cutting action, but the cutting action will stop here even though the saw blade continues to move. The remaining teeth will pass through the slot (kerf) the first tooth produced. However, if you move the board forward into the blade, new material will contact the teeth. Cutting action will continue. The movement of the board is the feed motion. Cutting takes place only when both cutting and feed motions are present.

electrical discharge sawing (EDS). **A process that uses a spark as a cutting tool between a moving knife edge band and the workpiece.**

electrical discharge wire cutting (EDWC). **A process that uses a spark as a cutting tool between a taut traveling wire and the workpiece.**

chemical machining. **A process that uses powerful chemicals to etch (eat away) excess material from the workpiece.**

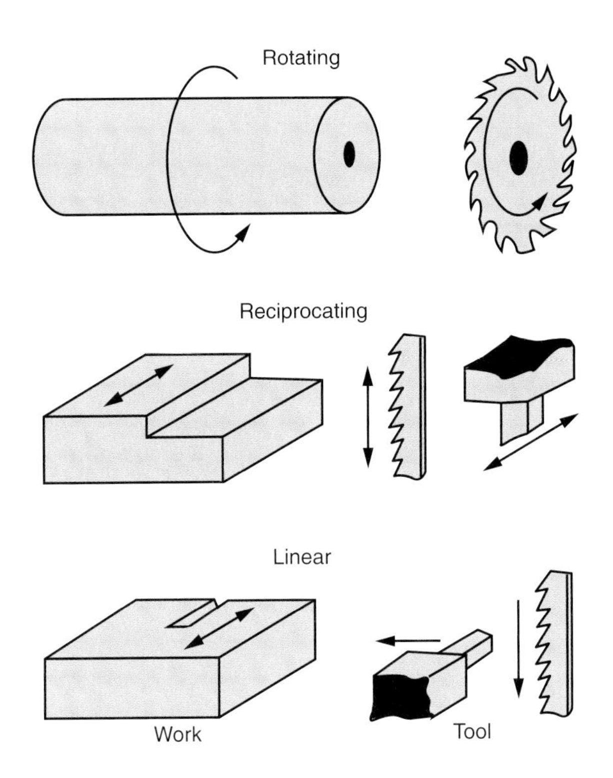

Figure 12-19. These are the types of motion present in separating operations.

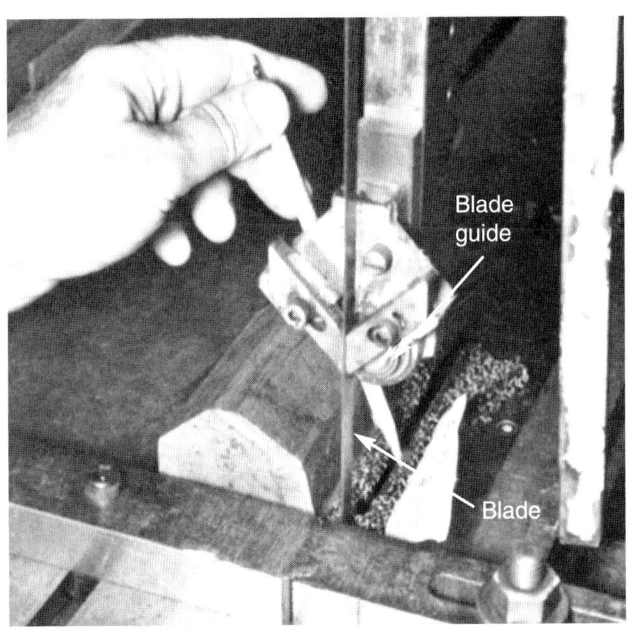

Figure 12-20. This diagram shows the cutting motion for a band saw in linear (downward) movement of the blade. The feed motion is linear movement (horizontal) of the workpiece.

Depth of Cut

The amount of material separated from the workpiece in one stroke or revolution is the *depth of cut*. It is the measured difference between the original surface and the new, machined surface.

The distance the cutter is fed into the workpiece or the distance the workpiece is fed into the cutter determines the depth of cut. In flat machining, the depth of cut and the reduction in the size of a material is equal. However, in rotating work operations, such as lathe turning, the reduction in the diameter of the work is twice the depth of cut.

Depth of cut is generally measured in thousandths (1/1000) of an inch for precision machining. Measurements are in fractions of an inch, however, for construction and woodworking processes.

Clamping Devices

Most separating operations use machines that hold or support both the tool or cutter and the workpiece. These are *clamping devices*. The type of clamping device used is directly dependent on the feed and cutting motions used.

The type of cutting motion must be considered in selecting clamping devices for both the tool and the workpiece. Rotating work and tools must be gripped and turned. ***Chucks,*** such as drill chucks and router collets, answer this need. Revolving cutters may be placed on shafts called ***arbors.*** Milling machines and grinders use this type of holding system. Finally, rotating workpieces can be supported between centers in lathes and cylindrical grinders.

laser beam machining (LBM). A cutting method that uses a beam of focused light to melt a path in the material and separate the excess from the workpiece.

cutting motion. Relative movement between the workpiece and the tool that causes material to be removed.

rotate. To turn about a central axis, or revolve. Either the workpiece or the tool rotates in many types of machining operations.

reciprocate. Move back and forth. Some types of machining are done with a reciprocating motion of either the tool or the workpiece.

linear. Movement in a straight line; often used in reference to a machining operation.

feed motion. In machining, the movement that brings new material in contact with the cutting element.

depth of cut. The amount of material that a tool separates from the workpiece in one stroke or revolution.

clamping devices. Machines that hold or support the workpiece or a cutting element.

Vises, clamps, and magnetic tables hold down the workpiece for flat, reciprocating, or stationary work. Work that is fed linearly is often clamped to a table or held in a machine vise. The table is then moved by gearing or hydraulic systems to provide feed motion. Woodworking operations often use guides (miter gauges and rip fences) so the work can be fed by hand.

You will see examples of all these clamping techniques in the next section of this chapter. Pay special attention to this feature in the figures that illustrate these separating operations.

Separating Machines

Six basic machines separate materials. These machines, as shown in **Figure 12-21,** are turning machines, drilling machines, milling and sawing machines, shaping and planing machines, grinding and abrasive machines, and shearing machines. In addition, there are separating machines and equipment that perform nontraditional machining. These machines also perform flame cutting.

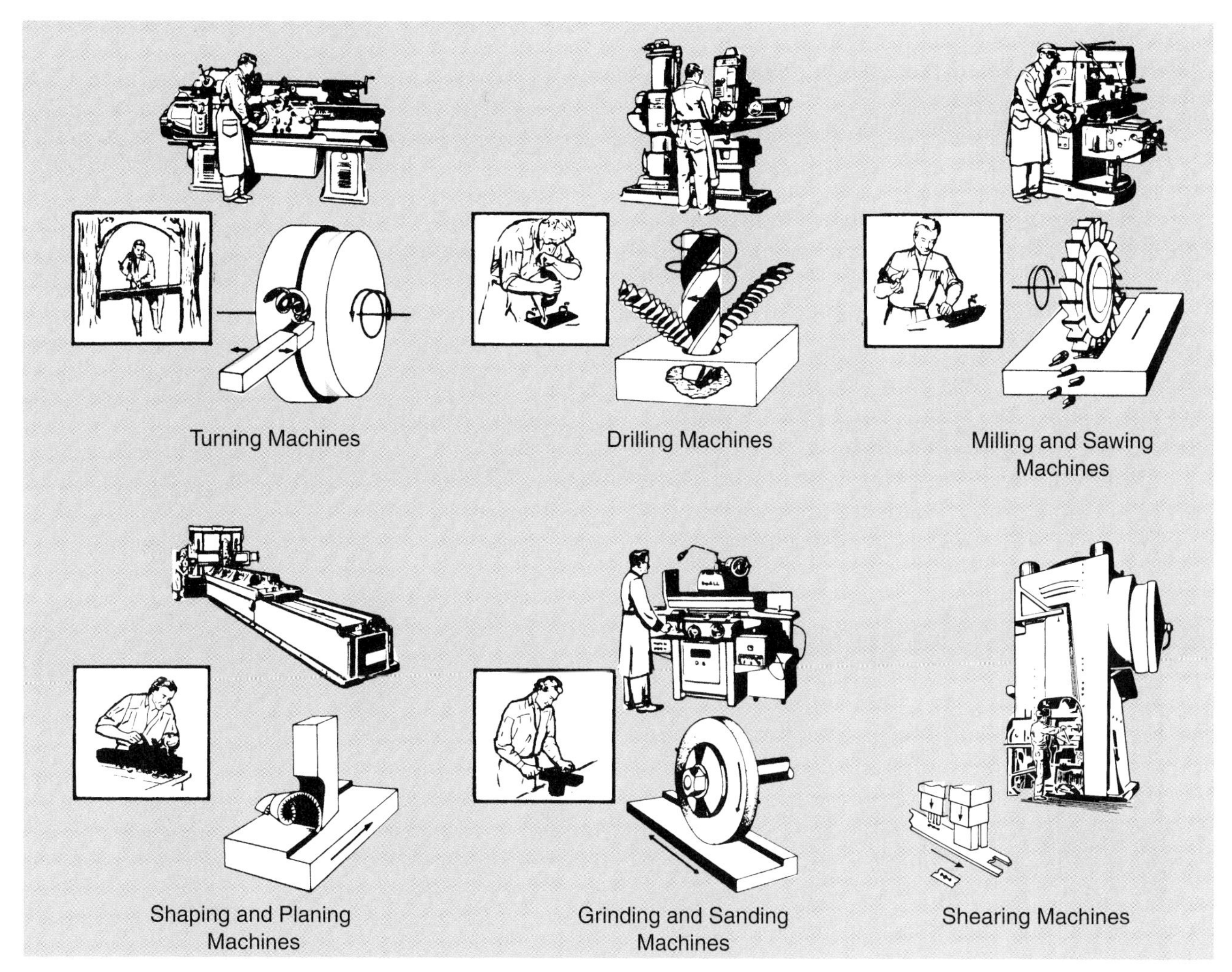

Figure 12-21. These are six basic machine tools. (DoAll Co.)

Figure 12-22. Turning machines evolved from the potter's wheel.

Turning Machines

Turning machines are almost as old as civilization itself. An early predecessor of this machine is the hand- or foot-operated potter's wheel. See **Figure 12-22.** Clay placed on the flat potter's wheel was rotated. As the clay rotated, the potter shaped the material using his or her hands. Vessels and other ceramic objects were produced using this machine. From this early start, came our modern turning machines that are generally called lathes.

A large number of processes are completed on turning machines. However, these machines can complete two unique operations. These operations are straight turning and face turning. See **Figure 12-23.** In both cases the work rotates to generate the cutting motion. A single-point tool is fed into and along the work. This action establishes the depth of cut, and produces the feed motion. The work may be held between centers or in a chuck. See **Figure 12-24.**

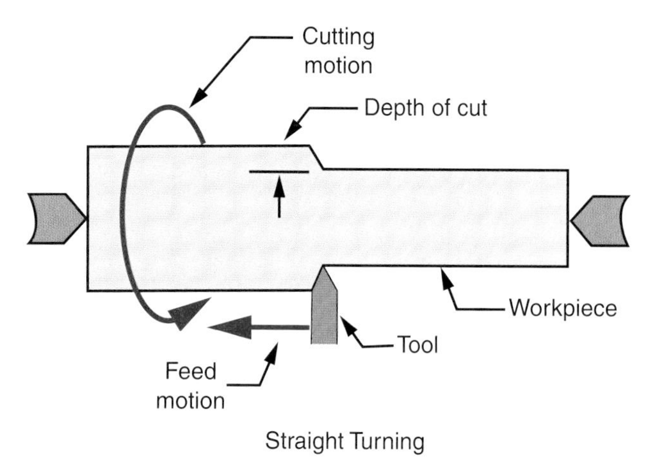

Figure 12-23. There are two basic types of turning.

Figure 12-24. This piece of stock is being held in a chuck on a lathe. (Inland Steel Co.)

Straight turning

Straight turning is performed on the external surface of a shaft and reduces the diameter of a material. The operation is performed by feeding the tool into the stock to establish the depth of cut. The tool is then moved along the length of the shaft parallel to its axis. This action produces a uniform diameter along the length of the part.

A modification of straight turning is *taper turning*. This operation produces a straight cut, but the diameter uniformly changes along the length of the part. Taper turning uses a straight tool path that is not parallel to the axis of the shaft.

Face turning

Face turning produces a straight cut across the face or end of a piece of material. The material can be round or rectangular. It is generally held in a chuck.

straight turning. Machining done by the cutting action of a single point tool to produce a uniform diameter on a rotating workpiece.

taper turning. Machining done by the cutting action of a single point tool to produce a uniformly changing diameter along the length of a rotating workpiece.

Special operations can also be completed on lathes. These include drilling and reaming holes, cutting threads inside holes and along shafts, and cutting grooves and angles.

Many types of turning machines can be used to machine metals. These include the engine lathe, turret lathes, automatic bar machines, and turning centers. Other lathes can be used to machine woods and spin metals (a forming process).

Drilling Machines

The first drills were produced about 40,000 years ago. A small, pointed stone was probably attached to a stick. This "drill" was rotated between the palms of both hands to cut a hole in a soft material. This same cutting principle is found in modern drilling machines. See **Figure 12-25.**

Figure 12-25. This drill press is being used to cut threads. (Inland Steel Co.)

Figure 12-27. This liquid oxygen fuel tank part for the space shuttle is being drilled in a special computer-controlled machining center. (Fansteel, Inc.)

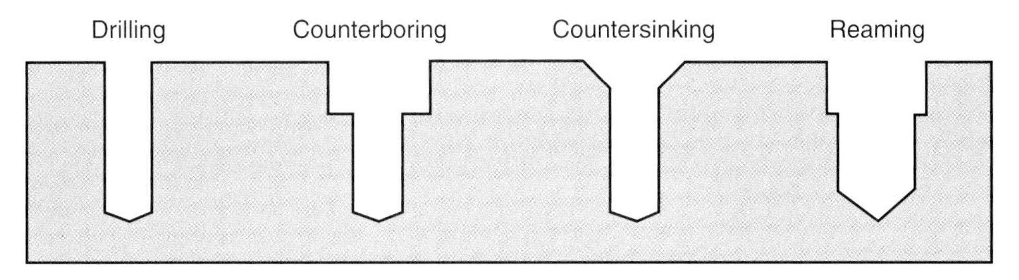

Figure 12-26. These are common operations performed with a drilling machine.

Drilling machines include drill presses (hole drilling), boring (hole diameter truing) machines, and tapping (thread cutting) machines. Simple drilling tools found in many home workshops might include the woodworking brace and auger bit, the hand drill and twist drill, and the portable electric drill with a drill bit.

Several operations can be completed on drilling machines. These, as shown in **Figure 12-26,** include:

- *Drilling.* Produces a straight, round hole in a workpiece. These holes can accommodate shafts and mechanical fasteners (screws, bolts, rivets, etc.).
- *Counterboring.* Produces two holes around the same axis. The lower hole is smaller in diameter than the upper hole. Counterboring allows the head of a fastener to be set below the surface or it may seat a part with a shoulder.
- *Countersinking.* Produces a bevel at the top of a hole. This is done to accommodate a flat head screw.
- *Reaming.* Shapes or enlarges a hole. It is often used to widen an existing hole.

Drilling machines generally grip the cutting tool in a chuck. The tool, which may be a drill, reamer, or other bit, is rotated to generate the cutting motion. It is then moved downward (linearly) to produce the feed motion. The workpiece is held stationary on a table, clamped in a vise, or gripped in a special jig as shown in **Figure 12-27.**

face turning. A machining process in which the cutter moves across the face, or end, of a rotating workpiece.

drilling. A machining operation that produces a straight, round hole in a workpiece.

counterboring. A machining operation that produces two straight, round holes on the same axis in a workpiece. The upper hole is larger, to allow the head of a fastener to be set below the surface.

countersinking. A machining operation that is similar to counterboring, but with a beveled upper hole. Countersinking is done to sink the head of a flathead fastener below the surface.

Academic Link

There was a time when people did not communicate as they do today. They had no books, newspapers, or any other method of recording what they did. What is referred to as recorded history is from the time humans first started recording what they did to present time. History—that is, the human past documented in some form of writing—began 5000 years ago in parts of southwestern Asia and as recently as the late 19th century AD in central Africa and parts of the Americas. *Prehistory* covers past human life from its origins up to the start of written records. Because there are no written records for prehistory, prehistorians rely entirely on material remains for evidence.

Archaeology is the science of the human past and their material remains. Throughout this book is discussed the development of manufacturing processes and the tools used for them. Many of the tools we used today had their beginings in this prehistory time period. We know this because archaeologists found the remains of these tools and trace the origins to these times. Archaeology is a science well suited for anyone who enjoys biology, botany, geology, chemistry, history, psychology, art, and solving a great puzzle.

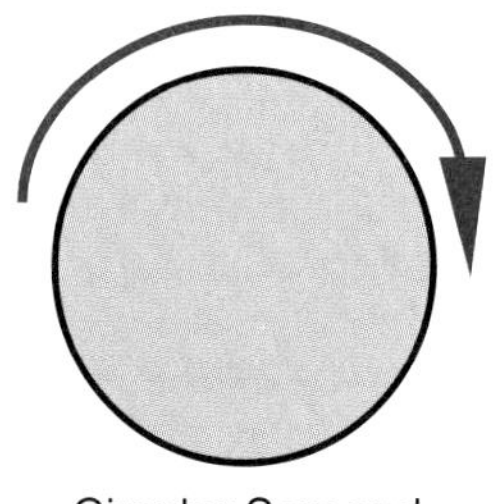

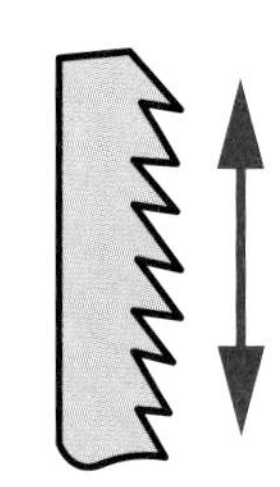

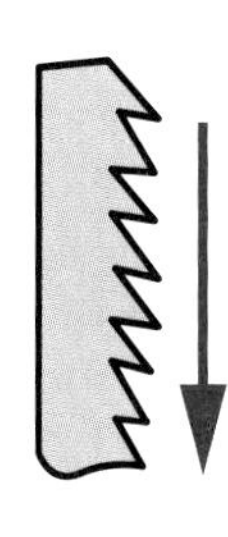

Figure 12-28. This diagram shows cutting motions for sawing and milling machines.

Milling and Sawing Machines

People began using saws in early history. Archaeologists have found stone saws that date back about 7000 years. Later, saws were made from bronze and iron. The metal file, which works on the saw principle, was invented almost 2000 years ago.

All of these tools had one thing in common. They had regularly spaced teeth on a straight piece of material. As machines were developed, teeth were also shaped on the round pieces similar to the circular saw blades and milling machine cutters of today.

reaming. **A machining operation that uses a rotating cutter to shape or enlarge a hole.**

By observing cutting motion, you can place milling and sawing machines into three basic groups. As shown in **Figure 12-28,** these machines generate their cutting motion by one of the following:

- Rotating the cutter
- Moving the blade linearly
- Producing a reciprocating blade motion

Rotating cutting machines

The largest group of sawing machines separates material by rotating the cutter on a shaft. This action produces the cutting motion. These machines produce their feed motion by moving the stock linearly into the cutter or blade. This feed motion may be generated either by manually feeding the stock or by using automatic table feeds. The depth of cut in this group of machines is established by either raising or lowering the cutter or the table upon which the workpiece rides.

The two most common machines in this class are the ***milling machine*** for metalworking and the woodworking *circular saw*. Observe the cutting and feed motion in these illustrations. See **Figure 12-29.**

Figure 12-29. Notice the cutting and feed motion for the milling machine (top) and the circular saw (bottom).

Variations on these machines are the radial saw and the cut-off saw. The ***radial saw*** is widely used to cut material to length on construction sites and in furniture and cabinet factories. Various ***cut-off saws*** are used to cut finish trim for buildings (motorized miter saw), metal to length (chop saw), and concrete block and brick.

Both of these machines have a rotating saw blade that generates the cutting motion. However, the work is stationary while the blade moves linearly downward or outward to produce the feed motion. See **Figure 12-30.**

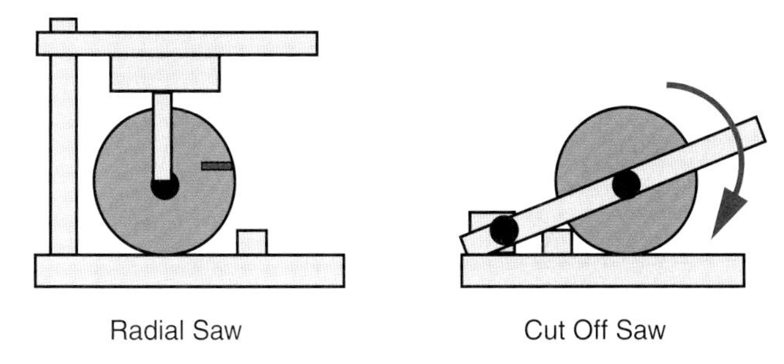

Figure 12-30. Feed motion for radial and cut off saws is shown in these diagrams.

Other machines that operate on milling principles are the woodworking jointer, shaper, surfacer, and router. Each of these machines has a rotating cutter that produces the cutting motion. The stock moves in a straight line past the cutter to develop the feed motion.

Other sawing machines

Power hacksaws and band saws are important saws that do not use rotary cutters. The blade is a strip of steel with teeth on one edge. The ***hacksaw*** uses a reciprocating motion with the cutting action taking place on the forward stroke. The feed motion of power hacksaws is similar to the cut-off saw. The saw unit is fed downward to produce the feed motion.

The band saw blade is a thin steel band that travels around two wheels. Standard ***band saws*** expose the right side of the loop where it passes through the table. See **Figure 12-31.** The work is placed on the table and fed linearly into the blade. The standard band saw is used to cut curves and arcs in wood products, plastic sheets, as well as metal plates and sheets.

The *horizontal band saw,* like the vertical band saw, has a linear cutting motion. However, the stock to be cut is clamped in a vise. The blade unit is fed downward, like the cut-off saw, to produce the feed motion. Horizontal band saws are widely used to cut metal bars, rods, and structural shapes to length.

circular saw. **A cutting device with a rotating blade. The workpiece may be fed into the blade, or the blade moved across the work.**

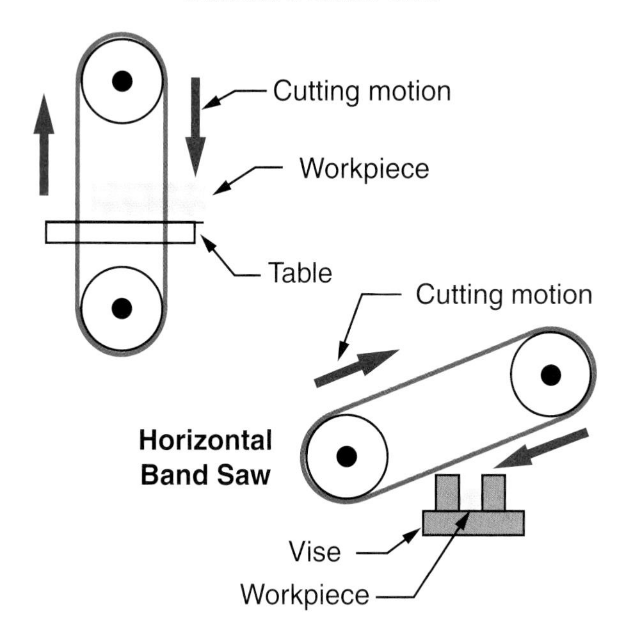

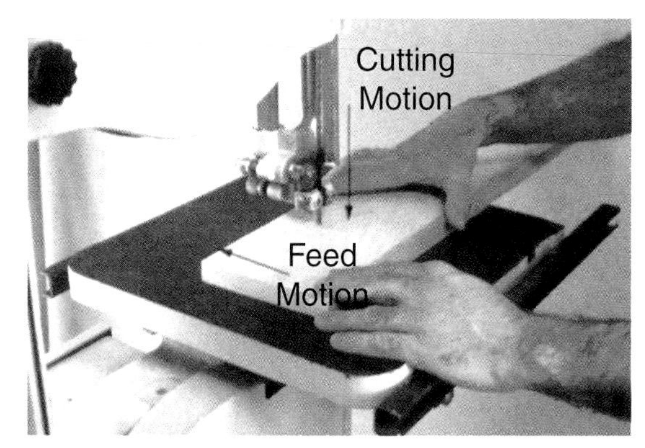

Figure 12-31. Note the types of band saws on the left. On the right look at the cutting and feed motions for a standard band saw.

Shaping and Planing Machines

The principle of shaping and planing is found in the old hand woodworking planes. Early carpenters used blocks of wood and an iron blade to shape parts. These tools are still found in antique shops and museums. The plane was moved forward to make a cut then backward to position it for the next cut.

The *metal shaper* cuts material by moving a single point cutting tool back and forth across a stationary workpiece. See **Figure 12-32.** The work is clamped in a vise and positioned in front of the tool. A ram pushes the tool forward and through the material. The tool cuts on the forward stroke. On the backstroke, the tool pivots up and the work moves over (feed motion) for the next cut.

The *metal planer* uses a reciprocating workpiece to develop the cutting motion. The single point tool can cut on the forward stroke of the work or on both the forward and backward strokes. The work is clamped on the planer table. The tool is positioned over the work for the first cut. The table travels under the tool to produce the chip. The tool moves over for a new cutting path at the end of the cutting stroke.

Metal shapers are generally smaller than planers. Both machines are only of limited use in modern industry.

Grinding and Abrasive Machines

metal shaper. A machine that removes excess material by moving the cutting tool back and forth over a stationary workpiece.

metal planer. A machine that removes excess material by moving the workpiece back and forth against a stationary cutting tool.

Our ancestors found that rubbing a strip of metal over a hard, rough material could form sharp edges. As early as 1500 BC, this practice was in general use. *Abrasive machining* has developed from this early separating technique. These machines are used for grinding and sanding operations that remove excess material. They also create smooth surfaces where needed.

Unlike the other types of machines discussed so far, grinding machines do not have specific cutting and feed motions. They are adaptations of other machine tools. What is common among these machines is their use of a natural or human-made abrasive grain. The material is bonded to a wheel or adhered (glued) to a flat paper or cloth backing material.

Abrasive wheels are used on several types of grinding machines. Two of these machines are cylindrical grinders and surface grinders.

In principle, ***cylindrical grinders*** are very much like lathes. See **Figure 12-33.** They hold and rotate the workpiece in a chuck or between centers. The rotating grinding wheel is fed into the work to generate the chip. Moving the work and its holder that is attached to a table produces the feed motion. The distance the wheel is fed into the work establishes the depth of cut.

Figure 12-32. This twin head metal shaper cuts on both the forward and back passes. The workpiece moves, while the tool is stationary during the cutting act. (Inland Steel Co.)

Figure 12-33. The cylindrical grinder is similar to the lathe in appearance and cutting action.

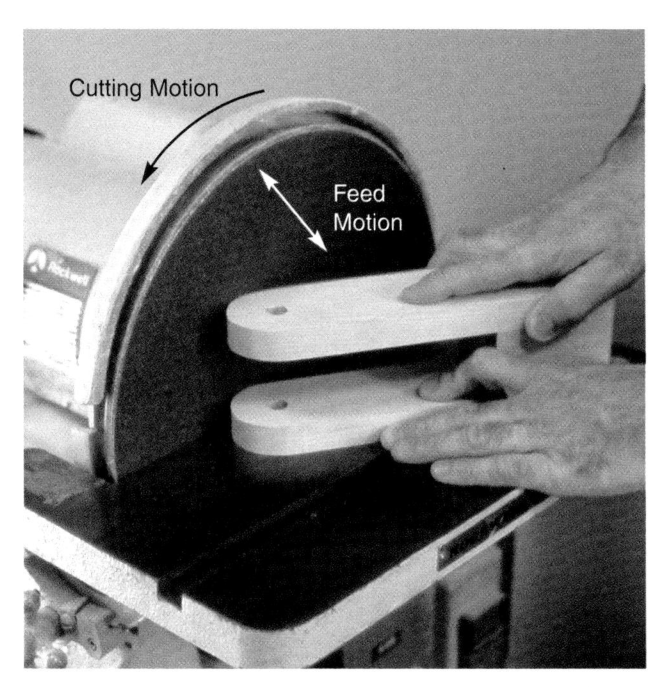

Figure 12-34. The disc sander, like the milling machine, uses a rotating cutter into which material is fed.

The ***surface grinder*** works on the milling machine principle. The workpiece is held on the machine table with bolts, clamping fixtures, or magnetic force. The rotating grinding wheel produces the cutting action. The work is reciprocated under the rotating grinding wheel to produce the feed motion.

abrasive machining. A process that uses an abrasive, in wheel or sheet form, to remove material from the workpiece.

Machines that use flat abrasives are called ***sanders***. These machines can use abrasive sheets, discs, or belts to sand and polish metals, woods, ceramics, and plastics.

The ***disc sander***, as shown in **Figure 12-34,** has an abrasive sheet attached to the machine's metal disc. This assembly forms a random, multipoint cutting tool. It operates like a milling machine. The work is fed into the revolving disc to produce the cut. Portable disc sanders, like the one shown in **Figure 12-35,** are often used to clean machines and cast parts.

The ***belt sander*** uses abrasives glued to a continuous band of material. See **Figure 12-36.** Its cutting principles are similar to the band saw. The abrasive travels horizontally or vertically past the work to generate the cutting motion. The work is fed linearly into the belt to produce the feed motion.

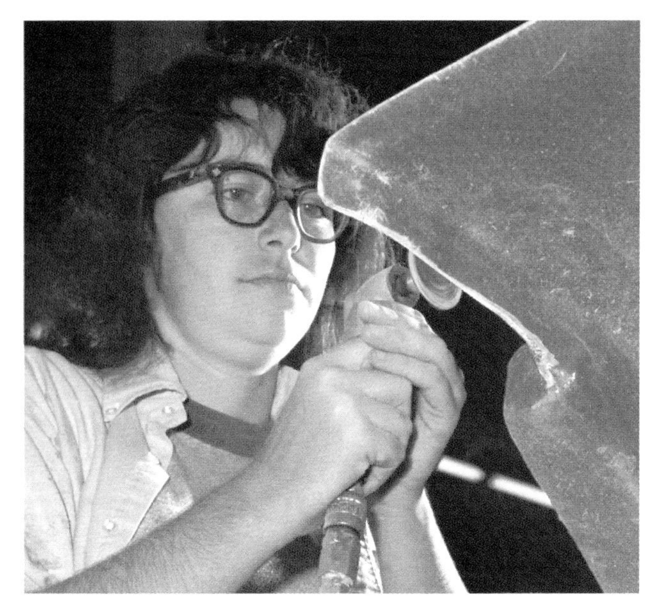

Figure 12-35. This worker is using a pneumatic-operated sander to touch up a fiberglass truck cab component. (ARO Inc.)

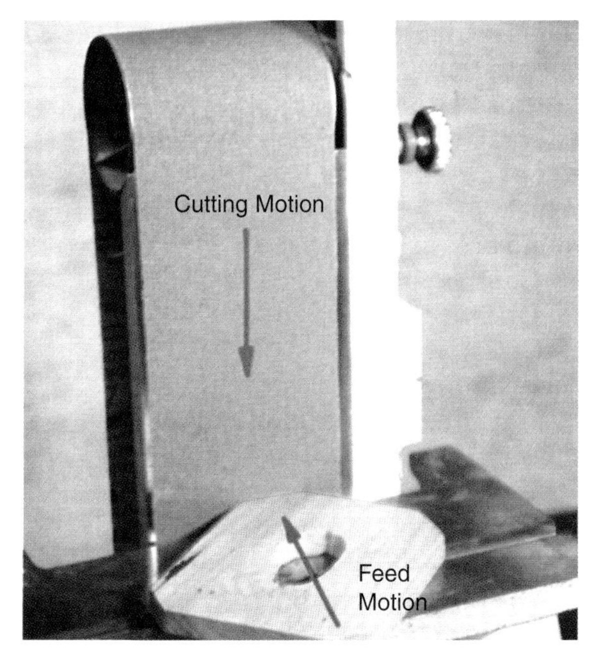

Figure 12-36. This belt sander uses a fast moving abrasive belt to cut away unwanted material.

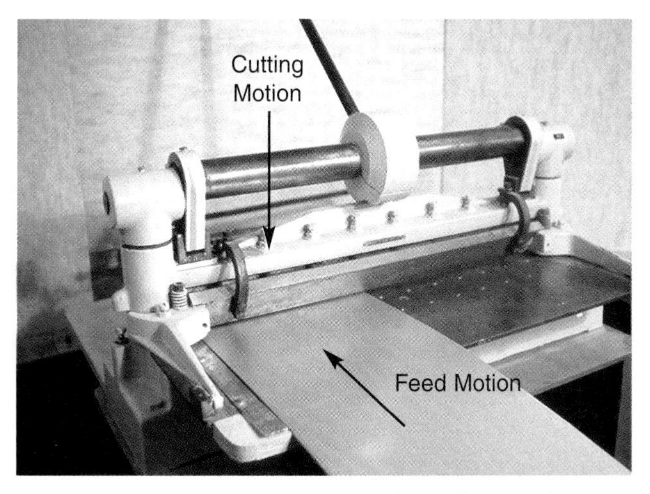

Figure 12-37. This is a squaring shear that can be used to cut metal strips to length.

Shearing Machines

The shearing machine is a major type of separating equipment. While other machines force tools against a workpiece to create a chip, *shearing machines* use two opposed edges to fracture or break the excess material from the workpiece. Shearing actions, unlike chip removal, do not lose materials. The length of the two parts will add up to the original length of the strip.

You have practiced shearing many times. Every time you cut paper or cloth with a pair of scissors, you are using a shearing technique.

Most shearing machines have a moving blade or die and a stationary edge. ***Squaring shear,*** shown in **Figure 12-37,** use two straight shearing blades to cut sheet metal to size. The upper blade moves downward to fracture the material over the lower blade.

Dies, like the one shown in **Figure 12-38,** often produce a shaped hole or outline. The movable upper die is mated to a stationary lower die. The die halves are mounted in a punch press. The material between the die halves is sheared as the punch press closes. Vast quantities of sheet metal and plastic are cut to size and shaped using dies.

shearing machines. **Devices that cut material by fracturing it between opposing tool edges.**

Other Separating Equipment

Nontraditional separating methods use special purpose machines such as those discussed earlier. Flame cutting is often done with oxyacetylene equipment similar to that, which will be introduced in Chapter 14.

Figure 12-38. Special shapes can be cut using dies. (Strippit, Inc.)

Summary

Separating processes are widely used in industry to size and shape metal, wood, ceramic, and plastic materials. The separation action is accomplished through the use of tools and forces generated by a combination of cutting and feed motions.

Hundreds of separating machines have been developed and used over time. These machines can best be understood if they are viewed in terms of:

- The type of tool used.
- The movement used to generate the cutting and feed motions.
- Methods used to clamp the work and tool.

Separating actions can be grouped into two categories: machining and shearing. Machining sizes and shapes material by removing excess materials in the form of chips or small particles. Shearing fractures the excess material away from the workpiece.

Safety with Separating Processes

- Do not attempt a process that has not been demonstrated to you.
- Always wear safety glasses or goggles.
- Keep your hands away from all moving cutters and blades.
- Use push sticks to feed small pieces of stock into woodcutting machines.
- Use all machine guards.
- Stop machines or equipment when making measurements and adjustments.
- Do not leave a machine until the cutter has stopped rotating.
- Clamp all work when possible.
- Unplug machines from the electrical outlet before changing blades or cutters.
- Remove all chuck keys or wrenches before starting machines.
- Remove all scraps and tools from the machine before turning on the power.
- Remove wood scraps with a push stick. Use a brush to remove metal chips and particles.
- Keep your hands behind the cutting edge of chisels and punches and behind screwdriver points.
- Obtain the instructor's permission before using any machine.
- Keep all work areas clear of scraps and unneeded tools.
- Use only sharp cutting tools for separating operations.

Key Words

All of the following words have been used in this chapter. Do you know their meanings?

abrasive machining
chemical machining
circular saw
clamping devices
counterboring
countersinking
cutting elements
cutting motion
depth of cut
drilling
electrical discharge grinding (EDG)
electrical discharge machining (EDM)
electrical discharge sawing (EDS)
electrical discharge wire cutting (EDWC)
face turning
feed motion
flame cutting
laser beam machining (LBM)
linear
metal planer
metal shaper
multiple point tools
rake
reaming
reciprocate
relief angle
rotate
shearing
shearing machines
single point tools
straight turning
taper turning

Test Your Knowledge

Please do not write in this text. Place your answers on a separate sheet.

1. What is the difference between machining and shearing?
2. *True or False?* One of the earliest power-driven metal-cutting machines was the lathe.
3. List the three essential elements of separating.
4. Name six basic machines that have been developed to separate materials.
5. *True or False?* Face turning is performed on the external surface of a shaft and reduces the diameter of a material.
6. Which of the following produces a straight, round hole in a workpiece?
   a. Countersinking.
   b. Drilling.
   c. Counterboring.
   d. Reaming.
7. List the three ways milling and sawing machines can generate their cutting motion.
8. *True or False?* Metal shapers and planers are used extensively in modern industry.
9. Abrasive machines are used for:
   a. Sanding operations that remove excess material.
   b. Creating smooth surfaces.
   c. Grinding operations that remove excess material.
   d. All of the above.
10. While other machines force tools against a workpiece to create a chip, _____ machines use two opposed edges to fracture or break the excess material from the workpiece.

Applying Your Knowledge

Note: Be sure to follow accepted safety practices when working with tools. Your instructor will provide safety instructions.

1. Manufacture a part or product using common separating processes. This might include a simple box for the injected molded checkers you made during the forming activity or a hot dish holder (trivet), as shown in **AYK 12-1.**
   a. If you decide to manufacture the box for the checkers, produce a design. (Be sure to survey your laboratory equipment to make certain you can produce the product you design with the equipment available.)
   b. Develop a procedure for making the box. Discuss the procedural steps with your instructor.
   c. Collect the materials needed.
   d. Secure the tools and begin the activity.
   e. If you decide to build the trivet instead, you can omit the design phase unless you wish to modify the design.
   f. See your instructor. He or she may suggest procedures or have you develop them.
2. List all the ways you use separating actions during one day of normal activities at home and at school. Describe each action in terms of the tool or machine used, the type of cutting element, and the material separated.
3. Analyze a simple product and make a list of the separating actions that were needed to produce the product.
4. Observe the cutting action of three machines in the technology education laboratory. Analyze them in terms of the type of cutter, the methods used to generate the cutting, and feed motions.

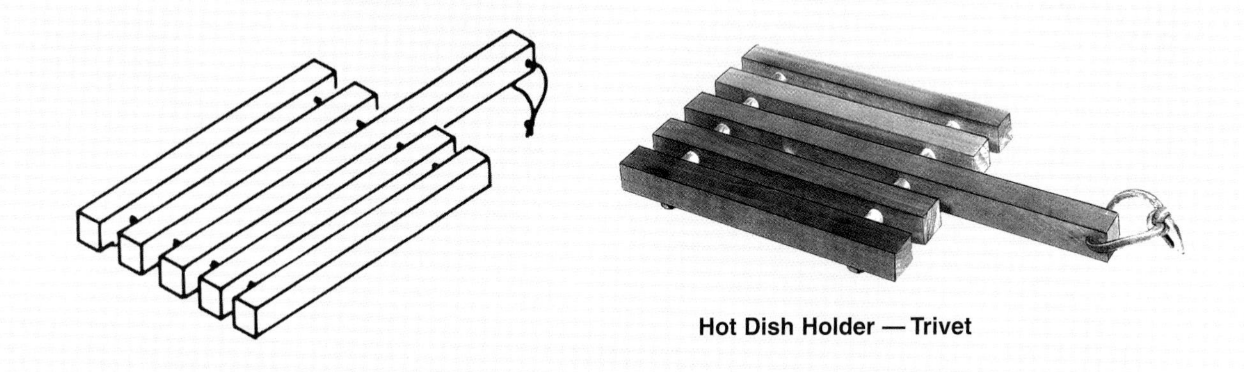

Hot Dish Holder — Trivet

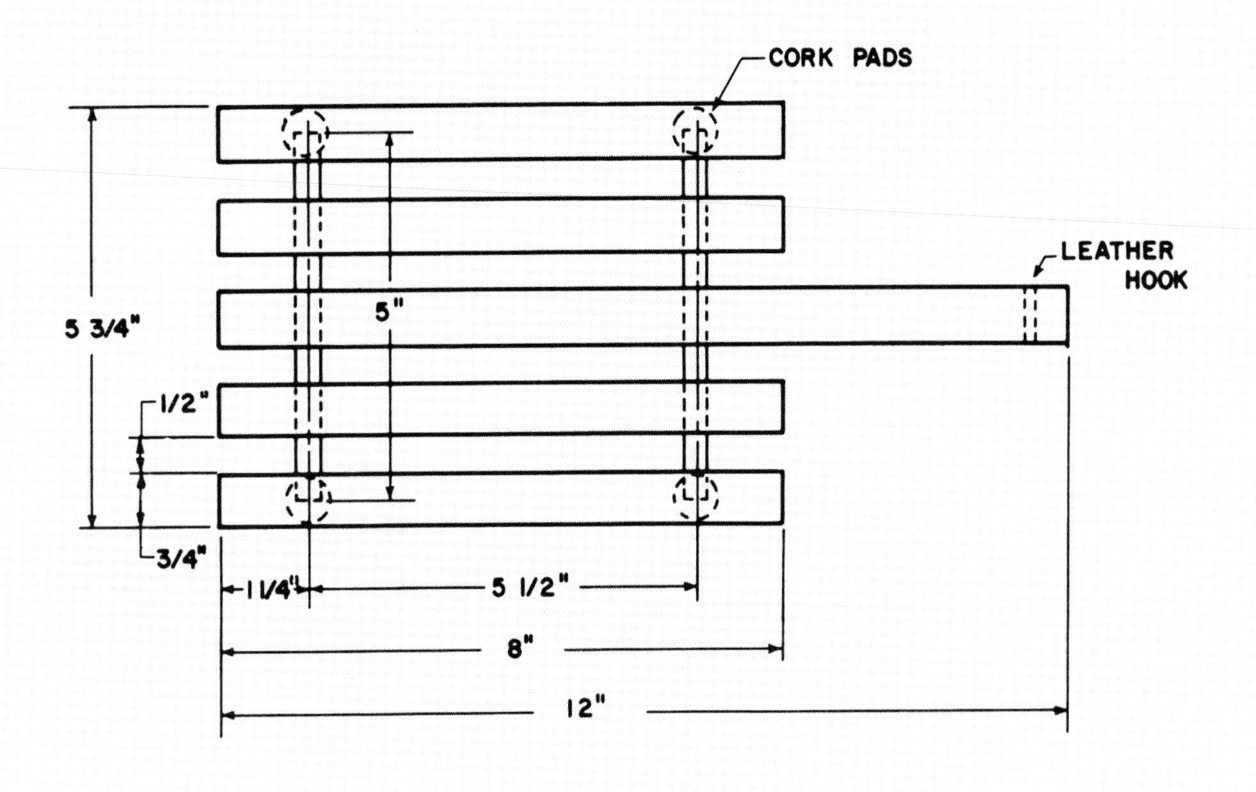

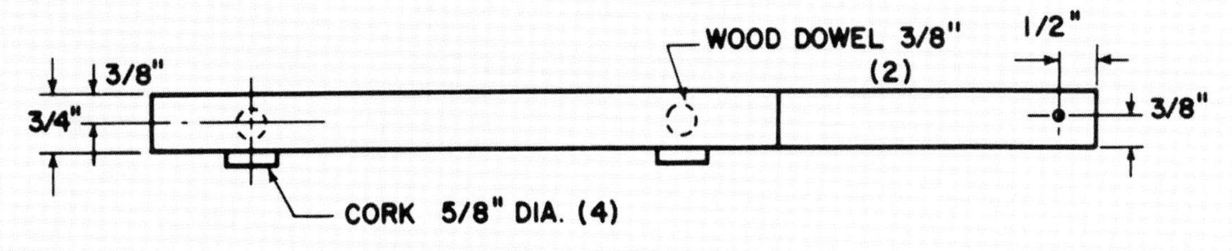

AYK 12-1. Several separating processes are involved in producing this hot dish holder (trivet). Can you name them?

Chapter 13 Conditioning Processes

Objectives

After studying this chapter, you will be able to:

✓ Describe how heat, mechanical force, or chemical action can be used to condition materials.
✓ List reasons materials are conditioned.
✓ Name and describe six common thermal conditioning practices.
✓ Give examples of how chemical conditioning is used.
✓ Explain mechanical conditioning and how it occurs.

In the last three chapters, you learned about processes that give parts their size and shape. These processes change the external appearance of the material. You will recall that in casting processes, the material receives a new size and shape by melting and pouring it into a mold. As materials go through forming processes, a shaping device is used to apply pressure to reshape the material. Separating processes remove excess material to size and shape a part. In many cases, these processes do not produce a totally usable part.

Sometimes, changing the outward, physical properties of a material is not enough. Often, the mechanical properties of a material must also be changed. In order to be suitable for a specific task or use, the strength, toughness, hardness, elasticity, or other mechanical properties of a material may need to be improved. See **Figure 13-1.**

Processes that use heat, mechanical force, or chemical action to change the mechanical properties of a material are called ***conditioning processes.*** In selecting and using conditioning processes, several factors must be considered. These, as shown in **Figure 13-2,** are that:

* The desired mechanical property must be established.
* The internal structure that produces this property must be determined.
* The procedure that will develop this structure must be selected.

Mechanical Properties of Materials

The first step in selecting a conditioning process is to determine the appropriateness of a material for a job. The material must be matched to the conditions

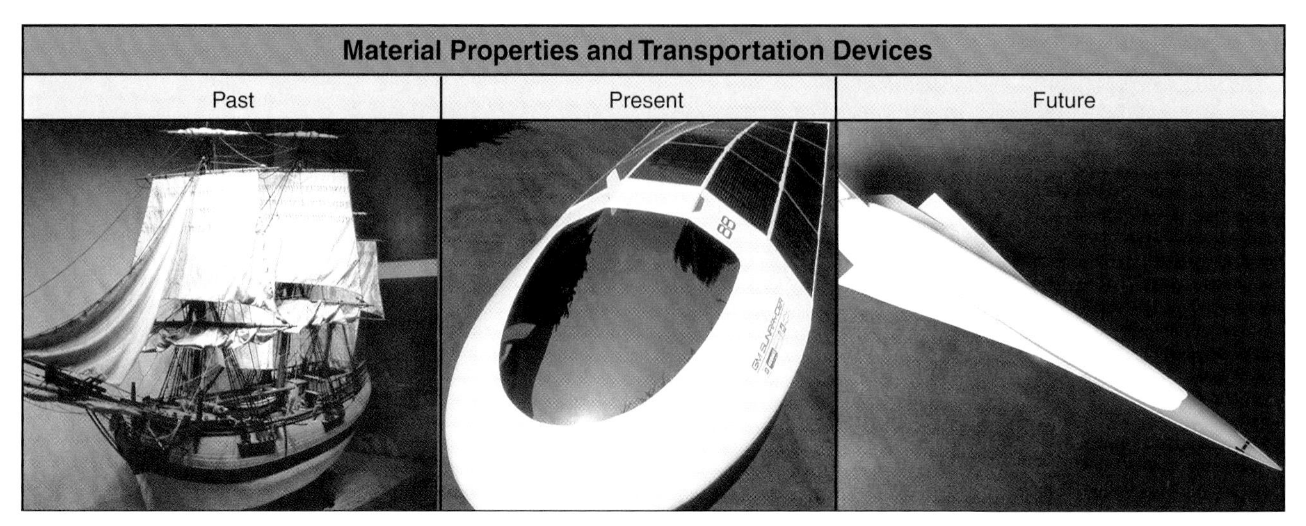

Figure 13-1. Look at these transportation devices. What material properties do you think were considered in designing: Left–The sailing ship? Center–The frame for the Sunraycer solar-powered car? (General Motors Corp.) Right–The hypersonic aircraft? (NASA)

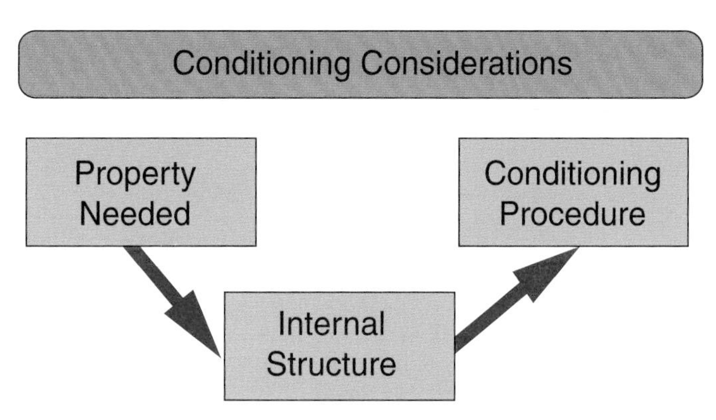

Figure 13-2. Certain considerations must be made before choosing a conditioning process.

under which it must function. Each of its many properties must be compared to the operating conditions. This will ensure that the product produced from the material will operate in the environment. It must also perform its task successfully.

In Chapter 5, you learned that all materials have a number of specific properties that can be grouped into seven categories: physical, mechanical, chemical, thermal, electrical and magnetic, acoustical, and optical. Some of the specific properties under each of these categories are listed in **Figure 13-3.**

Conditioning processes primarily deal with mechanical properties. These are the properties that resist mechanical forces or load. The major mechanical forces are:

- **Compression.** This is the force that squeezes or crushes a material.
- **Tension.** This is the force that pulls or tears a material apart.
- **Torsion.** This is the force that twists a material around an axis.
- **Shear.** This is the force where opposing forces fracture a material along a plane.

The strength of materials is measured by the amount of force needed to break or rupture them. Parts that need to withstand any or all of these forces must be strong. Chains, tractor hitches, and engine parts are examples of items that must have high strength.

Other mechanical properties include hardness, softness, ductility, brittleness, elasticity, and stiffness. Some of these properties can also be changed through conditioning.

A material is considered to be hard if it resists denting and scratching. ***Hardness*** is important for parts that will be subjected to friction or wear (rubbing, turning, or sliding). Bearings that fit around shafts and allow them to turn more freely must be very hard. Conditioning prepares them for this task. Likewise, cutting tools must be hard.

The opposite of hardness is ***softness.*** Soft materials will scratch and dent easily. Most solid wood products are considered to be soft. This quality allows wood to be easily cut and sanded.

Another important mechanical property is ***ductility.*** A material is ductile if it can be deformed (bent, compressed, or stretched) to a considerable extent without breaking. At the same time, the material structure must hold a fairly heavy load while being deformed. For instance, chewing gum would not be considered ductile even though it can be drawn out. It cannot hold up to a load. Ductility allows materials to be bent, drawn into wire, and formed using various mated dies. Generally, the harder a material is made, the less ductile it becomes.

Nonductile materials are described as ***brittle.*** A brittle material breaks with little or no elongation (stretching). Glass and many clay (ceramic) products are considered to be brittle.

Elasticity is the ability of a material to be deformed and return to its original shape when the force is removed. An elastic material can be flexed (stretched) many times without changing its shape. Elasticity is an important property for springs, vehicle tires, and basketballs.

Stiffness is the mechanical property that describes a material that resists being flexed or bent. This property is needed in diving boards, archery bows, and knife blades.

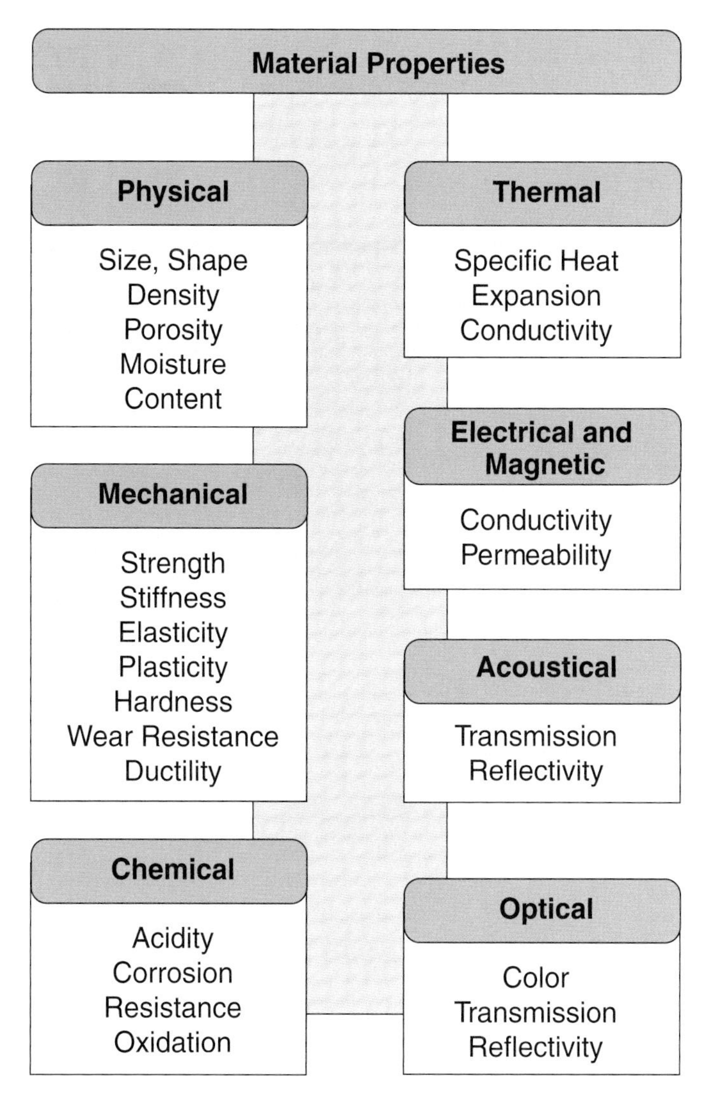

Figure 13-3. Properties of materials determine the choice of conditioning processes.

Structure and Material Properties

The specific mechanical properties a material exhibits are a direct result of its physical structure. Metals form crystal structures when they cool. Genetic factors and growing conditions determine the structure of wood. Ceramic materials are complex, crystalline structures. Composite materials are composed of a filler in a matrix. Unique structures account for many of the differences among these groups of materials.

The properties of a material can be altered only if the structure is changed. Therefore, once the desired property is established, the structure that will provide this property must be determined. A conditioning process is then selected to develop the appropriate structure.

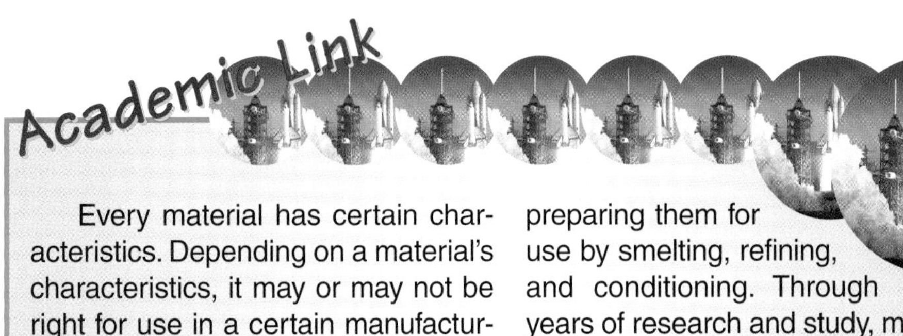

Every material has certain characteristics. Depending on a material's characteristics, it may or may not be right for use in a certain manufacturing process. These characteristics are determined by many factors. As stated in the previous section, the structure of a material accounts for many of the differences in materials.

Metallurgy is the science of separating metals from their ores and preparing them for use by smelting, refining, and conditioning. Through years of research and study, metallurgists have developed many methods of conditioning metals. These conditioning processes are used to change the structure of a metal so that select properties (hardness, ductility, brittleness, etc.) are right for a given manufacturing process.

Types of Conditioning Processes

Materials are conditioned for a number of reasons. The most important of these are to:

- Make materials easier to form or separate.
- Remove internal stress that has built up during casting, forming, separating, and assembling processes.
- Develop specific mechanical properties.

Several types of processes may develop changes in internal structure. These, as shown in **Figure 13-4,** are thermal (heat) conditioning, chemical conditioning, and mechanical conditioning.

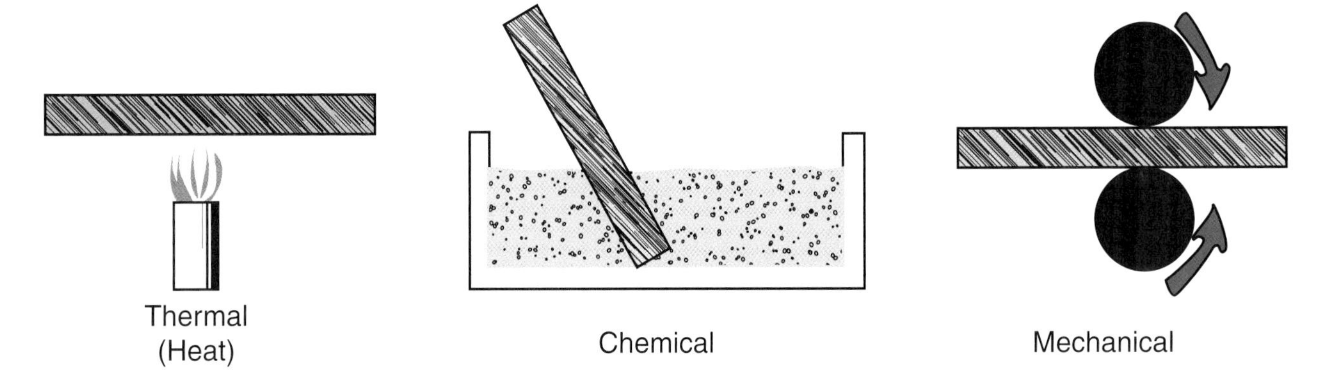

Figure 13-4. These diagrams illustrate methods of conditioning materials.

Thermal Conditioning

thermal conditioning. A process that changes the internal structure of a material through controlled heating and cooling.

Thermal conditioning changes the internal structure of materials through controlled heating and cooling. These processes can be used to change the hardness of a material, remove internal stress, or change the moisture content.

There is a number of different thermal conditioning techniques. Three important ones are drying, heat treating, and firing.

drying. A process that removes moisture from a material. Drying can happen naturally or can be aided by applying heat.

Drying

Drying removes moisture from a material. It is used to solidify clay slip by changing a liquid suspension into a solid. The physical property of the material is changed.

A widely used drying sequence reduces moisture in wood. Wood expands, contracts, and warps (twists or bends) as its moisture content changes. The wood is often stabilized by drying it using a process called ***seasoning.***

Two types of seasoning are used. One is natural, or air-drying. The lumber is carefully stacked. Stickers (spacers made from strips of wood) are placed between each layer so air can circulate. See **Figure 13-5.** The wood will naturally lose some of its moisture content this way. The lowest level that can be obtained with air-drying is about 15%.

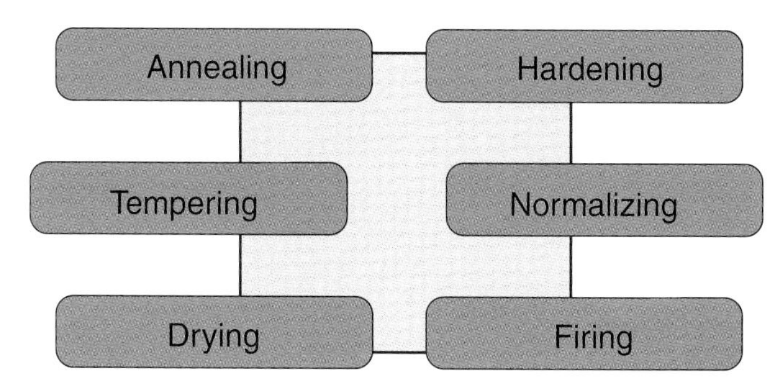

Figure 13-5. Lumber being air-dried uses spacers between boards to allow for good air circulation.

Air-dried lumber is suitable for some construction applications. However, it is not good for interior construction, cabinets, or furniture. Lumber for these uses is kiln dried. The stacks of lumber are placed in a large oven called a kiln. Air circulation, heat, and humidity are carefully controlled as the lumber is dried. Kiln-dried lumber will have moisture content of 6% to 12%.

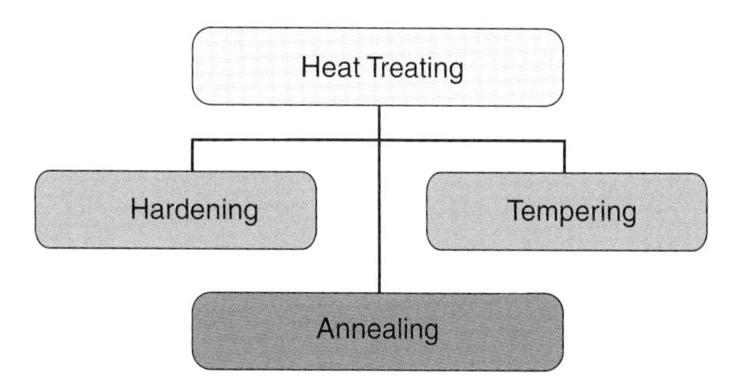

Figure 13-6. Types of heat treating.

Heat Treating

heat treating. The thermal conditioning of metals using a process of heating and cooling solid metal to produce certain mechanical properties. Heat treating includes the three major groups: hardening, tempering, and annealing.

Heat treating is thermal conditioning of metals. It is a process of heating and cooling solid metal to produce certain mechanical properties. Heat treating includes the three major processes shown in **Figure 13-6.** These are:

- *Hardening.* Increasing the hardness of a metallic material.
- *Tempering.* Removing internal stress.
- *Annealing.* Softening a material.

Steels (iron alloys) lend themselves to the widest variety of heat treatments. The carbon content determines the condition of steel after it has been heat treated. Generally, the higher the carbon content the harder steel becomes under proper heat treatment.

For Heat-Treating, a Steel Must ...

Have Enough Carbon

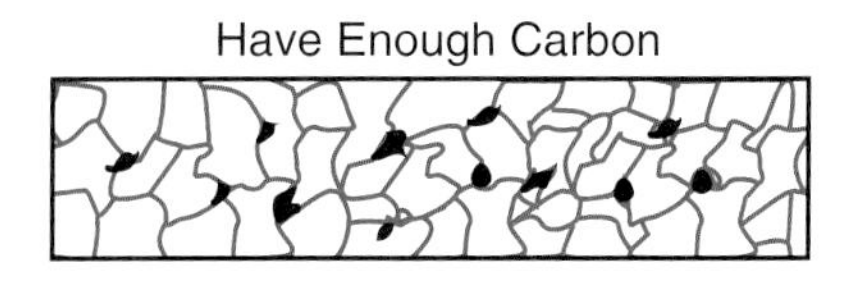

Be Heated to
Proper Temperature

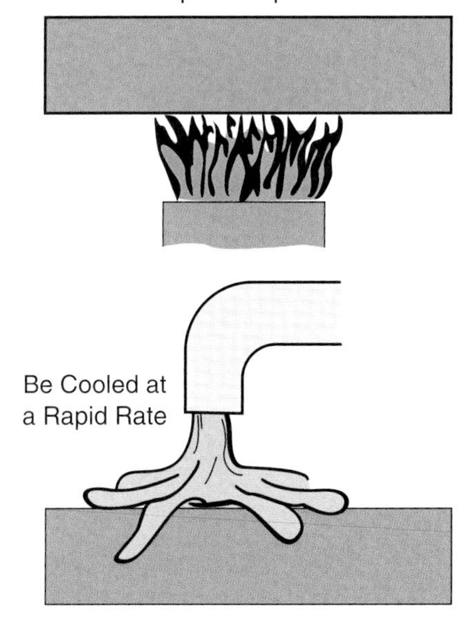

Figure 13-7. In order for steel to be heat-treated, these conditions must exist.

All heat-treating processes produce one of two results. They either harden the metal or they soften it. Often, during hardening, internal stress is produced. Softening processes (annealing and tempering) reduce this internal stress.

For many applications, such as tools, bearing surfaces, and forming dies, steel must be made hard. Hardening steel, as shown in **Figure 13-7,** requires that:

- The metal has sufficient carbon (0.80% to 1.50%). This steel is called ***tool steel.***
- The metal is heated to the proper temperature (1400° to 1500°F or 760° to 816°C). The temperature varies with the carbon content of the steel.
- The metal is cooled rapidly. Oil or water quenches are used to ensure uniform and rapid cooling.

In normal heat-treating, steel of proper carbon content is placed in a furnace. See **Figure 13-8.** Here the metal is carefully heated to the correct temperature. It is then allowed to soak so that all portions of the material reach an established temperature. The part is removed from the furnace and quickly quenched in water or oil. See **Figure 13-9.** This rapid cooling sets the internal grain structure.

The process hardens the entire part and is, therefore, classified as a full-hardening process. It produces a fine-grained steel that is hard, very brittle, and has internal stresses. For this reason, after hardening, the steel receives additional heat-treating to remove these stresses. This process is called tempering. In this stress-relieving process, the metal is heated to 600° to 1200°F (316° to 649°C). It is then allowed to cool slowly. The resulting part is somewhat less hard, but much less brittle.

hardening. A process used to make a material more resistant to denting, scratching, or other damage.

tempering. A thermal conditioning process that removes internal stresses in materials.

Tempering is also used to relieve stress in glass products. The newly cast or formed glass is slowly cooled in an oven to reduce the internal stress. These stresses can cause the product to break unexpectedly. See **Figure 13-10.**

Additional metal hardening processes produce a hard surface. These processes are called ***casehardening*** techniques. Carbon or nitrites are added to the surface of a mild steel workpiece. The part is then heated and quenched, producing a hard shell on a ductile core. The hard surface resists wear, while the soft core can absorb stress and strain.

annealing. A heat-treating process that relieves internal stresses in a part and softens any work hardening.

Metals can be softened by a heat-treating process called annealing. Annealing removes the hardness from a part that has occurred during other processing activities or that has occurred as a result of heat-treating. In annealing, the metal is heated to a specific temperature. It is then allowed to soak so that a uniform material temperature is developed. The part is then allowed to cool slowly. Often, the material is covered with sand to ensure a slow, uniform cooling cycle.

normalizing. A thermal conditioning process that relieves stress in steel and develops a fine, uniform grain structure.

Normalizing is a softening technique that removes internal stresses. It is a process that is similar to annealing. The major difference is that the metal is heated to 100° to 200°F (38° to 94°C) above the annealing temperature. As the metal cools slowly, a fine grain structure is developed.

Figure 13-8. The hot parts (top) are leaving the heat-treating furnace. They will be ground to become machine rolls (bottom). (Bethlehem Steel Co.)

Figure 13-9. These steel parts are being removed from a quench tank. (ARO Corp.)

Figure 13-10. This ribbon of glass is entering a tempering line. (PPG Industries)

Not all thermal conditioning is restricted to metals. Heat is often used to harden ceramic materials in a process called *firing*.

The ceramic material to be fired contains clay and a glass-like material called flux. When the material is heated, the flux becomes a "cement" that melts and fuses into a hard, rigid structure.

Firing is a thermal conditioning process that uses high temperatures that melts the gassy part of a ceramic. Upon cooling, the product will have ceramic particles that are bonded together with this glassy material. Firing is often done in large kilns, as shown in **Figure 13-11,** that have several zones that are held at different temperatures. The ceramic ware is loaded on carts. See **Figure 13-12.** The carts enter the kiln where the first zone heats the product and removes excess moisture. The center, hot zone fires (bakes and fuses) the clay. The third zone allows the clay to cool slowly. In this zone, the internal stresses are removed.

firing. A thermal conditioning process used on ceramics. Firing melts the glassy part of the ceramic. Upon cooling, the product will be particles held together by the glassy material.

Wood materials are also placed in kilns. They are heated slowly to remove excess moisture. When the wood cools, it is more stable and easier to work. Furniture, cabinets, and most homes are built with kiln-dried lumber.

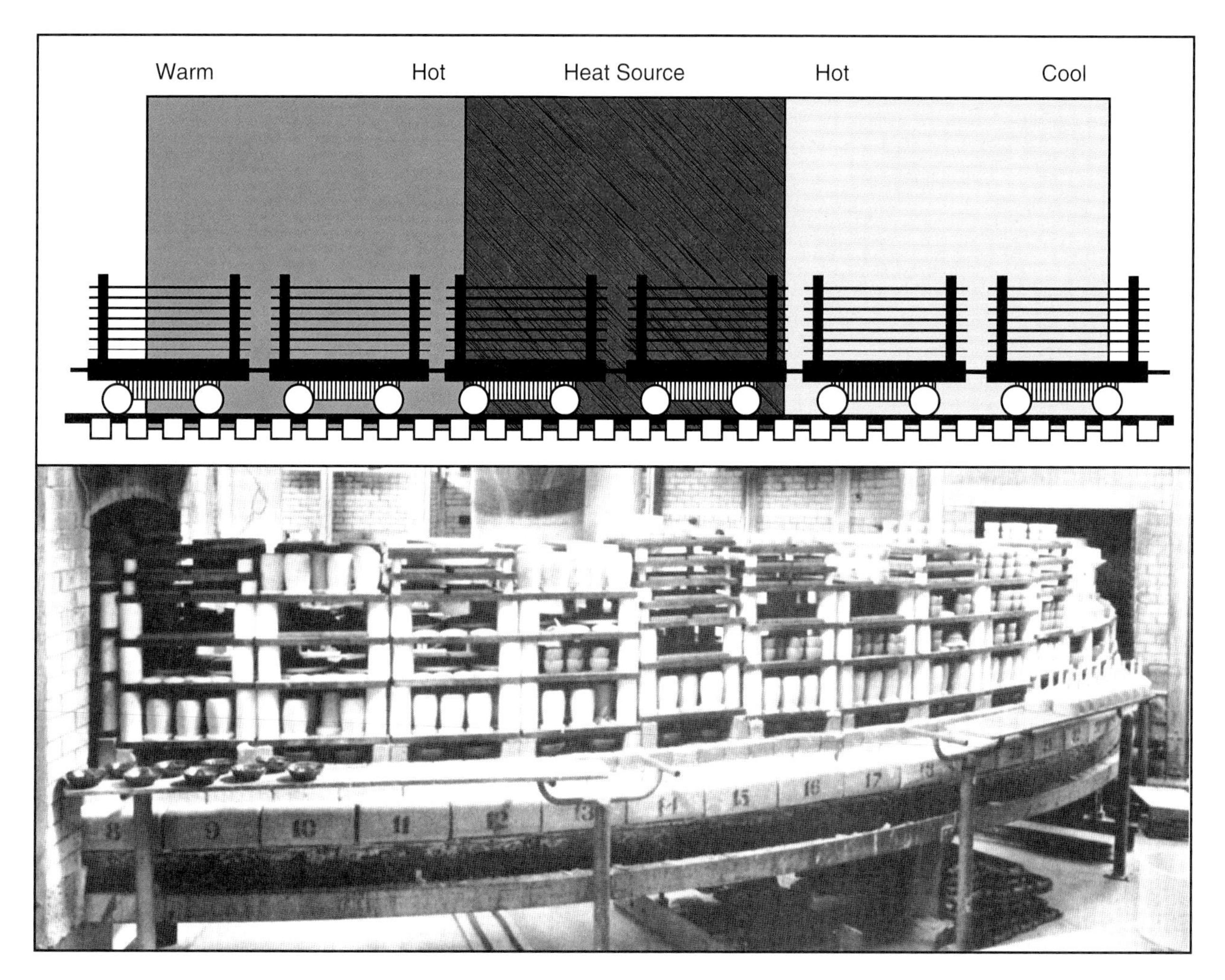

Figure 13-11. Tunnel kilns, like the one diagramed here, are used to fire ceramic products.

Figure 13-12. Loaded cars ready to enter the kiln. (Syracuse China)

Chemical Conditioning

Internal properties of materials may be changed by chemical action. This is called *chemical conditioning*. A catalyst (material that starts a chemical action) may be added to a liquid plastic material. The molecules in the plastic undergo change. They link together to form longer, more rigid polymer chains. This action is called *polymerization*. The material is changed from a liquid to a solid.

This action is often used with plastic casting operations. See **Figure 13-13.** The result is a harder, more durable product. In many cases, a fracture-resistant material is added to the plastic to improve its usefulness and durability.

Glass lenses for eyeglasses are treated with chemicals to make them more fracture-resistant. This property is essential for safety glasses.

Animal hides are also chemically conditioned. They are treated with a series of chemicals during the tanning processes. The result is a material called leather.

Figure 13-13. The plastic material that binds the fiberglass in these hot tubs was hardened (set) by chemical action. (Rohm and Haas)

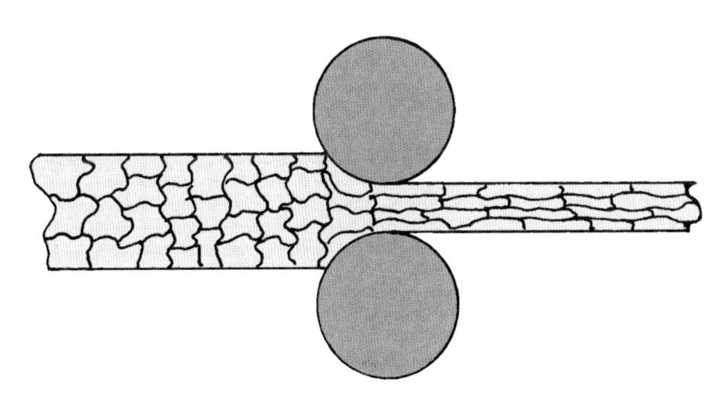

Figure 13-14. The effects of cold rolling on grain structure are shown in this diagram.

Mechanical Conditioning

Mechanical force is sometimes used to condition materials. This is called *mechanical conditioning*. Often, this action is an unwanted result of other manufacturing processes. During cold-forming operations, the material is subjected to rolling and hammering forces. See **Figure 13-14.** These forces change the structure of ductile metals into long, closely packed grains of hard metal. This change is called *work hardening*. Although this change is not primarily the purpose of forming and machining processes, it must be considered.

Shot peening is an intentional mechanical conditioning process that sprays small steel balls (shot) against parts. The action produces a series of small depressions (dents) in the metal. This increases the ability of the material to resist fatigue. Shot peening is widely used to condition springs. After shot peening, a spring can withstand a greater number of flexures (stretches) before it fails.

chemical conditioning. **A method of changing the internal properties of a material through chemical reaction.**

polymerization. **Linking together of chain-like molecules of a plastic, changing it from a liquid to a solid.**

mechanical conditioning. **Using physical force to modify the internal structure of a material.**

work hardening. **A structural change in metal caused by such processes as hammering or rolling.**

shot peening. **A process that sprays tiny steel balls against metal parts. The balls pit the surface of the metal and increase its fatigue-resistance.**

Summary

Not all manufacturing processes are designed to change the visible properties of a material. Conditioning processes use heat, chemicals, or mechanical force to change the internal properties of materials. These processes make materials:

- Harder so they can withstand wear.
- More ductile to allow for easy forming.
- Stronger, more elastic, or stiffer to meet specific product demands.

Safety with Conditioning Processes

- Do not attempt a process that has not been demonstrated to you.
- Always wear safety glasses or goggles.
- Wear protective clothing, gloves, and face shield when working around hot metal or other materials.
- Do not leave hot products or parts where other people could be burned.
- Constantly monitor material temperatures during conditioning processes in which heat is used.
- Use care when working with chemicals used for conditioning.
- Use a holding tool such as pliers or tongs to hold metals that are being heat treated.
- Use a spark lighter to light a heat-treating furnace.
- Stand to one side of the quench solution when quenching hot metals.
- Place a "Hot Metal" sign on any parts that are air cooling.

Key Words

All of the following words have been used in this chapter. Do you know their meanings?

annealing
chemical conditioning
drying
firing
hardening
heat treating
mechanical conditioning
normalizing
polymerization
shot peening
tempering
thermal conditioning
work hardening

Test Your Knowledge

Please do not write in this text. Place your answers on a separate sheet.

1. Processes that use heat, mechanical force, or chemical action to change the mechanical properties of a material are called _____.
2. List three factors that must be considered when selecting and using conditioning processes.
3. *True or False?* The specific mechanical properties a material exhibits are a direct result of its physical structure.
4. List three reasons materials are conditioned.
5. Name and describe six of the most common thermal conditioning practices.
6. *True or False?* Annealing and tempering increase internal stresses in materials.
7. Polymerization is an example of _____.
   a. thermal conditioning.
   b. chemical conditioning.
   c. mechanical conditioning.
   d. None of the above.
8. Shot peening is an example of _____.
   a. thermal conditioning.
   b. chemical conditioning.
   c. mechanical conditioning.
   d. None of the above.

Applying Your Knowledge

Be sure to follow accepted safety practices when working with tools. Your instructor will provide safety instructions.

1. Heat-treat a material. This can be a product you have cast or formed in earlier activities. For instance, you may want to heat-treat the screwdriver or tack puller blade produced in the forming activity.
   To harden a metal:
   a. Bring the heat-treating furnace up to the temperature desired. Caution: If the furnace is a gas-fired unit, follow the manufacturer's instructions in lighting it. Stand to one side and avoid looking into the fire box.
   b. Heat the metal to its critical temperature.
   c. Remove the part with preheated metal tongs. Caution: Wear protective clothing including face mask, gloves, and apron.
   d. Quench (cool) the part in the cooling solution (usually water). Twirl the part so it will cool quickly and evenly.
   e. Test the hardness with a file.
   To temper the metal:
   a. Polish about an inch of the point or edge of the part. (This will make it easier to see the change of colors as the part is heated.)
   b. Heat the metal slowly and evenly behind the polished surface as heating progresses. When the proper color appears, quench the part as in hardening. Observe the same safety safeguards as in hardening.
   There are other methods and procedures for heat-treating. Discuss the preferred method with your instructor and, if time permits, research other methods.
2. Using a common, inexpensive hacksaw blade, cut a piece of metal. Try to bend it. What happened? Anneal one piece of the metal. Using a file, cut the original piece and the annealed piece. Compare your results. Make a poster to show what you learned.

Statisticians

Statistics is the application of mathematical principles to the collection, analysis, and presentation of numerical data. Statisticians contribute to scientific inquiry by applying their mathematical knowledge to the design of surveys and experiments; collection, processing, and analysis of data; and interpretation of the results. Statisticians often apply their knowledge of statistical methods to a variety of subject areas, such as biology, economics, engineering, medicine, public health, psychology, marketing, education, and sports. Many applications cannot occur without the use of statistical techniques, such as designing experiments to gain federal approval of a newly manufactured drug.

In business and industry, statisticians play an important role in quality control and product development and improvement. In an automobile company, for example, statisticians might design experiments to determine the failure time of engines exposed to extreme weather conditions by running individual engines until failure and breakdown. At a computer software firm, statisticians might help construct new statistical software packages to analyze data more accurately and efficiently. In addition to product development and testing, some statisticians also are involved in deciding what products to manufacture, how much to charge for them, and to whom the products should be marketed. Statisticians may also manage assets and liabilities, determining the risks and returns of certain investments.

Chapter 14 Assembling Processes

Objectives

After studying this chapter, you will be able to:

✓ Describe how assembly processes have evolved.
✓ Explain how bonding takes place.
✓ List five categories of bonding methods.
✓ Describe how mechanical fasteners are used in assembling processes.
✓ Give examples of permanent fasteners, semipermanent fasteners, and temporary fasteners.
✓ List five basic forms of joints.

Think of the many products you use every day. Few are made from a single part or component. Other than nails, pins, paper clips, and buttons, few other products are made of a single part or component. Most products are composed of a series of individual parts. A common lead pencil, for instance, has at least five parts. See **Figure 14-1.** The body is made up of two wood parts bonded around the graphite "lead." The eraser is held with a metal band.

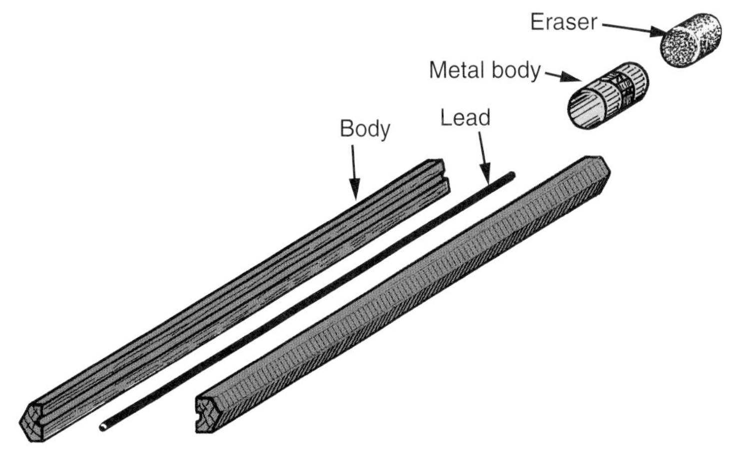

Figure 14-1. A simple lead pencil is assembled from at least five separate parts.

Assembled products come in all sizes, as illustrated in **Figure 14-2.** Some can be held in the palm of your hand, while others can carry thousands of passengers across the sea.

All multipart products are put together using assembling processes. ***Assembling processes*** are all practices used to temporarily or permanently attach parts to form assemblies or products.

Assembly uses two major types of processes. These, as shown in **Figure 14-3,** are bonding and mechanical fastening. Bonding permanently fastens parts using heat, pressure, and/or a bonding agent. *Mechanical fastening* temporarily or permanently holds parts together using mechanical fasteners or mechanical force.

mechanical fastening. **A method of holding parts together temporarily or permanently with mechanical fasteners or force.**

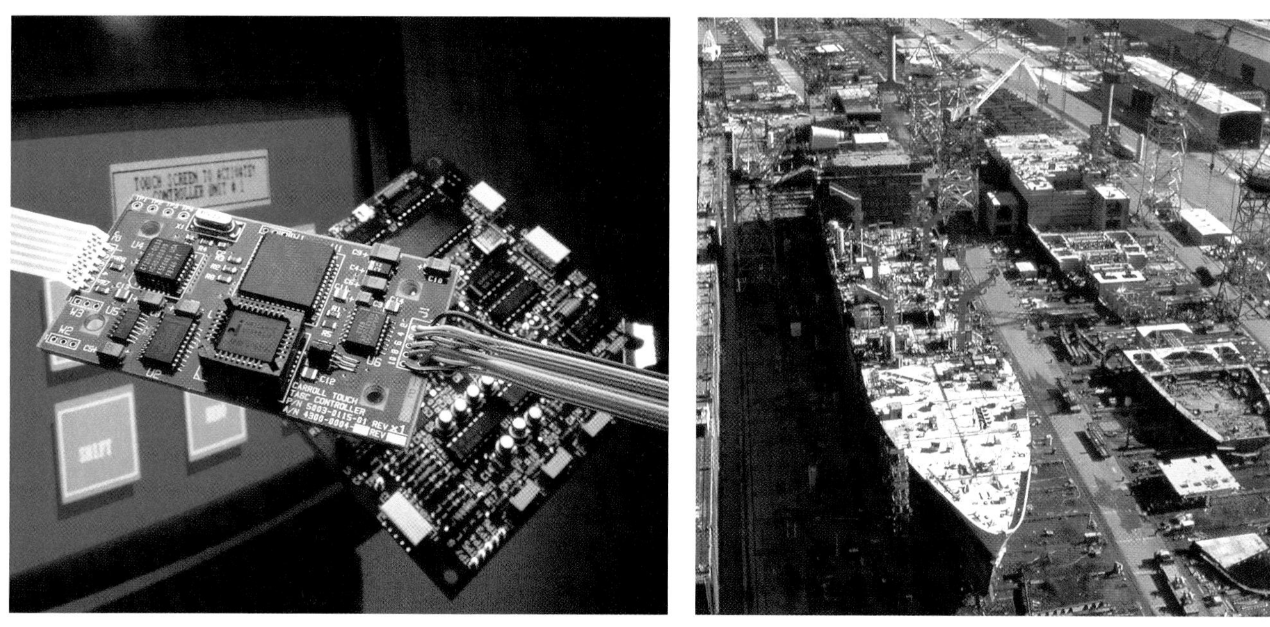

Figure 14-2. Left—Manufactured products may be small, complex assemblies like the computer circuits. (AMP Inc.) Right—They can also be huge ships that are made of tons of parts and subassemblies. (Colt Industries)

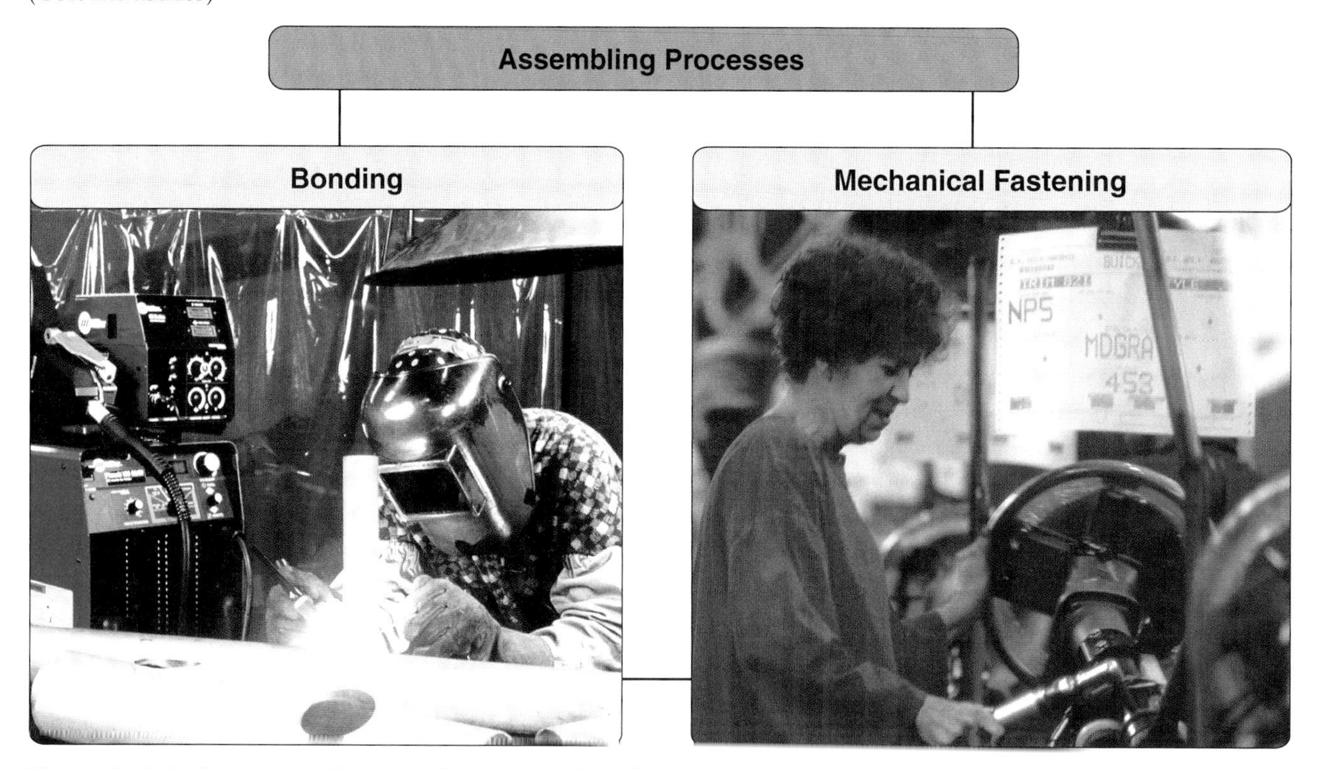

Figure 14-3. Left—Assembling practices can use bonding techniques such as gas tungsten arc welding. (Miller Electric Mfg. Co.) Right—They can also use mechanical fasteners, such as bolts and screws. (General Motors Corp.)

Development of Assembling

Assembling practices are about as old as civilization. At one time humans used animal tendons to sew animal skins together to form clothing and shelters. Vines were used to lash stones onto sticks to make axes. Early assembly practices involved techniques that were like present-day sewing and lacing.

As civilization developed, humans discovered new ways to improve assembled products. By Pharaoh Tutankhamen's (King Tut's) time, wood glues were in use by Egyptians. Made from animal parts, these adhesives were used well into the 20th century. Wood products have been found in which this glue is still holding after 3300 years.

By 600 B.C., the Greeks had developed several types of metal clamps. These were used to hold building stones in place. During the European Dark Ages, cabinetmakers were using complex joints to assemble wood furniture and cabinets.

By 1786, a machine was developed to mass-produce nails. Before this machine was invented, blacksmiths forged the nails one at a time. Later, machines that produced bolts, screws, and other mechanical fasteners were developed.

Early artisans found that metal could be assembled without a fastener. By heating two metal parts and hammering them together, the parts could be welded into one assembly. This forge-welding practice was used by colonial blacksmiths to produce wagon tires (the steel rim that went around the wood wagon wheel). Later, heat from burning gases and electric arcs was used to weld metal parts together. See **Figure 14-4.** From these and other earlier discoveries, a wide array of assembling practices has developed.

Figure 14-4. Electric arc welding was first attempted in 1890 with a bank of lead storage batteries supplying the electrical power. (Miller Electric Mfg. Co.)

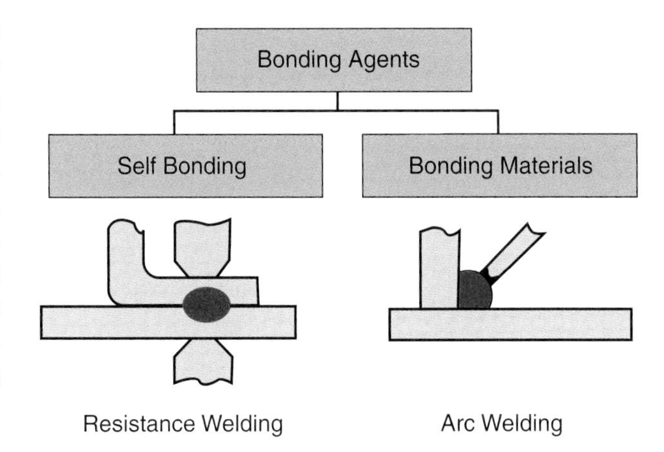

Figure 14-5. Self bonding or bonding materials are used as bonding agents.

Assembling by Bonding

Bonding techniques use either cohesive (self bonding) or adhesive bonds (bonding materials) to permanently hold parts together. See **Figure 14-5.** *Cohesive bonds* are the same forces that hold the molecules of a substance together. For example, cohesive bonds hold molecules of steel in crystals, and these crystals make up the solid metal.

Adhesive bonds use the tackiness or stickiness of a substance to hold parts together. Adhesive bonds are used to glue wood parts together to form furniture.

Three essentials are considered when parts are bonded into assemblies. These are:

- The *agent* (or substance) used to bond the parts together.
- The *method* selected for creating the bond.
- The type of *joint* needed to provide a strong bond between parts.

cohesive bonds. **A means of joining materials that uses the same forces that hold the molecules of the substance together.**

adhesive bonds. **A means of joining materials that makes use of the stickiness or tackiness of the bonding substance.**

Bonding Agents

atomic closeness. In cohesive bonding, atoms of each part must have the same atomic closeness (be as close together) across the joint as they do within the materials themselves.

fusion bonding. A bonding process that uses the same material as the base metal to create the weld. Thick parts require the use of a filler rod. This rod is made of the same material as the base metal. It provides more metal to produce a strong weld. The heat is provided by burning gases or by electric sparks.

Cohesive bonding requires *atomic closeness* between parts. This means that the atoms of each part must be as close to each other across the joint as they are within the parts. Under normal conditions, that closeness is not possible. Simply laying two parts together will not create a bond. Heat and/or pressure are used to produce the required atomic closeness for bonding.

Adhesive bonding requires that a film of a tacky substance be applied on the surfaces of parts that will contact one another. This material is usually an organic substance or a metal. When the parts are positioned and clamped, a bond will result. Organic adhesives form polymer chains that extend from one part to the other across the joint. The strength of the bond is related to the number and strength of these chains. Metal adhesives (solder and bronze) form a bridge between the two parts.

Most bonding processes add a bonding agent to improve the strength of the joint. These agents can be grouped under three classes:

- The *same material* as the material in the parts. For example, the parts, themselves, may be melted to produce a bond as in fusion welding.
- The *same general type of material* but different composition. For example, the agent may be a different metal as in soldering and brazing.
- A *totally different material* with special bonding properties. For example, woodworker's glue may be used to assemble a wood cabinet.

The type of bonding material (agent) and practice used will depend on the material being assembled. It will also depend on the conditions under which the assembly must function.

Methods of Bonding

A large number of bonding methods are used to assemble wood, plastic, ceramic, metal, and composite materials. Each technique differs in the bonding agent used and the methods used to apply the agent. See **Figure 14-6.** These techniques can be grouped under the five categories of fusion bonding, flow bonding, pressure bonding, cold bonding, and adhesive bonding.

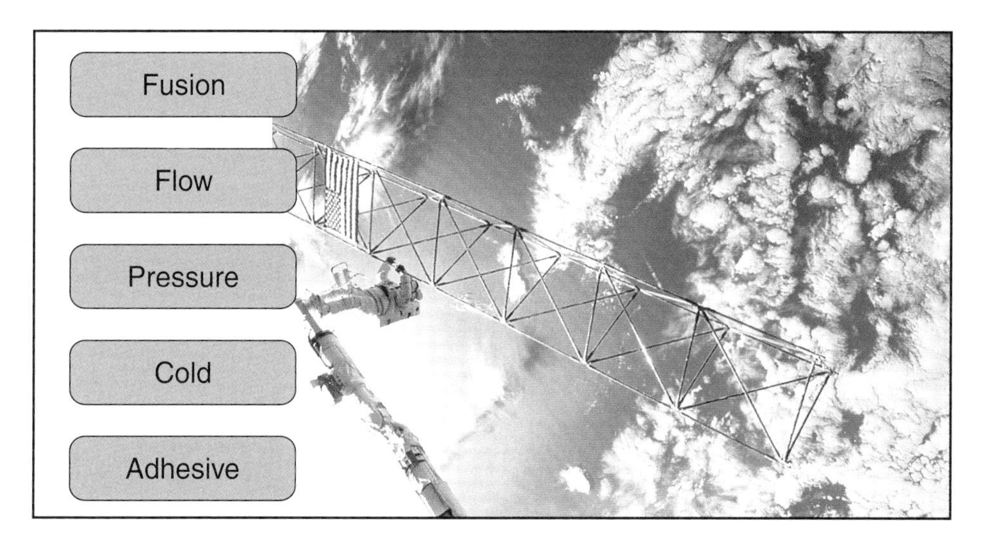

Figure 14-6. These are the five types of bonding. How many of these do you think were used on this space structure? (NASA)

Fusion Bonding

Fusion bonding melts the edges of parts to be joined. The molten material is allowed to flow between the pieces to create a bond. The result, as shown in **Figure 14-7,** is an assembly that looks and acts as a single piece of material.

Fusion bonding can be used to bond ceramics, plastic, and metal materials. This process is called *welding* when it is used on metals. However, other welding processes do not fit the fusion bonding category. Resistance welding and inertia welding are examples that fit other bonding categories.

The bonding agent, in metal fusion bonding (fusion welding), may be the part itself. The edges of the parts may be melted and caused to flow together. However, with materials thicker than 1/8", a filler material (rod) is generally used. The filler material is made up of metal that is similar to the base material.

The heat required to melt the base material and filler rods is provided by one of two basic sources, burning gases or electric arcs. The most common fusion welding techniques are named after these heat sources.

In most *gas welding*, heat is generated by burning a mixture of oxygen and acetylene. This is referred to as *oxyacetylene welding*. The gases, as shown in **Figure 14-8,** are kept in individual pressure tanks. These gases are allowed to flow through pressure-reducing valves to a torch. Inside the torch, the gases are mixed in specific proportions. Control valves on the torch set the percentage of each gas in the mixture. The gas mixture flows out of the torch tip where it is ignited and burns at about 6000°F (3316°C).

This temperature is sufficient to melt and fuse most common steels, aluminum, cast iron, and other metals. Most oxyacetylene welding is used for small quantity production, artist's work, and machine repair. The process is slow and requires considerable operator skill.

Most production fusion welding is done with electric arc systems. The simplest is called *arc welding*. See **Figure 14-9.** This process uses high amperage, low voltage, alternating (AC) or direct (DC) current.

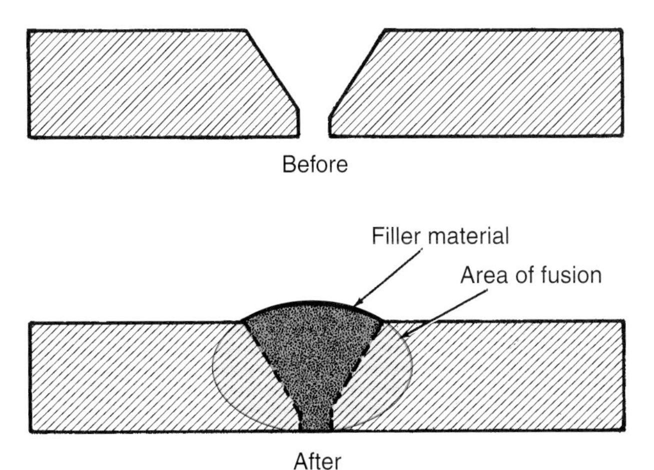

Figure 14-7. Fusion bonded metal appears to be one piece after the bonding operation.

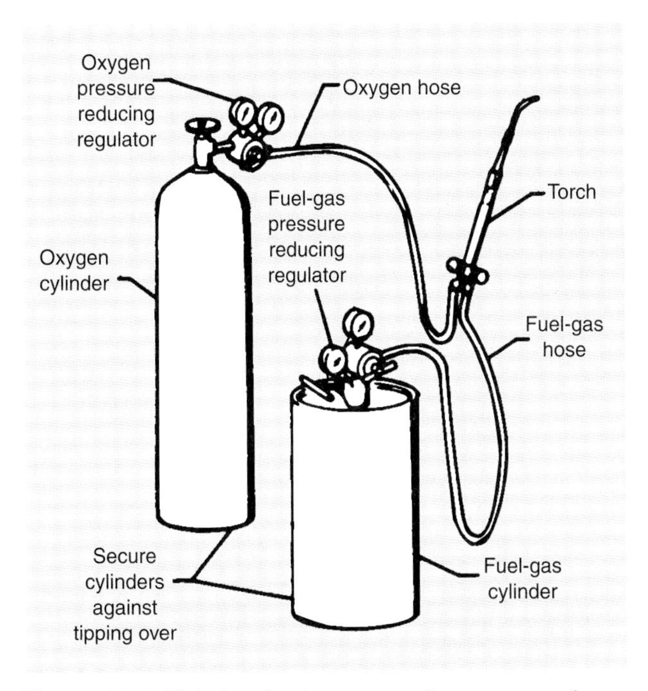

Figure 14-8. This is a basic oxyacetylene system for gas welding. (Linde Div., Union Carbide Corp.)

Whenever electricity jumps a gap, heat is generated. Arc welding uses this fact of physics to produce welding temperatures. One side of an electrical source is attached to the work or a metal table on which the work is placed. The other lead is attached to the welding rod in a rod holder. When the rod is brought near the work, an arc is developed. This electric spark produces one of the hottest sources of energy known—about 11,000°F (6093°C). (Only nuclear reactions are hotter.) This temperature can easily melt both the filler rod and the base metal. The metal from these sources flows together forming a permanent, cohesive bond between the parts.

Changes to arc welding processes have been developed. The most important of these is ***shielded arc welding.*** See **Figure 14-10.** Shielded metal arc welding uses an electrode that is a filler rod with a chemical coating. Some of the

gas welding. **A means of bonding metal, using the heat of an oxygen/acetylene flame to melt arid fuse the materials.**

arc welding. **A means of bonding metal, using heat developed by an electric arc to melt and fuse the materials.**

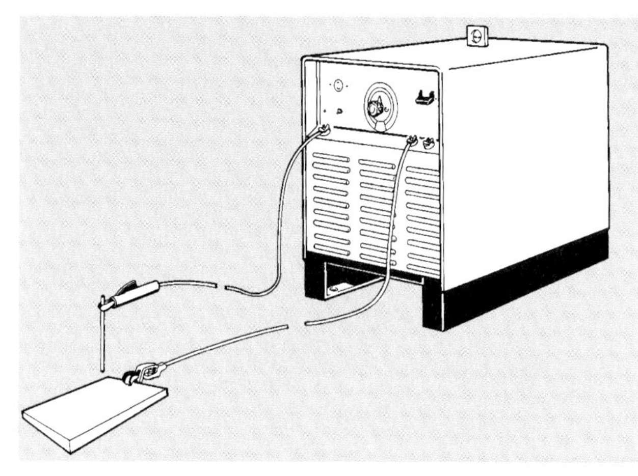

Figure 14-9. A basic arc welding system. (Miller Electric Mfg. Co.)

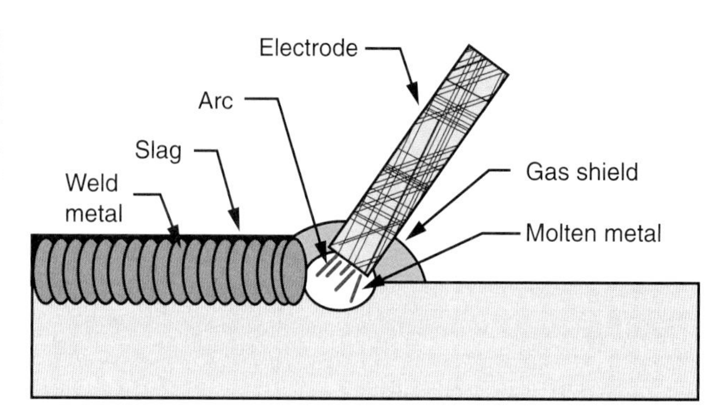

Figure 14-10. This diagram illustrates shielded arc welding. If the gas shield were removed, it would be normal arc welding.

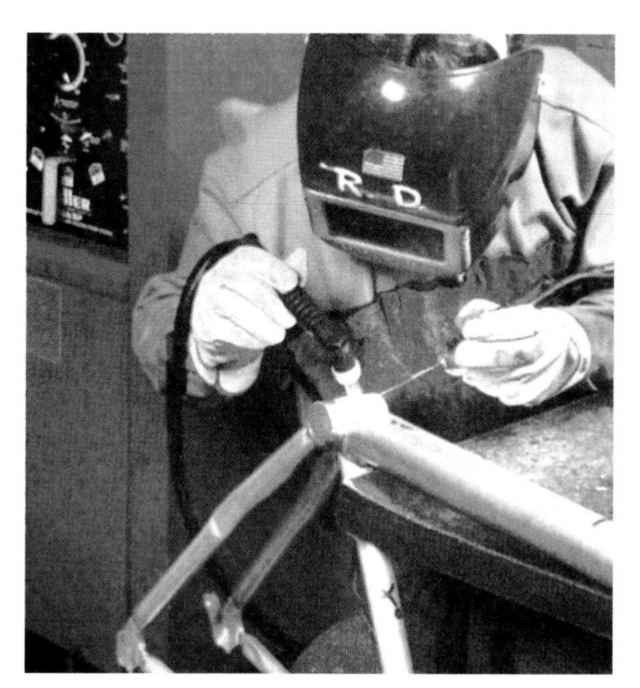

Figure 14-11. This worker is welding an aluminum with an inert gas system. (Miller Electric Mfg. Co.)

coating breaks into a cloud of protective gas when the arc is struck. This cloud provides a shield around the weld area.

Other processes use a gas that is carried from a storage cylinder to the weld area. This process is called inert-gas welding. The most common of these is ***gas tungsten arc welding* (GTAW)**. This process is also called tungsten inert gas (TIG) welding. See **Figure 14-11.** Gas tungsten arc welding produces an arc between the work and one or more tungsten electrodes. In both of these cases, the shield protects the weld area from impurities in the air. The weld area is also protected by an inert gas.

Arc welding is used extensively for repair and maintenance work and in assembling heavy components for agricultural and construction equipment. Shielded arc welding is used in assembling nonferrous metal and specialty steel parts.

A process that is very similar to fusion bonding is *solvent bonding*. In this process, a solvent, instead of heat, is used to cause materials to flow together. Acrylic plastic parts can be bonded in this manner. The edges of parts are coated with a solvent that softens the surface. The parts are pressed together and the surfaces flow together. The solvent is allowed to evaporate resulting in a rigid assembly.

solvent bonding. A joining process for plastics that uses a solvent to soften the edges of the pieces to be bonded.

Solvent processes can also assemble ceramics. For example, the handles of china cups are attached to the body using this process. The ends of the handles are dipped in a slip (the same clay water mixture used to mold the handle). The liquid slip softens the handle that is pressed to the cup body. The soft clay dries, leaving the handle permanently attached to the cup body.

flow bonding. A method of joining materials that uses a filler metal that melts onto a heated base metal. When cool, the filler metal acts as an adhesive.

Flow Bonding

Flow bonding heats, but does not melt the base metal. See **Figure 14-12.** Then a dissimilar filler metal is melted onto the hot base material. The filler metal forms a bridge between two parts with adhesive forces.

In flow bonding, a clean, close-fitting joint is produced. The parts are cleaned and often coated with liquid or paste flux. This is a material that cleans the metal to improve the flow and bonding characteristics of the filler material. The parts are assembled and heated. Then one of several filler materials is melted on the joint. Capillary action (the same action that causes water to move uphill through a paper towel) causes the filler material to flow between the parts. Upon cooling, the filler material forms a strong, adhesive bond with the base material.

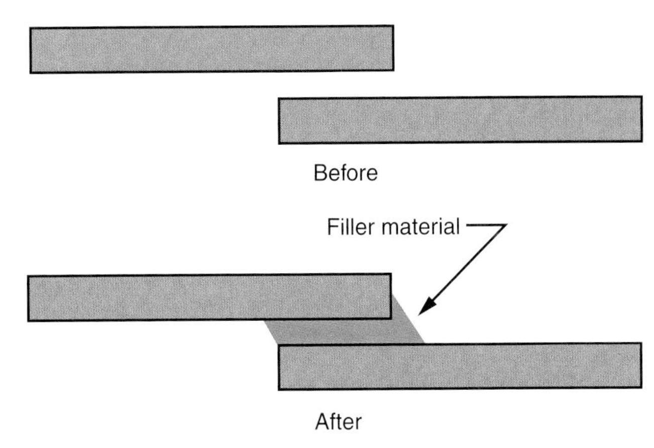

Figure 14-12. Flow bonding does not melt the metal in the parts being joined.

The two common flow bonding techniques are brazing and soldering. The difference between these processes lies in the temperatures required and in the bonding material used.

Temperatures between 1000° and 2500°F (538° and 1371°C) are used in brazing. This is a wide range of temperatures. The base metal and the filler material being used determine the specific temperature used. The most common brazing metals (filler materials) are copper, copper alloys (bronze), silver alloys, and aluminum alloys.

Heat for brazing processes is produced in three different ways:

- Placing the flame of an oxyacetylene torch on the parts.
- Placing the parts in a furnace or an oven until the filler metal melts.
- Dipping the parts into a hot bath of filler metal or flux.

The first method is similar to oxyacetylene welding. The flame heats, but does not melt, the parts. A hot filler rod is dipped in powdered flux. The rod is then heated with the flame. The flux melts and flows onto the joint. Then the rod melts and follows the flux into the joint. Upon cooling, the braze metal forms the bond between the parts.

When a furnace is used, the filler metal is preformed into discs or flat wafers. The base metal is cleaned and the filler metal is placed in the joint area. The assembly is clamped into position and put in the furnace. The heat melts the filler material that forms the bond.

The dip method is similar to the furnace method. A molten bath is used to heat the metal. In a *filler material bath,* the metal both heats the parts and forms the bond. In a flux *bath,* the parts and preformed filler are clamped together. The assembly is placed in the bath where it is heated and the bond is formed.

Brazing is used in repair and maintenance work, in fabrication of copper and brass products, and in assembling thin sheet and tubing parts. Artists also use brazing in developing metal sculptures.

Soldering is much like brazing except soldering uses an alloy of tin and lead, called *solder,* as a bonding agent. Soldering also uses a lower temperature, generally below 600°F (316°C).

The melting temperatures can be obtained with electrically heated guns and irons, **Figure 14-13,** propane torches, and solder baths. The guns and torches are widely used in electrical component repair while the baths are used in the production of electronic circuit boards.

brazing. **A method of joining metals that involves heating them and melting a filler material that will act as an adhesive.**

soldering. **A process similar to brazing, but generally performed at temperatures below 600°F.**

Figure 14-13. An electronic device is being repaired with a soldering iron. (Xerox Corp.)

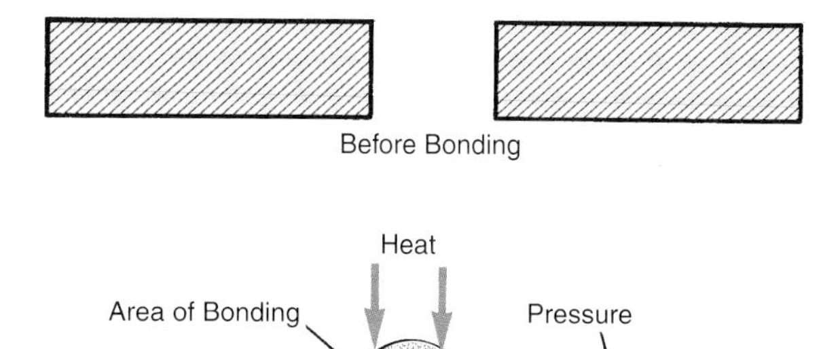

Figure 14-14. Pressure bonding uses heat and pressure but no filler material, to form the bond.

Pressure Bonding

Since *pressure bonding* uses both heat and pressure to develop a bond, the term pressure bonding can be somewhat confusing. See **Figure 14-14.**

Pressure alone can produce a bond. However, the great amount of force required reduces its commercial usefulness. By using heat, along with pressure, the weld can be made with less force.

In pressure bonding, the base materials (the parts) form the bonding agent. No additional filler material is added.

The most widely used pressure bonding practice is ***resistance (spot) welding,*** See **Figure 14-15.** In this practice, sheet, rod, or band steel is placed between two electrodes. Then a four-step cycle is started:

1. The electrodes are closed to apply pressure to the metal (squeeze time).
2. Electrical current is caused to flow between the electrodes and through the parts (weld time). During the weld time, a spot (kernel) of molten metal is formed between the two parts and in line with the electrodes.
3. The current is stopped and, while the pressure is maintained, the melted spot hardens (hold time).
4. The electrodes are released from the work (off time).

Resistance welding is widely used to assemble sheet metal parts. This process is common in the home appliance and automotive industries. See **Figure 14-16.** Probably more metal is bonded by resistance welding than in any other welding process.

pressure bonding. A method of joining materials that uses both heat and pressure.

Another type of pressure welding *is* ***forge welding.*** In this process, heated metal is pressed or hammered together, creating a welded joint. In forge welding, the metal is heated almost to its melting point. The surfaces are placed together. Then a hammering or pressing action creates the required atomic closeness. The force and the heat fuse the parts together. Links in common chain are assembled using this operation.

A final pressure welding technique is ***inertia welding,*** which is also known as ***friction welding.*** In this process, one part is spun against the other part that is held stationary. The friction between the parts generates a great deal of heat. The heat melts the contacting surfaces of the parts. Then the rotating part is stopped and forced into the stationary part. The result is a welded assembly.

cold bonding. A means of joining ductile metals, such as aluminum or copper, by applying very high pressure in a small area.

Cold Bonding

Cold bonding, involves applying great force to create the atomic closeness needed for bonding. See **Figure 14-17.** This process is generally limited to very ductile metals such as copper and aluminum.

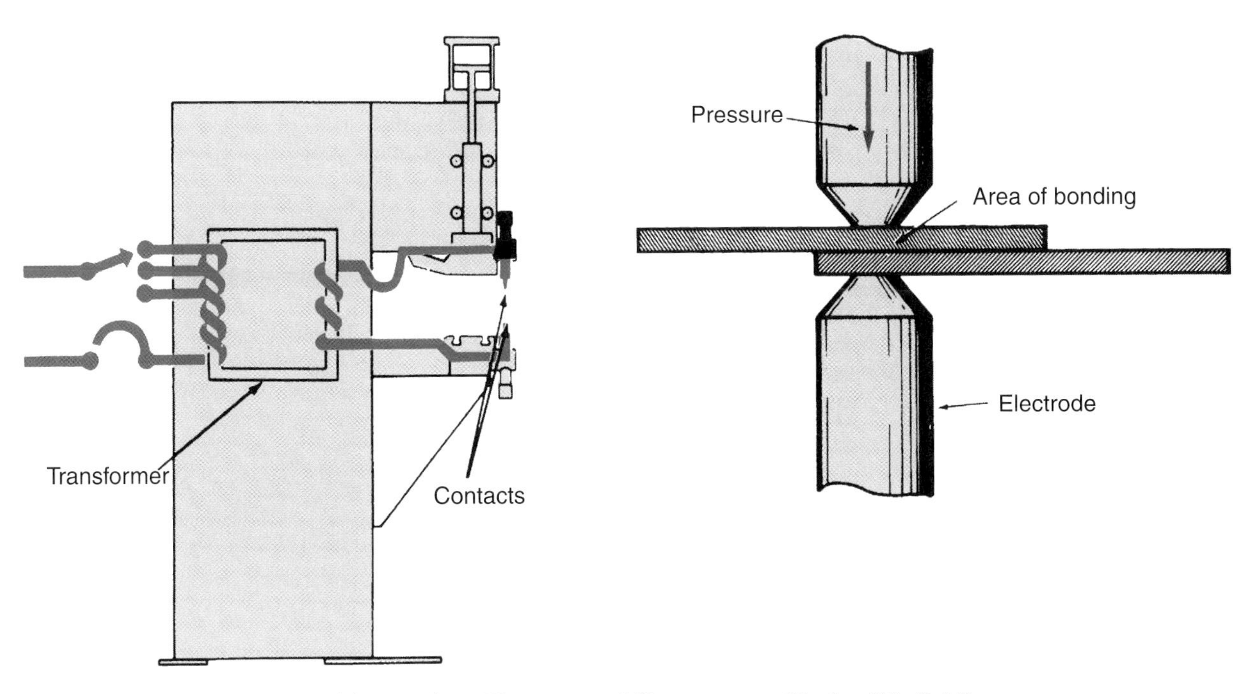

Figure 14-15. Left—A spot welder. Right—The spot welding process (Taylor Winfield)

Pressures from 50,000 to 200,000 pounds per square inch (psi) are needed to create the bond. The pressure is often developed between two aligned punches.

Pressure bonding is useful in joining light-gage tubes and splicing wires. It is also used for lap joints in thin sheet metal where indentations at the weld are acceptable.

A recent development in cold bonding process is ***explosive welding,*** which is also known as ***cladding.*** See **Figure 14-18.** In this process, two metal sheets are bonded by an explosive charge. The charge is detonated above the sheets. The shock of the explosion creates a cohesive bond between the parts. This process is used in specialized tasks such as in cladding expensive metal to inexpensive metal cores, assembling chemical process vessels, and forming tubing and channels.

Figure 14-16. A robot is welding automotive components. (Fanuc Robotics)

Adhesive Bonding

Adhesive bonding, involves permanently assembling parts using a material that has "stickiness" or "tackiness." See **Figure 14-19.** In adhesive bonding, adhesive, glue, or cement is applied to the parts to be assembled. The parts are brought together where adhesive bonds form

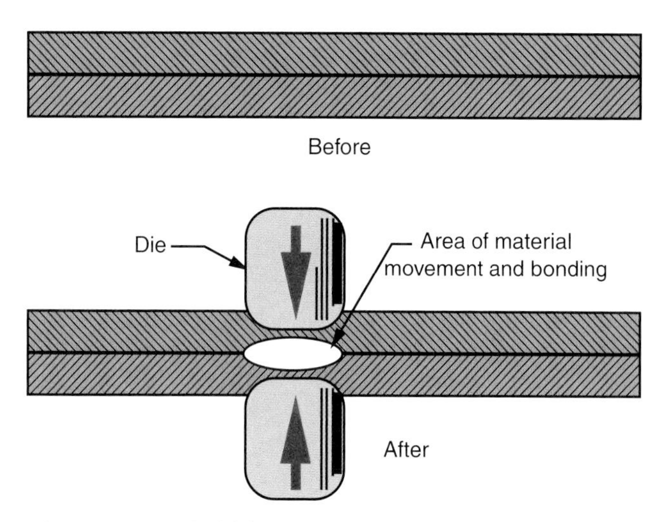

Figure 14-17. Cold bonding uses heavy pressure to bond materials together.

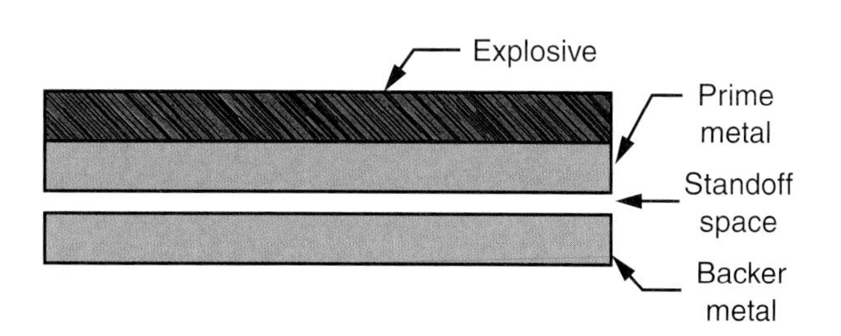

Figure 14-18. A typical explosive forming operation is illustrated above.

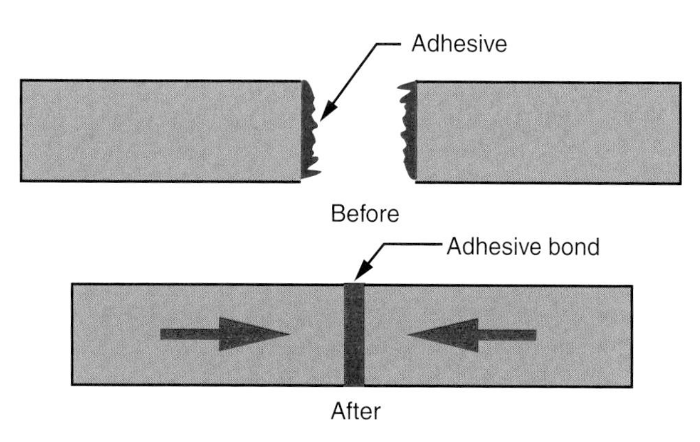

Figure 14-19. Adhesive bonding depends on a sticky material to hold the parts in position.

and are cured. Almost all adhesive assembling processes follow the same six steps. These are:

1. The adhesive material and joint are selected.
2. The material is prepared to ensure a good fit, and bonding surfaces are cleaned.
3. The adhesive is applied by brushing, rolling, or spraying.
4. The product is clamped with mechanical or pneumatic (air) clamps.
5. The adhesive is cured by drying, applying heat, or radio frequencies (RF curing).
6. The assembly is unclamped.

The bonding agents for adhesive bonding fall into two basic classes, inorganic and organic. The typical inorganic adhesives are metal alloys used in soldering and brazing. These processes were discussed under flow bonding. Organic adhesives can be grouped into the three major types of thermoplastic, thermosetting, and elastomers.

Thermoplastic adhesives

Thermoplastic adhesives are polymers that have a basic tackiness. They are usually suspended in liquids or are produced in hot melt sticks. They achieve their adhesion when the solvent evaporates or is absorbed or when the hot melt bead cools. Thermoplastic adhesives are not waterproof or heat-resistant.

The common woodworker's white glue is an example of a thermoplastic adhesive. The hot-melt adhesives used to fabricate and close corrugated boxes also fall into this class.

Thermosetting adhesives

Thermosetting adhesives are either powdered or liquid polymers that cure by chemical action. The curing action changes relatively simple, short polymer chains into complex, linked chains. A catalyst that is added to the adhesive or is activated when water is added often starts curing. Some thermoset adhesives do not cure by catalytic action, but are set with heat.

thermoplastic adhesives. Adhesives usually resins suspended in water (solvent). They form a bond when the solvent evaporates or is absorbed into the material.

Thermosetting glue joints are strong, but brittle. They can withstand heat and varying amounts of water. Most plywood, furniture, and high-quality cabinets are assembled with this type of adhesive.

Elastomers

Elastomers are rubberlike adhesives. They produce a flexible bond that cannot withstand heat. Typical elastomer adhesives are rubber and contact

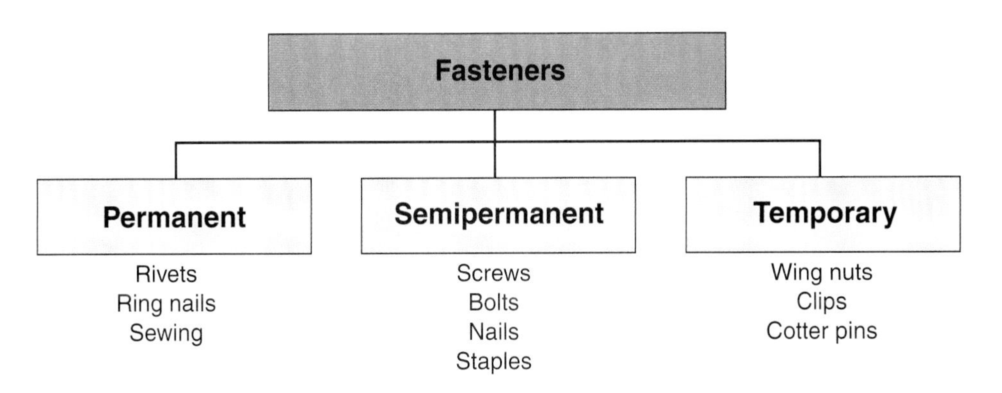

Figure 14-20. There are three types of fasteners.

thermosetting adhesives. Powdered or liquid polymers that cure by either chemical action or the application of heat.

cements. They find their widest industrial use in applying plastic laminates to cabinet faces and countertops or in pressure-sensitive labels and tapes.

Assembling with Mechanical Forces

Two major assembling techniques use mechanical force. These are mechanical fasteners and mechanical force.

Mechanical Fasteners

Mechanical fasteners are devices that hold two or more parts in a specific position in respect to each other. There is almost no limit to the number or variety of fasteners available. This is because many fasteners have been developed for limited special applications.

permanent fasteners. Joining devices designed to remain in place. To be removed, they must be destroyed.

Fasteners can be grouped in many different ways. One useful system groups them by the permanence of the assembly they produce. See **Figure 14-20.** In this system the three types of fasteners are permanent, semipermanent, and temporary fasteners.

Permanent fasteners

Permanent fasteners are meant to be installed and not removed. If removed, the fasteners are destroyed. Rivets are the most common permanent fastener used on engineering materials. When properly installed, the rivet is enlarged at both ends. See **Figure 14-21.** The force created when setting the rivet (forming the straight end) holds the parts. When correctly installed, the rivet is destroyed when removed.

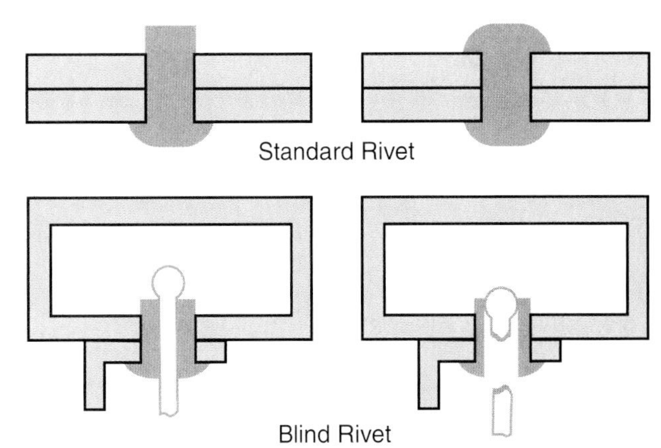

Figure 14-21. There are two major types of rivets: standard and blind.

Rivets are widely used for attaching sheet metal to frames. Aircraft and trailer companies are major users of rivets.

Semipermanent fasteners

Semipermanent fasteners are designed to remain in position. However, they can often be removed without destroying them or the materials they hold. These fasteners can be grouped into two major classes:

semipermanent fasteners. Devices designed to remain in place, but capable of being removed without being seriously damaged. A wood screw is an example.

Figure 14-22. Threaded fasteners are used to attach hardware to wood products. (Andersen Corp.)

Head Shape	Size	
	Penny	Length
Common	2d	1
	3d	1 1/4
Box	4d	1 1/2
	6d	2
Finish	8d	2 1/2
	10d	3
Casing	12d	3 1/4
	16d	3 1/2
	20d	4

Figure 14-23. Common nail length and head shapes.

threaded fasteners. Devices that use the friction between two threaded parts, or between threads and the material itself, to hold pieces together.

wire fasteners. Mechanical fasteners, formed from Wire, that make use of friction between the fastener and the material to hold pieces together.

temporary fasteners. Devices designed to allow easy removal for disassembly of a product or component. Cotter pins and wing nuts are examples.

- *Threaded fasteners.* These include nuts, bolts, and screws. They use friction between two threaded parts to grip the assembly. For example, a bolt and nut is used to hold the seat of a bicycle in position. As you turn the head of the bolt, the nut is drawn tightly against the bicycle frame. This creates friction between the head, nut, and frame. The friction keeps the bolt in place.

  In other cases, one of the parts of an assembly acts as the nut. Assume you are attaching two pieces of wood with a screw. As you turn the head of a screw, threads are cut in the bottom piece. The action draws the bottom board up against the top piece. This is the exact action that causes a nut to be drawn up when the head of a bolt is turned. Again, friction keeps the screw gripping the boards. See **Figure 14-22.** Therefore, the more threads per inch or the larger the diameter of the fastener, the stronger the assembly will be.

  The three major threaded fasteners are wood screws, machine screws, and bolts. Wood screws are used to attach a metal, wood, or plastic part to a wood member. Sheet metal screws are similar to wood screws except their threads extend the full length of its shank. A sheet metal screw, as the name implies, is designed to hold two pieces of sheet metal together. The screw should fit easily through the hole in the first member. It will cut threads (self-threading) in the second part. Machine screws are threaded fasteners used to assemble metal parts.

- *Wire fasteners.* These include nails and staples. They use friction to hold the parts together. Likewise, friction keeps the nail in place and makes it hard to remove. Typical wire fasteners are common and box nails, used by framing carpenters, casing and finish nails used by finish carpenters, wire nails and brads used in cabinets, and staples. Some typical nails and their common sizes are shown in **Figure 14-23.**

Temporary fasteners

Some applications require easy and frequent assembly and disassembly. In these cases, *temporary fasteners* are used. Cotter pins, spring clips, paper clips, and other similar fasteners make up the temporary mechanical fastener group. These fasteners are a vital part of assembling. For example, cotter pins keep the spindle nut on an automobile transaxle. Wing nuts are another example. They are often used to hold safety cover plates on equipment.

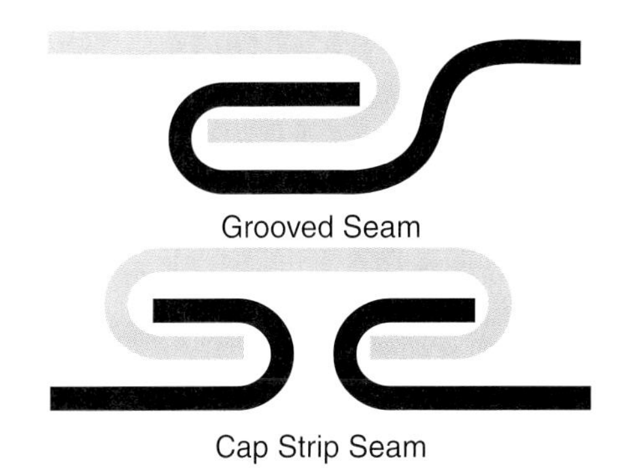

Figure 14-24. This common sheet metal seam uses mechanical force to hold the parts together. The strength of the seams depends on the ability of the metal to hold its shape after it is formed.

Fastening by Mechanical Force

Many parts are connected without the aid of bonding or a fastener. These techniques use mechanical force to position and hold assemblies together.

One common mechanical force assembling technique is *seaming*, see **Figure 14-24.** Sheet metal parts are assembled by bending and interlocking the components using various seams. The ability of the metal to retain its shape keeps the assembly together.

Another type of assembly uses *interference fits* such as press fits or shrink fits. For instance, if you watch football games, you have seen a backfield player try to gain yardage through the line. This player bumps and pushes against the line of other players. These players interfere with the progress.

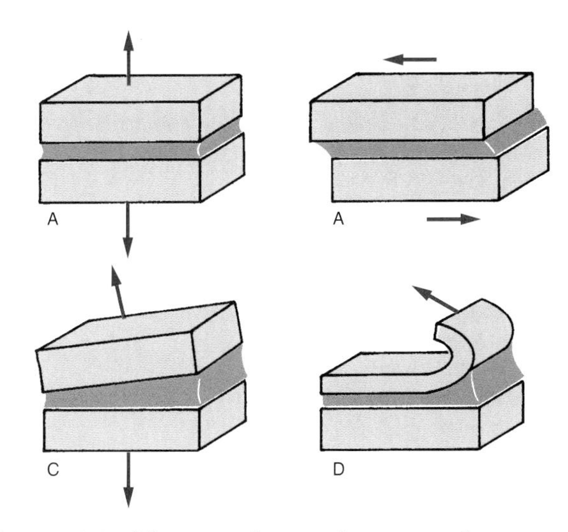

Figure 14-25. These are four major types of stress on glue joints. A—Tensile stress. B—Sheer stress. C—Cleavage stress. D—Peel stress.

Likewise, if you tried to put a round part into a hole of smaller diameter, you would encounter interference. You could put pressure (press or hammer) on the shaft to force it into the hole. This would result in a *press fit*. The friction between the larger part and the smaller hole would hold the assembly together.

In some cases, a metal part cannot be forced into a smaller opening. In a *shrink fit*, the smaller dimensioned part is heated. The expansion of the part allows easy assembly. When cooled, the part contracts (shrinks) around the other part. This creates a fit similar to a press fit.

seaming. **A means of joining sheet metal and similar materials by folding and interlocking the edges.**

interference fits. **A means of assembly that makes use of mechanical force to hold parts together.**

press fit. **A method of fastening that relies on the friction between closely fitting parts to hold them together.**

Joints

In all assembling operations, there is a point where two parts come together. This point is called a *joint*. Properly designed joints increase the strength of the assembly. They also make parts alignment easier. A good joint is designed to withstand the forces that are placed on the assembly. Four basic forces must be considered. See **Figure 14-25.** These include:

- ***Tensile stress.*** These are opposite forces that try to pull the bonding agent apart.
- ***Shear stress.*** This is a sliding force that tries to break or fracture the bond.

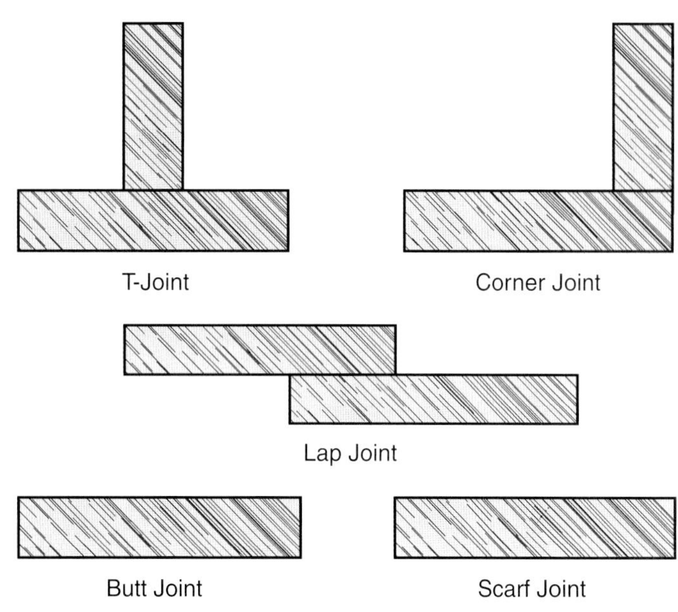

Figure 14-26. Five basic types of joints are shown above. Joints are chosen to meet strength and appearance requirements.

shrink fit. A fastening method in which one part is heated to make it expand. After the parts are assembled, the heated piece cools and shrinks, resulting in a tight fit.

joint. The point where two parts of an assembly come together and are fastened.

- ***Cleavage stress.*** These are angled forces that try to rupture and tear the bond.
- ***Peel stress.*** This is a curving force that tries to curve one part and fracture the bond.

Most common joints are variations of following five basic forms: T-joint, corner joint, lap joint, butt joint, scarf joint.

The T-joint and corner joint are used to join parts that intersect at an angle. See **Figure 14-26.** The lap, butt, and scarf joints are used to join parallel parts. They may be used to join the faces, edges, or ends of materials. The lap joint is strongest but does not provide a smooth joint. The butt joint forms a smooth intersection, but it is relatively weak. The scarf joint combines the advantages of both—strength and smooth intersection.

The matching of an appropriate joint with the proper bonding agent provides a serviceable assembly. Each joint is chosen to meet strength and appearance requirements.

Summary

Assembling is a method of permanently, semipermanently, or temporarily attaching parts to form products. This action may be done by bonding the parts together or by using mechanical forces.

All assembly operations must consider the agent that will be used, the method of creating the assembly, and the joint that will provide the best intersection of the parts.

Safety with Assembling Processes

- Do not try to complete a process that has not been demonstrated to you.
- Always wear safety glasses.
- Wear gloves, protective clothing, and goggles for all welding, brazing, and soldering operations.
- Always light welding torches with spark lighters, never matches or lighters.
- Handle all hot materials with gloves and pliers.
- Perform welding, brazing, and soldering operations in well-ventilated areas.
- Use proper tools for all mechanical fastening operations. Be sure screwdrivers and hammers are the proper size for the work being performed.
- Carefully follow instructions for lighting welding torches. Light the gas (acetylene, etc.) first; then turn on the oxygen.
- When shutting off a welding torch, first turn off the oxygen, then the gas.

Key Words

All of the following words have been used in this chapter. Do you know their meanings?

adhesive bonds
arc welding
atomic closeness
brazing
cohesive bonds
cold bonding
flow bonding
fusion bonding
gas welding
interference fits
joint
mechanical fastening
permanent fasteners
press fit
pressure bonding
seaming
semipermanent fasteners
shrink fit
soldering
solvent bonding
temporary fasteners
thermoplastic adhesives
thermosetting adhesives
threaded fasteners
wire fasteners

Test Your Knowledge

Please do not write in this text. Place your answers on a separate sheet.

1. *True or False?* All multipart products are put together using assembly processes.
2. Distinguish between bonding and mechanical fastening.
3. _____ bonds are the same forces that hold the molecules of a substance together, while bonds use the tackiness or stickiness of a substance to hold parts together.
4. List five methods of bonding.
5. *True or False?* In most arc welding, heat is generated by burning a mixture of oxygen and acetylene.
6. Pressure bonding uses which of the following to develop the bond?
   a. Heat.
   b. Pressure.
   c. Both a and b.
   d. None of the above.
7. List the three major types of organic adhesives.
8. *True or False?* A nail is an example of a semipermanent fastener.

9. If you hammer a shaft to force it into a hole of a smaller diameter, you are creating a _____:
   a. seam.
   b. press fit.
   c. shrink fit.
   d. None of the above.
10. Name the five basic forms of joints.

Applying Your Knowledge

Be sure to follow accepted safety practices when working with tools. Your instructor will provide safety instructions.

1. Identify products that were assembled by mechanical fastening and by bonding.
   a. Identify three products that use mechanical fastening. Identify the assembly technique used to produce each product and briefly describe the process.
   b. Identify three products that were assembled by bonding. Again, identify the assembly technique for each and briefly describe the process.
   c. Organize all of your data on a chart similar to the one in **AYK 14-1.**
2. Assemble parts to form. Use common assembling processes you have studied in this chapter. This could be a product of your own choice, or you could select one of the following:
   a. Make a screwdriver or tack puller from the handle produced in the casting activity (Chapter 10) and the blade produced in the forming activity (Chapter 11).
   b. Assemble the parts of the hot dish holder (trivet) produced in the separating activity (Chapter 12).
3. Develop a transparency or chart on fusion, flow, pressure, cold, and adhesive bonding.
4. Produce several of the types of joints described in this chapter. Design a technique to test their strength. An idea for a strength test is shown in **AYK 14-4.**

Bonding	Technique	Description
Mechanical fastening		
1.		
2.		
3.		
Bonding		
1.		
2.		
3.		

AYK 14-1. On a separate sheet of paper, prepare a chart similar to this one on which to record your responses.

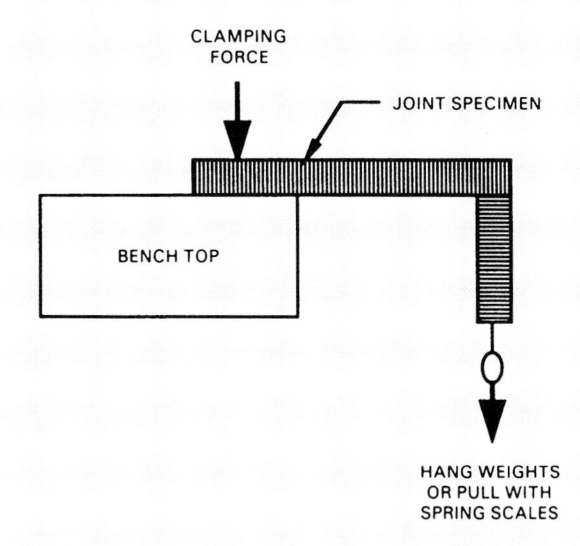

AYK 14-4. Once you have assembled a joint, test its strength by using a test similar to this one.

Chapter 15 Finishing Processes

Objectives

After studying this chapter, you will be able to:

✓ Describe how finishing processes have developed.
✓ List the three basic steps involved in the application of a finish.
✓ Name and describe the two basic groups of finishing materials.
✓ Explain how materials are prepared for finishing.
✓ Describe how different finishes are applied to materials.

Most consumer products—from coffee cups to wooden chairs—have a surface finish. A surface treatment applied to a material is called a finish. It includes all processes that convert or coat the surfaces of materials to beautify or protect them.

Many people are drawn to a product by its beauty. A surface finish is often applied to improve the appearance of a product. See **Figure 15-1.** The result is a more attractive item that will have an increased sales appeal.

However, appearance is not the only reason surface finishes are applied to products. Surface finishes increase the durability of a product. Most industrial materials are attacked by elements in the environment. For instance, copper and aluminum can corrode. Iron and steel can rust. Wood can crack (check) and warp from absorbing and losing moisture. Water and dirt can penetrate ceramic materials. Almost all materials are subject to wear—scratching and denting. A finish is often applied to protect a material from chemicals in the air and from moisture. See **Figure 15-2.** The coating helps extend the life of a product.

Figure 15-1. Many of the items shown in this room have had surface finishes added to them to improve their performance and appearance. (Drexel-Heritage Furnishings, Inc.)

Development of Finishing

From earliest times, people have tried to make their surroundings more pleasant. Cave dwellers painted the walls of their caves. These paintings have

Figure 15-2. The yellow paint on this earthmover probably did not cause the owner to decide to buy the product. However, the paint does protect the steel parts from environmental attack that could result in rusting. (Deere and Co.)

Figure 15-3. This modern Southwest Indian Pueblo pottery was hand formed and decorated.

been discovered in a number of locations around the world. The pictures made the walls of caves more attractive and recorded personal accomplishments. Modern finishing processes are an outgrowth of these early finishing techniques.

Nearly 10,000 years ago, the Egyptians developed painting into an art. The freshness of color and the beauty of the art found in Egyptian tomb paintings have amazed archaeologists.

Decorating Pottery

The art of making pottery (earthenware vessels) is among the oldest of all crafts. Decorating pottery began early in human history. It is believed that the first designs were used to identify the owner and the contents of the container. Later, beautification of the vessels became important.

The Greeks of 500 B.C. were noted for their black pottery decorations. Black figures on reddish clay added beauty to functional items. Similar practices were developed in civilizations throughout the world. One of these is purely American. The natives of the Southwest (Arizona, New Mexico, and southern Colorado) developed it long before Columbus arrived. The Pueblo Indians of that region became very sophisticated in making and decorating pottery. See **Figure 15-3.**

Protecting Wood and Metals

The use of wood and other fibrous materials led to the development of additional finishing materials. Early in history, wood was a primary material for all types of products. Because wood is greatly affected by changes in humidity, it can warp and check. Therefore, people devised ways to protect it from the environment. Early boat builders used reeds that grew in shallow water to make simple vessels. They found that their boats could be protected with a paste made from clay and bitumen. Later, wood boats were sealed with tar and resin collected from trees.

Early humans had what they made, ate what they hunted and gathered, and lived where it was convenient. As people banded together and communities grew the demand for materials for housing, tools, utensils, and art grew as well. Early humans used many materials for these purposes. However, without proper conditioning and finishing the usefulness of these items were short lived. Adding protective (and decorative) finishes to materials assured that they could be used longer.

The art of covering pottery with enamel was invented by the Egyptians at a very early date. They applied it to stone as well as to pottery. Egyptians made pottery before the building of the Pyramids. This is evident from the presence in older hieroglyphic (Egyptian language) writing of pottery. Pictures of pottery and small pieces of pottery have been found in Egyptian tombs around the time of the building of the Great Pyramid (about 2350 B.C.).

From these early beginnings, natural varnishes and shellac evolved. Both materials provided protection and brought out the natural beauty of wood in items such as cabinets and furniture. These materials led to the development of paint that was used to decorate household items and walls of homes. See **Figure 15-4.**

The protection and decoration of metals is a more recent development. As metals, such as iron and steel, became popular in everyday products, techniques were developed to prevent them from rusting and to improve their surfaces.

Figure 15-4. Wall decoration, like this reproduction of a mural found in Awatovi, near Hopi First Mesa, Arizona, has a long history. (Maxwell Museum of Anthropology)

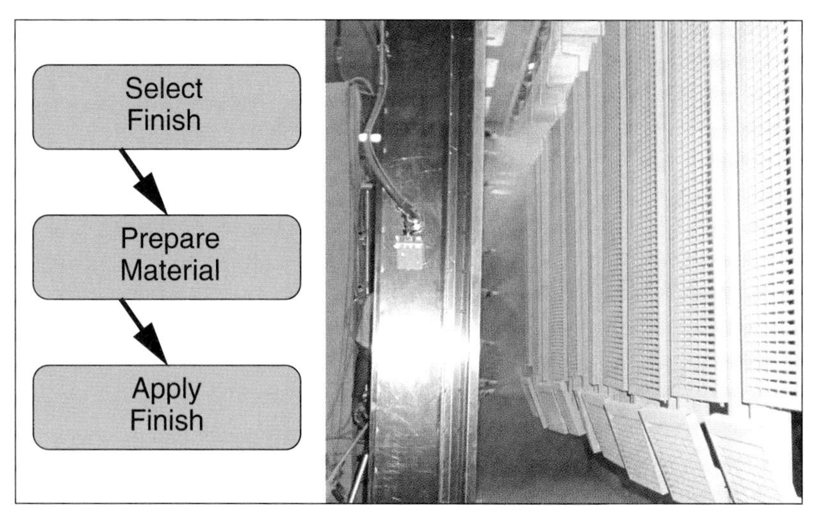

Figure 15-5. The three basic steps in applying finishes are outlined here. (Ransburg-Gema)

Essential Elements of Finishing

The materials for most manufactured products are first sized and shaped using casting, forming, separating, and/or conditioning techniques. After these processes, the parts require one or more additional operations to clean, protect, and decorate the surfaces.

These finishing tasks may be completed before or after assembling. The appropriate sequence of these operations is determined by the material and processes selected.

conversion finish. A process that chemically alters the surface of a material into a protective layer.

anodizing. A process that oxidizes the surface of a part, giving a layer of oxide that will resist corrosion. Anodizing is often used on aluminum and magnesium parts.

The application of a finish requires three basic steps, as shown in **Figure 15-5.** These steps are:

1. A finishing material is selected.
2. The part or base material is prepared to accept the finish.
3. The finish is applied.

Finishing Materials

Finishing materials may be divided into two basic groups. These groups consist of two types of finishes—conversion finishes and coating finishes. See **Figure 15-6.**

The surface of a material may be converted into a protective layer. This action is usually carried out through a chemical process that produces a layer that is chemically different from the base material. The new layer is called a *conversion finish.* It generally has better wear properties, resistance to atmospheric chemicals, and appearance than the base material.

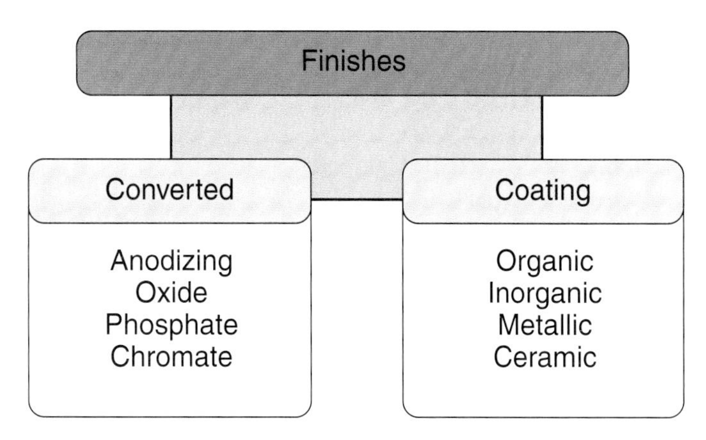

Figure 15-6. These are the types of finishes used in manufacturing.

Conversion finishing

Colored aluminum tumblers, ashtrays, and similar products have a converted surface. This finish was developed using a process called *anodizing.* This term describes an electrolytic treatment of any metal to develop a stable film or surface layer. The film is an oxide of the base metal.

All metals tend to develop an oxide coating. This coating resists further environmental attacks. Anodizing uses an electric current passing through an acid electrolyte (a solution that will conduct electricity). This action accelerates the natural oxidation action. The result is a uniform, hard, thicker oxide layer on the part or product. This layer accepts dye so it can also be decorative.

Other conversion finishes that are chemically developed include:

- *Phosphate conversion coating*. This is a conversion finish that is used as a prepaint coating on steel and as anti-scuff coatings on gears and engine parts.
- *Chromate conversion coating*. This is a conversion finish that can be a clear coating over cadmium and zinc-plated steel parts.
- *Oxide conversion*. This is a conversion finish that provides corrosion and abrasion resistance (blackening, nitriding).

phosphate conversion coating. **A finishing process used as a prepaint coating on steel.**

chromate conversion coating. **A finishing process that provides a clear coating on cadmium- and zinc-plated steel parts.**

oxide conversion. **A process that provides a finish that resists corrosion and abrasion.**

Surface coatings

Parts and products can also be coated with a layer of a second material. See **Figure 15-7.** The result is a *surface coating*. The coating is made of one of two basic types of materials:

- *Organic coatings*. These often come from petroleum and natural gums or resins from trees.
- *Inorganic coatings*. These come from mineral-based metallic and ceramic materials.

Organic coatings are the most widely used. As with all coatings, they are designed to produce a surface layer of material that adheres to the base material. This layer seals the part or product against the outside environment. See **Figure 15-8.** If color is added, it can add beauty to the base material.

The term "organic coatings" covers a wide range of materials. Most of these coatings are made up of a vehicle or binder and pigments that are suspended in a solvent. The vehicle is the material that adheres to the part to provide the protection. The solvent carries the vehicle to the part. The pigment provides color to the final coating.

Organic finishes are applied as liquids. Then they must form their coating film through one of three basic actions. Some coatings are formed through polymerization. The rather simple polymer chains in the finish combine to form more rigid, complex chains. Other finishes harden through *solvent evaporation*. As the solvent evaporates, the protective coating is left behind. A few finishes are *hot melts*. They are heated and applied to the part. As they cool, a durable coating is developed.

Figure 15-7. Coatings are a protective layer applied to a base material. This dishwasher case is receiving a dip coating of porcelain that will be fired in an oven to cure it. (White Consolidated Industries)

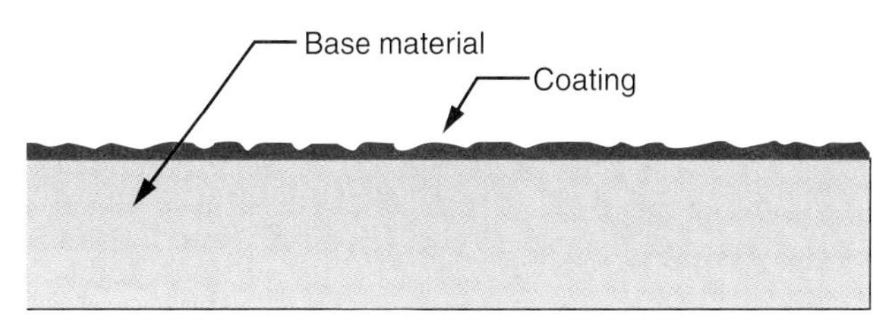

Figure 15-8. Coating forms a protective layer of new material on top of a base material.

Figure 15-9. Enamel is a colored mixture of vehicles and solvents. (PPG Industries)

Figure 15-10. This large steel structure is being dipped into molten zinc to produce an inorganic coating. (American Galvanizers Association)

surface coating. A finishing process that adds a layer of a different material to the surface of a base material.

organic coatings. Finishing products derived from living or once-living sources. Natural gums or resins from trees are widely used.

inorganic coatings. Coatings made up of metals or ceramic materials.

solvent evaporation. The means by which many organic coatings change from a liquid to a solid, state.

hot melts. A type of finish that is made liquid by heating, then forms a coating as it cools to a solid state.

Organic coatings include any number of specific finishes. The most common are:

- *Paint*. This is any coating that dries through polymerization. Examples include varnish and enamel.
- *Varnish*. This is a clear paint made from a mixture of oil, resin, solvent, and drier.
- *Enamel*. This is a varnish to which a pigment has been added. See **Figure 15-9.**
- *Lacquer*. This is a solvent-based synthetic coating that dries through solvent evaporation.

Inorganic coatings are layers of metallic or ceramic material deposited on the part. See **Figure 15-10.** Most metallic coatings are deposited on metal parts and products. However, coating plastics with metals is becoming more popular as plastics replace metals for automotive and appliance trim. Typical metal coatings are chromium, zinc, and tin. Decorative shapes coated with silver and gold are used to produce jewelry.

Ceramic materials make a hard, chemical-resistant finish. These coatings are all the glass and glasslike materials used on metal and ceramic items. Two common ceramic coatings include porcelain enamel and glazes. ***Porcelain enamel*** is used on better appliances and bathroom fixtures. ***Glazes*** are used to coat dinnerware, decorative objects, wall and floor tile, and other ceramic products. Both of these finishing materials consist of finely ground glass suspended in water. When heated (fired), they melt and fuse into a uniform, colorful finish.

Preparing Materials for Finishing

Most materials cannot be finished as they leave casting, forming, and separating processes. They generally require some preparation. Scratches and dents must be removed. Dirt, grease, oil, surface oxides, and other impurities must also be removed.

Material preparation activities use two basic techniques. They are mechanical preparation and chemical preparation.

paint. A coating that changes from a liquid to a solid by means of a polymerization reaction. (This is a linking of molecules into strong chains.) Many paints have a coloring agent added.

varnish. A clear coating material made from a mixture of oil, resin, solvent, and drier.

enamel. A varnish to which a pigment has been added to produce a colored coating.

Mechanical preparation

Mechanical preparation involves using a rubbing action to prepare a material for finishing. This usually means using abrasive particles, wire brushes, or metal slugs to wear off impurities and roughness. The most common mechanical preparation processes are:

- *Abrasive cleaning*. Involves using a coated abrasive paper, belt, or pad. See **Figure 15-11.** The abrasive is used to sand, polish, or buff the material. *Sanding* is the coarsest of these operations. It uses coated abrasives to remove scratches, oils and grease, and other impurities. *Polishing* uses powdered abrasives and a pad to remove oils, greases, and minor surface defects. *Buffing* uses a pad and fine abrasives to remove the scratches produced by sanding and polishing.
- *Media cleaning*. Involves using fine shot or ceramic shapes to polish and remove unwanted material on the surface of the part. The media may be propelled by air against the surface (sand blasting), tumbled in a container with the parts, or vibrated against the material.
- **Brushing.** Involves using a wire brush to clean the surface of a material. The rubbing action of the bristles of the brush dislodges impurities and cleans the surface.

Figure 15-11. The surface of this metal part is being prepared to accept a finish. (ARO Corp)

Chemical preparation

Chemical preparation involves using chemical action to prepare a material for finishing. The material may be dipped in a chemical or may be sprayed with it. Often, the chemicals in a dip tank are agitated to increase the speed of the cleaning action. Likewise, spraying hot solvents will be faster than using cold liquids.

Chemical cleaning is often used to prepare metals to accept a metallic coating. See **Figure 15-12.** If the chemical is an acid, the process is called *pickling*. Alkaline solutions are used in *caustic cleaning*. *Steam cleaning* uses blasts of steam and detergent to remove oils and greases.

Figure 15-12. This row of tanks contains chemicals that clean parts before a finish is applied.

Applying Finishes

Applying something often means using a process to meet a goal. You can apply a formula to a math problem to reach a solution. The problem-solving process can be applied to solve a technological or scientific problem. Likewise, you can *lacquer.* A material containing a polymer coating and a solvent. A lacquer dries as the solvent evaporates.

mechanical preparation. Use of an abrasive or similar material to prepare a surface for finishing.

apply a finishing process to protect or beautify a material. As you studied earlier, finishing processes can be divided into two groups in terms of the results obtained—conversion finish or a coating. Each of these results has its own technology.

Applying conversion finishes

Conversion finishes, as you learned earlier in this chapter, are developed through chemical means. Their goal is to change the chemical composition of the outer layer of the part or material.

One common surface conversion technique, anodizing, has already been discussed. You will recall that it is used to convert the surface of aluminum objects into aluminum oxide.

Another similar process is phosphate conversion coating. This technique involves treating the metal in a bath containing phosphoric acid and the phosphate of the base metal (iron phosphate, zinc phosphate, etc.). As shown in **Figure 15-13,** the metal is dipped in a number of chemical and water baths. The process converts the surface layer of the base metal into a metallic phosphate. This layer provides an excellent base for paint, resists oxidation (rusting for iron and steel), and reduces friction.

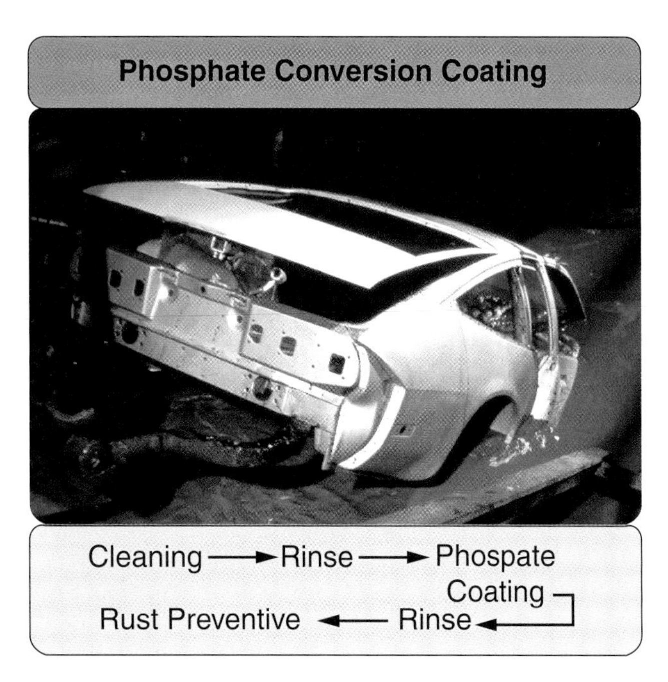

Figure 15-13. A phosphate conversion coating is used to rustproof an automobile body. (PPG Industries)

Applying coatings

Coating finishes make up the majority of all finishes. These finishes are layers of dissimilar materials added to the surface of the base material. The coating adheres to the material to provide the desired protection and beautification. As discussed earlier, coating materials can be grouped as inorganic (metallic coatings and ceramic coatings) and as organic coatings.

Coatings can be applied using several common methods. The method selected depends on the coating material, shape of the part, and the desired appearance. See **Figure 15-14.**

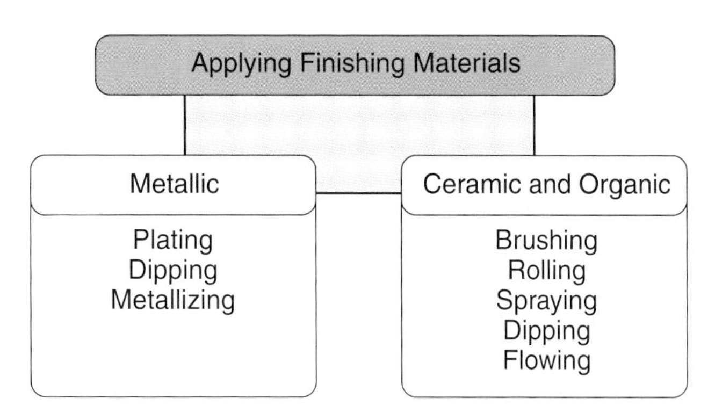

Figure 15-14. Methods of applying finishing materials are listed above.

Metallic coatings must deposit metal particles on the base material. The common processes used to accomplish this task are:

- Plating
- Dipping
- Metallizing

The most common plating process is *electroplating*. This process, as shown in **Figure 15-15,** involves using three components:

- ***Cathode.*** The part that has a negative electrical charge.

- *Anode.* A piece of plating metal that has a positive charge.
- *Electrolyte.* A liquid that conducts electricity.

In electroplating, the part (cathode) and the plating metal (anode) are placed in a vat of electrolyte. See **Figure 15-16.** Electrical leads are attached to the cathode and anode. The direct current is turned on causing charged atoms from the cathode to enter the electrolyte. These atoms, called ions, have a negative charge. Since opposite electrical charges attract, the ions move to the positively charged anode. There they adhere to the surface of the part. When the desired thickness is reached, the current is turned off and the part is removed. Metallic plating is widely used for jewelry, automotive trim, and hundreds of other products.

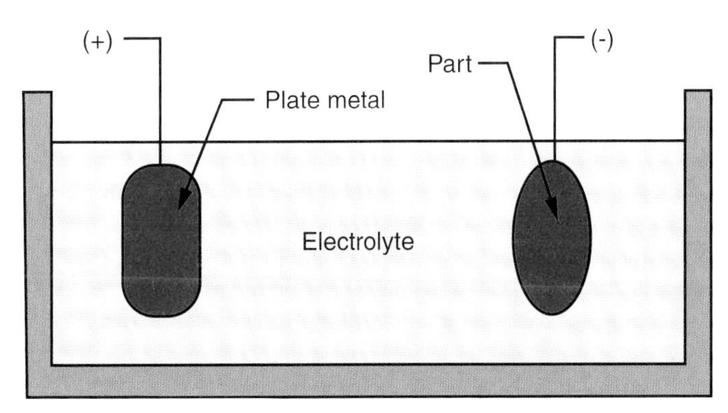

Figure 15-15. A basic electroplating system is diagramed above.

Figure 15-16. Electroplated parts are shown leaving a tank of an electroplating unit.

Electroplating is also used to produce coatings on industrial materials. Steel is electrocoated to produce galvanized (zinc coated) and tin plated sheets. See **Figure 15-17.**

A second method of applying metallic finishes is *dipping*. In this process, a vat of molten metal is used. The part or material is placed in the vat. The cooler base material solidifies a thin coating of metal onto its surface.

Electrocoating is a type of dipping process. See **Figure 15-18.** It uses unlike electrical charges for the parts and the finish. The charged part is dipped into a tank with finishing materials that have the opposite charge. The material is attracted to the part. When the part is removed from the tank it is rinsed with water. Then the paint is baked (dried) in a continuous oven.

Dipping may be used to coat individual items or batches of parts. In addition, industrial materials are often continuously fed through the vat to produce zinc (galvanized) coated sheets. The third method of applying metallic coatings is called *metallizing*. In

Figure 15-17. Galvanized sheets are shown leaving a continuous plating unit. (American Iron and Steel Institute).

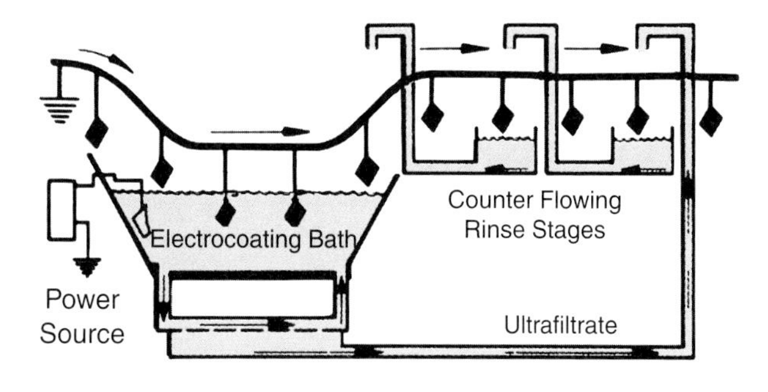

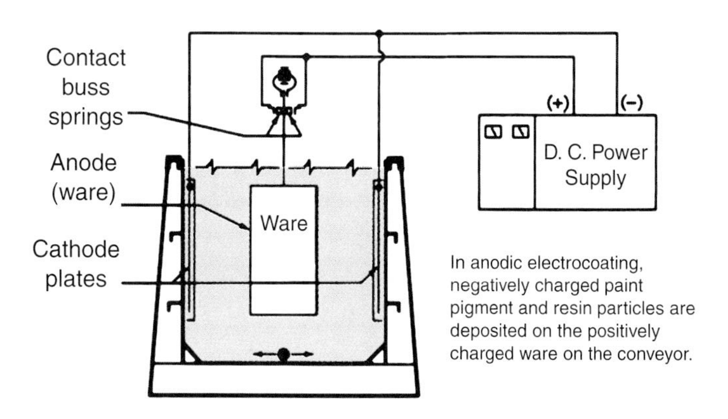

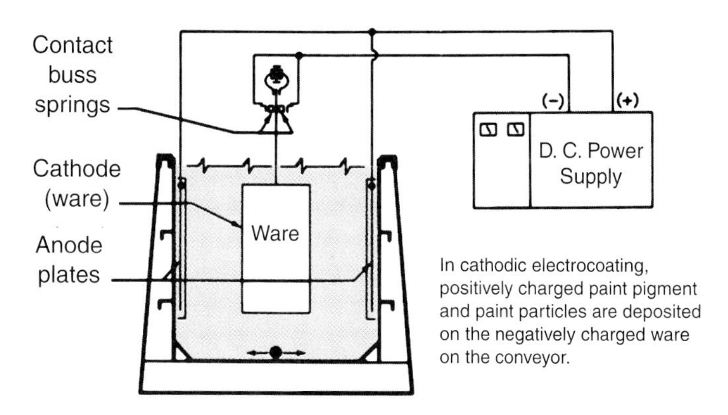

Figure 15-18. A diagram of the electrocoating process. (George Koch and Sons)

this process, fine, metallic particles are applied in one of three ways:

- *Wire metallizing* or *flame spraying*. The coating material is melted using an oxyacetylene flame and sprayed on the part with a blast of compressed air.
- *Plasma arc spraying*. Powdered metal is fed into a stream of superheated gas, usually argon that is directed onto the part. See **Figure 15-19.**
- *Vacuum metallizing*. Metal is vaporized in a vacuum chamber where the particles float and become deposited on slowly rotating parts.

Organic and many ceramic finishes are applied using one of several techniques. These include brushing, rolling, dipping, spraying, and flow coating.

Brushing is seldom used in finishing products on a commercial scale. It is the slowest finishing technique and produces the least uniform coating. Also, skill is needed to produce a high quality brushed finish. Occasionally, brushing is used to decorate or coat unique products.

Rolling techniques find wide use in assembly activities. Roll coating machines are often used to apply adhesives to furniture and cabinet parts and in plywood manufacturing. However, roll coating finds only limited use as a finishing technique in manufacturing. It is used in printing processes that produce the wood grain patterns of hardboard wall paneling and to coat some metal sheets.

Dip coating organic materials is basically the same process as dip coating metals. The parts or products are submerged in a vat of finishing material. The items are then removed and suspended over the container. Excess material runs to the lowest part of the product and drips off of the material. The coating is cured in the air or in an oven. Often, the finished part will have a buildup on the lowest edge or corner. This characteristic is unsuited for many applications where a very uniform surface layer is needed.

Dip coating may be used for individual parts, or it may be a continuous process. In continuous coating, parts are suspended on a conveyor. The moving conveyor dips them into successive tanks of cleaners, rinses, and coating materials. See **Figure 15-20.**

Dip coating is effective when a large number of irregular parts are to be coated with the same color. Changing colors is difficult or expensive because a

abrasive cleaning. **Rubbing a surface with sandpaper or a similar material to smooth it and remove grease or other contaminants.**

media cleaning. **Surface preparation methods that propel, vibrate, or tumble some form of abrasive against the surface of the material to be cleaned.**

separate tank is needed for each color. Dip coating is used to prime coat metal parts, apply glaze to ceramic ware, and surface coat small metal products, such as toys.

Many metal, ceramic, and wood products have a surface finish applied by *spraying*. The common spraying techniques, as shown in **Figure 15-21,** are:

- **Hand spraying.** One of the oldest spraying techniques. See **Figure 15-22.** The spray system is composed of a container of finish, a source of compressed air, and a gun that delivers the finish through the stream of air. Hand spraying is often used in low volume production settings, such as in many cabinet and furniture operations, and in repair and maintenance work. It finds limited use in modern high-volume manufacturing.
- **Automatic spraying.** Propels a finish onto the part in the same manner as hand spraying. The difference lies in two factors. First, a robot or other automatic device operates the spray head. Secondly, parts are moved automatically past the spray heads.
- **Electrostatic spraying.** A specialized automatic spraying operation. It uses one of the physical laws of electricity to increase the efficiency of spraying operations.

In hand spraying, a large portion of the finish material is lost to overspray. This is because the finish material misses the object at the start and end of each pass of the spray gun. Additional material bounces off the part because of the velocity it is traveling. Automatic spraying eliminates some overspray. It carefully controls the start and end of the spray pass. The introduction of airless spraying reduces the speed the finish material travels to the part. This reduces the amount of finish material that fails to stick to the surface. However, automatic spraying still presents some problems. For instance, overspray still occurs. Also, round parts have the heaviest buildup directly in front of the spray nozzle, while the top and bottom of the parts receive only a thin coat of coating material.

chemical preparation. Use of solvents or steam and detergents to remove grease and other contaminants from a surface prior to finishing.

pickling. A cleaning method that uses acid to prepare the surface of a metal.

caustic cleaning. A surface preparation method employing strongly alkaline solutions.

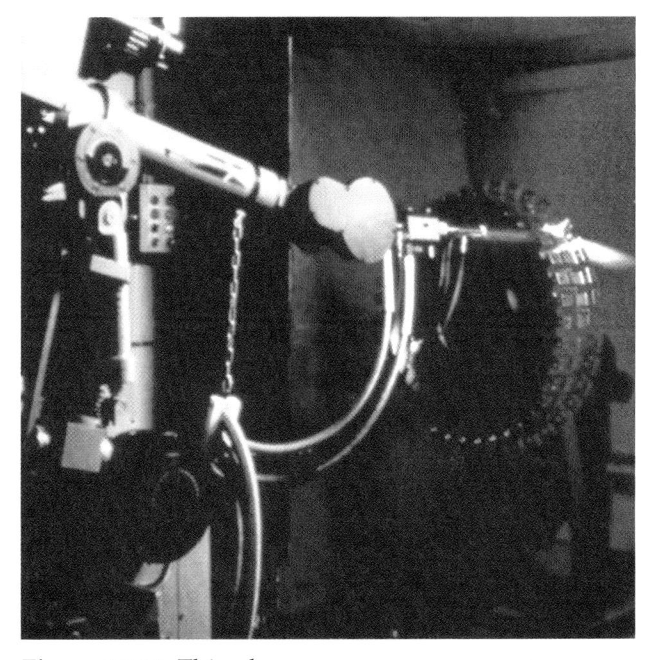

Figure 15-19. This plasma spray gun can spray vaporized metals or ceramic materials. (METCO)

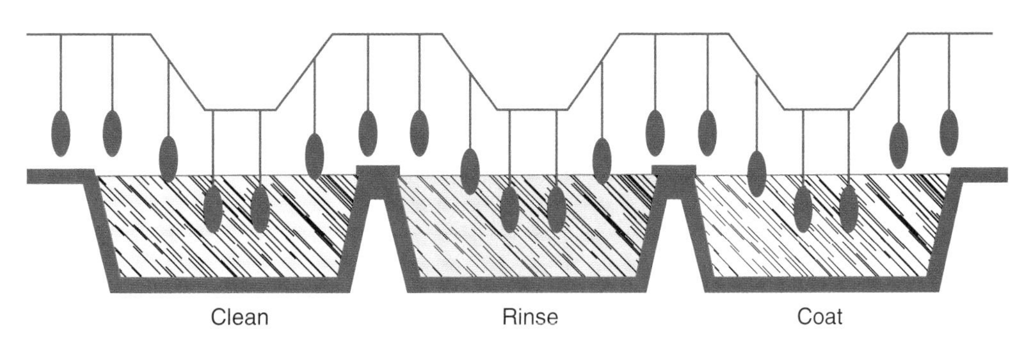

Figure 15-20. A schematic of a continuous dip coating operation is shown here.

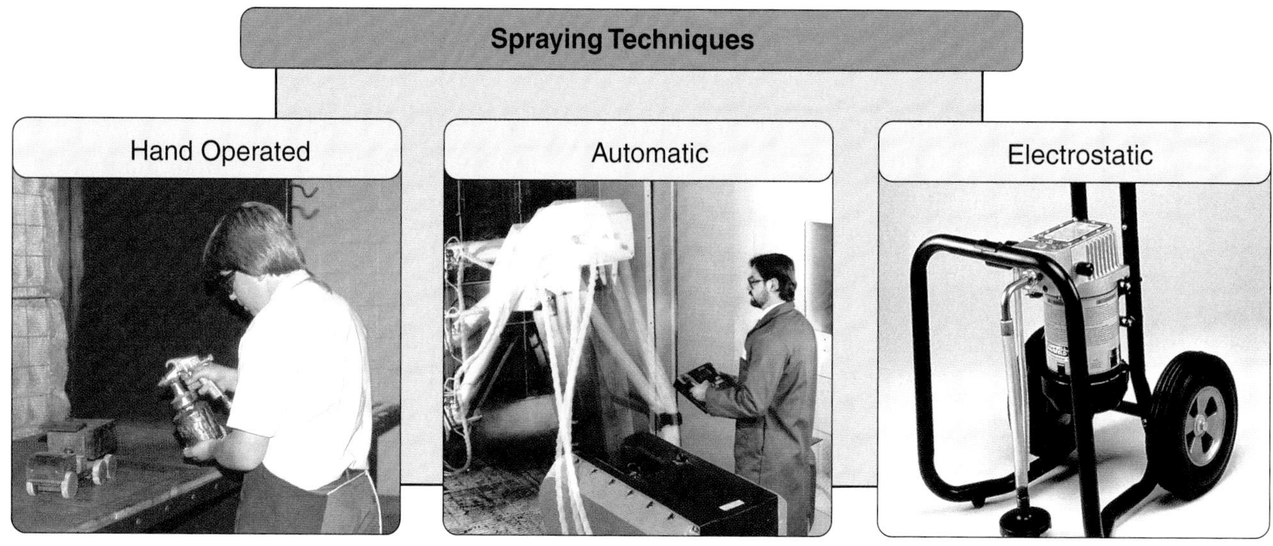

Figure 15-21. Spraying techniques used to apply finishes. (DeVilbiss Co., Ransburg Corp., and Campbell Hausfeld)

steam cleaning. The use of steam and detergents to remove grease and oily deposits from the surface of metal.

metallic coatings. Finishes that consist of metal particles forming a surface layer on the base material.

electroplating. The use of an electric current to deposit a thin, uniform metal layer on the surface of a base material.

dipping. A finishing method in which the material is dipped into a vat of molten metal.

metallizing. Application of a very thin finish of metal particles by means of spraying or vaporizing in a vacuum chamber.

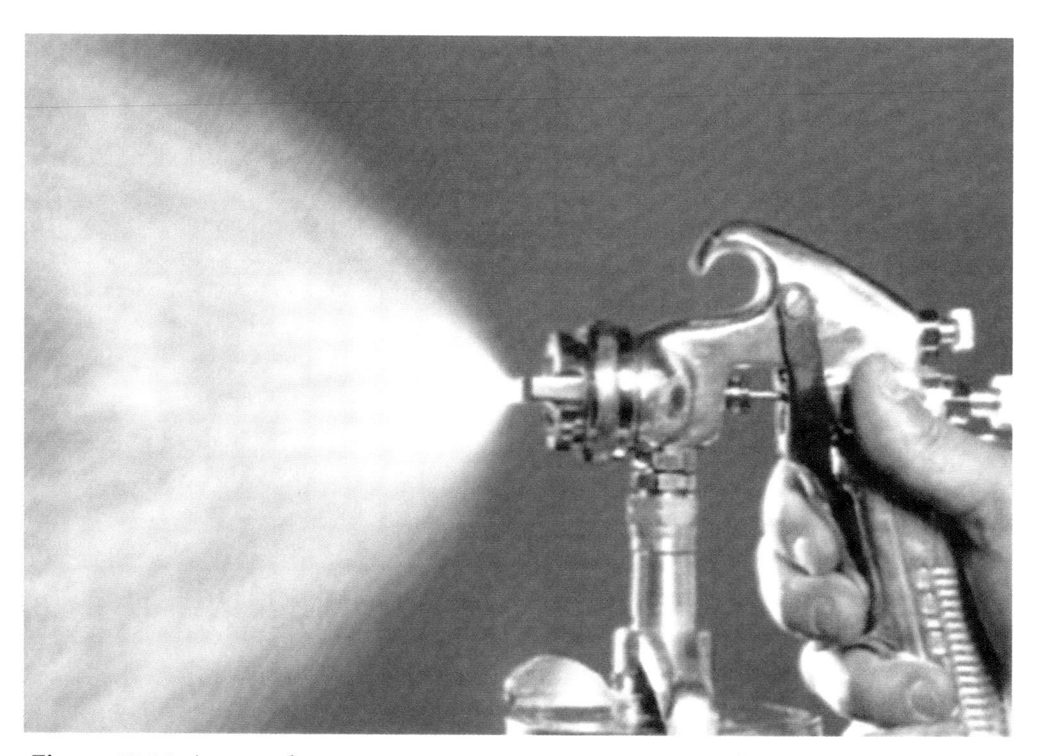

Figure 15-22. A general-purpose spray gun uses air to propel finish onto a surface. (DeVilbiss Co.)

Electrostatic spraying solves both of these problems by giving the paint and the part opposite electrical charges. Since opposite charges attract, the paint is drawn to the part as it is sprayed. In fact the paint will actually wrap around the part giving a uniform coat on all surfaces, see **Figure 15-23.**

Another coating technique is *flow coating*. This process floods the surface with finishing material. The process is very similar to dip coating, except the

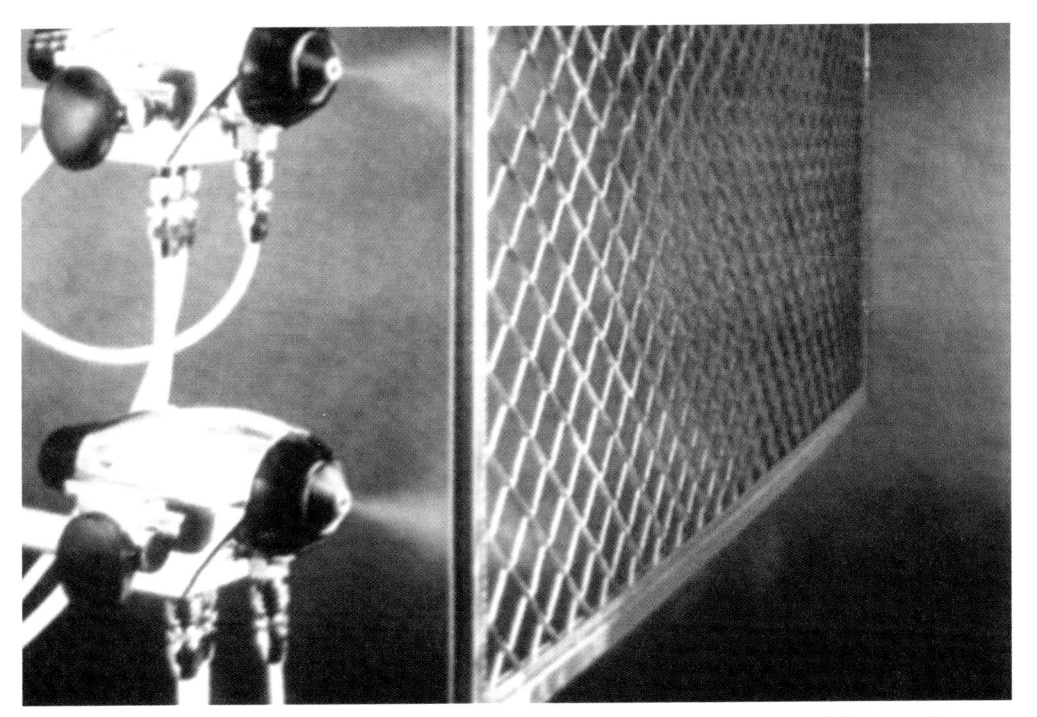

Figure 15-23. This automatic electrostatic spray unit coats both sides of the metal grate in one pass. The paint is attracted to the part by magnetic charges. (DeVilbiss Co.)

rolling. A forming process that uses a rotating applied force to change the thickness of a piece of steel or other material. Also, a finishing method in which a coating is applied by a roller.

dip coating. A finish obtained by submerging the material in a container of coating material.

spraying. A finishing method in which paint or similar material is vaporized and deposited on the surface of the material.

flow coating. A process that floods the surface with a finishing material. The excess is allowed to drip off the material.

part is not immersed in a tank. Instead a stream of paint flows over the surface as the part moves under a nozzle. The material or product is tilted as it passes under the nozzle. The excess will flow off the part's lower edge.

Flow coating is useful in coating an exterior surface of a product or the top of sheet material. It does not work well on interior surfaces.

Summary

Finishing processes are designed to protect products from environmental elements and wear. They can also be used to beautify materials by adding color and luster to surfaces.

Finishing operations must prepare the surface to accept the material. Then the appropriate finish application process must be selected and the finish applied.

A finish may be a layer of the base material that has be converted to be more attractive and corrosion resistant, or it may be a separate organic or inorganic coating material that is applied to the surface of the material or product. Coating materials are applied using plating, metallizing, brushing, rolling, dipping, spraying, and flow coating operations.

Careful attention to the selection, preparation, and application of a finish will add to the beauty and durability of products. The finish will also make the product more marketable for the company and more pleasing to the customer.

Safety with Finishing Processes

- Do not try to complete a process that has not been demonstrated to you.
- Always wear safety glasses when performing finishing processes.
- Always apply finishes in well-ventilated areas.
- Do not apply finishing materials near an open flame.
- Always use the proper solvent to thin finishes and clean finishing equipment.
- Avoid splashing or flipping finish onto other people or surfaces when you are using a brush.
- Dispose of all waste finishes and solvents in the proper manner.

Key Words

All of the following words have been used in this chapter. Do you know their meanings?

abrasive cleaning
anodizing
caustic cleaning
chemical preparation
chromate conversion coating
conversion finish
dip coating
dipping
electroplating
enamel
flow coating
hot melts
inorganic coatings
lacquer
mechanical preparation
media cleaning
metallic coatings
metallizing
organic coatings
oxide conversion
paint
phosphate conversion coating
pickling
rolling
solvent evaporation
spraying
steam cleaning
surface coating
varnish

Test Your Knowledge

Please do not write in this text. Place your answers on a separate sheet.

1. *True or False?* Surface finishes are applied only to improve the appearance of products.
2. List the three basic steps involved in the application of a finish.
3. Name the two basic groups of finishing materials.
4. *True or False?* Organic coatings come from mineral-based metallic and ceramic materials.

5. Which of the following finishes is a solvent-based synthetic coating that dries through solvent evaporation?
   a. Paint.
   b. Lacquer.
   c. Varnish.
   d. Enamel.
6. List three types of mechanical preparation processes for finishing.
7. List three types of chemical preparation processes for finishing.
8. Anodizing is an example of a(n) _____ finish.
9. _____ finishes are layers of dissimilar materials added to the surface of the base material.
10. Which of the following is used in applying metallic finishes?
    a. Dipping.
    b. Metallizing.
    c. Electroplating.
    d. All of the above.
11. *True or False?* Brushing is often used in finishing products on a commercial scale.
12. In which of the following coating techniques is a part or product submerged in a vat of finishing material?
    a. Rolling.
    b. Dip coating.
    c. Spraying.
    d. Flow coating.

Applying Your Knowledge

Be sure to follow accepted safety practices when working with finishes. Your instructor will provide safety instructions.

1. Apply a common finish to a product.
   a. Select a product and a finish. Suggestion: dip finish the trivet produced during the separating activity in Chapter 12.
   b. Examine the surface of the product. Is it ready for finishing? It may require cleaning, smoothing, or some other preparation.
   c. Prepare surface as may be required.
   d. Procure finishing materials.
   e. Select tools and equipment needed.
   f. Apply the finish. (Note: if you use volatile materials or spraying on finishing material, be sure to use proper ventilation and wear a mask or respirator. Check safety equipment with your instructor.)
2. Apply three different wood finishes to sample wood blocks. Wrap a piece of 120 grit abrasive paper to a brick or other weight. Attach a string to the weight and drag it across the wood samples 1, 5, 10, and 20 times. Observe and record your results for each test on a chart similar to the one in **AYK 15-2.**
3. Collect five samples of materials that have a finish applied to them. Prepare a label for each, similar to the one in **AYK 15-3.**

Times	Sample	Results	Conclusions
1	1		
5	1		
10	1		
20	1		
1	2		
5	2		
10	2		
20	2		
1	3		
5	3		
10	3		
20	3		

AYK 15-2. On a separate sheet of paper, prepare a chart similar to this one on which to record your responses.

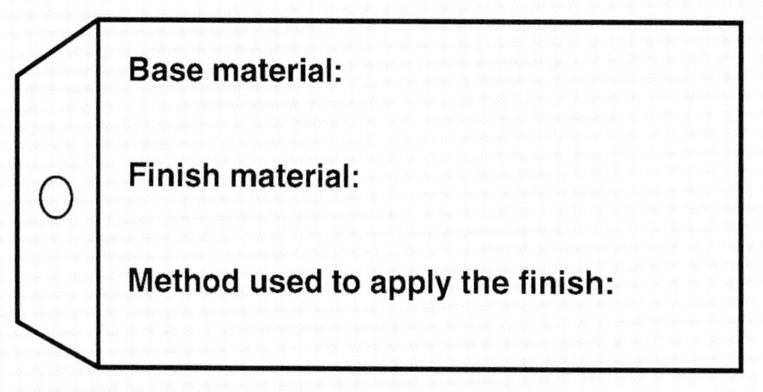

AYK 15-3. Prepare labels similar to this one. Analyze each sample and fill in the labels.

Technology Link

Medical Technology

Manufacturing is a unique and broad technology. In basic terms it is the process that changes resources into products. Manufacturing affects many other technologies. Other technologies, such as medical technology, have an effect on manufacturing as well. The link between manufacturing and medical technology is a two-way street.

Have you ever thought about how lucky we are to be living in a time when medical treatment for just about everything is just a doctor's visit away? We live in a time when damaged human body parts and organs can be replaced with new manufactured ones. Also, think about all the instruments that a surgeon would use during an operation. From the electronic monitors to the scalpel, and from the medicines to the surgical gloves, all of these items are products of manufacturing processes.

What most people don't realize is the direct effect other technologies have on manufacturing. With the advancements in medical technology, we now have a healthier workforce. Today, having healthier workers results in a more experienced, attentive, and efficient workforce. An added effect that medical technology has had on manufacturing is safety in the workplace. Through medical research, many unsafe procedures and materials are no longer being used. For example, at one time asbestos brake linings were used in the manufacture of automobiles. Because medical technology and research discovered that asbestos could be harmful if inhaled, stringent restrictions of use have been implemented.

Unit 5 Establishing a Manufacturing Enterprise

Chapter 16 Introduction to Enterprises and Management

Objectives

After studying this chapter, you will be able to:

- ✓ Identify the roles of managers in companies formed to manufacture products and construct structures.
- ✓ Define and describe management.
- ✓ List the functions of management.
- ✓ Explain and give examples of management functions.
- ✓ List five kinds of activities carried on by management.
- ✓ Explain the kind of work done in each management activity.

Have you ever wondered how a company comes into existence? It starts with someone who has a new idea. See **Figure 16-1.** For instance, the idea for Federal Express started with a graduate school paper on the need for quick package delivery. Apple Computers was based on the belief that individuals would use computers as a tool. Just think about the impact that these two ideas have had. See **Figure 16-2.**

Entrepreneurs

Some people have ideas for new products and services. They also have the personal drive to develop them. People with these special talents are called entrepreneurs. They are action-oriented people who are willing to take financial risks. They are challenged by organizing and watching their companies prosper. Entrepreneurs are often called "enterprising" people because they see opportunities and are willing to pursue them. See **Figure 16-3.**

Figure 16-1. Entrepreneurs of all ages and backgrounds start companies with bright ideas and the drive to develop them.

Figure 16-2. The personal computer industry has grown to the point that nearly everyone has access to a computer. (AT&T)

Figure 16-3. If these were enterprising people, what type of business do you think they would start?

Enterprises

Many efforts of entrepreneurs result in forming a business or enterprise. There are enterprises of all sizes and types. See **Figure 16-4.** An enterprise may be a local dry cleaner or pizza parlor, or it may be an economic giant such as General Motors or Wal-Mart. Enterprises are referred to by a number of different names, including companies, firms, establishments, and corporations. Of these enterprises, fewer than five percent are devoted to manufacturing. Though few in number, manufacturing companies produce nearly all the products made in America.

You will note that the word, industry, is not used to describe an enterprise. This term, when properly used, describes a series of enterprises that make similar products. Examples of these include the automobile industry, the forest products industry, and the food processing industry. Each of these industries is made up of a number of competing business enterprises. See **Figure 16-5.** Their products are similar or can be substituted for one another.

Free Enterprise

The key to most successful companies is free enterprise. This is an economic system that is common in many Western societies. Free enterprise believes that business should be conducted through privately owned companies. In this system, profit-centered companies conduct the vast majority of economic activities. They must compete for business, employees, and natural resources. The marketplace establishes the selling price for their products, the wage rate for their workers, and the purchase price for their material inputs. Free enterprise allows companies to operate with a minimum amount of governmental control. Only selected segments of the economy are government owned or regulated. These segments provide services that are so essential that public ownership is desirable. They also provide services when the costs make unregulated private ownership unwise. Consider the police force. What would happen if it were privately owned? See **Figure 16-6.** It would have to show a profit at the end of each year. It could

Figure 16-4. Enterprises come in all sizes and shapes.

Figure 16-6. Imagine what would happen if the police department was run for profit.

Figure 16-5. An industry is made up of a series of companies that make similar products and compete with one another.

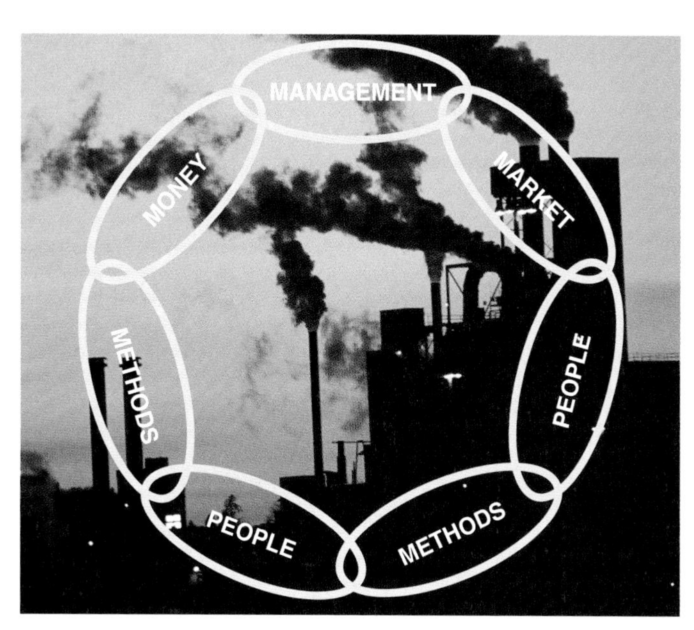

Figure 16-7. These are the parts of successful manufacturing. Each "link" of the company must be strong to produce useful products.

only serve those customers who could afford the services.

Management

There are many companies that operate many types of manufacturing systems. They produce a great variety of products. However, these systems have the same seven components. See **Figure 16-7.**

Every manufacturing system must be managed. *Management* coordinates (brings together to do one thing) the basic inputs of *people, machines,* and *materials.* These inputs are paid for with *finances.* Materials and machines are purchased. People receive wages or salaries for their work. These resources are brought together by *methods* (ways of doing things). Workers use machines to change the form of material as they produce products. The products are sold in the *market* where people pay money to own these products.

management. **The group who supervises and brings together the inputs of people, machines, materials, finances, and people to produce and sell products.**

expenses. **Money that a company spends to purchase goods and services.**

profit. **The amount of money left over after all the expenses of a business have been paid.**

retained earnings. **The portion of profits that a company uses to expand its operations or to develop new products.**

dividends. **Share of profits that is paid to a company's stockholders.**

planning. **Setting goals and the major course of action to reach them.**

The money the company gets from sales is called income. The income must pay for the *expenses,* or the cost of producing the products. Any income above the expenses and taxes is called *profit.*

The average manufacturing enterprise earns about 6% profit. This means that for every dollar of sales, the company earns about 6 cents.

Profit is used by the company in two basic ways. Some profit is kept by the company to invest in additional productive capital. It pays for such things as new machines and new buildings. This money is called *retained earnings.* Profit is also used to pay *dividends,* which are paid to the owners. These payments are the rewards for investing their money in the company. Dividends are usually paid to the owners quarterly (every three months). How income and profits are used in a company is shown in **Figure 16-8.**

Management is the process of guiding and directing company activities. The goal is efficient use of company resources. See **Figure 16-9.**

Studying three factors can develop an understanding of management. These are the functions of management (what managers do), levels of management, and areas of activity.

Functions of Management

Anyone who manages has four functions to carry out. See **Figure 16-10.** These can be described as follows:

- *Planning.* Setting goals and the course of action to be followed.
- *Organizing.* Dividing tasks into jobs and establishing lines of authority (who gives orders to whom).

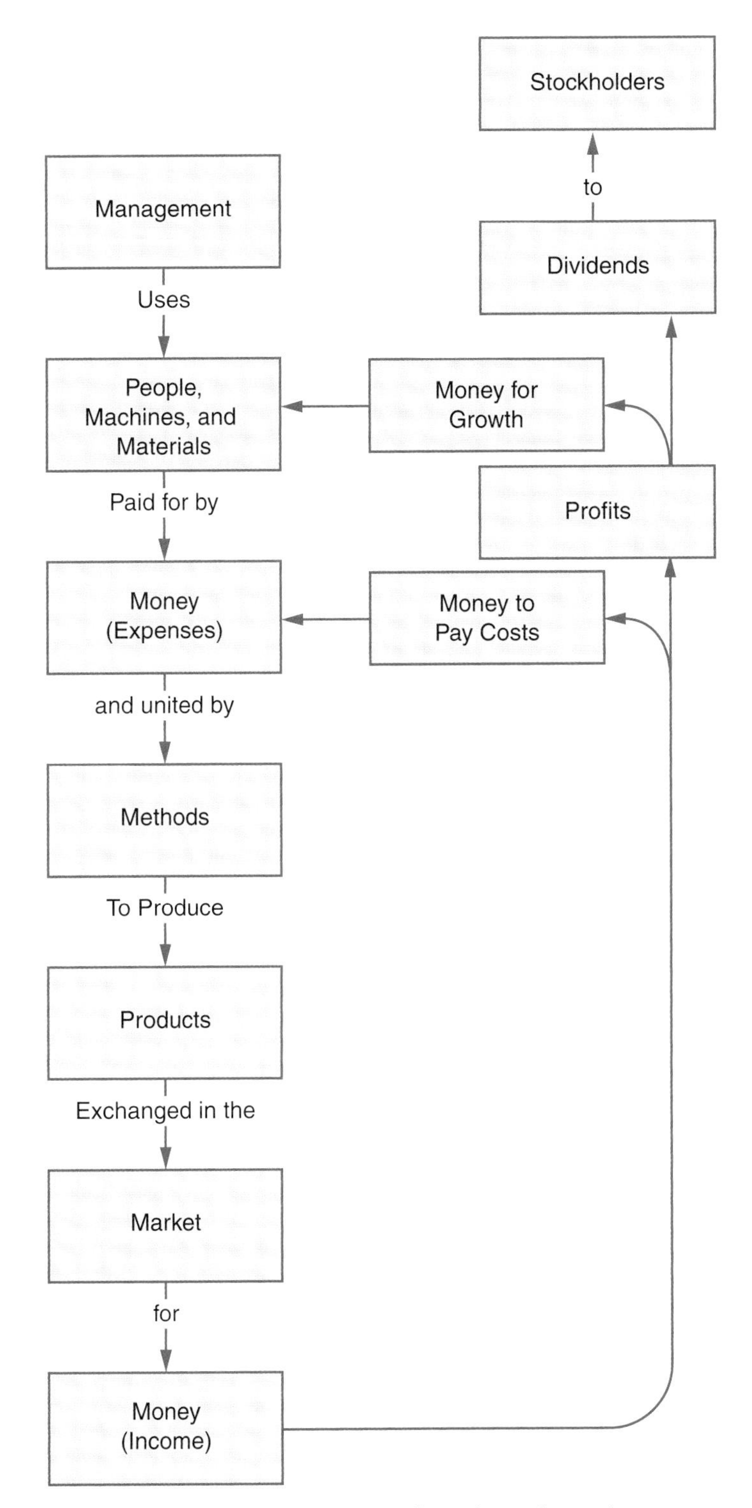

Figure 16-8. The left side of the flowchart shows how management uses resources to produce products. The right side shows the ways income is used.

Figure 16-9. Managers use many techniques to guide and direct company activities. (Hon Industries)

Figure 16-10. Managers plan, organize, direct, and control company activities. (Rohm & Haas)

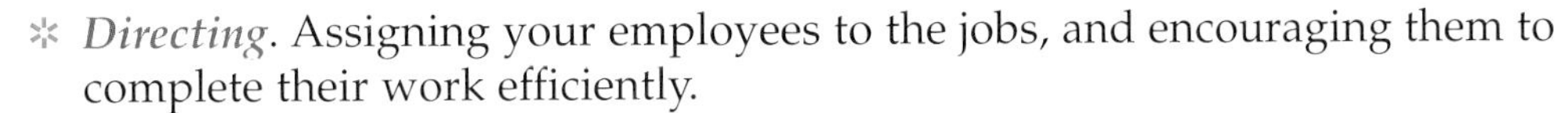

- *Directing*. Assigning your employees to the jobs, and encouraging them to complete their work efficiently.
- *Controlling*. Comparing the results of employees' work with the company plan.

organizing. Developing A structure to reach goals that have been established.

directing. Assigning employees to jobs and encouraging them to complete their work efficiently.

Figure 16-11. Planning follows four logical steps.

controlling. Comparing actual results to goals and the plans developed for reaching them.

Planning

Planning is the first function of management. See **Figure 16-11.** Planning sets the goals for the company or for one of its activities. Basically, planning means to:

- *Gather* information about the task or problem.
- *Arrange* the information so you have a "picture" of the task or problem.
- *Identify* several solutions or courses of action.
- *Decide* on the best solution or course of action. Refer to Figure 16-6.

Planning is used in all parts of the company. See **Figure 16-12.** It may be done to establish goals for the entire company. It can also be used to set smaller goals, such as: production goals, financial goals, or training goals. These goals can be short-term—daily or weekly. They may also be part of long-term goals that cover one or more years.

Organizing

Organizing is the process of ranking the importance of and assigning resources to complete tasks. See **Figure 16-13.** Each task uses human, material, and capital resources. People must also be given authority to complete the

Figure 16-12. A product planning session allows employees to exchange and evaluate product plans. (Motor Vehicle Manufacturers Assoc.)

work. They must understand how their job fits within the overall company activity. Individuals need to know who answers to them and, likewise, to whom they must answer. In short, people should understand who works for them and who is their boss.

Organizing involves three decisions:

- Who is to do each task.
- How much authority is needed to complete each task.
- How many resources are needed to complete each task.

Figure 16-13. Through organizing activities, managers decide who is to do what tasks, with what authority, and with what resources. (Nevada Power Company)

Directing

Directing begins when goals are set and tasks are organized. Now people must be assigned to do these tasks. However, more is needed. They must be trained and motivated (given a reason) to work efficiently. See **Figure 16-14.**

To be successful, the directing phase must let employees know:

- *Why* each task is important.
- *How* to complete the task.
- The *rewards* (pay and recognition) for doing the job well.

Managers who direct employees will provide proper training. They will also let each worker know he or she is important to success. *Each task, large or small, is a step toward moving a product idea to the marketplace.*

Figure 16-14. These employees receive training as part of management's directing role.

Figure 16-15. This worker uses a computer control system to help mange the output of a manufacturing process. (AMP Inc.)

Controlling

As work is completed, it must be checked and measured. See **Figure 16-15.** The results of human effort must be compared to the company's goals. This task is called ***controlling.*** It means several things. The quality of the product must be controlled.

The manager must see that the results are in balance with the goals. To do this, managers:

- Gather performance data (such as records of sales, production, and payroll).
- Compare performance to the plan.
- Determine if changes are needed.
- Decide what action should be taken.
- Begin the right action to correct any problem.

A plan may work fine the first time with the performance matching the goals. However, management does not stop there. Plans are constantly changed as people work to make the company more efficient. New organizing and directing activities are encouraged. These improvements help a company increase its productivity (output per amount of labor). A more productive company can pay its workers more or sell its products for less. It can compete better with other companies and finance new products and plants. It can grow larger and employ more people.

Levels of Management

levels of authority. **Responsibility and decision-making paths within a company.**

Managers must organize the company and establish *levels of authority*. Some people are given more responsibility than others. Some employees have greater decision-making powers than other people in the company. There is a "pecking order" within the company. This is true of all kinds of organizations.

Think about the school you are in. There are many citizens in your school district. They probably do not have the time or ability to run the schools. Thus, the voters elected a school board who hired a superintendent. He or she manages the day-to-day operation of the school district. Larger districts have several assistant superintendents who manage areas such as curriculum, personnel, and business affairs.

Principals are hired to manage individual schools. Often, people are assigned to oversee (supervise) departments such as technology education, mathematics, art, science, and history. Teachers conduct the classes offered by the departments that give students a chance to learn.

Academic Link

It is said that humans are the most complex beings that roam Earth. Many animals have methods of communicating with each other. Whether it is a rattlesnake using its rattle, a dog barking a warning, a cat arching its back, or crickets chirping, animals are constantly sending messages to each other. With innovation in the communication world, humans have taken the skill of communicating to a higher level.

A manager must be able to communicate with their bosses, other managers, and workers. Communication innovations have changed the way people communicate in the manufacturing world. At one time most managers were on-site. They could go out on the facility floor and talk to the workers. Today, with telecommuting, teleconferencing, web casting, e-mail, faxes, and telephone, a manager may be hundreds of miles from the workplace. Through communication innovations a manager can still have direct communication with what is happening minute-by-minute.

School		Business
Title	Task	Title
Voter	Final Control	Stockholder
School Board	Form Policy	Board of Directors
Superintendent	Day-to-Day Control	President
Assistant Superintendent	Control of Major Function or Area	Vice President
Principal	Manages Single Facility	Plant Manager
Department Head	Manages an Area	Department Head
Teacher	Assigns and Supervises Workers	Supervisors
Students	Does Work	Workers

Figure 16-16. A comparison between the levels of authority in a school and a business corporation.

How Companies are Organized

A company is organized much like a school. See **Figure 16-16.** There are levels of responsibility. Let us look at a corporation, a type of business organization. (The different types of businesses will be discussed in Chapter 17.)

Most corporations have stockholders who are like the citizens of the school district. *Stockholders* are the owners of a corporation. They buy a portion of the

stockholders. **Those who hold shares of stock, representing partial ownership of a company.**

board of directors. A group of people elected by the stockholders to represent their interests. The board sets company policy.

president. A full-time manager hired by the board of directors. The president is the top manager in a company.

vice presidents. People who are in charge of a major part of the company such as sales, marketing, engineering, manufacturing, or personnel.

plant manager. An individual is in charge of an entire production facility. This person manages all activities at a single plant.

department head. A person who runs one department in a plant, such as accounting or assembly.

company; however, they do not run the company. They probably live all over the country and have their own jobs. They may not know how to manage an enterprise so they elect a group to represent them. This group is called the *board of directors*. The board gives the company direction and form policy. The board hires a full-time manager called a *president* who is the top manager. She or he has several *vice presidents* who are in charge of a major part of the company. There may be vice presidents for sales, marketing, engineering, manufacturing, personnel, and so on.

Let us look at the manufacturing side of the company structure. The next level is the *plant manager*. This individual is in charge of an entire production facility and manages all activities at a single plant. The plant is usually divided into departments such as machining, shipping, accounting, assembly, or welding departments. *Department heads* aided by *supervisors* run them. These individuals assign and supervise the production *workers*.

Employees at each level have work to do. Each person is important. One job should not be viewed as better than another job. They are only different.

Production workers are essential. A company cannot survive without them. However, they have different responsibilities than do the vice presidents. This division of responsibility and authority is essential. It allows a company to be managed efficiently.

Areas of Activity

Managers have functions to carry out with a certain amount of authority to do their job. They also work in certain areas that move a product from the idea stage to completion. When all the work is done, an idea becomes a product. It is sold to customers for a profit. There are five major managed areas of activity, which are shown in **Figure 16-17:**

- **Research and development.** Discovers, designs, develops, and specifies new and improved products.
- **Production.** Engineers, designs, and sets up manufacturing facilities that produce products to the company's quality standards.
- **Marketing.** Identifies the people who will buy the products. Then marketing promotes, sells, and distributes the products.

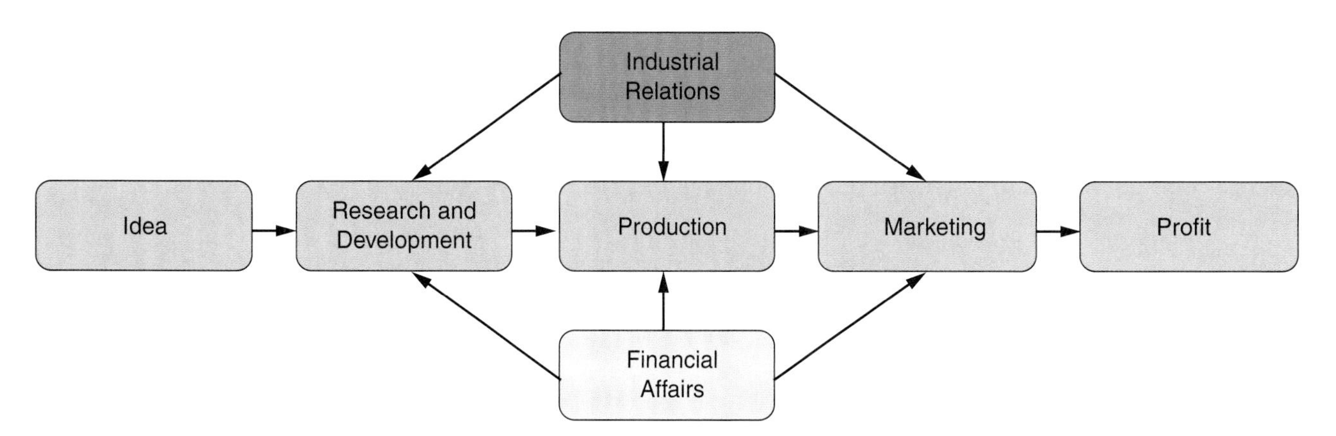

Figure 16-17. The five managed areas of activity involved in changing ideas into products.

Figure 16-18. The better a company is organized, the better it will run. The employees must work together for a company to be a success. (General Electric)

- **Industrial relations.** Operates programs to find and train the company's workforce. It also does things that make the public and workers feel good about the company.
- **Financial affairs.** Raises and controls the company's money. These five areas of activity cause a product to evolve. These activities will be discussed more fully in Chapters 17 through 33.

supervisors. Individuals who assign jobs to and supervise production workers.

workers. The people who do the actual production.

Summary

Management directs and controls company activities. See **Figure 16-18.** A view of management requires some basic knowledge that involves an understanding of functions, levels of authority, and areas of activity for managers.

Managers plan, organize, direct, and control as they manage single tasks or entire companies. They perform these functions in a structured way. The structure extends all the way from the owners down to the workers. Managers work in five areas of activity: research and development, production, marketing, industrial relations, and financial affairs.

Key Words

All the following words have been used in this chapter. Do you know their meaning?

board of directors
controlling
department head
directing
dividends
expenses
levels of authority
management
organizing
planning
plant manager
president
profit
retained earnings
stockholders
supervisors
vice presidents
workers

Test Your Knowledge

Please do not write in this text. Place your answers on a separate sheet of paper.

1. What are the three factors you must study to understand management?
2. From the following list, select those activities that are the functions of management.
   a. Planning (setting goals and courses of action).
   b. Identifying markets.
   c. Organizing (assigning tasks to certain jobs and establishing lines of authority).
   d. Directing (assigning employees to jobs).
   e. Controlling (comparing employees' work with company plan).
   f. All of the above.
3. Stockholders (do, do not) run the company.
4. List the five major managed areas of activity. Describe what is done in each area.

Applying Your Knowledge

1. Study the organization of a church, school club, trade association or other group. List the officers (managers) of the group according to the level of authority.
2. Organize a group to perform a community service such as picking up litter. Plan the activity, organize the group, direct the group, and control the activity.

Chapter 17 Organizing and Financing an Enterprise

Objectives

After studying this chapter, you will be able to:

✓ List different ways companies are owned.
✓ Describe three methods of ownership.
✓ Discuss advantages and disadvantages of each type of ownership.
✓ Identify steps required by law to form a company.
✓ Describe three kinds of company organization.
✓ Describe how a company determines its money needs.
✓ List ways of financing (getting money) for starting a company.

Enterprises are organized by a person or group of people. As the company is organized, decisions are made. The main ones are:

- What type of ownership will be used?
- What local or state rules must be met?
- What type of management will work best?
- How will the finances (money or capital) be raised?

Each new company faces the same questions. Let us look at these elements one at a time.

Forms of Ownership

Most companies are publicly owned. That is, one or more individuals own them. The government does not own them. Most companies are formed to make money for the owners. They are said to be *profit-centered*.

profit-centered. **Companies that are formed to make money for the owners.**

There are three forms of public, profit-centered ownership as shown in **Figure 17-1.** The type of ownership that is selected for a company will depend on several factors.

Figure 17-1. These are the three forms of ownership.

Figure 17-2. A farmer is often the proprietor of the farm he or she operates. (Deere and Company)

Proprietorship

The proprietorship is the simplest and oldest form of ownership. A *proprietorship* is a business enterprise owned by one person, called the *proprietor*. Many service stations, antique shops, farms, and retail stores are owned by a single individual. See **Figure 17-2.**

The proprietorship is used for enterprises that are small and need little capital. They have several advantages. See **Figure 17-3.** Proprietorships are easy to form because a simple business license is all that is usually required. The management structure is simple. The owner directly controls all operations. This gives the company flexibility (ability to change easily) to react quickly to changes in the market. Finally, the owner has the right to all after-tax profits.

However, proprietorships have certain disadvantages. See **Figure 17-4.** Often the enterprise cannot easily raise more money. Thus, its growth is sometimes held back.

proprietorship. **A business enterprise owned by one person.**

Also, few individuals have all the talents needed to run a company. Limited management talent can cause the enterprise serious problems.

Finally, the owner is responsible for all debts of the company. If the company fails, the owner must pay the debts with his or her own money. This is called *unlimited liability*.

Partnership

A partnership overcomes some of the disadvantages of a proprietorship. The *partnership* is an association of two or more people to run a legal business.

Figure 17-3. A proprietorship has some advantages.

Such businesses usually are easy to start and end. They can offer more management talent and more ways to raise money for the company.

The partners must still accept unlimited liability for the company debts. Also, the owner-managers might have arguments. They must share responsibilities as well as profits. See **Figure 17-5.**

Figure 17-4. Proprietorships also have disadvantages.

Corporation

Most manufacturing companies are organized as corporations. A *corporation* is a legally created business unit that is an *artificial being* in the eyes of the law. See **Figure 17-6.** It is created in one state and can operate in all states. Like all beings, corporations can own property, sue or be sued, and enter into contracts.

Corporations have long lives because their owners (stockholders) can sell their holdings without causing a change in management. Generally, the owners are not the managers. Instead, individuals with special skills are hired to manage the enterprise.

Corporate owners have a definite advantage. They have limited liability. The corporation, not the owners, is responsible for all debts. Owners can lose their original investment if a corporation fails. They do not, however, have to furnish additional money to pay outstanding debts.

unlimited liability. Exists when the owner of a business is responsible for all debts of the company (if it is a proprietorship or a partnership). The owner must pay the debts with his or her own money.

Of course, corporations have some disadvantages. The owners generally have little interest in the daily operations of the company. Their main interest is in dividends, which often causes management to work toward high short-term earnings. Long-term growth may not be given proper attention by the management.

Also, corporations must file many government reports that make their operations more public. This exposure makes competing more difficult. Corporate profits are heavily taxed and the stockholders' dividends are also taxed.

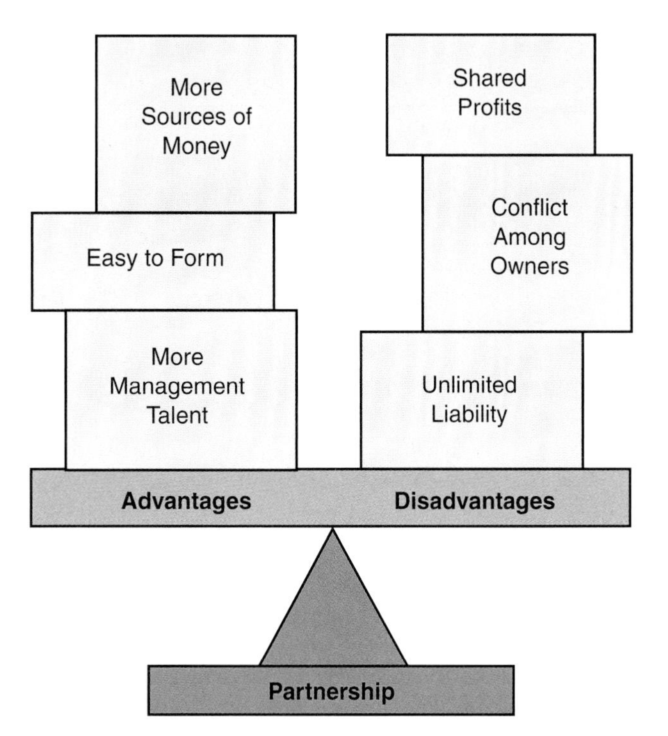

Figure 17-5. Two or more people can form a partnership. It has both advantages and disadvantages.

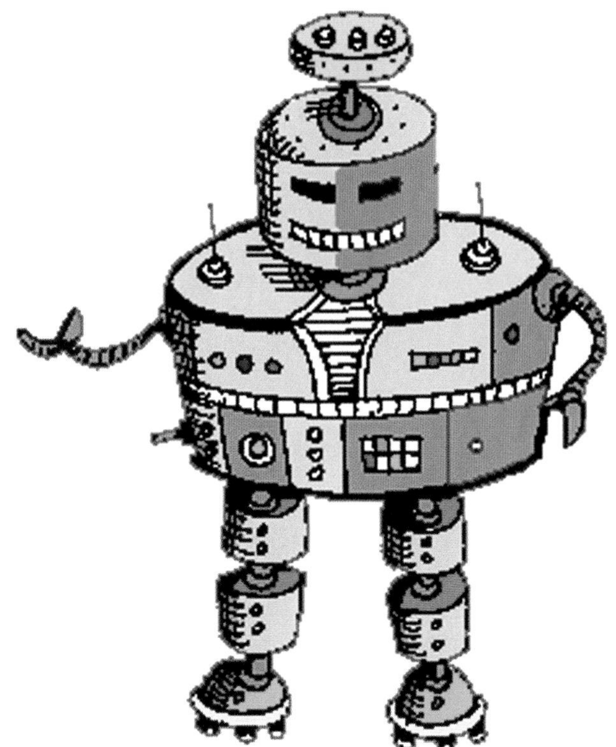

Figure 17-6. A corporation is an artificial being in the eyes of the law.

partnership. A form of business ownership involving two or more persons. Usually, partners will have equal shares of ownership of the business.

corporation. A legally created business unit. It is an "artificial being" in the eyes of the law.

articles of incorporation. The application for a corporate charter, containing information on the activities of the proposed corporation and the name of persons involved in it.

Forming a Company

All enterprises must become a legal company. Most manufacturing companies are corporations. Therefore, we will limit our discussion to forming a corporation.

As we said before, a corporation is an artificial being. Therefore, it must be born through a specific process shown in **Figure 17-7.**

Articles of Incorporation

Three major things control a corporation. These are:

- The laws of the state where it is formed.
- The corporate articles of incorporation.
- The corporate bylaws.

The laws vary from state to state. However, all provide basic rules for ownership and financing of a company.

The company must select a state where it wants to incorporate. It prepares an application form. This form is called the *articles of incorporation* or application for a charter. It is filed with the proper state official. The certificate usually requires:

- Name of the company.
- Purpose for forming the company.

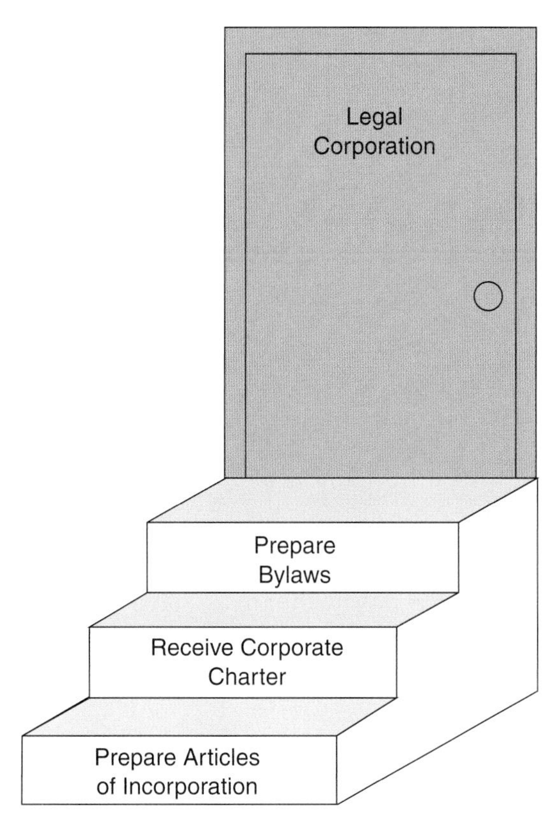

Figure 17-7. There are three steps in forming a corporation.

Figure 17-8. Articles of Incorporation mark the "birth" of a corporation. (Boise Cascade Corp.)

- Names and addresses of those forming it.
- Location of the company office in the state.
- Type and value of the stock to be sold.
- Names of principal officers of the company.

Other information, which may vary from state to state, may be requested.

Corporate Charter

The state officials review the articles of incorporation and determine the filing fee. This fee is generally based on the number and value of the shares of stock to be sold.

When the articles are approved and the fees are paid, the state issues a *corporate charter*. See **Figure 17-8.** This charter authorizes the company to do business in the state. It is the corporate *birth certificate* that all other states recognize. A charter from one state allows a company to conduct business in all states.

corporate charter. A document issued by the state that legally authorizes the company to do business in the state.

BYLAWS

OF

BOISE CASCADE CORPORATION

Offices

Section 1. The registered office of the corporation in Delaware shall be in the City of Wilmington, County of New Castle.

Section 2. The corporation may also have offices at such other places both within and without the State of Delaware as the board of directors may from time to time determine or the business of the corporation may require.

Meetings of Stockholders

Section 3. All meetings of the stockholders for the election of directors shall be held in Boise, Idaho, at such place as may be fixed from time to time by the board of directors, or at such other place either within or without the State of Delaware as shall be designated from time to time by the board of directors and stated in the notice of the meeting. Meetings of stockholders for any other purpose may be held at such time and place, within or without the State of Delaware, as shall be stated in the notice of the meeting or in a duly executed waiver of notice thereof.

Figure 17-9. Bylaws provide general rules for operating a corporation. (Boise Cascade Corp.)

Corporate Bylaws

The corporate charter is very general. It does not give many directions for running the company. More detailed information is spelled out in the company *bylaws*. See **Figure 17-9.** Most bylaws outline:

- Date and location of the annual stockholders meeting.
- Date and place of periodic board of directors meetings.
- List of corporate officers with their duties, terms of office, and method of appointment.
- The number, duties, and terms of office of the directors.
- Types of proposals that must have stockholder approval.
- Method to be used to change the bylaws.

The stockholders have the power to develop and change the bylaws. Often they delegate this power to the board of directors. In the end, the stockholders have little control over the daily operations of the company. This power rests with the major corporate managers and the board of directors.

Developing a Management Structure

bylaws. Detailed information that outlines how a company will be run. The stockholders have the power to develop and change the bylaws.

management structure. The way a company is organized, how responsibility flows, and how decision making is divided.

Running a corporation properly takes a team of persons who will manage it. For each important function (task) of the company there will be a manager. There may be several levels of managers each with additional managers and staff. This is called a *management structure*. So that everyone knows who is her or his boss, the lines of authority are sketched into a chart. A management structure must be developed early in a company's life.

The main officers, managers, and workforce were presented in Chapter 16. These included:

- **President.** Responsible for all company activities.
- **Vice president.** Manages a major segment of the company.
- **Plant manager.** Manages a single production facility.
- **Department head.** Manages a major activity within the production facility.
- **Supervisor.** Directs the work of production workers.
- **Workers.** Complete the work outlined by the management.

This listing follows the production activity right down to the production worker. However, there are other areas of activity. Marketing, industrial relations,

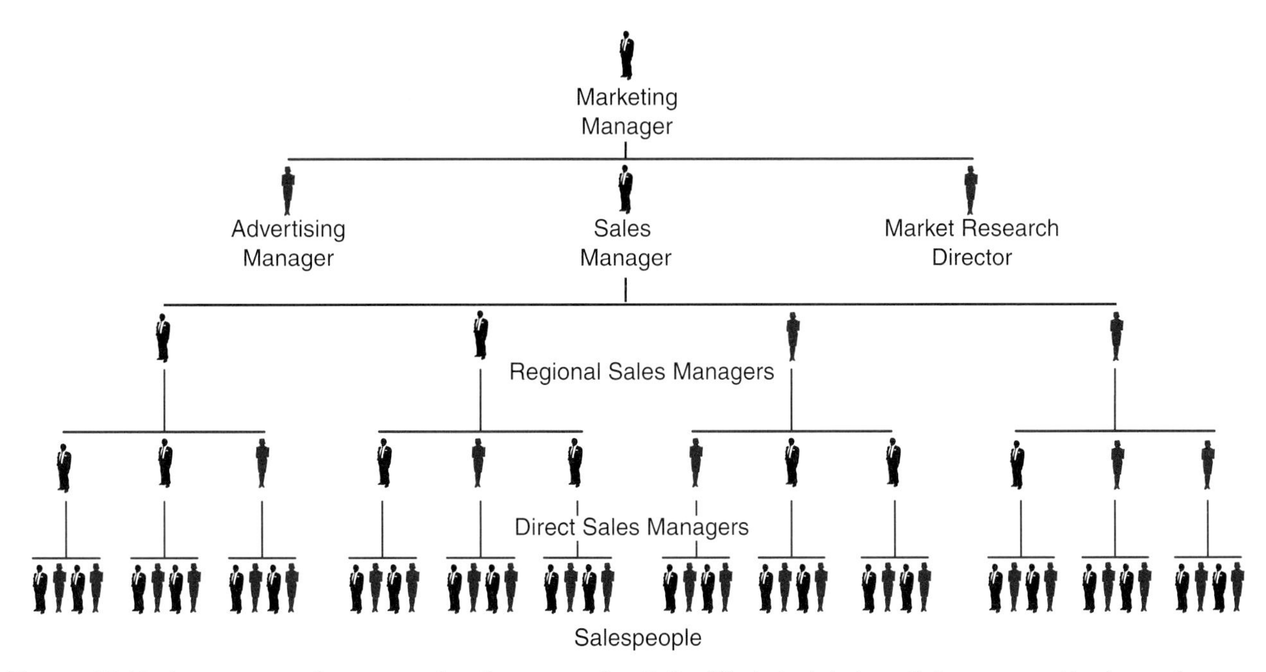

Figure 17-10. A structure of a nonproduction area of activity. Their task is to sell the corporation's products.

research and development, and financial affairs also have levels of authority. See **Figure 17-10.**

These various areas of activity and levels of authority must be organized, too. Responsibilities must be fixed and lines of authority are drawn. See **Figure 17-11.** The result is an organizational structure. There are a number of organization models (charts) but the three most common are:

- Line organization.
- Line and staff organization.
- Line and function staff organization.

Figure 17-11. Everyone in the company should know who is responsible for each job.

Line Organizations

Line organization is the simplest. A single line of authority flows from the president to the vice presidents. See **Figure 17-12.** From there, authority flows directly through various levels to the workers.

All information and direction flows vertically up and down the structure. There are no horizontal connections between different tasks. This type of organization is also called a military structure. It can work well only in very small organizations.

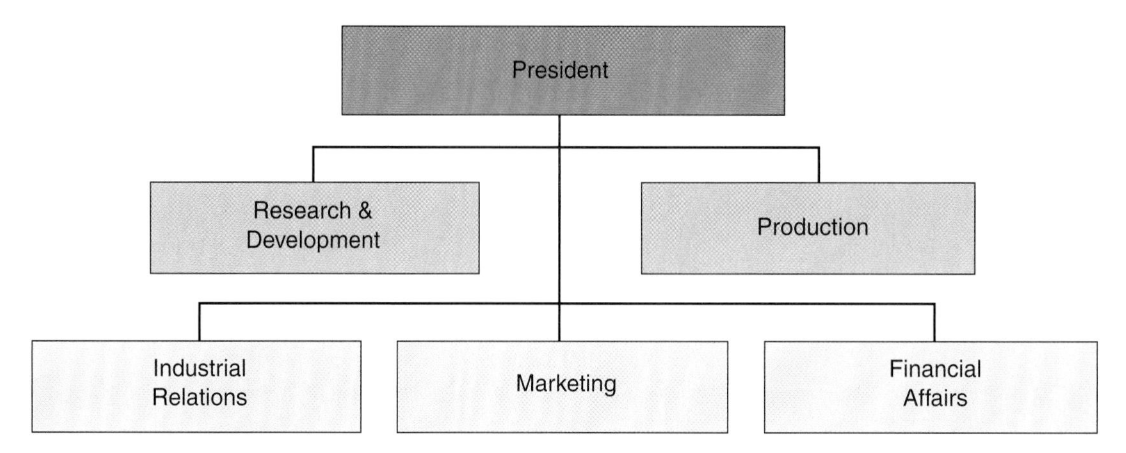

Figure 17-12. Typical areas under the control of the company president. She or he is the immediate superior of managers of all these departments.

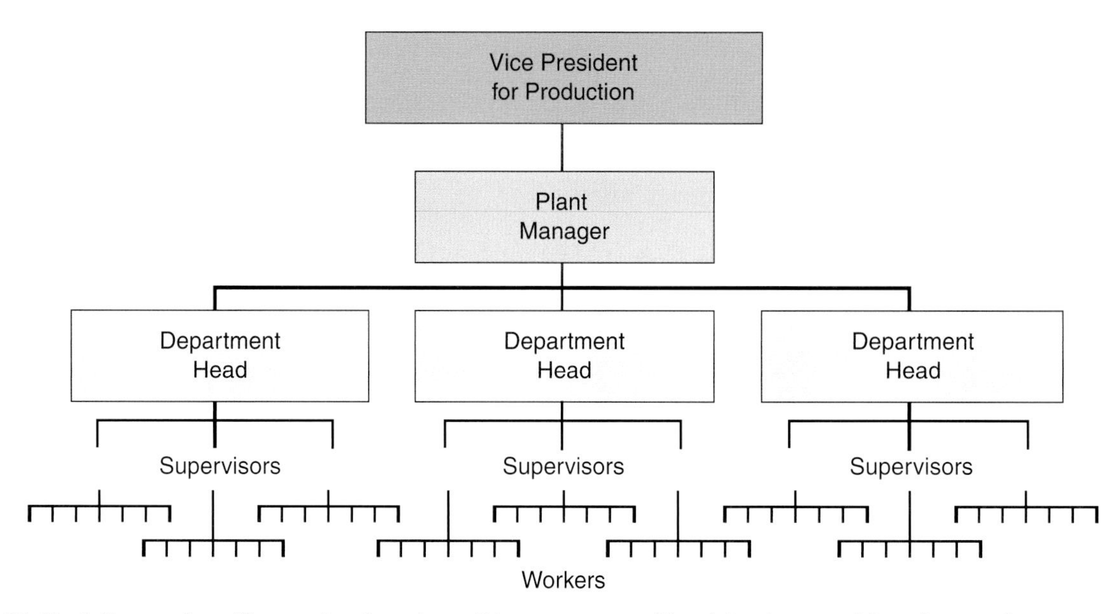

Figure 17-13. A line and staff organization chart. Line managers like this vice president have other managers to advise them.

Line and Staff Organizations

In line organization a single person oversees all operations. This person manages sales, public relations, research, distribution, manufacturing, and many other operations. In larger companies, this becomes an impossible task. Therefore, the line and staff organization is used. See **Figure 17-13.** The major line managers have staff people to advise them. The line managers still have the authority over operations while staff advice helps them manage more effectively.

Line and Functional Staff

The third organizational structure is line and functional staff. The main change is in the function of the staff, which is given authority over specific areas.

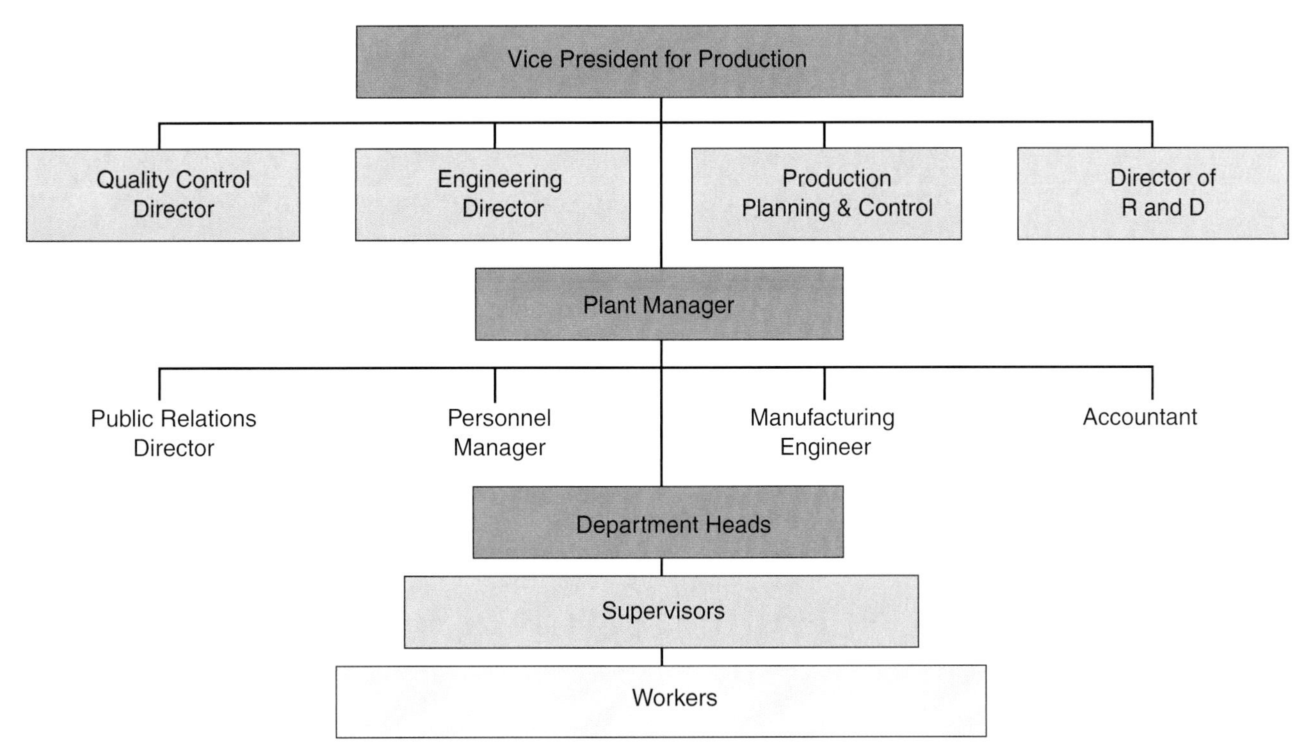

Figure 17-14. A line and functional staff organization chart. Officers in each area make the decisions in their own departments.

The staff members move from advice giving to decision making. The staff, such as quality control, has authority over its activities until conflict occurs. Then higher management must resolve the differences. For example, the quality control staff may not agree with the production-planning director. The vice president for production must then decide who is right. See **Figure 17-14.**

Determining Financial Needs

All companies need money to operate. Established companies can obtain most of their money from sales. New and expanding companies must raise finances from outside sources.

Companies find out how much money they need by making up a budget. *Budgets* are estimates of income and expenses. They detail the costs of operation and sources of income. There are six major types of budgets, **Figure 17-15:**

budget. **Plan that a business uses to forecast income and expenses.**

- A sales budget estimates sales for a specific period.
- A production budget estimates the number of products to be produced to meet the sales budget.
- A production expense budget estimates production costs. They include material, labor, and overhead (equipment, utilities, rent, etc.) costs.
- A general expense budget estimates cost not directly related to the manufacture of products. This includes marketing and administrative costs. Research and development, financial affairs, industrial relations, and top management's expenses are often considered administrative costs.
- A master budget summarizes all other budgets.

Academic Link

With every business, accounting practices and budgeting can be a make-or-break situation. Accounting is a field of study that uses numbers, mathematics, and organizational skills. It is the art of recording, classifying, and summarizing a business and its actions in terms of money, transactions, and events. Figuring budgets is done by analyzing previous accounting reports, finding income and expense trends, and determining estimates. As stated in the previous section, a budget is an estimate of future income and expenses.

Let's say the accounting reports indicate that over the last four years the sale of grass seed has increased 10% each year for the month of April. There may be many other factors that influence accounting and budgeting decisions. However, with this example simple estimating would suggest that 10% more seed than was sold this April should be budgeted for next April.

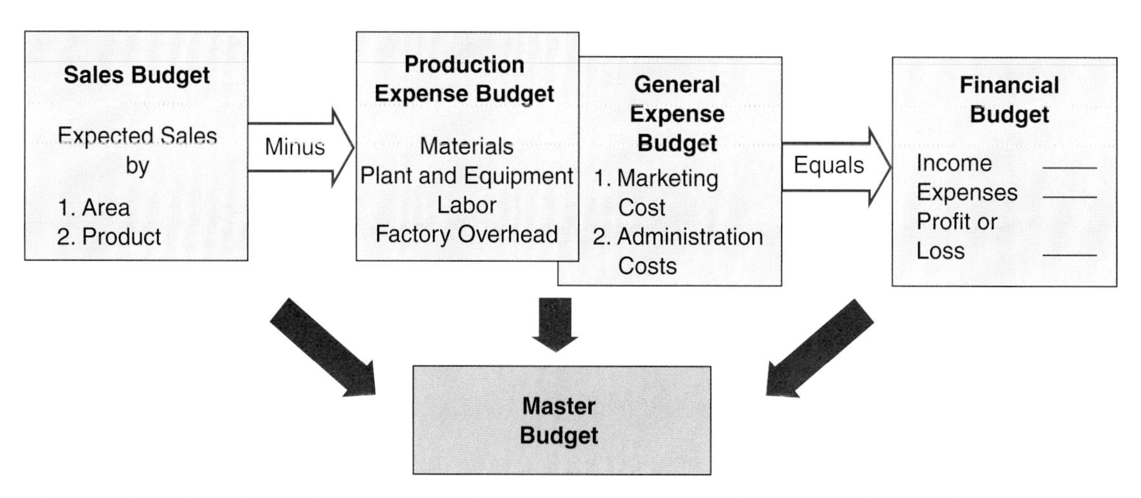

Figure 17-15. Flowchart shows how a master budget draws information from other budgets.

Getting Financing

equity financing. Raising money for a company by selling shares of ownership.

stock. Ownership rights to the company, bought by people investing in the company.

Budgets help managers determine their company's need for money. Often this money must be raised from outside the company.

There are two main ways to raise outside finances. Ownership rights to the company can be sold. This is a technique called *equity financing*. People are sold *stock* and become part owners, or shareholders, of the company. See **Figure 17-16.**

Companies can also borrow money using *debt financing*. Basically they agree to "rent" some money from someone who will provide money for a period of time. This money is called a *loan* that the company must repay plus interest (rent).

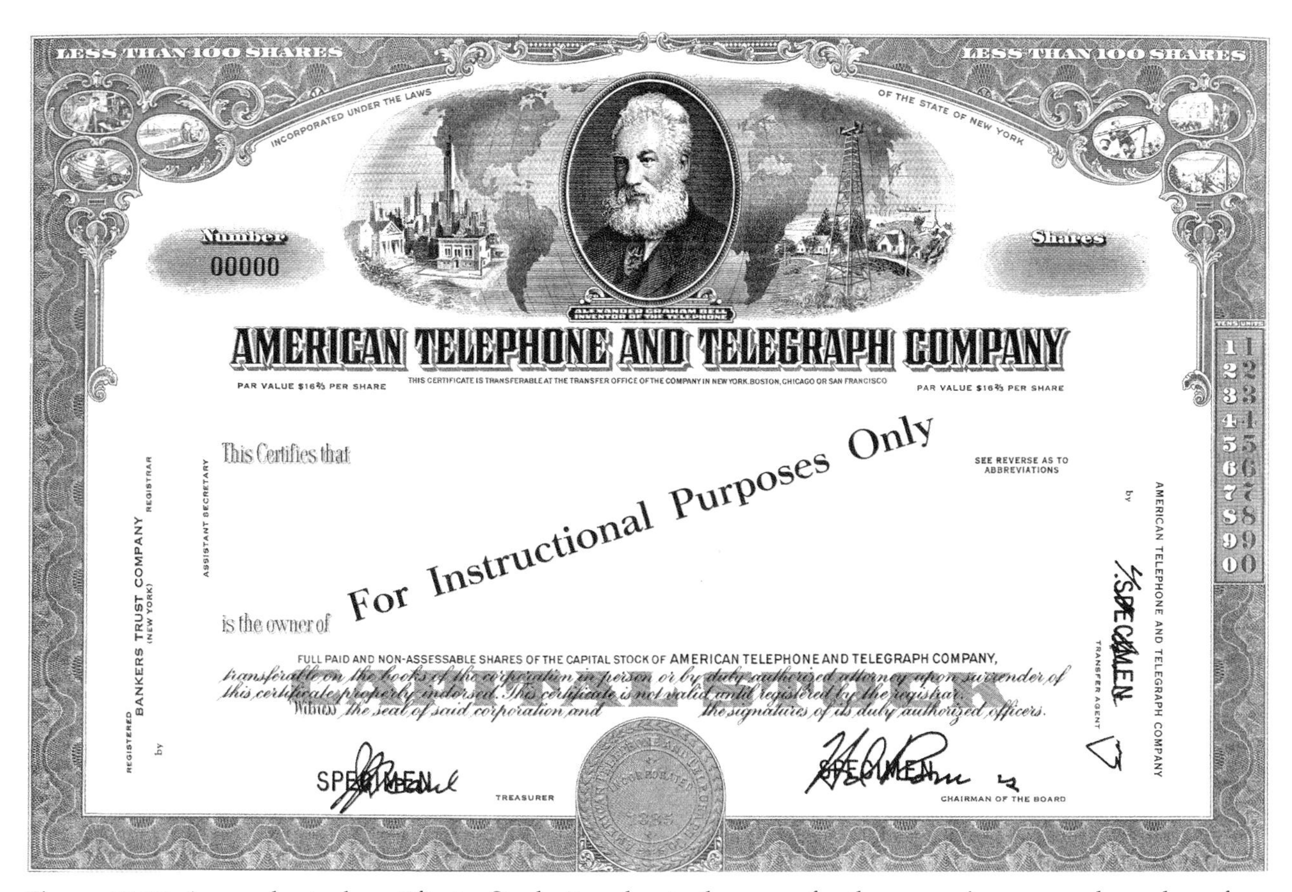
LESS THAN 100 SHARES

LESS THAN 100 SHARES

INCORPORATED UNDER THE LAWS OF THE STATE OF NEW YORK

Number

00000

Shares

ALEXANDER GRAHAM BELL
INVENTOR OF THE TELEPHONE

AMERICAN TELEPHONE AND TELEGRAPH COMPANY

PAR VALUE $16⅔ PER SHARE

THIS CERTIFICATE IS TRANSFERABLE AT THE TRANSFER OFFICE OF THE COMPANY IN NEW YORK, BOSTON, CHICAGO OR SAN FRANCISCO

PAR VALUE $16⅔ PER SHARE

This Certifies that

SEE REVERSE AS TO ABBREVIATIONS

For Instructional Purposes Only

is the owner of

FULL PAID AND NON-ASSESSABLE SHARES OF THE CAPITAL STOCK OF AMERICAN TELEPHONE AND TELEGRAPH COMPANY,
transferable on the books of the corporation in person or by duly authorized attorney upon surrender of this certificate properly indorsed. This certificate is not valid until registered by the registrar.
Witness the seal of said corporation and the signatures of its duly authorized officers.

SPECIMEN

TREASURER

SPECIMEN

CHAIRMAN OF THE BOARD

REGISTERED BANKERS TRUST COMPANY (NEW YORK) REGISTRAR by

ASSISTANT SECRETARY

AMERICAN TELEPHONE AND TELEGRAPH COMPANY TRANSFER AGENT by

SPECIMEN

1 2 3 4 5 6 7 8 9 0

Figure 17-16. A sample stock certificate. Study it and note the space for the owner's name and number of shares owned.

debt financing. Raising money for a company by borrowing from a bank or selling bonds.

loan. Money borrowed by a company from a bank or other lender for a specific term at a stated rate of interest.

Summary

All beginning companies must be organized and financed. They become a legal business by meeting certain state requirements. They may be formed as proprietorships, partnerships, or corporations.

The company must establish a managerial structure. Responsibilities and lines of authority for managers must be developed.

Finally, money must be raised to run the company. Budgets are developed to determine financial needs. Money is obtained through equity or debt financing.

Key Words

All of the following words have been used in this chapter. Do you know their meaning?

articles of incorporation
budget
bylaws
corporate charter
corporation
debt financing
equity financing
loan
management structure
partnership
profit-centered
proprietorship
stock
unlimited liability

Test Your Knowledge

Please do not write in this text. Place your answers on a separate sheet of paper.

1. Name and describe the three profit-centered types of company ownership.
2. The _____ _____ _____ is an application form for a state charter.
3. Most bylaws outline six basic directions by which companies run. What is included in these directions?

Matching questions: Match the definition (4 through 8) with the correct term (a through e).

4. _____ Manages a major activity within the production facility.
5. _____ Manages a major segment of the company.
6. _____ Directs the work of production workers.
7. _____ Single line of authority flows from president to vice presidents; from them to workers.
8. _____ Major line officers have staff officers to advise them.
   a. Line organization.
   b. Supervisor.
   c. Line and staff organization.
   d. Department head.
   e. Vice president.
9. _____ are estimates of income and expense.
10. *True or false?* A master budget estimates production costs including labor, material, and overhead.

Applying Your Knowledge

1. Study the organization of a church, school, or other group. Prepare an organization chart for the group.
2. Visit a local company and obtain their organization chart. Share it with your class.
3. Work with your parents to set up categories for a household budget. *Do not* include dollar values for each category.
4. Assume you and two friends are going to start a lawn care service. What type of ownership would you use and why?

Technology Link

Energy and Power

Manufacturing is a unique and broad technology. In basic terms, it is the process that changes resources into products. Manufacturing affects many other technologies. Other technologies, such as energy and power, have an effect on manufacturing as well. The link between manufacturing and energy and power is a two-way street.

When you walk into a dark room and switch on the lights, have you ever wondered how the light is made, how it got the power to light up, or how that power is generated? The plastic for the switch went through a forming process, the wires were extruded, and the light fixture and light was manufactured. All of the components that made the light turn on are products of a manufacturing process.

What most people don't realize is the direct effect other technologies have on manufacturing. Energy and power has a significant role in the world of manufacturing. There has always been a need for transportation of resources to manufacturing facilities and transportation of final products from the facilities to the marketplace. With the desire of reducing inventory costs and the innovation of just-in-time manufacturing, demands of efficient and timely transportation sources are that much greater. In a world that wants its products "yesterday," transportation of products to the marketplace is that much more important.

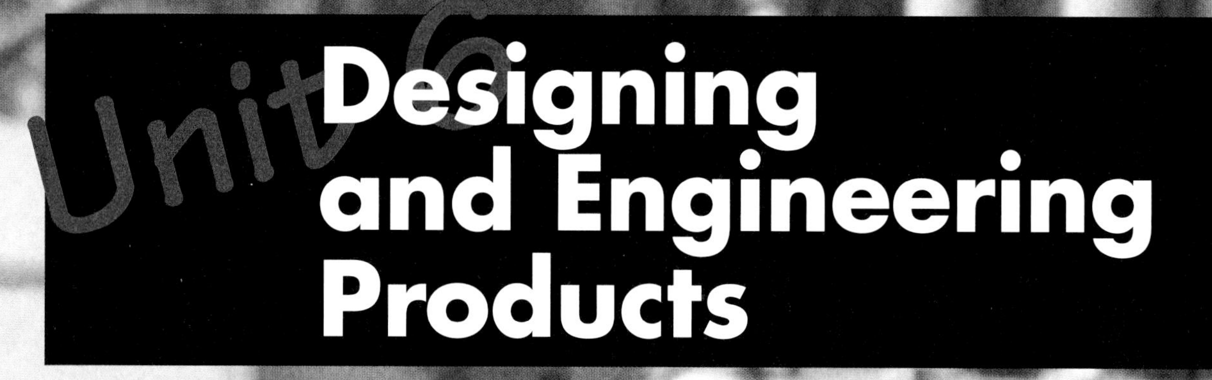
Unit 6
Designing and Engineering Products

Chapter 18 Establishing Customer Needs and Wants

Objectives

After studying this chapter, you will be able to:

- ✓ Define and describe two approaches manufacturing companies use in developing new products.
- ✓ Explain how companies can identify good product ideas.
- ✓ Describe several processes companies might use in developing new product ideas.
- ✓ Appreciate the role product development plays in the success of a manufacturing enterprise.
- ✓ Develop some methods of your own for finding products your class can mass produce and sell.

All products must start with an idea that usually is based on someone's needs. See **Figure 18-1.** Someone thinks of a product that people need. Companies use two basic systems in developing products. These are a production approach or a consumer approach. See **Figure 18-2.**

Figure 18-1. All products start with an idea.

Production Approach

The production approach to product design stresses producing products. A design staff develops a product, and then it is produced in quantity. See **Figure 18-3.** Major advertising campaigns try to convince people they need the product.

Many high-volume consumer goods are developed by the production approach. Cosmetics, toiletries, soaps, toothpaste, and designer clothing are examples of products using this approach. It is doubtful that large numbers of people were asking for CD players or electric pencil sharpeners to be developed. They were first designed and produced, and then we were told, through advertising, that we needed them.

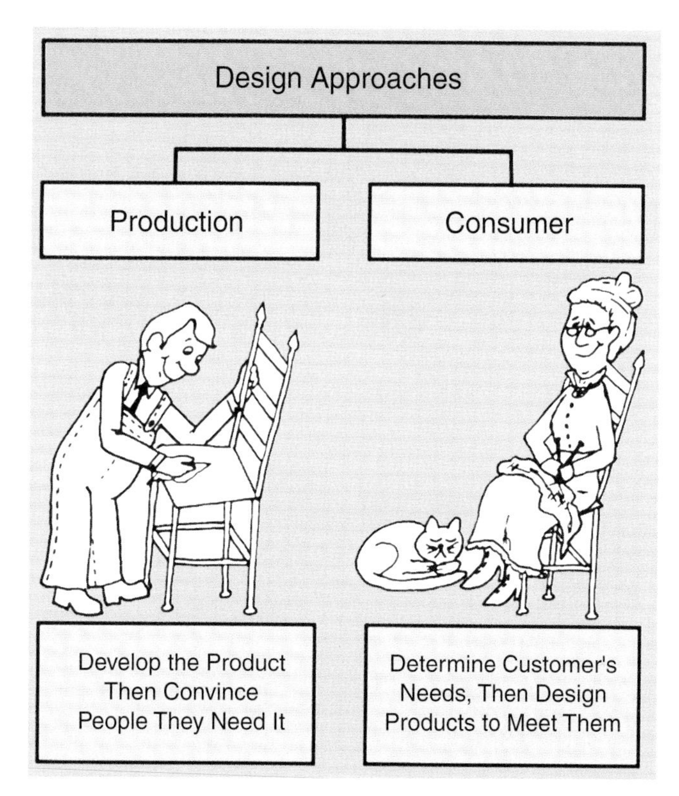

Figure 18-2. These common design approaches are used by companies.

Figure 18-3. Many high-volume, low-cost goods, such as this toy, are produced using a production approach. (Ohio Art Company)

Figure 18-4. This aircraft was developed using a consumer approach. (Gulfstream)

Consumer Approach

The consumer approach first identifies products people need. Then the products are designed and produced. This approach is widely used in developing industrial goods. See **Figure 18-4.**

For example, The Boeing Company carefully collects data from airlines before they design a new aircraft. They try to include all the features requested by the many different airlines.

Another company asked boat builders and casket makers "What kind of sander do you need?" Their answer was "A small machine that will sand curves and small surfaces." This research led to the hand-held oscillating sander.

Until recently most products you bought were designed using the production approach. Growing competition has caused more companies to use the consumer approach.

Identifying Product Ideas

Developing products is hard. New products must balance the needs of the customer with the strengths of the company. See **Figure 18-5.** Customers must need or want the products. Likewise, the company must have the resources to design, produce, and market the product.

A company cannot produce a product just because it is a good idea. It must fit the company's area of operation. A metal machining company will not be very interested in an idea for a wood desk.

Determining Consumer Needs and Wants

Each of us has needs and wants. You may want a new bicycle but your best friend may want a stereo system. This difference causes a problem for companies. They cannot manufacture a product just for you. They need to produce products many people want (mass appeal).

The company must decide what product a group of people will buy. The way this is done depends on how the company develops products. See **Figure 18-6.**

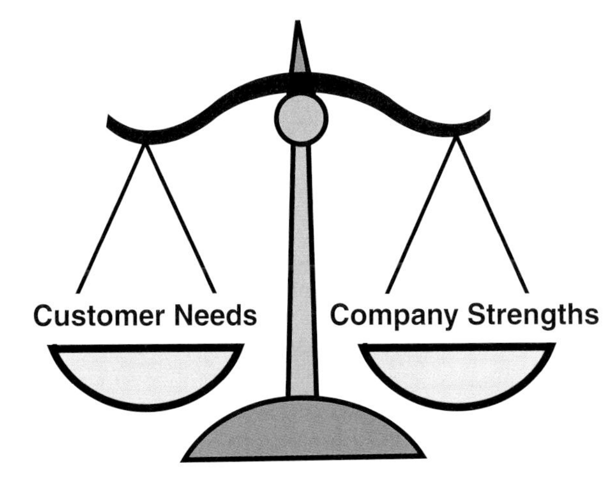

Figure 18-5. Products must be designed to meet customer needs and use company strength. (BEA Systems)

Some companies are imitators that produce products much like those other companies make. They let someone else identify and build the market. The basic information they need is sales figures. Assume Company D (developer) is selling lots of widgets. Then Company I (imitator) will also want to make widgets. *Imitation* is a common product development technique. Think of products that are widely imitated. How about home computers, stereo receivers, clothing styles, and toys?

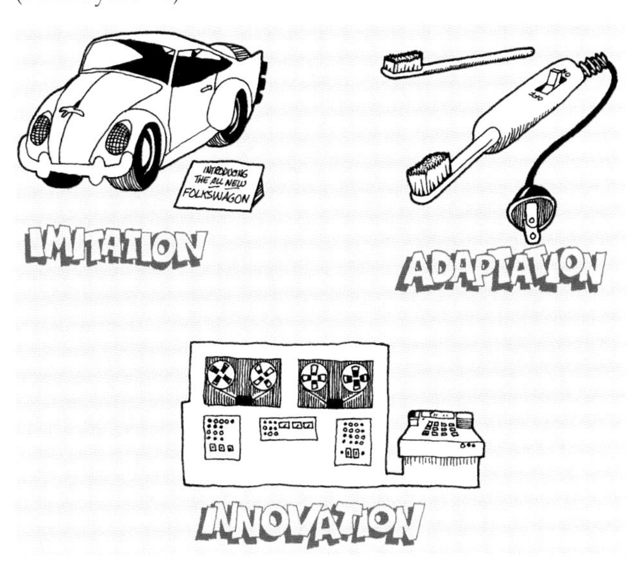

Figure 18-6. Product development may imitate, adapt, or innovate.

Another technique is *adaptation*, which means developing an "improved" product by changing its operation. The manual typewriter was adapted to be an electric typewriter. The electronic typewriter was a further adaptation, replacing the electric typewriter. Today's personal computer loaded with word processing software has replaced the typewriter in most workplaces and many homes. See **Figure 18-7.**

The last technique is *innovation* (creating something new) where a totally new product is developed. The video recorder, CD player, microchip (**Figure 18-8**), and polyester fibers are recent innovations.

imitation. A common product development technique. A company will produce a product much like those of other companies, letting someone else identify and build the market.

adaptation. Changing a product by improving it. For example, the electric typewriter is an adaptation of the manual typewriter.

Sources of Product Ideas

Most companies are constantly seeking new products. See **Figure 18-9.** Each new product starts with ideas that come from both inside and outside the company. They basically arise from studying three sources shown in **Figure 18-10.**

Market Research

Market research is used to study people's thinking about products. It may tell the company what products people want. It may test people's feelings about a product that the company already makes. The information gathered is used for either product development or product improvement.

Figure 18-7. A—The electronic typewriter was the precursor to the personal computer. (IBM Corp.) B—Today's personal computer allows the user to go beyond typing to integrate text and graphics to produce high-quality documents.

Figure 18-8. The silicon wafer is an example of an innovation. Each small square on the wafer contains a custom designed integrated circuit. (AT&T)

Figure 18-9. Successful companies generate a constant flow of new products. (Hon Industries)

innovation. **A product development technique where a totally new product is developed. The videocassette recorder, CD player, and the microchip are examples of innovations.**

Market research gathers information in three major areas:

- Information about customers
- Information about the product
- Information about the marketing approach

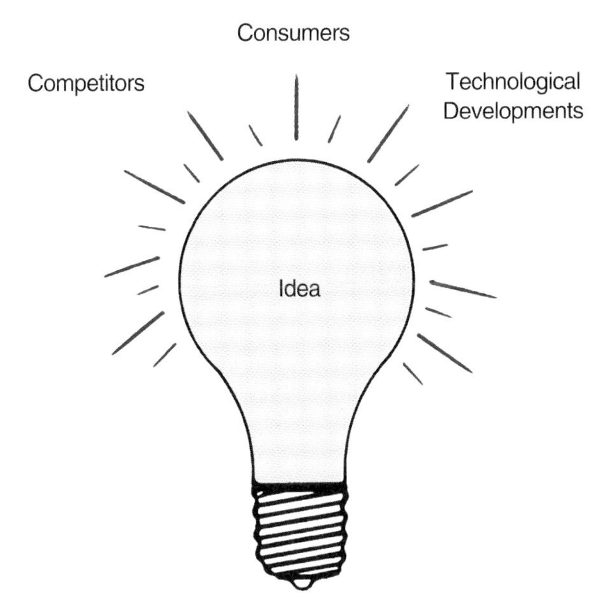

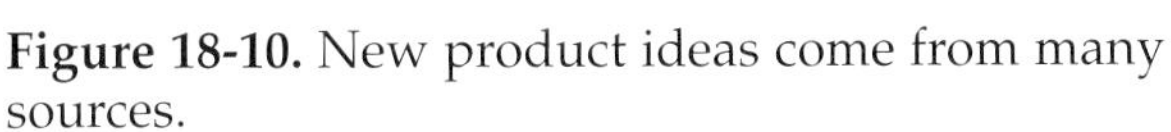
Figure 18-10. New product ideas come from many sources.

Figure 18-11. Some product ideas are tested through consumer surveys. (American Woodmark)

The results give the maker information about people's choice of product size, color, style, and function. This information is the starting place for good product design or redesign.

market research. **The process of gathering and analyzing information about customers' desires, competing products, and sales results.**

Some kinds of research collect information about the market itself. A company learns things about its customers: their age, gender, where they live, their occupation, and income.

Finally, a company gathers data on how good its marketing activities are. Information is obtained on types of stores where the product is purchased.

More detailed information about the product is obtained by surveys. Individuals are often asked to use or taste the product. See **Figure 18-11.** Then they are questioned about the product. The actual questions try to bring out the following information:

- Feelings about the product
- Evaluation of product quality
- Whether the person would like to own or use the product
- Reaction to product color, size, and function
- Expected selling price
- Number of products that would be purchased in a year
- Use for the product—gift or personal
- Improvements that could be made
- Comparison to other similar products on the market
- Type of store in which the product would be expected to be found

Often the warranty card, **Figure 18-12,** is used to gather this data. Basic information is usually needed on:

- Name and address of purchaser
- Gender and age of the new owner

TYCO R/C CONTROL REGISTRATION CARD

WE CARE!
We are continually striving to please our customers by making quality products that satisfy their needs and interests. As a Customer, your opinion is very important to us. Please take the time to fill out this brief questionnaire, which will help us better serve you.

Thank you for your help.

1. **Date of Purchase:** ____/____/____ Day Mo. Yr.
2. **Name of your R/C vehicle** ____
   Product Stk. No. ____
3. **Age of person for whom R/C vehicle was purchased:** ____
4. **Who purchased this product? (check one)**
   ☐ Mother ☐ Father ☐ Friend ☐ Relative ☐ Yourself
5. **If a gift, was it requested by name?** ☐ Yes ☐ No
6. **Do you or anyone else in your household own other Tyco R/C Products?** ☐ Yes ☐ No
   Which Products? ____
7. **Which one of the following most influenced your decision to buy or request this R/C vehicle?**
   ☐ Tyco R/C name ☐ Newspaper/Magazine Ad
   ☐ "Package caught my eye" ☐ A television commercial
   ☐ A friend ☐ Other____
8. **What real cars would you like to see us make in an R/C vehicle?**
   ____
9. **Please comment on how we might improve our R/C vehicles.**
   ____

Your Name ____ Phone No. (__) ____

Address ____

City ____ State ____ Zip Code ____

Figure 18-12. Study the questions on this warranty card. How would the information help a product designer? An advertising person?

- Income of customer
- How the customer heard about the product
- Type of store in which the product was purchased
- Whether the product was purchased for a gift or personal use

Competitive Analysis

All companies carefully study the products of their competitors. From this study a company can determine the need to improve its own products. This is called *competitive analysis.*

Product imitators can determine trends in product development. They can identify areas where they want to develop similar products.

Of course, each company must ensure that they do not break patent laws. They must either:

- Develop their own technology
- Pay to use someone else's ideas
- Use technology that is not patented

There are thousands of ideas not patented. The patents of many other products have expired.

competitive analysis. Companies carefully study the products of their competitors. From this study a company can determine the need to improve its own products.

Academic Link

An important part of marketing research is communicating with the consumer. Knowing their likes and dislikes is essential when trying to improve an old product or develop a new one. Marketing surveys give the manufacturer a good idea of consumer likes and dislikes.

Developments in communication technology have made it easier to gather marketing information. For example, developments in the telephone system make it easy for a company in New York to survey customers in Illinois. The Internet has made it even easier to reach consumers worldwide. Being able to develop and put into practice Internet surveys is an important part of communicating with customers.

Technological Developments

The last source of product ideas is *technological developments*. Advancements in science and technology can give designers product ideas. New materials may suggest new products. The development of composites gave us fiberglass fishing poles and lightweight aircraft parts. The invention of the laser and glass fibers gave us fiber-optic communication systems.

Companies must never stop gathering information about new developments. They may produce the advancement in their own research labs. See **Figure 18-13.** Other data are obtained by outside sources. These sources include:

- Other companies
- Government agencies
- Universities
- Private research centers
- Private inventors

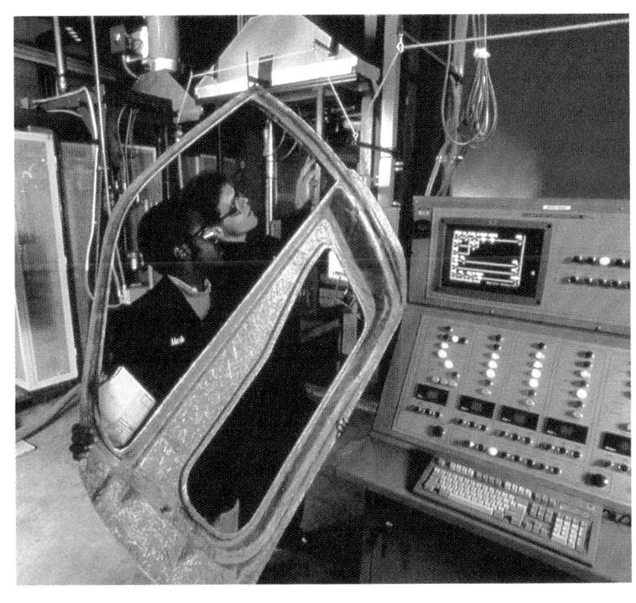

Figure 18-13. These engineers are developing a new composite doorframe. (Du Pont)

technological developments. **Advancements in science and technology that give designers new product ideas. Companies must never stop gathering information about new developments.**

Company Profile

Product ideas, as mentioned, must be matched to the company strength. These strengths are often contained in a list called a company profile. Five main elements are included in a company profile. It describes the market the company knows and presently serves. The types of products the company produces are listed. The sales volume the company expects from each product is presented. Also, the cost to develop and engineer the product is estimated. Finally, the financial resources (money) that the company has available for the project is identified.

Summary

Companies may develop products using either production or consumer approaches. They start the development process with product ideas. The ideas may be imitations (copies) of other products. Some ideas adapt or improve on existing products. A few ideas are truly new and are called innovations.

Studying consumers, competitors, and technological developments generates product ideas. All ideas must fit the company. They must fall within a profile of the company's strengths.

Key Words

All of the following words have been used in this chapter. Do you know their meaning?

adaptation
competitive analysis
imitation
innovation
market research
technological developments

Test Your Knowledge

Please do not write in this text. Place your answers on a separate sheet of paper.

1. Explain the difference between a consumer approach and a production approach to product development.
2. Adaptation means producing a new product by _____. (select best answer)
   a. improving an old design so the product does the job better.
   b. changing a product so it looks different.
   c. making a product of cheaper materials to make more profit.
   d. stealing another company's product ideas.
3. _____ _____ is studying people's thinking about products.
4. List the three sources of ideas for product development.
5. Describe two methods of getting information from users of a company's product.
6. When a company studies the products of its competitors to see how it can or should improve its own products, the method is called _____.
   a. comparison shopping.
   b. competitive analysis.
   c. technological development.
   d. patent search.

Applying Your Knowledge

1. Working in a group, identify ten new product ideas that people in your age group need or want.
2. Visit a company and interview a product designer. Find out how she or he finds new product ideas.
3. Interview a retail store manager. Ask him or her to discuss products which:
   a. Were designed to meet a basic customer need.
   b. Were first designed and then the need was developed by advertising.
4. Do you have a new product idea? Make a sketch of the product and market it to your friends. Do they want to buy it? Why or why not?

Chapter 19 Designing Products

Objectives

After studying this chapter, you will be able to:

- ✓ List and describe the objectives of design.
- ✓ Explain factors of industrial design.
- ✓ List and describe steps in a design process.
- ✓ Explain different types of models used in presenting designs.

Product ideas must be developed to meet product needs. Product designers must change words into products. They convert statements of need into product ideas.

Designers use a process called *ideation*. They sketch many ideas and then the best ones are selected. The ideas are refined and designers move ideas-in-mind to ideas-on-paper. Product design has objectives and a process. It also has goals and a method (way of doing).

ideation. **A process designers use to move from ideas in their mind to ideas on paper.**

Design Objectives

All product design activities are aimed at meeting a goal or purpose. The major goals are to develop products that customers want, meet or beat the competition, and are profitable to make.

Successful designers always keep three major factors in mind. They design for function, manufacture, and selling. See **Figure 19-1.**

Designing for Function

All products are designed to do a job. A train or truck must move freight. See **Figure 19-2.** A picture must decorate a wall and a washer must wash clothes.

The ability to do a job is called *function*. Designers consider the product's purpose, operation, and safety. These are the factors in designing for function.

function. **The ability of a product to do a job.**

For example, a designer of toy trucks must consider many things. Several of these are related to function.

- Where is the toy going to be used? Will it be an indoor or outdoor toy?
- Is it designed for educational purposes?

Figure 19-1. Products are designed to meet three factors. They must fill a need (function), be easy to make (manufacture), and customers must want them.

Figure 19-3. Designers considered manufacturing when they designed this airplane. Design for manufacture allows the skin to be easily mated with the frame. (The Boeing Co.)

Figure 19-2. Trucks and trains are designed to move freight. This is their function. (Freightliner Corp., Norfolk Southern)

- What type of surface must it roll on?
- What age of child will use it? What safety features are required for this age group?

Every product's worth is measured in terms of function. It must operate under typical conditions.

Designing for Manufacture

It is not enough to have a functional product. The company must be able to build it efficiently. Designers consider ease of manufacture. See **Figure 19-3.**

Designers will also consider the number of parts needed. Usually, the fewer parts the better. Moreover, they will use standard parts and materials whenever possible. Why make a bolt if you can buy it cheaper?

In addition, the number of different parts are considered. A designer would not want to use a 1" × No. 8 screw in one place and a 1-1/4" × No. 8 screw in another. They are the same diameter and number of threads per inch is the same. There is only a slight difference in their length. Using only one size reduces inventory costs.

The ability to process the part is another consideration. Look at **Figure 19-4.** The part on the left would be more difficult to cut out and sand. The shape on the right is a better design for manufacture.

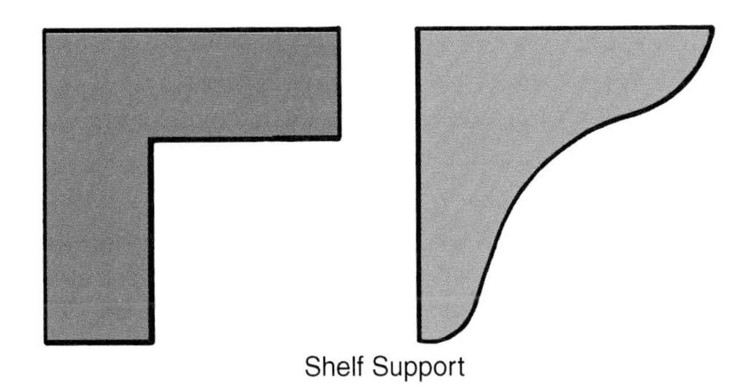

Figure 19-4. The part on the right would be easier to manufacture than the one on the left.

Designing for Selling

Product function and manufacture are important but the product must also sell. Customers must want to buy it because it meets their needs. These needs include function, appearance, and value. See **Figure 19-5.**

Figure 19-5. Look at these toys. Do you think they were designed for selling? Why? (Ohio Art Co.)

We buy products because they do a job for us. But we also want them to be attractive. Even a table saw is designed to look good. Finally, the product must have value. We must feel that it is worth the price.

Generally we are first attracted to a product by its looks. We then decide if it will do the job. The final decision is related to its cost. Is it worth the money it costs?

Design Process

Product designers follow a few basic steps in developing product ideas. These steps, shown in **Figure 19-6,** include:

- Developing preliminary (beginning) designs
- Refining designs
- Preparing models
- Communicating designs
- Obtaining approval for designs

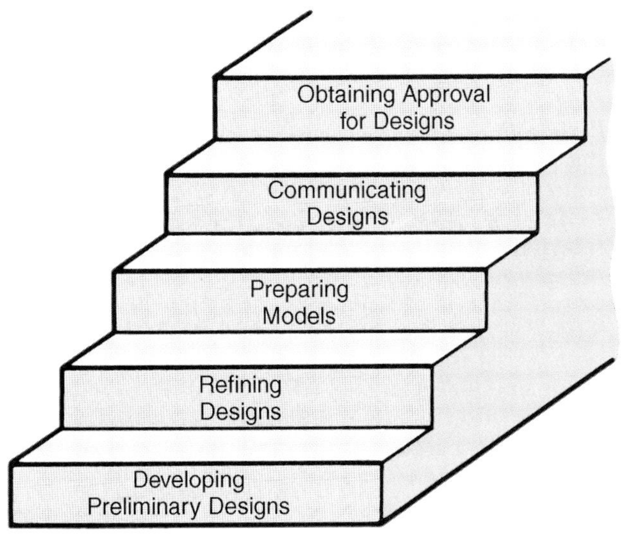

Figure 19-6. There are five steps in product idea development.

Figure 19-7. Designers develop rough sketches for their "library" of product ideas. (RCA)

Figure 19-8. This designer is working on a sketch for a new toy. (Ohio Art Co.)

Preliminary Designs

thumbnail sketch. Preliminary drawing made by a designer to show a design idea. Also referred to as *rough sketch.*

rough sketch. Preliminary drawing made by a designer to show a design idea. Also referred to as *thumbnail sketch.*

refined sketch. A sketch that shows shape and size of a product idea. It gives a fairly accurate view of the designer's ideas.

CAD. A computer-based system used to create, modify, and communicate a plan or product design.

Generating ideas is the first step in product design. Designers quickly sketch as many ideas as they can. See **Figure 19-7.** Often these sketches are simple "doodles" that record what the mind dreams up.

These first sketches are often called *thumbnail* (or *rough*) *sketches.* They serve the same purpose as notes do for writers. They are assorted pieces of information that can be sorted and organized into more complete pictures.

You might think of the rough sketches as a "library of ideas." A large "library" is more likely to have good ideas. Therefore, it is important for a designer to develop many rough sketches.

Refining Designs

During refining, the designer selects the best ideas from the many rough sketches. These ideas are improved and details are added. Several sketches may be fused (put) together to form a better idea. These sketches, called *refined sketches,* are more complete. See **Figure 19-8.**

The refined sketch shows shape and size. It gives a fairly accurate view of the designer's ideas.

Many designers now use computers to develop sketches. See **Figure 19-9.** Computer systems allow the designer to quickly change lines and details. The stroke of a "wand" changes the computer picture. These pictures can be stored for later use or be used to create a drawing. The computer will direct a pen to draw a sketch. These systems are called *CAD* (**c**omputer **a**ided **d**esign) systems.

Preparing Models

Many times sketches do not show enough detail. They are generally two-dimensional (have width and height only) and do not show depth. People may have a hard time imagining this third dimension. Therefore, models are often built. *Models* are three-dimensional representations of a system or product.

There are two major types of models: mock-ups and prototypes. A *mock-up* is an appearance model. See **Figure 19-10.** It shows what the product will look like. Mock-ups are usually made of easily worked materials. Cardboard, balsa wood, clay, Styrofoam™, and plaster are commonly used.

Prototypes are working models that generally show the product in full size. Prototypes use the same material the product will use. Their purpose is to check the operation of the final product. **Figure 19-11** shows a typical prototype.

Models allow the design to be viewed more completely. It can be seen from all angles. The appearance and operation of the product can be checked carefully.

Figure 19-9. This 3D CAD drawing allows the designer to see a car's engine. (Ford Motor Co.)

Figure 19-10. This designer is working on a scale model of a new vehicle. (Ford Motor Co.)

Communicating Designs

Developed designs must be communicated to management. Often, managers want to study the ideas for size, shape, color, and decoration. Typically, completed designs are shown in two ways: models and renderings.

The models are usually final prototypes or mock-ups. Earlier models were used for the designer to check ideas. Now, new models are made that include all design changes. See **Figure 19-12.** These models will look exactly like the finished product.

Renderings are colored pictorial sketches used to show final designs. They show the overall detail of the design. **Figure 19-13** shows a rendering of a new product.

Getting Management's Approval

The final step for a design idea is getting approval. Management must give its "OK" to continue developing the product.

models. A three-dimensional representation of a product. The two major types of models are *mock-ups* and *prototypes.*

mock-ups. Models designed to simulate the appearance of a new product.

prototype. A working model. Prototypes are used to check the operation of the final product.

rendering. A colored pictorial sketch used to show the final appearance of a product design.

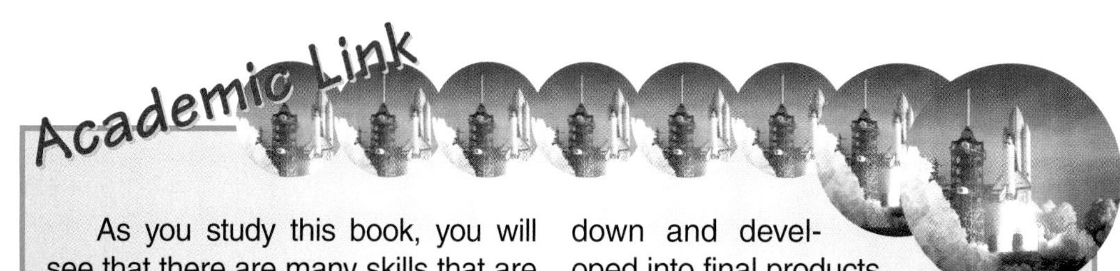

As you study this book, you will see that there are many skills that are significant partners with manufacturing. For example, artistic ability is something that one might not consider being related to manufacturing. However, have you wondered why a beverage container is shaped and colored the way it is? Why are the headlights on your family's car shaped as they are? All of these things at one time were an idea that was sketched down and developed into final products.

The study of the art and design has many avenues that lead to the world of manufacturing. Sketching, airbrushing, clay modeling, and computer designing that is discussed in this chapter are all artistic skills. They are used by many manufacturers for the development of product designs, mock-ups, and prototypes.

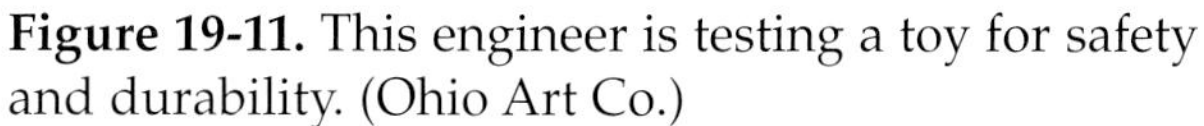
Figure 19-11. This engineer is testing a toy for safety and durability. (Ohio Art Co.)

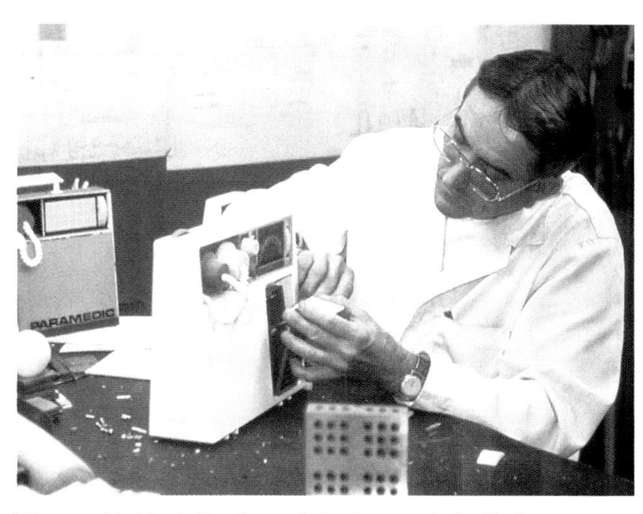

Figure 19-12. A final model of a toy is built by a model maker. (Ohio Art Co.)

Management evaluates the product design against several factors. These factors include: strength of competing products, the cost to manufacture the product, the size of the market, the money that can be earned, and the resources required.

The managers usually receive preliminary cost estimates. They also review the design sketches and models. See **Figure 19-14.** They decide the fate of the product ideas. If approved, the designs will be sent to product engineering where the product is refined, specified, and tested. These activities are the subject for the next chapter.

Figure 19-13. This designer is developing a rendering of a possible new product. (Ohio Art Co.)

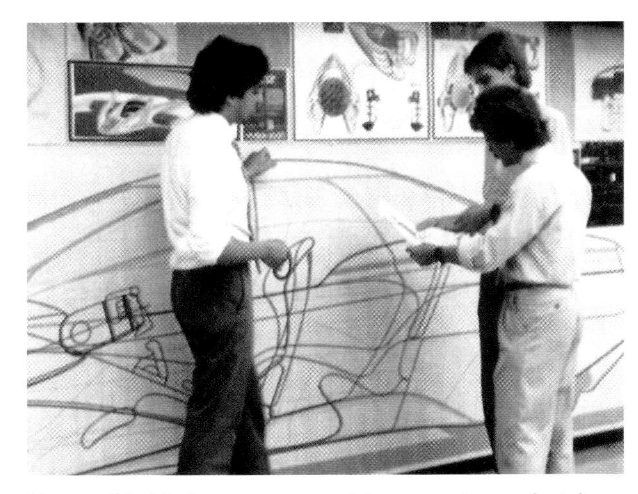

Figure 19-14. A management team reviews sketches for a new automobile. (Ford Motor Co.)

Summary

Designing products involves careful consideration of function, manufacture, and selling. Products must be designed to do a job. They also must be efficiently produced and they must sell.

Creating a successful design involves several steps. Ideas must be generated through rough sketching activities. These ideas are further developed by preparing refined sketches and models. The refined ideas are communicated to management for approval. Approved ideas are turned over to product engineering.

Key Words

All of the following words have been used in this chapter. Do you know their meaning?

CAD
function
ideation
mock-ups
models
prototype
refined sketch
rendering
rough sketch
thumbnail sketch

Test Your Knowledge

Please do not write in this text. Place your answers on a separate sheet of paper.

1. Give the steps for the process of ideation.
2. Tell what is meant by:
   a. Designing for function.
   b. Designing for manufacture.
   c. Designing for selling.
3. Arrange the following steps for design in their proper order:
   a. Obtaining approval for designs.
   b. Developing preliminary designs.
   c. Communicating designs.
   d. Refining designs.
   e. Preparing models.
4. What are the two types of models used by product designers? What is the difference between them?
5. Using computers to draw up designs is known as _____. The term stands for _____ _____ _____.
6. A rendering is a:
   a. Final prototype.
   b. Colored pictorial sketch.
   c. Final mock-up that has been painted.

Applying Your Knowledge

1. Visit a product designer to see samples of product sketches. Ask about the design process (steps) used. Find out about the way products are designed and approved for manufacture.
2. Select three simple products in your home. Evaluate their designs in terms of function, manufacture, and selling.
3. Sketch three to five new products you would like someone to design.

Chapter 20 Engineering Products

Objectives

After studying this chapter, you will be able to:

✓ Explain how products are engineered.
✓ Define the terms: engineering modification and engineering specification.
✓ Recognize different types of engineering drawings and tell how they are used.
✓ Describe a bill of materials and explain its use.
✓ Discuss specifications and describe their form and contents.

Product designs show how the product will look and work. But the product is not ready to be produced. The design must be refined through a process called *product engineering*.

Product engineering modifies and specifies the design. *Specifying* means to give the size, material, and quality requirements for the product. Product engineers also test the product's operation and safety, as well as other important features.

product engineering. The process of preparing a design for production. The design is specified and tested for operation and safety.

specifying. To determine the size, material, and quality requirements for a product.

Specifying Designs

Most products have several parts that must fit and work together. See **Figure 20-1.** Often the product has parts from several sources. Standard items are bought from suppliers while other companies may build special parts. Suppliers may bid to produce parts of the product. The company, itself, may produce many different parts. All of them must fit together to make the product. See **Figure 20-2.**

Each supplier must know the exact size, shape, and properties of the components they make. This information is found in the specifications. Product engineers specify product characteristics (features) in three ways using:

Figure 20-1. The automobile is a product with many individual parts. (Ford Motor Co.)

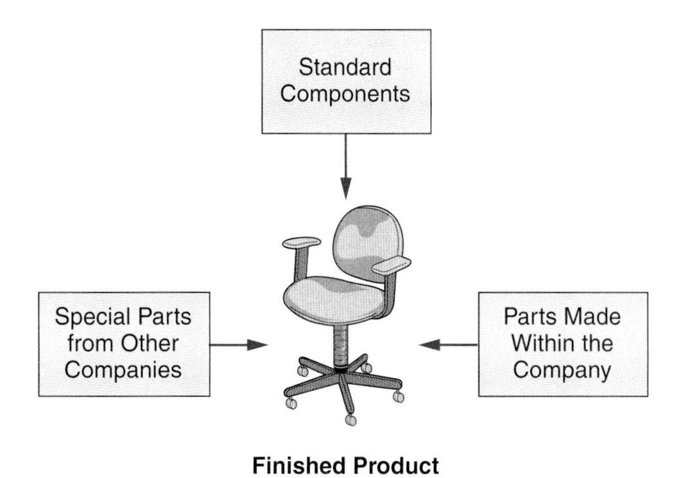

Figure 20-2. Parts from several sources must fit together to make a typical product.

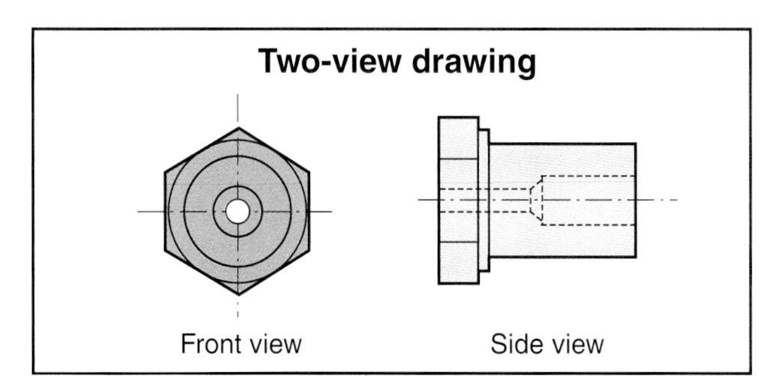

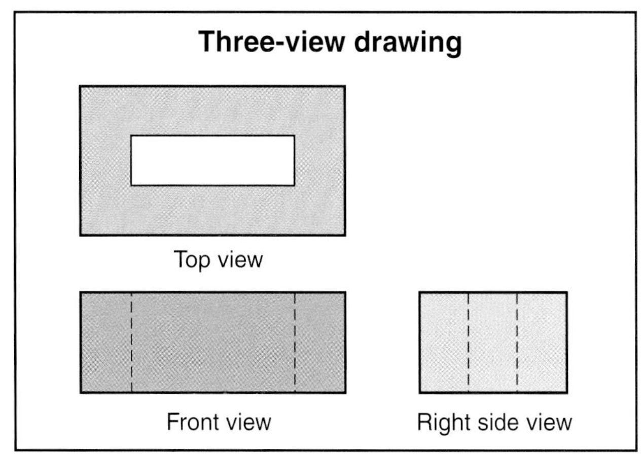

Figure 20-3. Orthographic projections are "straight on" drawings of parts. Two types are shown.

- Engineering drawings
- Bill of materials
- Specification sheets

Engineering Drawings

Engineering drawings tell how to make the product. The drawings include specifications for individual parts and give information needed to assemble the product. This basic information is placed on three types of drawings:

- Detail drawings
- Assembly drawings
- Systems drawings

Detail drawings

Engineering will prepare a detail drawing for each different part. *Detail drawings* give the exact size of the part as well as the size and location of all features. These features may include holes, notches, curves, and tapers. These features give the parts their final form.

Detail drawings usually show the part from several sides. These drawings use a system called *orthographic projection*. This system generally shows the part in two or three views. See **Figure 20-3.** Round parts are shown by an end view and a side view. Other shapes are shown in three views: front, top, and right side. You see each view in two dimensions, height, and width. It is as though you are looking directly at that side of the part.

Many detail drawings are now prepared using CAD (computer-aided design) systems. These systems allow the drafter (a person who prepares drawings) to quickly draw and change a drawing while working at a computer. **Figure 20-4** shows a CAD drawing on a computer screen.

A detail drawing must give all the information needed about a part. The manufacturer (maker of parts and products) must be able to make the part from the drawing. Refer to **Figure 20-5.** Could you make the table leg from the information given? What additional information would you need? Remember we are talking about making many parts that are alike. Could you make the curves identical on each part? Does the drawing give you that information? You can easily see why the drawing must be complete. Without good detail drawings, parts cannot be accurately produced.

detail drawing. **A drawing that gives the exact size of a part as well as the size and location of all features. These features may include holes, notches, curves, and tapers.**

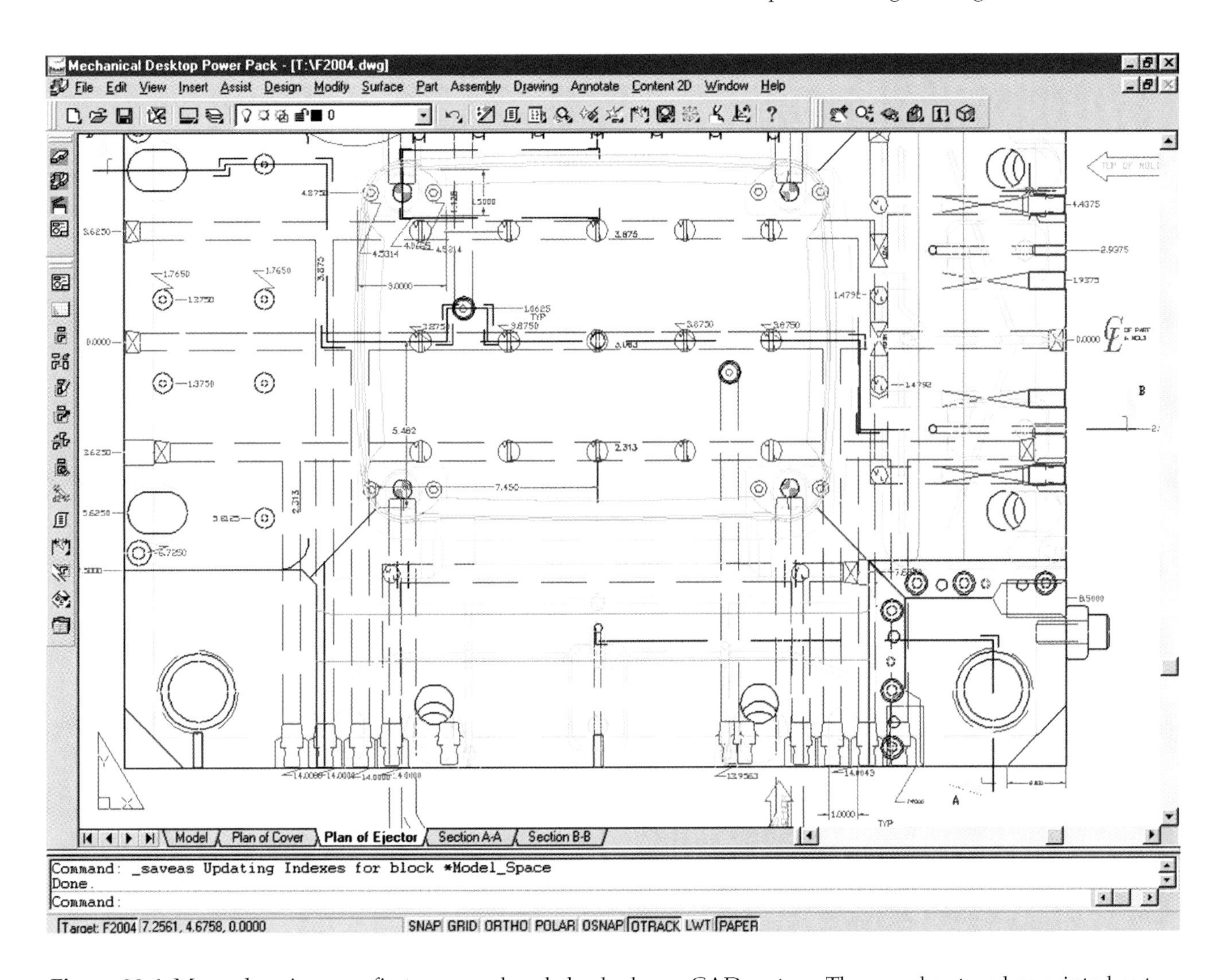

Figure 20-4. Many drawings are first prepared and checked on a CAD system. They can be stored or printed out.

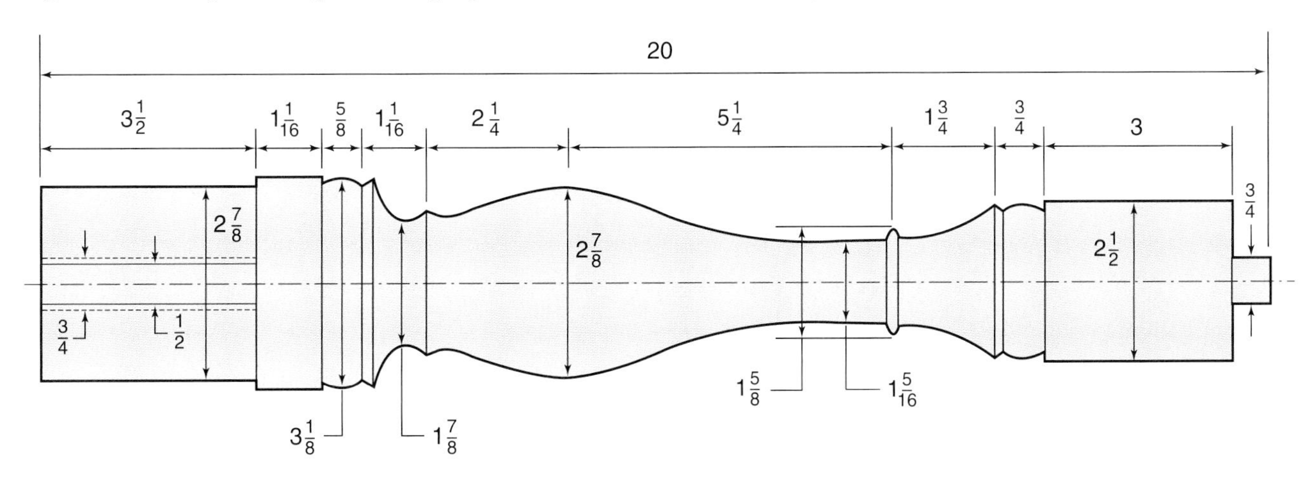

Figure 20-5. This is a simple detail drawing.

Assembly drawings

Parts must be put together to make many products. How these parts mate is shown on *assembly drawings*. These drawings identify the parts by a code (number, letter, etc.) and then show where each part goes.

orthographic projection. A method of presenting a product in drawings by showing the top, front, and right side views.

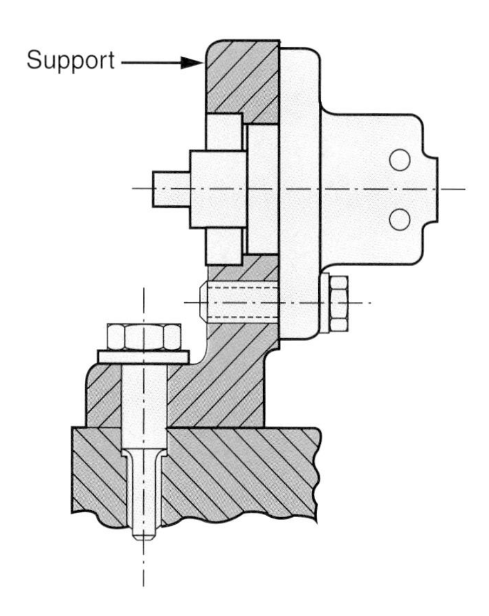

Figure 20-6. A simple assembly drawing shows where each part belongs.

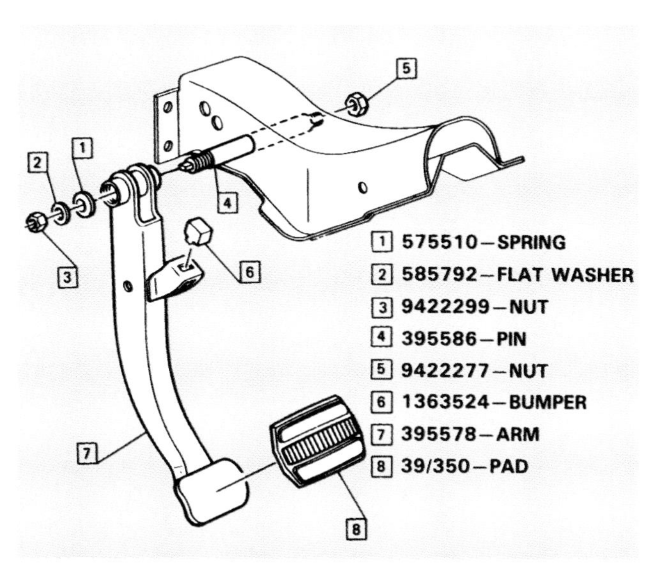

Figure 20-7. An exploded assembly drawing is pulled apart so you can see where each part belongs.

Figure 20-8. This drafter is carefully checking a drawing for accuracy.

Assembly drawings may be two-dimensional (width and height) like the one in **Figure 20-6.** This type of drawing shows the parts in their final assembled position.

Pictorial (picture) drawings are also used. These drawings represent the parts in a single three-dimensional (has depth as well as height and width) drawing. Often the drawing is an exploded view in which parts are pulled apart to show how they fit together. See **Figure 20-7.**

Assembly drawings, like detail drawings, must communicate. They must give all information needed to put the product together. Each assembly drawing must be checked, **Figure 20-8,** to be sure that it:

- Identifies the parts by name or number.
- Shows the location of each part.

assembly drawing. **Engineering drawing that provides information to show assemblers how parts of a product fit together.**

systems drawing. **Drawings that are used to show electrical, pneumatic (air), and hydraulic (fluid) systems. They show the location of parts in the system and connections.**

Systems drawings

Systems drawings are used to show electrical, pneumatic (air), and hydraulic (fluid) systems. They show the location of parts in the system and how they are connected. See **Figure 20-9.**

Systems (also called schematic) drawings give information needed for assembly and servicing. Workers can easily see how components fit in the system. Most systems drawings do not show distances between components. Instead they show how the components relate to each other.

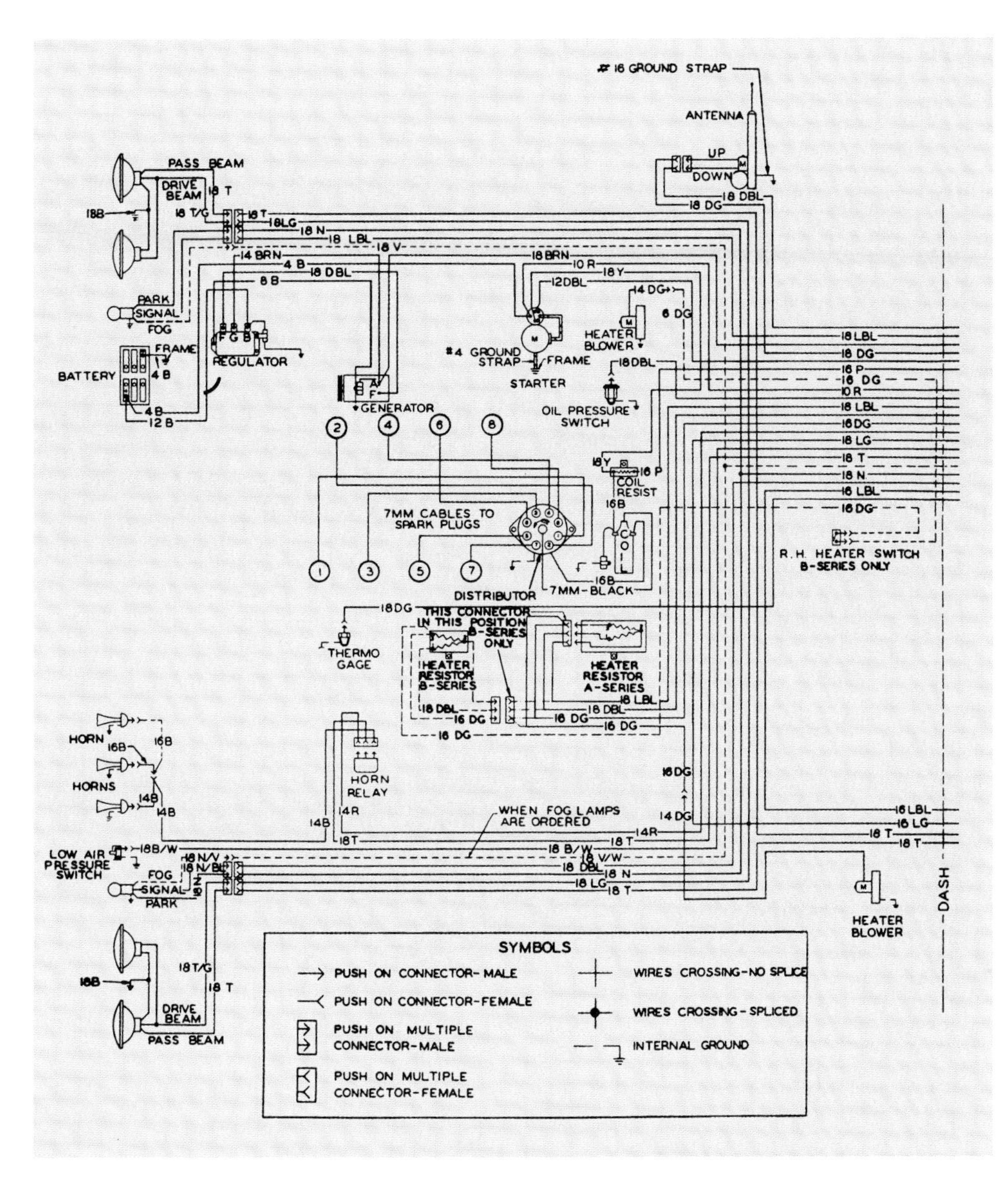

Figure 20-9. An electrical systems drawing. (General Motors Corp.)

bill of materials. A list of all the parts and hardware items needed to make one product.

Bill of Materials

A second tool of the product engineer is the bill of materials. It is a list of all materials needed to make one product. A *bill of materials*, shown in **Figure 20-10,** includes:

- Part number
- Part name
- Quantity of each part needed
- Size of each part
- Material to be used

MASTER BILL OF MATERIALS

PRODUCT: ____________________

PRODUCT CODE NUMBER: ____________________

Part No.	Part Name	Material	Qty.	Size T	Size W	Size L

Figure 20-10. A bill of materials lists sizes and quantities of materials in orderly fashion.

The sizes are given in a logical order. The thickness is listed first, then the width and, finally, the length.

A bill of materials is a valuable form that can be used to determine the material to order. It is also used to estimate the cost of manufacture. However, a bill of materials, itself, does not list costs. It is used to determine quantities of materials. These are then multiplied by current prices. If a bill of materials did include costs, it would soon be out-of-date since prices change often.

Specification Sheets

Some items and certain qualities cannot be shown on a drawing. How would you show adhesives on a drawing? A drawing of sheet steel would be of little value.

The important characteristics for these materials are not size and shape. The detail drawings show the final size and shape. They could describe size and shape of a note holder made from wood. However, the wood is chosen because of its properties that are described on a specification sheet.

Product engineers need to know or specify these various properties. The material's strength, weather resistance, and other qualities must be determined.

Small manufacturers cannot afford to have materials developed for them. An adhesive (glue) may be needed to bond an aluminum sheet to plywood. A small manufacturer would call an adhesive manufacturer and describe the need. The adhesive company would provide *technical data sheets* that describe adhesives that would do the job. See **Figure 20-11.** They would give:

- Properties of the adhesives
- Information on its application, clamping, and curing

PLASTIC SHEET LAMINATING GLUE #7700

Ready to Use
No Staining
High Solids: 52-54%
Storage: up to one year
Quick Setting: 20-30 minutes

DESCRIPTION
USE
SPREADING
ASSEMBLY TIME
PRESSURE
PRESS TIME
STORAGE

Figure 20-11. This outline is for a specification sheet for a common cabinetmaking adhesive. (National Casein Co.)

technical data sheet. **Information sheets that describe the characteristics of a product to a designer or engineer.**

The product engineer reviews the data sheets and picks the best adhesive for the job.

Larger companies may specify the material characteristics they want. They prepare a "specification sheet" for the material. Suppliers offer to supply the material that is produced to meet the customer's specifications.

Testing Products

Product engineers also test designs to make sure that the product works. See **Figure 20-12.** Just as important, they will test it for safety.

The product is put under actual conditions of use and closely watched. See **Figure 20-13.** The data gathered helps engineers to decide if the product works. Also, testing information is one source of ideas for product improvement.

One more use of product testing is to answer customer complaints. Broken products may be returned. Testing can determine if the product design has flaws or if customer misuse caused the failure. If so, new instructions for use may be needed.

Figure 20-12. This engineer is testing a cylinder head. (General Motors Corp.)

Figure 20-13. These technicians are testing the design of automobile. (Ford Motor Co.)

Summary

All products move from design to engineering. Designers only develop products. Product engineers must specify their characteristics.

An engineering staff prepares detail drawings for each part. They describe the size, shape, and surface finish for the part. Assembly drawings show how the parts go together. Systems drawings describe electrical, pneumatic, and hydraulic systems.

A bill of materials lists all parts needed to make a product. The list is a guide for purchasing and cost estimating.

Finally, the properties of a material or product are specified. They are contained in technical data sheets and specification sheets.

Product engineers also test a product. They check its operation and safety. After these activities are completed, the product design is released for manufacture. The product has been designed and engineered. Now it needs to be produced and sold.

Key Words

All of the following words have been used in this chapter. Do you know their meaning?

assembly drawing
bill of materials
detail drawing
orthographic projection
product engineering
specifying
systems drawing
technical data sheet

Test Your Knowledge

Please do not write in this text. Place your answers on a separate sheet of paper.

1. Three ways to specify characteristics (features) of a product are _____, _____, and _____.
2. Indicate which of the following types of drawings you would use to give the exact size of a part:
   a. Systems drawing.
   b. Assembly drawing.
   c. Detail drawing.
3. If you wanted to show all the features of a drawing (including features like holes, notches, curves and tapers), you would use a(n) _____ drawing.
4. An orthographic projection (select all correct answers):
   a. Allows you to see height, width, and depth in a single view.
   b. Generally shows parts in two or three views.
   c. Shows each view in two dimensions, height and width.
   d. Shows round parts in two views.
   e. Often shows three views: front, top, and right side.
5. A drawing that presents the parts as though it were a picture is called a _____ drawing. It shows the part in _____ dimensions, _____, _____, and _____.
6. List the information included on a bill of materials.
7. If a small company asked an adhesive manufacturer to supply information on a glue to meet the company's needs, would the adhesive manufacturer supply the information on a technical data sheet or a specification sheet?
8. When are specification sheets used?

Applying Your Knowledge

1. A maple bookrack has three major parts—a large end which is $3/4 \times 6 \times 8$, a small end which is $3/4 \times 6 \times 3$, and a shelf which is $3/4 \times 6 \times 14$.
   a. Prepare a bill of materials for the product.
   b. Prepare a drawing for the large end.

2. Interview a drafter to find out:
   a. Types of drawings he or she prepares.
   b. Methods used to prepare the drawing.
   c. Basic requirements for the job.
3. Take apart something that has several parts and reassemble it.
4. Make a dimensioned sketch of a simple product you own.
   The use of the computer as a design tool has allowed human creativity to be extended.

Unit 7 Developing Manufacturing Systems

Chapter 21
Choosing a Manufacturing System

Objectives

After studying this chapter, you will be able to:

- ✓ Recognize the differences between custom, intermittent, and continuous manufacturing.
- ✓ Compare these three types of manufacturing.
- ✓ Define terms used with each type.

At some point a new product has been developed. Drafters have completed and released detail and assembly drawings. The bill of materials and specification sheets are available. The product is ready to be manufactured, or is it? How is it going to be produced? Where is it going to be made? Where are the factory, machines, and people? What operations will be used to make the parts and assemblies? How will product quality be maintained? These are just a few of the manufacturing challenges that must be met before the product can be actually produced. Every manufacturing enterprise faces these "where" and "how" questions when a new product is produced. The success of the company will rest, to a great extent, upon the answers to these critical questions. The key to success is careful planning of manufacturing systems before making the product. The "six P's of successful manufacturing" summarize the importance of planning: Proper prior planning prevents poor performance (and poor profits). See **Figure 21-1.**

Proper Prior Planning Prevents Poor Performance

Figure 21-1. Products like this plane are the result of carefully planned manufacturing operations. (Boeing Co.)

The proper planning starts with selecting a manufacturing system. Three basic types of manufacturing systems are used today. These are outlined in **Figure 21-2.**

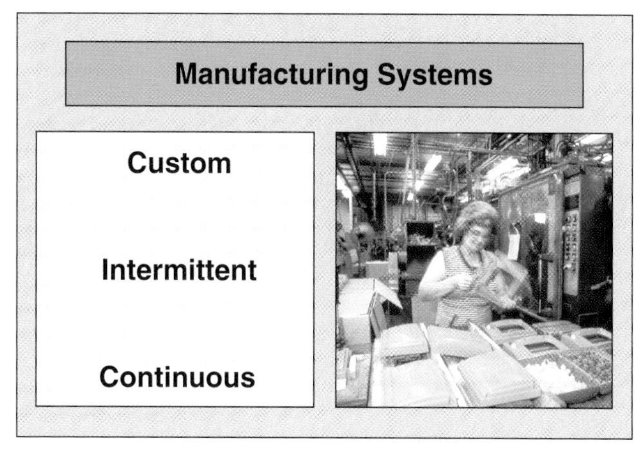

Figure 21-2. All manufacturing is done by one of these systems.

Figure 21-3. This passenger ship was custom manufactured to the customer's specifications. (National Steel and Ship Building)

Figure 21-4. Most tooling for manufacturing processes, like this press break tooling, is custom manufactured. (Accurate Manufacturing Co.)

Custom Manufacturing

Custom manufacturing is used to make small numbers of products. Often they are one-of-a-kind items. See **Figure 21-3.** The company produces them to a customer's specifications. The buyer decides the features of the product.

Some people have clothing manufactured to fit them. Kitchen and bathroom cabinets are sometimes built to fit one house. Tooling, like that shown in **Figure 21-4,** is custom manufactured. *Tooling* is the equipment used for production. The customer's specifications are used so that the tooling is built to order.

Custom manufacture generally requires skilled workers. They must be able to read plans (drawings and specifications) plus set up and operate their own equipment. Each worker checks the quality of his or her own work. See **Figure 21-5.**

Custom manufacture is the most expensive system for manufacture. Workers are more skilled and are, therefore, paid more. Setting up and checking machines takes a long time. Only a part of the work time is spent in actual production.

Equipment use is also low because part of the time it is idle and at other times it is being set up. The cost of each machine is charged against products. If it is not used often, more cost is charged to each product.

For these and other reasons, custom manufacture is not often used. It is chosen only when demand for the product is low but users are willing to pay the added cost.

Figure 21-5. Custom manufacturing requires highly skilled workers who are responsible for their own work. Here a potter forms a bowl.

Intermittent Manufacturing

In *intermittent manufacturing*, products are mass produced in sizeable quantities. Intermittent means starting and stopping at intervals (periods of time). The process starts, is completed, and then stops. At a later time the cycle is repeated to produce more of the product.

You may have seen intermittent manufacture in your home. A member of your family may have baked a batch of cookies. Perhaps a week later another batch was produced and the same process was used. The two baking activities were separated by a period of time.

Intermittent manufacture can be used for both primary and secondary processing. Steel is an example of intermittent manufacturing. The mill produces and pours one melt of steel, **Figure 21-6,** then another batch is produced. Intermittent manufacture in primary processing is called *batch processing*.

Figure 21-6. A batch of steel is being poured. (American Iron and Steel Institute)

In secondary processing, intermittent manufacturing processes groups of items. A number of parts move from station to station until they are finished. At each station the entire group, called a lot, is processed. For example, a hole may be drilled at one station. The entire lot of parts are drilled before they are moved to the next station. There the hole may be reamed. This type of manufacture is called *job-lot manufacture*. See **Figure 21-7.**

Figure 21-7. These cabinets are being produced in job lots. Note the stack of parts. (American Woodmark)

A company may do intermittent manufacturing for two purposes. Products may be made for the company's own use or the system may be used to make products for other companies. The specifications, therefore, may be its own or another company's.

Management must do more planning for intermittent manufacture. The job must be scheduled through the plant. Machines and workers must be assigned for each task. The lot must, somehow, be moved from station to station. Inspections must be scheduled at critical points in the manufacturing process.

Also, machines may need to be set up for each operation. Tooling must be installed and checked. The first parts produced have to be carefully inspected.

Intermittent manufacture is more efficient than custom manufacture. While skilled machine set-up people are needed, less skilled machine operators can run the equipment. They receive lower pay since they are less skilled. Also, the equipment can produce any number of like parts; therefore, equipment is not idle as much.

custom manufacturing. **Used to make small numbers of products. The company produces them to a customer's specifications. Custom manufacture is the most expensive type of manufacturing; often used to make large products.**

Figure 21-8. A series of workers assemble a new toy. (Ohio Art Co.)

Figure 21-9. Like these dishwasher cases, many products are moved from station to station during manufacture. (General Electric)

Figure 21-10. Large products may be assembled without being moved. Workers move to them to complete their jobs. (McDonnell Douglas)

Continuous Manufacturing

Continuous manufacturing produces products in a steady flow. Materials enter the system to start the process. Parts are made from the materials and products are assembled from the manufactured parts. The process continues at a steady rate without stopping.

This method depends heavily on *division of labor*. This means the task of making the product is divided into *smaller* jobs. Each worker is trained to do one job in the total process. Parts are produced as workers complete their individual jobs. See **Figure 21-8.** The product slowly takes shape as it moves through the system.

Continuous manufacture is based on the movement of resources. Most often the worker stays in one place while the product moves to him or her. See **Figure 21-9.** As a worker completes one job, the product moves on to the next worker.

Not all products are easy to move. In these cases the product is not moved but the workers move to the product. Each worker completes a job then moves to the next product where they do their job again. Different employees work on the product at each stage of manufacture. Large air conditioners, electric generators, locomotives, and aircraft are examples of products that are assembled in one spot. See **Figure 21-10.**

Continuous manufacturing saves time. Since workers are trained to do one job, each person becomes skilled in completing that one task. See **Figure 21-11.**

tooling. Devices such as jigs, fixtures, patterns, and templates that help workers make products better and faster. Tooling is designed for three purposes: to increase speed, accuracy, and safety.

Figure 21-11. This worker is trained to see defects in kitchen cabinets. (American Woodmark))

Figure 21-12. This robot is automatically cutting a steel part. (Cincinnati Milacron)

Equipment is built or set up for a single operation and is used over a long period of time.

Special tooling and trained workers waste less material. Also, equipment can be developed to perform routine tasks such as spraying finishes, machining parts, or welding assemblies. See **Figure 21-12.**

A new continuous system, called *flexible manufacturing*, is now being used. This system allows companies to produce small numbers of products using intermittent manufacturing techniques at continuous manufacturing costs.

Flexible manufacturing uses computers to operate machines and material handling devices. The key to this system is rapid set-up and product changeover. With this system a small number of one part or product can be produced. Then in a short time, the tooling and machine-set-ups can be changed to run an entirely different part or product. Flexible manufacturing is used with just-in-time manufacturing systems. These systems have materials arrive just in time for production. The products are produced to just meet shipping schedules. The two systems (flexible manufacturing and just-in-time production) allow companies to reduce inventories, reduce costs, and maintain product flexibility.

intermittent manufacturing. A system in which parts are produced in groups or job lots, in contrast to continuous manufacturing.

batch processing. The intermittent manufacture of raw materials in groups called batches.

job-lot manufacture. Intermittent manufacture that produces finished products.

continuous manufacturing. A type of manufacturing that produces products in a steady flow. Materials go in and finished products come out at a steady rate.

division of labor. The task of making a product is divided into small jobs. Workers are trained to do one job.

flexible manufacturing. A system of computer controlled manufacturing that permits production of small quantities without increasing costs.

Summary

The system of manufacture is selected because of its nature. Custom manufacturing is suited for producing a few special products. Its cost of operation is high. A high degree of skill is required of workers in this type of manufacture.

Intermittent manufacture is used to produce set quantities of materials. The materials move through manufacture as a batch or lot. Materials measured by volume or weight, such as steel, are batch processed. Products that can be counted are processed in groups called job lots.

Continuous manufacture is generally the cheapest way to produce a product. However, a large quantity must be needed and unsold products mean lost money. Flexible manufacturing takes advantages of the low cost of continuous manufacture and the higher flexibility of intermittent manufacture.

The three major systems are compared in **Figure 21-13.** These comparisons are general and may not be true for all cases.

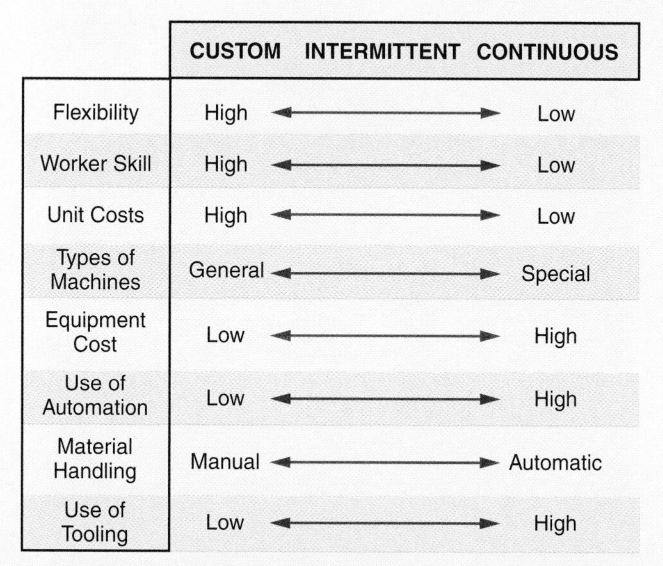

	CUSTOM	INTERMITTENT	CONTINUOUS
Flexibility	High	←→	Low
Worker Skill	High	←→	Low
Unit Costs	High	←→	Low
Types of Machines	General	←→	Special
Equipment Cost	Low	←→	High
Use of Automation	Low	←→	High
Material Handling	Manual	←→	Automatic
Use of Tooling	Low	←→	High

Figure 21-13. A comparison of manufacturing systems. As you can see, each has advantages.

Key Words

All of the following words have been used in this chapter. Do you know their meaning?

batch processing
continuous manufacturing
custom manufacturing
division of labor
flexible manufacturing
intermittent manufacturing
job-lot manufacture
tooling

Test Your Knowledge

Please do not write in this text. Place your answers on a separate sheet of paper.

1. The three major manufacturing systems are:
   a. Assembly line manufacture
   b. Custom manufacture
   c. Job-lot manufacture
   d. Batch processing
   e. Intermittent manufacture
   f. Continuous manufacture
2. Describe each of the major manufacturing systems.
3. Suppose that you were employed in a factory where you received a part for a product at your work area in lots of a dozen. You made a saw cut in each part and then moved them to another person's station for different operation. What kind of manufacturing would you be doing?
4. If there is a steady and high demand for a product, what type of manufacture would normally be best?

Applying Your Knowledge

1. If you had an order for two bookracks of the same design, which manufacturing system would you use? Which system would you use if your order were for 25 racks per month for two years?
2. Visit a local industry. Describe the manufacturing system being used.

Chapter 22 Engineering Manufacturing Facilities

Objectives

After studying this chapter, you will be able to:

✓ List the major engineering tasks in organizing a manufacturing operation.
✓ Describe how the manufacturing engineer performs these tasks.
✓ Demonstrate the use of forms such as the operation process chart and the flow process chart.

Manufacturing takes place in a carefully engineered facility called a factory. A factory's design and equipment permit efficient production.

Manufacturing engineers design and develop manufacturing facilities for approved products. They are responsible for five major tasks that are shown in **Figure 22-1.**

The tasks help the company produce its product at a *competitive price*. (This means it can be produced cheaply enough to sell at a price similar to like products on the market.)

Operation Selection
Material Handling Study
Tooling Design
Plant Layout
Efficiency Studies

Figure 22-1. These tasks are completed by manufacturing engineers in designing manufacturing facilities.

Selecting and Sequencing Operations

Hiring good workers, buying materials at the right price, and getting the right kind of equipment all help a company compete with other manufacturers. However, these activities alone cannot do the job. The equipment and workers must be employed wisely. The work must be done in an orderly way. The right processes and machines must be used to shape and build the product.

Seeing to this is the job of plant engineering. The first task is called ***selecting and sequencing operations.*** Simply put, it means deciding what processes must be done (selecting) and putting them in the order that they must be done

competitive price. A price at which a product can be produced cheaply enough to sell at a price similar to like products on the market.

manufacturing engineer. **A professional person who organizes manufacturing operations.**

(sequencing). The *manufacturing engineer* is a person trained to do this. She or he will use certain forms and methods to organize the manufacturing operations.

Operation Sheet

operations. **The processes that shape and assemble a product.**

operation sheet. **A form used to record the sequence of operations needed to produce a product.**

One of these forms is an operation sheet. Before she or he fills in the sheet, the manufacturing engineer must analyze (study parts to understand the whole) the product drawings. First the engineer decides what operations (processes) are needed to shape the product. See **Figure 22-2.** *Operations* are processes that shape the product and are recorded on the *operation sheet.* See **Figure 22-3.** This record gives the following information:

- Operation name
- Machine to be used
- Tooling needed

Shaping of the product is not the only thing done during manufacture. Parts and whole products must be inspected. They are also moved from place

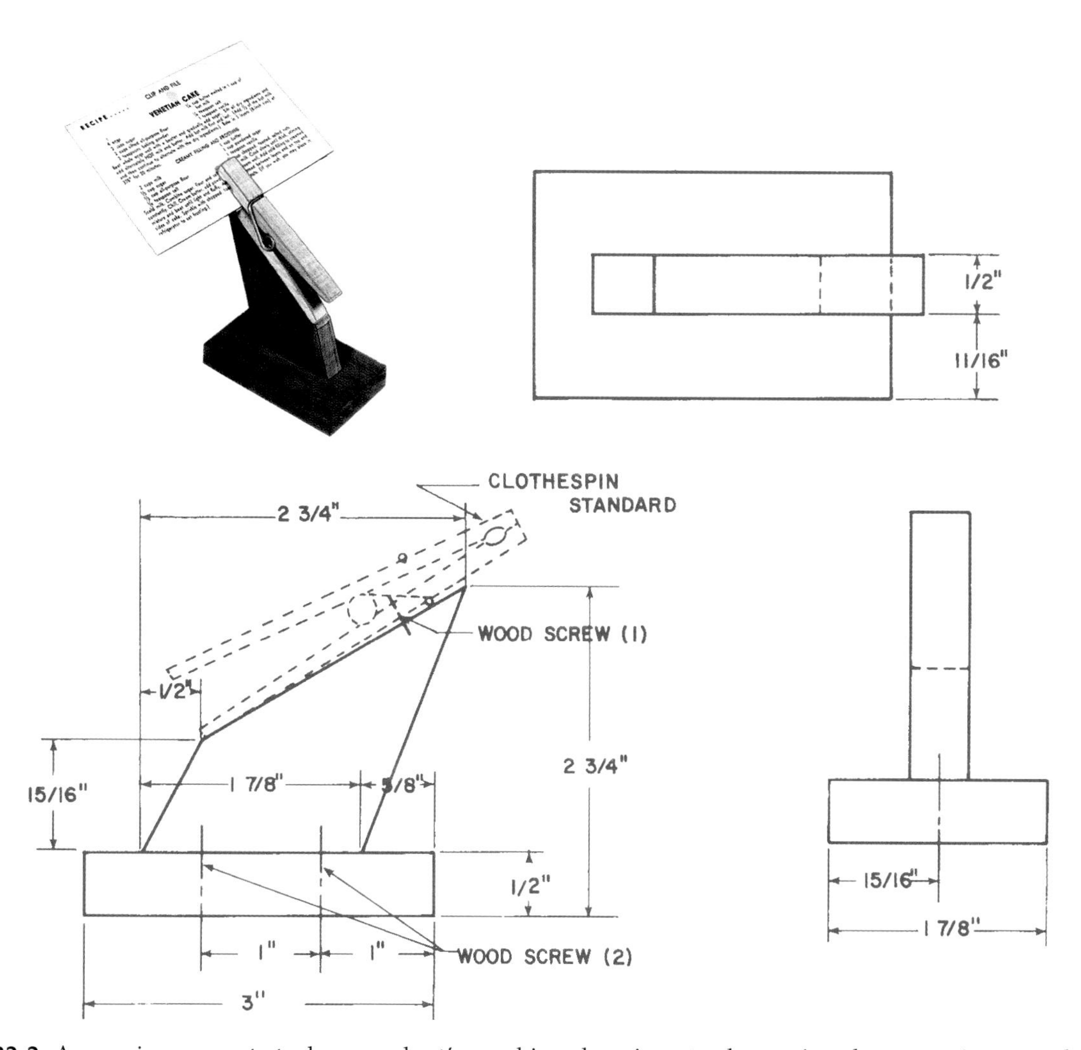

Figure 22-2. An engineer must study a product's working drawings to determine the operations needed for making it.

to place. Sometimes they are held or delayed for further processing. Materials, parts, and finished products may even be stored for a while.

All of these actions must take place in orderly steps. These related tasks make up the total manufacturing process.

Placing tasks in logical order is done with two major forms. These are the:

- Flow process chart
- Operation process chart

Figure 22-3. Products are usually made from a number of parts. The total manufacturing system is often shown on an operation sheet. (Zero Corp.)

Flow Process Chart

Look at the flow process chart in **Figure 22-4.** It shows at a glance the sequence of tasks for producing a single part. (Sequence means the step-by-step arrangement of tasks to complete a part.)

A *flow process chart* often:

- Describes the tasks
- Provides a code number for each task

flow process chart. **A graphic means of showing all the steps and processes a single part goes through as it is manufactured.**

FLOW PROCESS CHART

PRODUCT NAME	FLOW BEGINS	FLOW ENDS	DATE
RECIPE HOLDER	Upright 0-1	Upright T-2	
PREPARED BY: A.B. COMBS		APPROVED BY: D.E. FRY	

PROCESS SYMBOLS AND NO. USED: OPERATIONS 4; INSPECTIONS 1; TRANSPORTATIONS 2; DELAYS ___; STRORAGES ___

Task No.	Process Symbols	Description of Task	Machine Required	Tooling Required
0-1		*Cut top angle*	*Back saw*	*Jig U-1*
0-2		*Cut base angle*	*Back saw*	*Jig U-1*
T-1		*Move to sanding*		
0-3		*Face sand*		
0-4		*Edge sand*		
I-1		*Inspect*		*Gage I-1*
T-2		*Move to assembly*		

Figure 22-4. The flow process chart for the upright part of the product shown in Figure 22-2.

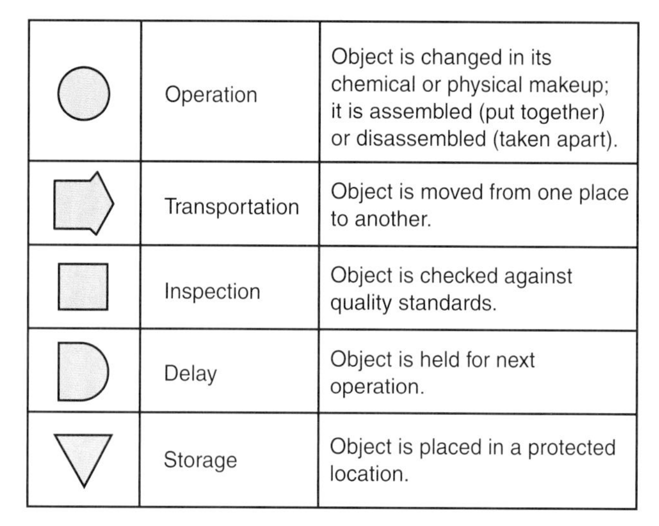

Symbol	Activity	Meaning
○	Operation	Object is changed in its chemical or physical makeup; it is assembled (put together) or disassembled (taken apart).
⇨	Transportation	Object is moved from one place to another.
□	Inspection	Object is checked against quality standards.
D	Delay	Object is held for next operation.
▽	Storage	Object is placed in a protected location.

Figure 22-5. A process flow chart uses symbols to indicate the type of manufacturing activity.

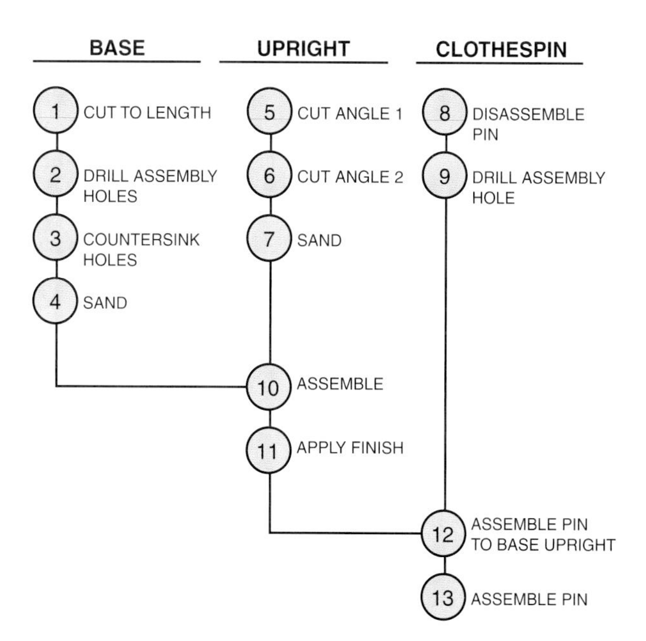

Figure 22-7. This operation process chart was made for the recipe holder shown in Figure 22-2.

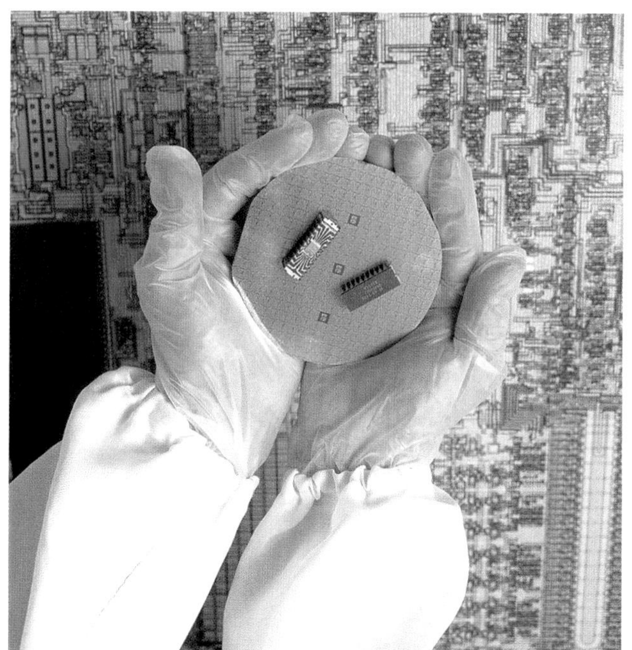

Figure 22-6. Efficient production of parts like these computer chips can be studied with flow process charts. (Hewlett-Packard)

- Identifies the machines to be used
- Lists tooling needed for the task

A standard symbol that is used throughout industry identifies each task. These symbols were developed by the *American Society of Mechanical Engineers (ASME)*. The five common flow chart symbols and their meanings are shown in **Figure 22-5.**

The flow process chart lets engineers study the sequence of operations. It helps them find the best way to make a part. See **Figure 22-6.**

Operation Process Chart

A flow process chart that shows how the individual parts are made is prepared for each part. However, flow process charts do not let you view the whole system. The operation process chart does. See **Figure 22-7.**

It shows how each part is made and then shows how they are assembled into a product. The operation process chart also shows where operations and inspections fit into the sequence. Transportation, delay, and storage tasks are not included. With the operation process chart, the manufacturing engineer can analyze the overall manufacturing process. See **Figure 22-8.**

The operation process chart is also used in scheduling production. It shows which parts must be finished first and the parts that can be produced later. This reduces the need to store some parts. The scheduling information is needed for a new manufacturing system: just in time (JIT). In this system each part is made or arrives at the plant "just in time" for assembly. The part does not have to be

Academic Link

Communication is the sending and receiving of messages. This book is a message sent to you. As you read this book you are receiving a message. We use the alphabet to form words. Before a true alphabet was created, humans used easily recognized symbols to communicate ideas. These are called pictograms and are still used today. For example, a school crossing sign may have a picture of children walking.

Another method of communicating is using ideograms. These are symbols that have a meaning that must be learned. The flow process chart mentioned in the previous section has a number of these symbols. A worker can look at the chart and recognize that the circle symbol means an operation is to be performed and arrow symbol means transportation step is to be performed.

stored, thus reducing costs. The product is also made "just in time" to meet a customer's order.

Figure 22-8. The total system of manufacture of products like these computer devices can be studied with an operation process chart. (Hewlett-Packard)

Designing Tooling

As you learned in a previous chapter, many operations cannot be done easily with standard machines. Often, special cutters are needed. In other cases, holders and clamps must be built. Sometimes a unique (one-of-a-kind) machine is required. At other times, molds and patterns must be made. All of these devices are called tooling.

Tooling helps the machine operator make parts better and faster by reducing the number of machine adjustments. Tooling often fixes the position of the material. See **Figure 22-9.** The part will be held, worked on, and released.

Tooling is designed to increase the operation's:

- Speed
- Accuracy
- Safety

tooling. Devices such as jigs, fixtures, patterns, and templates that help workers make products better and faster. Tooling is designed for three purposes: to increase speed, accuracy, and safety.

Speed of manufacture increases when the operations run smoothly. The tooling can be designed to operate at a set speed.

Tooling also increases accuracy of the operations. Operators do not have to position materials. Lines and layout marks are not needed. Look at **Figure 22-10.**

Figure 22-9. These cabinet parts are produced with the help of tooling. (American Woodmark)

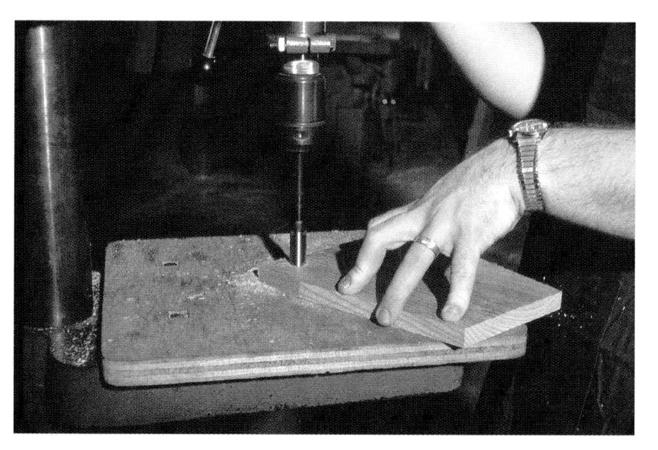

Figure 22-10. Tooling is used to increase the speed and accuracy of various operations. This fixture ensures that holes are drilled in the correct place in each part.

Figure 22-11. This machine will only close if the operator presses both switches at the same time. The orange boxes to each side of the press chamber produce a light beam that must be unbroken if the press is to operate. (Minster Machine Co.)

Notice how each part can be positioned in the fixture. The drill will produce a hole in the same place in part after part.

Finally, tooling increases worker safety. If the tooling holds the part, the operator's hands are out of the way and cannot be injured by the machine. Special devices may add more safety. Some machines will operate only when the operator holds switches with both hands. See **Figure 22-11.** Guards may come down and straps may pull hands away from dangerous positions.

Well designed tooling helps people and machines produce accurate parts. Quality is improved and scrap is reduced, which in turn lowers production costs.

Preparing Plant Layout

Selecting operations and designing tooling is not all that manufacturing engineers do. They must also be concerned with *resource flow*. This means moving materials and people through the factory efficiently.

resource flow. The flow of materials and people inside a plant.

Workers need to get to their workstations easily. Movement to and from restrooms, cafeterias, and other support areas must also be considered. There must also be rapid movement out of the plant in case of fire, storm, earthquake, or accidents.

All of these elements are considered in designing a factory. For good plant layout, manufacturing engineers must plan:

- Where to place machines
- How to move material
- Where to have aisles
- Where to locate utility systems (electricity, water, gas, etc.)

There are two basic types of plant layouts: process layout and product layout.

Process Layout

Process layout groups machines by the process they perform. A furniture plant may group machines by the process they perform. Joint cutting machines may be grouped in one location, assembly operations in another place. All finishing equipment could be in still another spot.

Process layout is used for custom manufacturing and in many intermittent manufacturing plants. These plants can produce a number of different products using the same equipment. Each product would have its own set of operations. Departments like machining, assembly, and finishing are created to process materials for a number of different products. See **Figure 22-12.**

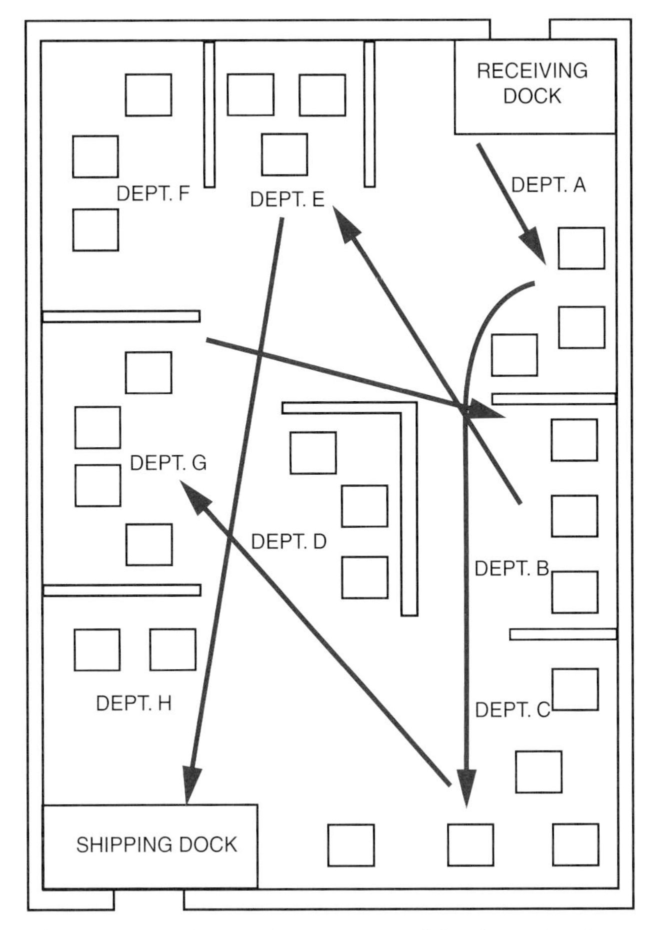

Figure 22-12. Process layout is used for factories that make a number of different products. Each product is moved to different departments as needed for its manufacture.

Product Layout

Product layout places machines according to the sequence of operations. It is used in most continuous manufacturing plants. The machine needed for the first operation is placed nearest the material storage area. The machine for the second operation is placed next, and so on. See **Figure 22-13.**

Continuous process lines can be laid out in several different patterns. The most common patterns are straight, S-shaped, circular, U-shaped, and random. See **Figure 22-14.**

process layout. A type of plant layout where machines are grouped by the process they perform.

product layout. A manufacturing arrangement in which equipment is arranged in the sequence of operations needed to produce the product.

Communicating Layouts

The manufacturing engineer must document their designs for the plant. He or she often communicates their plant designs using plant layout drawings on paper or a model. This allows plant engineers to locate equipment in the correct places and install the required electrical power, air, water, and other utilities.

Drawing the location of all equipment and features in two dimensions. However, in some cases, these drawings are confusing. Lines showing different elements (machines, conveyors, utility lines) may cross. Overhead clearances are difficult to judge. To overcome these disadvantages, a model may be used. See **Figure 22-15.** All features are shown in three dimensions. Sizes are easier to judge and clearance over aisles and under overhead conveyors are more visible.

Complex plant layout may use both drawings and models. Separate drawings may show equipment placement, utility runs, and conveyor locations. Models put all the features together to provide a miniature view of the plant.

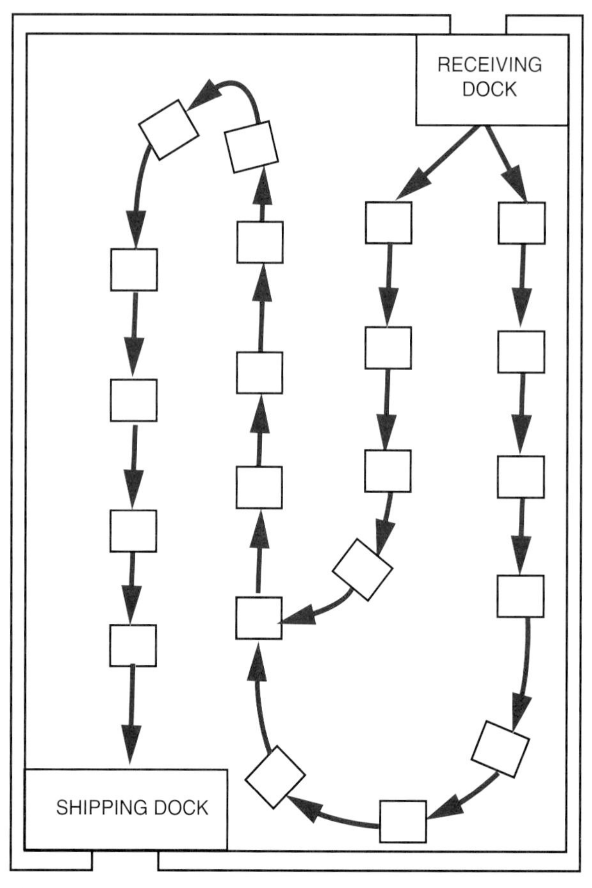

Figure 22-13. Product layout uses the sequence of operations for a product to determine machine location.

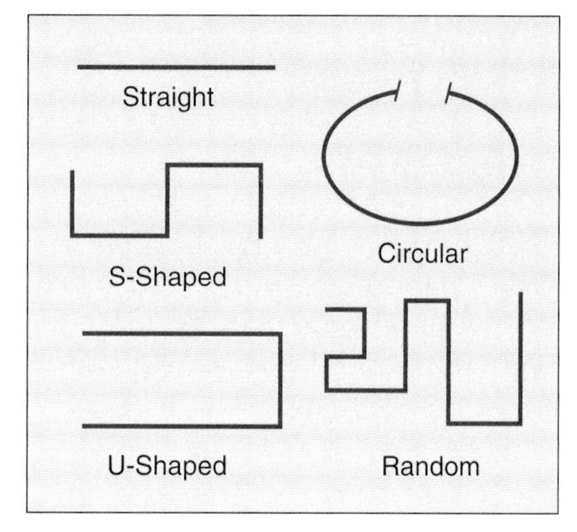

Figure 22-14. Plants can be laid out in one of five basic patterns.

Figure 22-15. A model of a plant layout uses blocks and parts that are scaled down to size. (Oldsmobile)

Designing Material Handling Systems

material handling. **The methods used to move material around a manufacturing plant.**

The fourth part of a manufacturing system is a way to move materials and products throughout the plant. This is called *material handling*.

The materials must be moved from storage to the manufacturing area. During processing the material must be moved from workstation to workstation. Finally, finished parts and products must be moved into storage or to transportation (trucks, trains, ships, planes, etc.).

Material handling devices are of two major types: fixed path and variable (steerable) path. See **Figure 22-16.**

Fixed Path Devices

Fixed path devices move a product from one fixed point to another. The item always travels on the same path much like a train on railroad. Once it starts down a main line, there is only one way to go. The train cannot decide to take a detour or make a turn.

Often fixed path systems are used in assembly lines. In these applications, feed lines supply parts to the main line. These lines can be arranged in one-sided or two-sided patterns. See **Figure 22-17.**

Figure 22-16. Material handling devices may be fixed path (left) or variable path (right) systems.

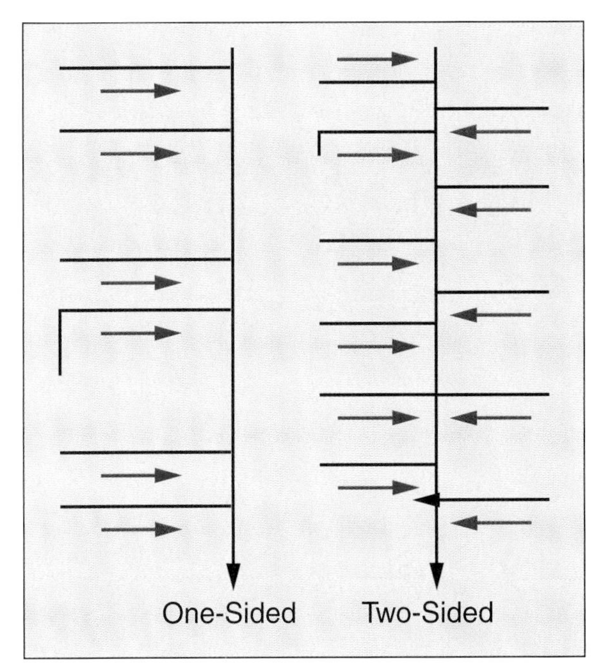

Figure 22-17. Assembly lines are one type of fixed path layouts. They use one-sided or two-sided layouts.

Figure 22-18. A "pick and place" robot is being tested for an assembly operation. It picks up a part in one location and delivers it to another. (Cincinnati Milacron)

Typical fixed path devices are conveyors, pipes, chutes, and elevators. A new fixed path device is the "pick and place" robot. See **Figure 22-18.**

Variable Path Devices

Variable path devices can be steered so that they move in a number of directions. Common variable path devices are forklifts, overhead cranes, tractors, and hand trucks. They require an operator to direct them.

Figure 22-19. An automated guided vehicle. (FMC Corp.)

Variable path robots generally follow wires buried in the floor. A computer can direct the robot to follow any path. **Figure 22-19** shows a robot vehicle that can move through a 77-acre plant along paths directed by over 19,000 feet of buried wire.

Variable path devices are used for two main purposes. They move materials and products in intermittent manufacturing plants. They also are used to load materials onto continuous manufacturing systems.

Improving Manufacturing Systems

The last task of a manufacturing engineer is improving the manufacturing system. Like products, production lines can be redesigned and improved. Operations can be refined. New ones can replace older, inefficient ones. Better material flow can be introduced and material-handling devices can be installed. More efficient machines and tooling can be developed or purchased.

Summary

Efficient manufacturing systems must be designed and engineered. This task is the responsibility of manufacturing engineers. They must select and sequence operations. Flow process and operation process charts are often used in this work.

Tooling is also designed. It includes special cutters, holding devices, molds, patterns, and machines. The tooling increases the speed, accuracy, and safety of operations.

The manufacturing plant must also be arranged for efficiency. Equipment is organized in process or product arrangements. The easy, safe flow of people and materials is always considered.

A material handling system is also needed. Materials, parts, and products must be moved through the factory. Fixed and variable path devices are used.

Finally, the system is always open for improvement. It is often redesigned and improved.

Key Words

All of the following words have been used in this chapter. Do you know their meaning?

competitive price
flow process chart
manufacturing engineer
material handling
operations
operation sheet
process layout
product layout
resource flow
tooling

Test Your Knowledge

Please do not write in this text. Place your answers on a separate sheet of paper.

1. What are the major engineering tasks?

Matching questions: Match the definition (2 through 7) with the correct term (a through f).

2. _____ Selecting process tasks to be done and putting them in order they must be done.
3. _____ Person who organizes people and machines for efficient manufacture.
4. _____ Form listing operations, machine to use, and tooling needed.
5. _____ Shows at a glance the sequence of tasks for producing a single part.
6. _____ Movement of part through manufacture.
7. _____ Represented by a circle on the flow process chart.
   a. Operation sheet.
   b. Flow process chart.
   c. Manufacturing engineer.
   d. Selecting and sequencing operations.
   e. Transportation.
   f. Operations task.
8. _____ refers to special tools or devices which make manufacturing operations more efficient.
9. What does the term "resource flow" mean?
10. _____ layout groups machines by the process the machines perform.
11. _____ layout is used in most continuous manufacturing plants.
12. List and describe the two major handling devices.

Applying Your Knowledge

1. Visit a manufacturing plant. Describe:
   a. Sequence of operations
   b. Tooling being used
   c. Type of layout used
2. Prepare a flow process chart for a simple task such as washing a plate.
3. With a partner, design a piece of tooling for a drill press that will drill:
   a. One hole in the center of a 4" × 4" part.
   b. Two holes along a center line of a 2" × 6" part. The holes should be spaced 2" apart.

Machinist

Machinists use machine tools, such as lathes, milling machines, and spindles, to produce precision parts. Although they may produce large quantities of one part, precision machinists often produce small batches or one-of-a-kind items. They use their knowledge of the working properties of the materials and their skill with machine tools to plan and carry out the operations needed to make machined products that meet precise specifications.

Machinists must first carefully plan and prepare the operation. They review blueprints or written specifications for a job, and calculate where to cut or bore into the workpiece (the piece is being shaped). They select tools and materials for the job, plan the cutting and finishing operations, and mark the workpiece to show where cuts should be made.

The machinists perform the necessary machining operations. During the machining process, they must constantly monitor the feed and speed of the machine. Machinists adjust cutting speeds to compensate for harmonic vibrations, which can decrease the accuracy of cuts, particularly on newer high-speed spindles and lathes.

Some machinists, often called production machinists, may produce large quantities of one part, especially parts requiring the use of complex operations and great precision. Production machinists work with complex computer numerically controlled (CNC) cutting machines. Frequently, machinists work with computer-control programmers to determine how the automated equipment will cut a part.

Chapter 23 Obtaining Human and Materials Resources

Objectives

After studying this chapter, you will be able to:

- ✓ List steps for hiring workers and describe how hiring is done.
- ✓ List the six major steps for buying materials.
- ✓ Describe various personnel and materials forms and state their uses.

The product has been designed and the engineering drawings have been prepared. The product is approved for manufacture. The manufacturing facility has been engineered with equipment and tooling is in place. Operations have been selected and sequenced. The material handling system is in place. Now what?

The product cannot be produced without more resources. See **Figure 23-1.** Workers must be hired and materials must be ordered. Securing these resources will complete the preparations for manufacture.

Figure 23-1. The plant is ready to produce products. All that is needed are workers. (American Iron and Steel Institute)

Employing Workers

Employees are people hired to do a specific job. See **Figure 23-2.** The company matches workers' abilities with its needs. Placing people in jobs is called *employment*. It requires the six major steps listed in **Figure 23-3.**

Determining Needs

The employment or human resources office must be told to hire a worker. Generally an employee requisition (request) form is prepared. This form tells the employment office:

employment. **Placing qualified people in jobs.**

- The job title
- The skills required
- The date the employee is needed

Figure 23-2. People are the most important resource a company has. (Hewlett-Packard)

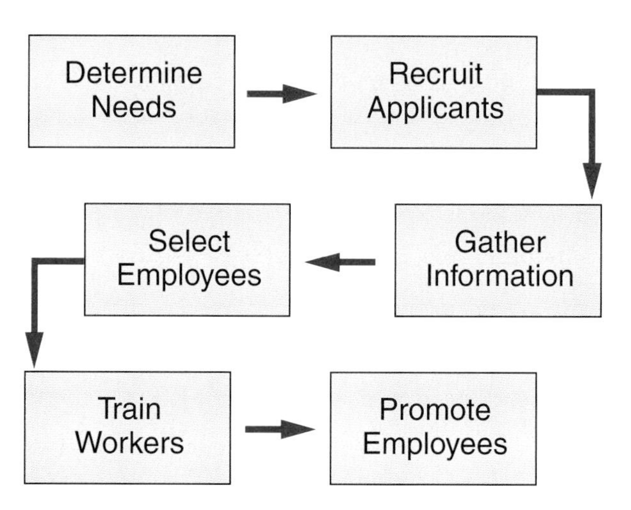

Figure 23-3. Six steps are followed in the employment and advancement of workers.

TYPE OF POSITION	SOURCES OF JOB APPLICANTS
PRESIDENT VICE PRESIDENT UPPER MANAGEMENT	• EXECUTIVE SEARCH (EMPLOYMENT) AGENCIES • ADVERTISEMENTS
SCIENTIFIC AND TECHNICAL	• UNIVERSITIES, TECHNICAL SCHOOLS • ADVERTISEMENTS • EMPLOYMENT AGENCIES
SKILLED WORKERS	• TRADE SCHOOLS • BUSINESS SCHOOLS • ADVERTISEMENTS • EMPLOYMENT AGENCIES • WALK-INS
SEMI SKILLED WORKERS (MACHINE OPERATORS, INSPECTORS, ETC.)	• ADVERTISEMENTS • STATE EMPLOYMENT AGENCIES • WALK-INS
UNSKILLED WORKERS	• WALK-INS

Figure 23-4. Sources for applicants from jobs at various levels.

Receiving a job requisition starts the employment process by providing basic hiring information.

Recruiting Applicants

People must be found to fill job vacancies (openings). See **Figure 23-4.** The employment office then begins to search for individuals who can do the job. This process is called *recruitment.*

There are four basic ways to recruit employees, which are shown in **Figure 23-5.**

The technique used will vary with the job opening. General factory and office jobs may be filled by walk-ins. These are people who come to the company looking for a job. They seek out the company and the job.

Jobs requiring more skill are filled by other methods. These jobs may include managers, technical staff, and skilled operators. Recruiters may visit schools or attend professional conferences to interview prospective employees.

The company may advertise in newspapers and special technical magazines for trained people. They may also use employment agencies, which are often called headhunters.

Gathering Information

recruitment. The process of attracting potential workers to a company.

Employers need information about an applicant (person seeking a job). They will use the information to decide how well suited the applicant is for the job. These facts may be gathered in several ways, **Figure 23-6:**

Figure 23-5. These four methods may be used to recruit employees.

Figure 23-6. Information about applicants may be gathered in several ways.

- Application blank. This is a form that provides basic personal data (facts) and work records.
- Previous employers. Information about the applicant's work habits and qualifications can be obtained from people he or she worked with at a former job.
- Interview. The applicant answers questions about his or her previous work experience, goals, and training. See **Figure 23-7.**
- Test. An applicant takes a written or performance test that can measure her or his ability to do the job.

Figure 23-7. These people are participating in a new employee interview.

Selecting Employees

It is in a company's best interest to choose the best applicant for each job. Information from the application blanks, interviews, and tests are used to arrive at the best choice. The best applicant is one who:

- Can do the job well
- Will fit into the company
- Is willing to accept training
- Wants to advance (get ahead)

Suppose that the person selected is hired. What happens next? He or she will receive a job notice or an employment letter. It will tell the worker the job title and rate for the job and when to report for work.

Training Employees

training. The process of preparing workers for jobs.

Few people can start a new job without some training. *Training* is the process of preparing workers to do their jobs. All newly hired workers need some basic information about the company itself. Knowledge of the company's products, competitors, and plant locations is also important. Basic rules about hours, pay rates, and work practices must be presented. This information is often given in an induction session for new employees.

They then receive any special training they need. Production workers may receive training through:

- **On-the-job training.** This is training at the workstation. Actual products on the production line are used. A supervisor or another worker will do the training.
- **Vestibule (off-line) training.** This is training in a special training area. Workers produce actual products, too. However, the training location is away from the manufacturing line. See **Figure 23-8.** A special instructor is often used.
- *Apprenticeship.* Such work preparation combines on-the-job with classroom training. See **Figure 23-9.** This type is used to prepare skilled workers. An experienced worker provides the on-the-job training. Often a special teacher provides the classroom instruction.
- **Cooperative education.** The new worker gets training by attending school part-time and working part-time.

Figure 23-8. This worker is receiving training on new manufacturing equipment. (AC Rochester)

Figure 23-9. These workers are receiving classroom training. (General Motors Corp.)

Employees also receive additional training during their working life. Individuals attend conferences and workshops. Special classes are offered. See **Figure 23-10.** Employees are trained in new developments, methods, and equipment. This ongoing employee development is essential for company growth.

apprenticeship. Training that combines on-the-job and classroom instruction. The on-the-job portion is provided by a skilled worker. A special teacher provides the classroom portion.

Advancing

Employees often change jobs as they develop new skills and knowledge. They are given more responsibility and advance to better jobs. These jobs are often more secure and pay higher wages.

Managers prefer workers who want to get ahead. Workers can show this by their willingness to accept new job and training opportunities.

Figure 23-10. Employees attend classes on new technologies. (Ford Motor Company.)

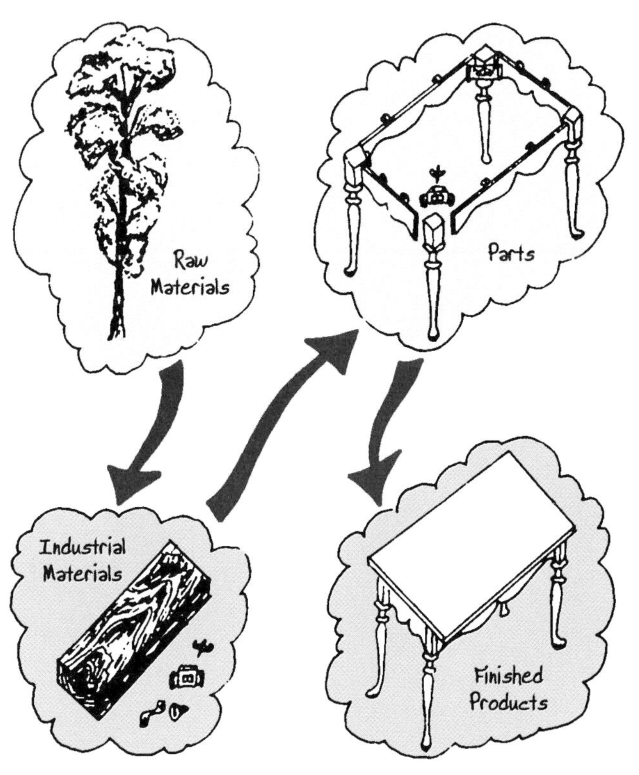

Figure 23-11. Purchasing takes place at each step in manufacturing.

Purchasing Materials

Making products requires many parts and materials to be purchased. For example, steel mills must purchase raw materials. Appliance manufacturers buy their steel. Power tool manufacturers purchase finished parts, like motors. Purchasing takes place at each step of manufacture. See **Figure 23-11.**

Purchasing brings suppliers and users together. This effort involves six major steps that are carried out by various people. These, shown in **Figure 23-12**, are:

purchasing. **The practice of buying the materials needed to manufacture products; it brings the suppliers and users together.**

- Requisition (request) materials
- Get bids or price quotes
- Issue a purchase order
- Receive shipment and invoice
- Accept shipment
- Pay for materials

Each of these steps involves filling out forms that record each activity. These forms are important for controlling purchasing activities. They ensure that the right material or equipment is ordered and received.

Material Requisition

The purchasing office buys materials when told what is needed. This may be done in a material requisition that lists:

- The material needed
- Quantity required
- Date needed
- Delivery location (plant, department, etc.)

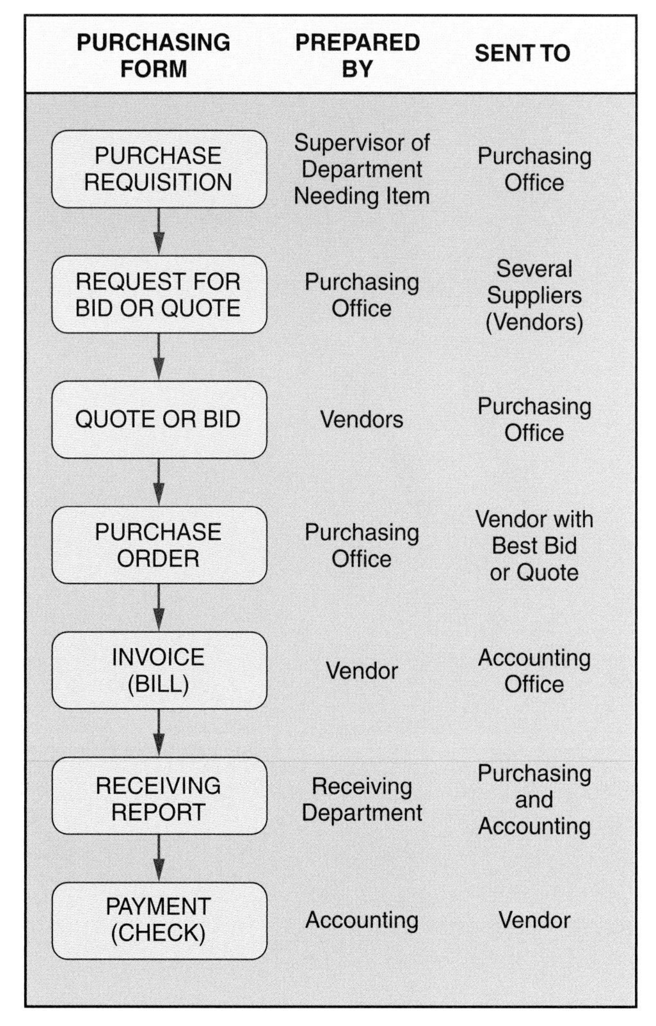

Figure 23-12. Steps in the purchasing procedure. Filing of proper forms keeps track of each purchase.

Often production planners fill out the requisition form. It is their job to determine when material and human resources are needed.

Bids and Quotes

The purchasing staff must find out the cost of the requested materials. First, they must find suppliers (also called vendors). Then a request for a price is sent to each vendor selected.

The vendors submit their prices for the materials. Some prices must be guaranteed for a period of time. These are called ***bids.*** Other prices are current prices that indicate the purchase price for that day. These prices are called ***quotes.***

Purchase Orders

The purchasing staff reviews prices by studying the bids or quotes. Then they will choose the best supplier. The choice depends on several factors including:

- Price
- Material quality
- Delivery date
- Vendor's reputation

Then the purchasing people prepare a purchase order that is sent to the selected vendor. The purchase order, **Figure 23-13,** lists the quantity needed, material description, and price. It also tells where and when the order is to be sent.

The vendor signs a copy of the purchase order and returns a copy to the company. The purchase order has become a legal, binding contract. The vendor must supply the materials and the purchaser must pay for them.

Invoice

When the ordered materials are shipped, a bill is also sent. This form, called an invoice, **Figure 23-14,** indicates that:

- The order has been shipped.
- The company owes the vendor for the price of the materials.

Receiving Report

Often, materials ordered from one location are received in another. For example, an order may come from Detroit to deliver materials in Kansas City. The order and receipt of materials must be coordinated.

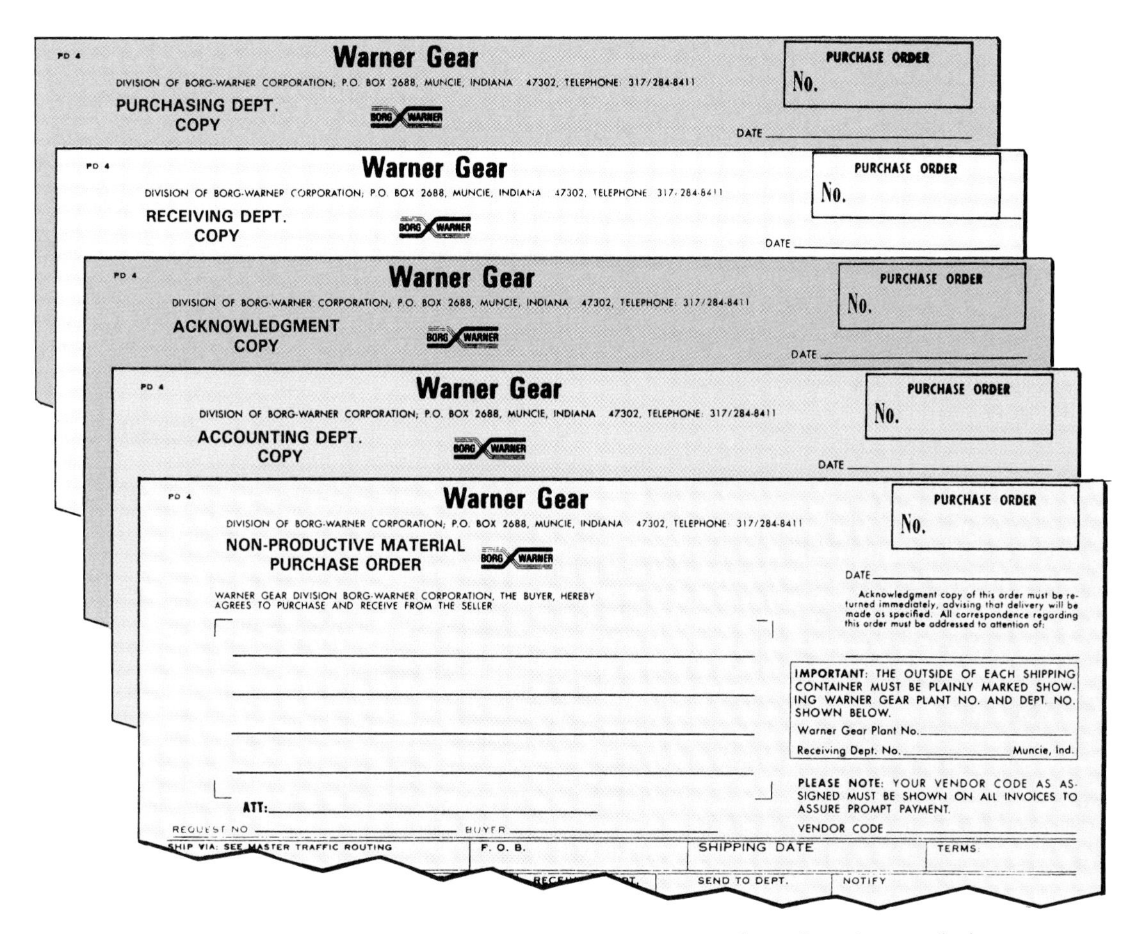

PD 4 **Warner Gear** PURCHASE ORDER No.
DIVISION OF BORG-WARNER CORPORATION; P.O. BOX 2688, MUNCIE, INDIANA 47302, TELEPHONE: 317/284-8411
PURCHASING DEPT. COPY
BORG WARNER
DATE ______

PD 4 **Warner Gear** PURCHASE ORDER No.
DIVISION OF BORG-WARNER CORPORATION; P.O. BOX 2688, MUNCIE, INDIANA 47302, TELEPHONE: 317/284-8411
RECEIVING DEPT. COPY
BORG WARNER
DATE ______

PD 4 **Warner Gear** PURCHASE ORDER No.
DIVISION OF BORG-WARNER CORPORATION; P.O. BOX 2688, MUNCIE, INDIANA 47302, TELEPHONE: 317/284-8411
ACKNOWLEDGMENT COPY
BORG WARNER
DATE ______

PD 4 **Warner Gear** PURCHASE ORDER No.
DIVISION OF BORG-WARNER CORPORATION; P.O. BOX 2688, MUNCIE, INDIANA 47302, TELEPHONE: 317/284-8411
ACCOUNTING DEPT. COPY
BORG WARNER
DATE ______

PD 4 **Warner Gear** PURCHASE ORDER No.
DIVISION OF BORG-WARNER CORPORATION; P.O. BOX 2688, MUNCIE, INDIANA 47302, TELEPHONE: 317/284-8411
NON-PRODUCTIVE MATERIAL PURCHASE ORDER
BORG WARNER
DATE ______

WARNER GEAR DIVISION BORG-WARNER CORPORATION, THE BUYER, HEREBY AGREES TO PURCHASE AND RECEIVE FROM THE SELLER

Acknowledgment copy of this order must be returned immediately, advising that delivery will be made as specified. All correspondence regarding this order must be addressed to attention of:

IMPORTANT: THE OUTSIDE OF EACH SHIPPING CONTAINER MUST BE PLAINLY MARKED SHOWING WARNER GEAR PLANT NO. AND DEPT. NO. SHOWN BELOW.
Warner Gear Plant No. ______
Receiving Dept. No. ______ Muncie, Ind.

ATT: ______

PLEASE NOTE: YOUR VENDOR CODE AS ASSIGNED MUST BE SHOWN ON ALL INVOICES TO ASSURE PROMPT PAYMENT.

REQUEST NO ______ BUYER ______ VENDOR CODE ______

SHIP VIA: SEE MASTER TRAFFIC ROUTING | F. O. B. | SHIPPING DATE | TERMS

RECE... | SEND TO DEPT. | NOTIFY

Figure 23-13. Sample of a large company's purchase order. Note number of copies needed.

The receiving area receives, from Detroit, copies of the original purchase order. This tells the receiving area: "These items are ordered. Be ready to receive them." When the order arrives, Kansas City personnel check the shipment. They compare the purchase order with the materials received. If they match, Kansas City signs a copy of the purchase order. It becomes a receiving report that is sent to the purchasing office. The signed order tells purchasing officers in Detroit that the materials have arrived safely.

Payment

The receiving report and the invoice are compared to see if the materials and prices are correct. The invoice and the purchase order should match. If they do, the invoice is approved for payment. A check is written, the vendor is paid, and the purchasing cycle is complete.

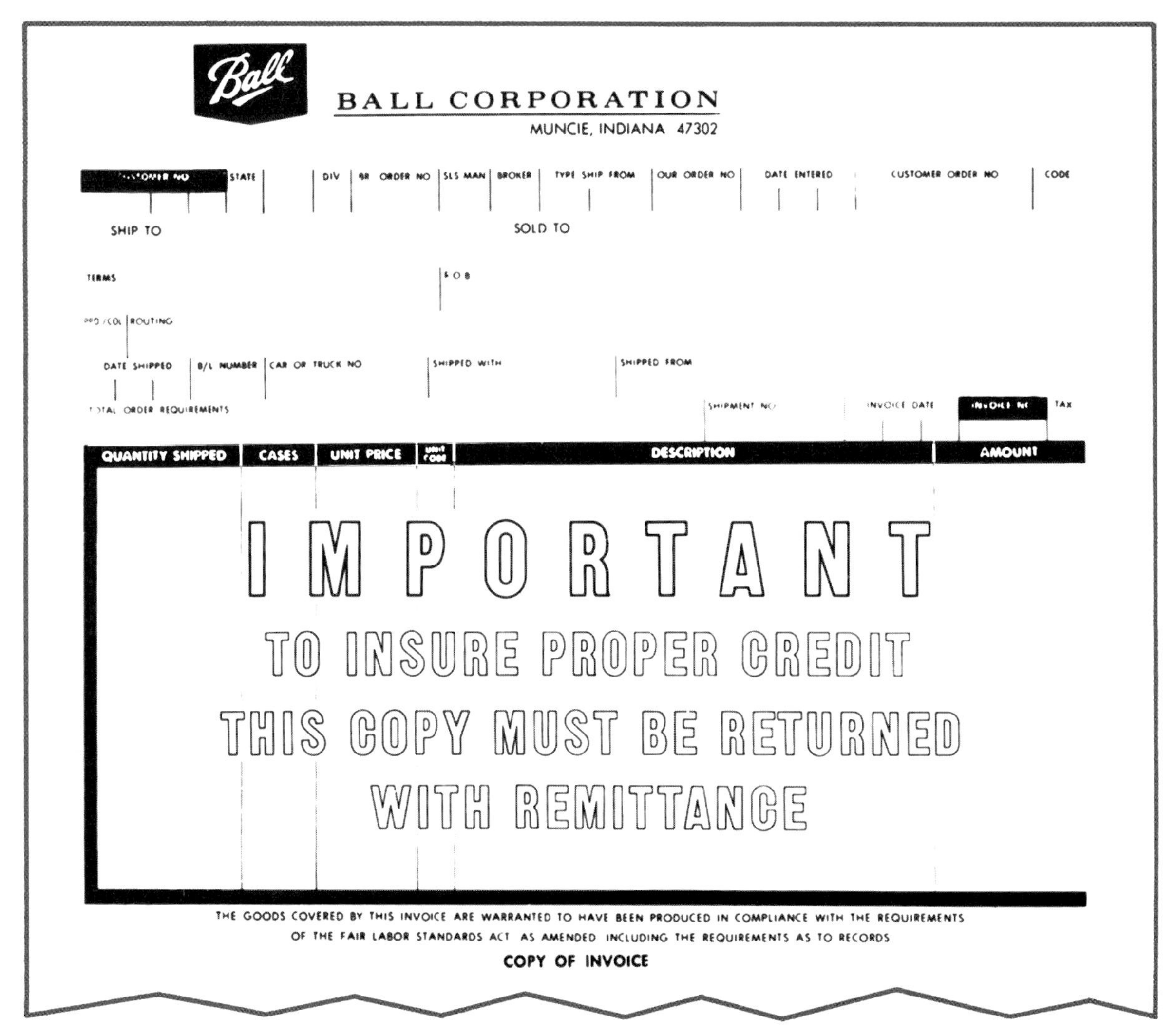

Ball

BALL CORPORATION

MUNCIE, INDIANA 47302

CUSTOMER NO | STATE | DIV | BR ORDER NO | SLS MAN | BROKER | TYPE SHIP FROM | OUR ORDER NO | DATE ENTERED | CUSTOMER ORDER NO | CODE

SHIP TO

SOLD TO

TERMS

F O B

PPD/COL | ROUTING

DATE SHIPPED | B/L NUMBER | CAR OR TRUCK NO | SHIPPED WITH | SHIPPED FROM

TOTAL ORDER REQUIREMENTS

SHIPMENT NO | INVOICE DATE | INVOICE NO | TAX

QUANTITY SHIPPED	CASES	UNIT PRICE	UNIT CODE	DESCRIPTION	AMOUNT

IMPORTANT

TO INSURE PROPER CREDIT

THIS COPY MUST BE RETURNED

WITH REMITTANCE

THE GOODS COVERED BY THIS INVOICE ARE WARRANTED TO HAVE BEEN PRODUCED IN COMPLIANCE WITH THE REQUIREMENTS OF THE FAIR LABOR STANDARDS ACT AS AMENDED INCLUDING THE REQUIREMENTS AS TO RECORDS

COPY OF INVOICE

Figure 23-14. A sample invoice. This is the bill that is included with delivery of an order. (Ball Corp.)

Summary

Manufacturing depends upon human and material resources. People, whose abilities match the jobs, are hired to do work. This is the task of employment. First the need for workers is determined. Then, applicants are recruited and information about them is gathered and studied. The best applicants are hired. New employees receive training about the company and job. Sometime during their employment they are likely to receive additional training.

Materials for production are purchased from the best supplier. The material is ordered and received. The vendor receives payment for the materials. These tasks are the job of people in purchasing.

Obtaining the best human and material resources is very important. Qualified workers and managers are keys to success. Proper material must be on hand to produce quality products. Without these resources, a company will surely fail.

Key Words

All of the following words have been used in this chapter. Do you know their meaning?

apprenticeship
employment
purchasing
recruitment
training

Test Your Knowledge

Please do not write in this text. Place your answers on a separate sheet of paper.

1. What are the major steps used to employ workers?
2. There are four basic ways workers can be recruited. Name them.
3. Which of the following are methods companies use to gather information about people looking for jobs with them?
   a. Have people demonstrate skills by giving them a test.
   b. Ask questions about their work experience, education, training, and goals.
   c. Have persons fill out a job application.
   d All of the above.
   e. None of the above.
4. Training a person at an actual workstation is called _____ _____ _____ training.
5. *True or false?* Apprenticeship training combines on-the-job training with special classes.
6. List the steps in a typical purchasing system.

Applying Your Knowledge

1. With a partner, design an application that could be used to gather information for job openings on your school newspaper (or other student position).
2. Brainstorm several questions to ask the applicant for the job in Activity 1. The questions should determine if the applicant can do the job well, will fit into the working environment, is willing to accept training, and wants to advance.
3. Role-play an interview between the supervisor and applicant for the position in Activity 1. Ask the questions you formulated in Activity 2.
4. Do some comparison shopping. Go to a store and choose an item, such as a CD-radio-cassette recorder. Compare the price, quality, stock availability, and manufacturer's reputation of three different models. Which do you think is the best purchase and why?

Technology Link

Construction

Manufacturing is a unique and broad technology. In basic terms it is the process that changes resources into products. Manufacturing affects many other technologies. Other technologies, such as *construction*, have an effect on manufacturing as well. The link between manufacturing and construction is a two-way street.

Maybe you have seen a new house or a new office building being constructed. Have you ever wondered how the materials used to construct the new structure were made? From the nails that hold the structure together to the window and door assemblies—all are the result of a manufacturing process. Almost every part of the new structure will have been through some manufacturing process. The framework of a window may be extruded and cut to size. The glass is formed and cut to size. All parts are then assembled to produce presized, preframed window assembly that is ready for installation.

What most people don't realize is the direct effect other technologies have on manufacturing. Just what effect does construction have on manufacturing? What about the buildings that make up manufacturing facilities? Without developments in the construction field, these manufacturing facilities would not exist. The offices that house the managerial teams that direct the flow of manufacture would not exist.

Chapter 24
Establishing Control Systems

Objectives

After studying this chapter, you will be able to:

✓ Name the four phases involved in developing and using a resource control system.
✓ List the major resources used to manufacture a product.
✓ Discuss the three major factors that affect labor costs.
✓ Name and discuss the two tasks in a total quality control system.

Good managers know how to use resources to get the best results. Machines that the company owns must be used as much as possible. People need to be kept busy doing productive work. Materials should not be wasted. The word for this management task is *resource control*.

resource control. **Making sure that resources are being fully utilized, and not sitting idle or being wasted.**

The outputs of the system must also be controlled. Waste, scrap, and pollution must be kept down. Product quality must be assured, too.

Control Systems

As we can see, all manufacturing activities need controlling. They must be carefully managed through established systems that have four phases as shown in **Figure 24-1.**

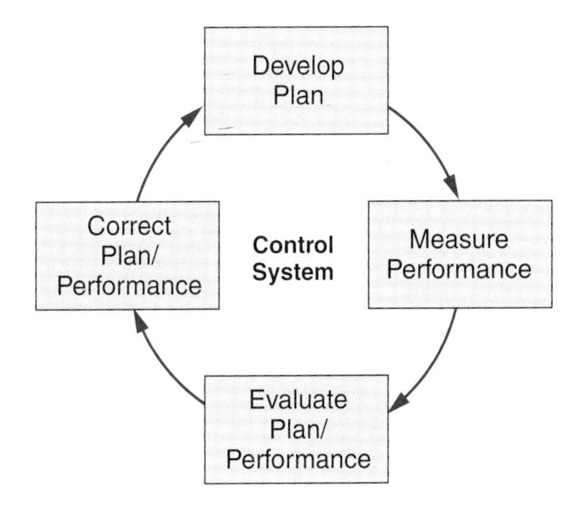

Figure 24-1. Phases of a control system.

Planning Activities

There are two basic types of manufacturing plans: long-range and short-term. Long-range plans are usually three to five year projections (forecasts or guesses) that direct overall business activities. Long-range product development, marketing, and finance plans are common. Long-range plans are general in nature. They only give guidelines for policy decisions.

Short-term plans are more concrete (real) statements that outline performance goals (work to be done) for a set period. Short-term plans can be for a day, week, month, or year.

Typical short-term plans include production schedules, budgets, and sales quotas. (A quota is a goal or amount to be done.) Goals guide day-to-day operations.

Manufacturing activities are guided by short-term plans. They include plans for using labor, materials, and machines. Also included are plans for controlling product quality.

Measuring Performance

Controlling requires measurement. The heat in your home is an example. Most houses have heating systems to keep them warm while some homes have air conditioning to cool them. These systems are controlled by thermostats that measure room temperatures and turn furnaces and air conditioners on and off.

Likewise, manufacturing activities are controlled. That is, their performance is measured. Some method is used to record the use of machines, labor, and material. See **Figure 24-2.** Records are kept of time worked, products produced, and material used. These measurements provide the basis for management decisions.

Figure 24-2. The use of labor, material, and machines must be controlled. (Reynolds Metals)

Evaluating Performance

Collecting data is not enough. To say, "It sure is hot today," is not very meaningful. "How hot is it?" compares today's temperature with some average and adds meaning. Maybe it is 10 degrees hotter than normal. This means more to us than just saying it is hot.

Manufacturing performance is also rated against some base. This base is often the company's plan. The number of products produced is measured against production plans. Scrap rates and rejected products are evaluated. Product quality, **Figure 24-3,** and worker output (amount of work done) are always measured. Product function, reliability, and appearance are evaluated.

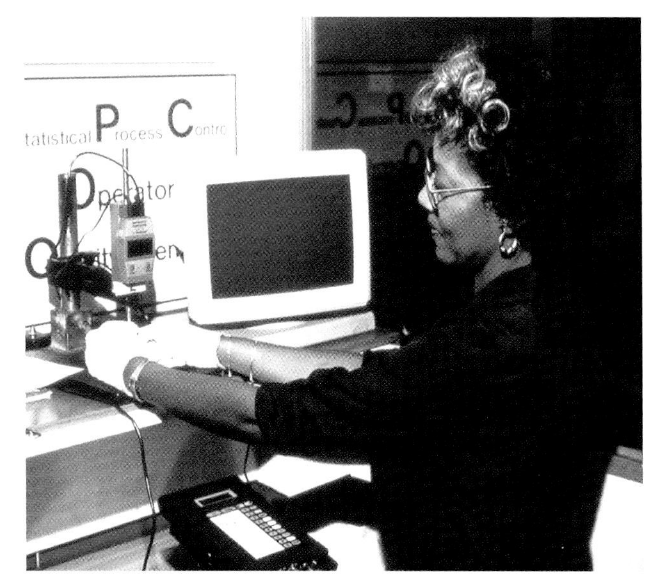

Figure 24-3. This inspector is measuring the performance of automobile components. (AC Rochester)

Corrective Action

Such measurement and evaluation often point out shortcomings. Performance may not live up to the plan requiring the company to correct the problems. Management must decide what action to take.

A number of things could be done. Manufacturing engineers may redesign operations. See **Figure 24-4.** They can change the workstations to make them more efficient. These actions could increase the worker's efficiency. Scrap rates could go down and productivity could rise.

Materials may be changed. Another standard size could provide better cuts. The size of scrap at the ends of the material could be reduced. A different material may work better.

Whatever the corrective action, product quality must be maintained. Managers try to find the most efficient system to produce good products.

Figure 24-4. A new process is being tested to see if it produces better parts. (Crouse-Hinds, Div., Cooper Industries)

Factors to Control

Everything a company does is controlled. Sales are controlled. Raw material and finished goods inventories (stores) are controlled. Income and expenses are controlled. Control is the basis for success. Two major concerns of any company are use of resources and quality of its products. These can be controlled.

Controlling Use of Resources

The major resources used to make products, **Figure 24-5,** are:

- Human labor
- Machine time
- Materials

Figure 24-5. This worker is using a machine to form a steel part. The part, the machine, and the worker are resources of the company. (Cincinnati, Inc.)

These resources must be managed. They must be used efficiently.

Controlling Human Labor

Each job takes time to do; therefore, there is a labor cost for the job. This cost adds to the total product cost. Management must see that the labor cost stays within the plan's limits.

Several things affect the actual labor cost. These include:

- How well the operations are planned
- Worker efficiency. (How well the workers do their job)
- Wage rate paid to the workers

Operations

These things can be changed through management action. Manufacturing engineers can study flow and operation process charts. Operations can be simplified and combined. Product design changes can do away with some assembly

Figure 24-6. Note how the worker sits comfortably while working. All machine controls and parts are within easy reach. This is an efficient workstation. (AC Rochester)

Figure 24-7. This automatic spraying machine coats panels without putting a human operator at risk from paint fumes. (White Consolidated Industries)

steps. New technology may make manufacturing more efficient. Any of these actions can reduce labor costs.

Worker Efficiency

Operations can be studied from the worker's point of view. The task can be made more efficient. Machines can be modified (changed) to make them easier to use. Tooling can be improved. Machine controls can be located for greater convenience. The amount of reaching can be reduced by handier location of materials. See **Figure 24-6.**

Often, the efficient use of labor is improved by *automation*. Simple, routine jobs are given to machines, robots, and computers. See **Figure 24-7.** Depending on company policy, workers can be retrained and reassigned to more challenging jobs.

Wage Rate

The final item in labor cost is *wage rates* (what the worker is paid). Generally, labor cost and worker skill are related. Difficult jobs require skilled workers and skills take time and ability to learn. This ability and training is rewarded with higher wage rates. *Deskilling* (making the job simpler) can control labor costs.

The total skill of any operation, **Figure 24-8,** includes:

- Skill built into the machine
- Skill of the worker

Manufacturing engineers always try to build skill into the machine. They can do this in several ways: better tooling, easier-to-use or self-adjusting controls. As a result, the operator needs less skill to run the machine. She or he can be hired at a cheaper wage.

Labor costs can only be controlled by studying results. Manufacturing information must be gathered.

automation. Replacement of human control for a machine, process, or system, with control by mechanical or electronic devices.

wage rates. A type of labor cost; the amount of money the worker is paid.

Measuring Labor

There are three major measurements used to control labor costs. These are:

- Time worked
- Products produced
- Productivity

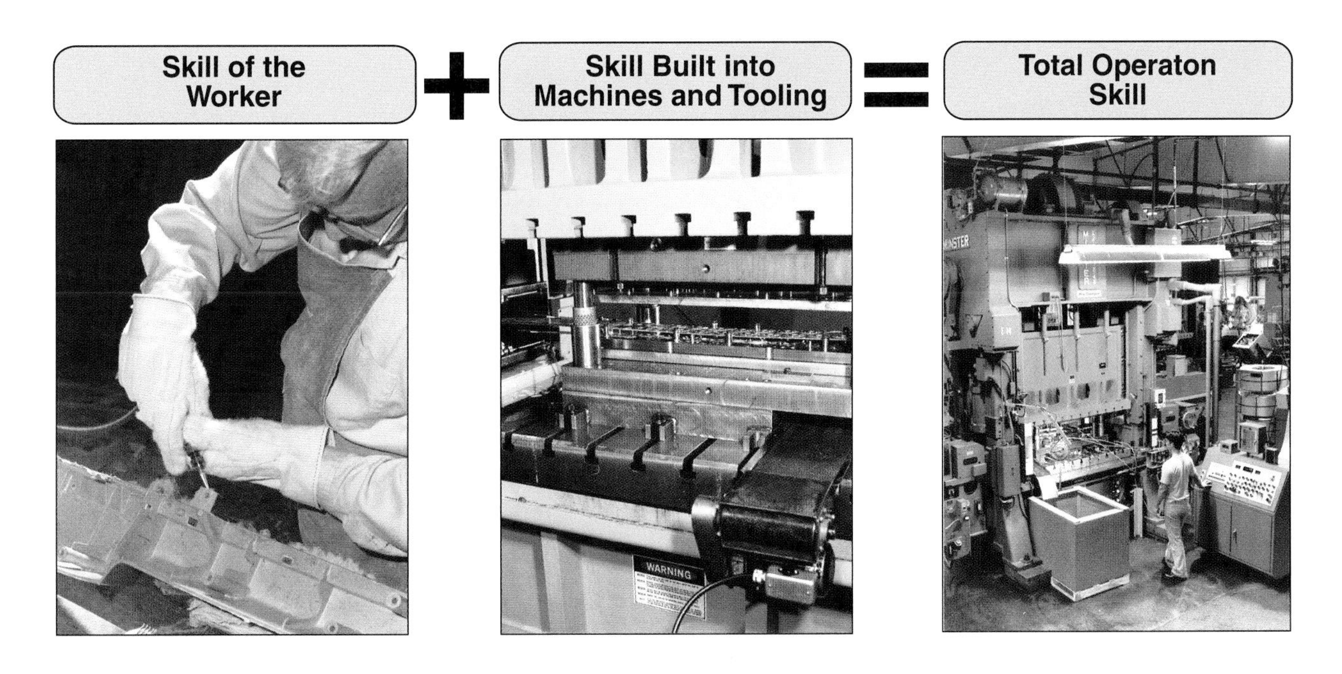

Figure 24-8. The skill of the job equals the skill of the machine plus the skill of the operator. (ARO Corp., Minster Machine Co.)

The simplest measure for labor is time worked. Everyday, employees record the times they start and end work. This time span is the workday. Often, a *time clock* and time card are used, **Figure 24-9,** to automatically print the employee's starting and ending times on the card. At the end of the week, total time worked is added together.

Work completed at each station is another way to measure labor. This technique records quantities of work rather than time.

The time worked is important to the worker because it is basis for many people's pay. See **Figure 24-10.** But, the number of good products made is the important measure for a company. Profits are made by producing quality products. Production records are often kept to record the number of products produced. The time it took to produce them is also recorded.

The production and time data are the basis for the third measure of labor efficiency. It is called *productivity*, which is a measure of the output per unit of labor. Typically, it is measured in terms of products per worker hour (one person working one hour) or worker day (one person working one day).

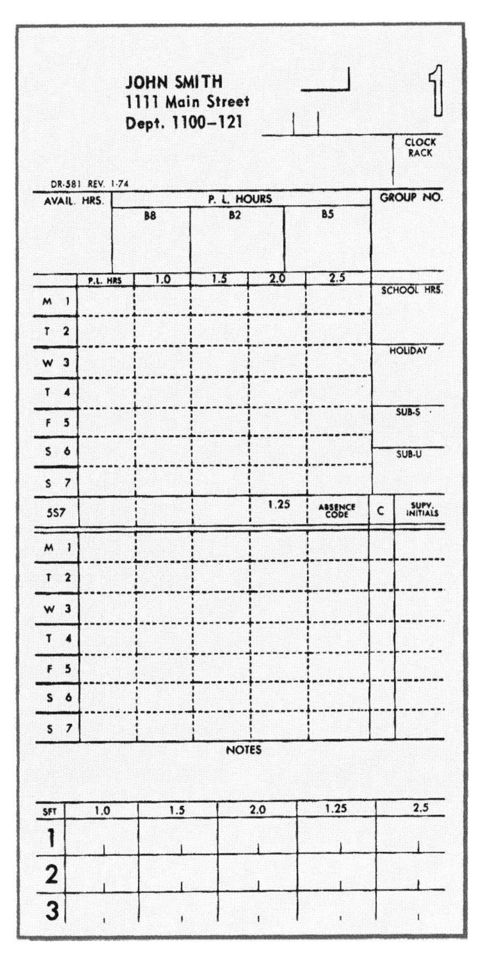

Figure 24-9. A typical time card. When card is inserted into a time clock, time will be recorded.

Controlling Machine Time

Proper use of machines has an important effect on cost of production. Machines must be carefully scheduled for use. Production planners must see that machines are not idle for long. See **Figure 24-11.** Similar operations should be scheduled on the same machine. This reduces set-up and tear-down

Figure 24-10. This worker is "punching" a job card. It records both the time she starts and completes a specific job. (ARO Corp.)

Figure 24-11. This production planner schedules the use of machines in an intermittent manufacturing system. (AC Rochester)

deskilling. A method used to control labor costs by making a job simpler.

time clock. A device that automatically prints an employee's starting and ending times on a card.

productivity. A measure of the output per unit of labor.

times. Also, long runs should be scheduled whenever possible. Efficiency rises as the length of the run increases.

Machine use can also be increased through careful operation design. All machine operations have at least three phases:

- Loading parts or materials into the machine.
- Processing the parts or materials.
- Unloading parts or materials from the machine.

Only the second phase is productive (makes products). It changes the form or shape of materials. The other phases contribute nothing to the form change. Loading and unloading time should be kept to a minimum.

At the same time, actual processing time should be studied. Correct machine speeds and feeds should be used. Drilling a hole at a speed slower than required is wasteful.

Controlling Material Use

inventory control. Controlling the amount of material on hand.

The use of materials must be controlled through an activity that is called *inventory* (materials on hand) *control.*

There are several materials that can be controlled. These, shown in **Figure 24-12,** are:

- **Raw materials.** Materials that will be processed during manufacture.
- **Purchased parts.** Hardware and parts bought from other companies. They will become part of the product.
- **Work-in-process.** Products that are being built but are not finished.

- **Finished products.** Completed products ready for shipment to customers.
- **Supplies.** Materials, such as abrasives or paint that are used up in the manufacture of the product.

These inventories are carefully kept. Often, computers are used to help keep the records. The record will list material by six categories:

- **On order.** Items ordered but not yet received.
- **Received.** Items received since the last update of the records.
- **On hand.** Items in storage.
- **Issued.** Items released from the stockroom to production or shipping.
- **Allocated.** Items being held that are marked for production or shipping.
- **Available.** Items in inventory that are available for use.

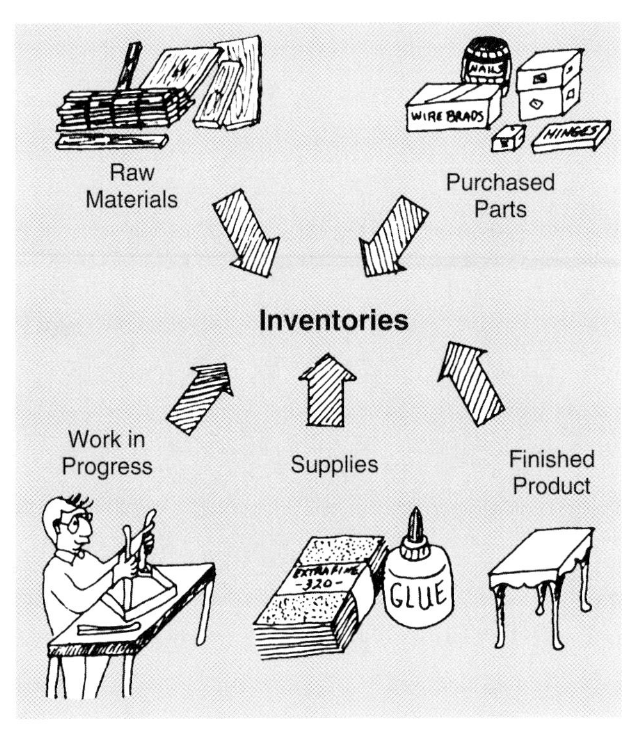

Figure 24-12. Types of inventories. Companies keep careful records of these items.

These records allow inventory control workers to keep track of material use. Also, the materials used can be compared with products produced. This will allow the manager to find areas of waste or theft.

Controlling Quality

Quality is important to everyone. Customers want products that work and retailers want to sell good products.

Manufacturers also want quality because returned products cost money to process. Also, creating dissatisfied customers can lose sales.

Quality is the business of everyone at every stage in manufacturing. Products must be designed with quality in mind. Facilities (space and equipment) must be engineered to turn out quality. Training programs must stress the importance of producing quality products. Customer service must be alert to quality problems.

The success of a company will depend on its image. Customers must see it as a quality manufacturer.

The key to producing a quality product is *quality control*. It ensures that the product meets standards. This function of a company has two tasks, as seen in **Figure 24-13:**

quality control. **The process of ensuring that a product meets specifications and standards.**

- Motivating workers to produce quality products.
- Inspecting products to remove substandard items.

Figure 24-13. Quality control programs have two functions: motivation and inspection. (General Motors)

Academic Link

Mathematics is used in many areas of the manufacturing world. One of the important uses of mathematics is with quality control and the measurement of the number of good products that are produced compared to the number of bad ones. In manufacturing processes, the sigma (a Greek letter) value is a measurement that indicates how well that process is performing—the higher the sigma value, the better.

More specifically, sigma measures the capability of the process to perform defect-free work. A defect is anything that results in customer dissatisfaction. With "Six Sigma" the common measurement index is "defects per unit," where a unit can be virtually anything—for example, a component part, piece of material, line of code, administrative form, time frame, or distance.

The sigma value indicates how often defects are likely to occur. As sigma increases, cost and cycle time go down while customer satisfaction goes up. Six sigma is the goal, which means products and processes will experience only 3.4 defects per million opportunities, or 99.99966% good. When it is said a process is at six sigma, we are saying it is the best in its class.

Motivation

Humans generally do what they think is important. For many years, companies emphasized production. Numbers of products manufactured was considered the most important factor. Quality was important, but took a backseat.

Workers reacted as expected by producing large numbers of products. Not all of them were good. But times have changed and now quality is very important. A number of programs have been developed to encourage quality. A few years ago "zero defects" was popular. Slogans, badges, and banners encouraged workers to produce good products. Managers developed these programs and they presented them to the workers.

These programs had mixed success because many workers did not believe their work was important. To them, quality was just a slogan. Foreign competition has changed workers' minds. Many companies and entire industries were seriously hurt by the image of poor quality.

Now workers are asked to help improve manufacturing systems. They are invited to join *quality circles* and "quality of work life circles." These are voluntary groups that meet often to discuss ways to improve the company and its operations. See **Figure 24-14.** New production methods, management activities, and other ideas are discussed.

Figure 24-14. These workers take time from their work to discuss ways to improve the company. (Brush-Wellman)

The goal of quality circles is to improve quality of both products and work life. Workers who join these groups tend to feel they are important. They can help the company improve and compete. Also, they have a better understanding of the importance of their job. They talk over problems with people in management.

Inspection

An old joke, known as "Murphy's Law," suggests that "if something can go wrong, it will." Of course this is not exactly true. But no person or machine is perfect because both can produce a poor product or a bad part. Items are sometimes built that fail to meet quality standards. These items must be taken out of the manufacturing line through an activity called *inspection*.

Figure 24-15. This computerized inspection system checks automobile engines before they are installed. (Ford Motor Co.)

Inspectors ensure that only good products leave the plant by checking the quality:

- Materials entering the plant
- Purchased parts
- Work-in-progress
- Finished products

Each of these items is compared with engineering standards. See **Figure 24-15.** They are often tagged to indicate the condition. Good items receive an "accepted"

quality circle. **Groups of workers that meet to discuss ways to improve the company and its operations.**

inspection. **A process that ensures only quality products leave the plant. Includes checking: materials entering the plant, purchased parts, work-in-progress, and finished products.**

Figure 24-16. The inspector enters inspection data into a computer. The computer will prepare an inspection report. (Whirlpool Co.)

rework. **The process of correcting defects.**

tag while other parts are marked for either rework or scrap. A *rework* corrects defects. Parts are then inspected again before they leave the plant.

Inspectors also prepare inspection reports, **Figure 24-16,** on their activity for a set period of time (hourly, daily, etc.). They record the number of parts:

- Inspected
- Accepted
- Rejected and the reason

This information will identify problems in the manufacturing system. Managers can then take proper corrective action.

Summary

Resources must be used wisely. Materials and machine time must not be wasted. People need to be assigned to appropriate work. Production planners must schedule the use of these resources. They then need to evaluate their use.

Product quality must also be maintained. Employees need to understand their role in producing quality products. Each worker must be motivated to produce good products.

In addition, inspection activities must remove defective items. Materials and purchased parts must meet standards. The product must also meet standards. Only through control of resources and quality can a company compete.

Key Words

All of the following words have been used in this chapter. Do you know their meaning?

automation
deskilling
inspection
inventory control
productivity
quality circle
quality control
resource control
rework
time clock
wage rates

Test Your Knowledge

Please do not write in this text. Place your answers on a separate sheet of paper.

1. List the four phases in developing and using a control system.
2. The number of products or parts spoiled during manufacture is called the _____ _____.
3. Which three of the following are resources for making a product?
   a. The building that houses the factory.
   b. The workers' labor.
   c. Machine time.
   d. Materials.
   e. Capital (money for running business).
4. List the three factors (things) that affect labor costs.
5. Making an operation easier to perform to lower a wage rate is known as _____.
6. Industry uses three methods to measure labor costs. Indicate which of the following are included:
   a. Quality of the work.
   b. Time worked.
   c. Productivity.
   d. Number of products produced.
   e. Number of rest periods in a day.
7. Products that are being built are known as _____.

Matching questions: Match the definition (8 through 12) with the correct term (a through e).

8. _____ Items in inventory available for use.
9. _____ Items being held in inventory that are marked for production or shipping.
10. _____ Items released from stockroom to production or shipping.
11. _____ Items in storage.
12. _____ Items ordered but not received.
   a. On order.
   b. Issued.
   c. On hand.
   d. Available.
   e. Allocated.
13. List the two tasks in a total quality control.
14. Indicate which of the following materials are checked by quality control inspectors:
   a. Materials coming into the factories.
   b. All parts purchased from other factories.
   c. Work in progress.
   d. Finished products.
   e. All of the above.
   f. None of the above.

Applying Your Knowledge

1. Design a poster to encourage quality work.
2. You have decided to manufacture track hurdles. What are the various elements that you will have to control? How would you control them?
3. With a partner, develop an inspection gage to determine if a board is 3/4" (±1/32") thick.
4. With a partner, develop an inspection gage to determine the location of a hole in the center of a 4" × 4" block. The hole should be within ±1/32" of the center.

Career Link

Public Relations Specialist

The reputation, profitability, and even continued existence of a company can depend on the degree to which its targeted users support its goals and policies. Building and maintaining positive relationships with the public is an important part of a public relations specialist's job. Public relations specialists may be advocates for businesses, nonprofit associations, universities, hospitals, and other organizations. They may inform the general public, interest groups, and stockholders of an organization's policies, activities, and accomplishments. Their work also involves keeping management aware of public attitudes and concerns of the many groups and organizations they must deal with from day-to-day.

In large organizations, the key public relations executive, who is often a vice president, may develop overall plans and policies with other executives. In addition, public relations departments employ public relations specialists to write, research, prepare materials, maintain contacts, and respond to inquiries.

People who handle publicity for an individual or who direct public relations for a small organization may deal with all aspects of the job. They contact people, plan and research, and prepare material for distribution. They may also handle advertising or sales promotion work to support marketing.

Unit 8 Manufacturing the Product

Chapter 25 Producing Products

Objectives

After studying this chapter, you will be able to:

✓ List and explain the three major steps in production.
✓ Define technical terms used in controlling production.
✓ Describe, in general terms, the method and sequence used in managing production.

Finally, when the manufacturing system is ready, actual production can begin. Manufacture of products involves three major steps. These, shown in **Figure 25-1,** are:

1. Scheduling production.
2. Producing products.
3. Controlling production.

Figure 25-1. There are three major steps in producing products.

Scheduling Production

Scheduling production means organizing a manufacturing production line so products or parts are produced on time. It must be done efficiently: high quality products made with the least use of time and materials. People who do this kind of work are called *production planners*. To plan properly they must know:

- The number of parts or products that must be built.
- The deadline (when the parts and products must be ready).

production planners. People who schedule and organize the manufacturing system to make sure that products are produced on time and at the lowest cost.

Using this information, they organize the workforce, equipment, and materials to do the job.

Figure 25-2. Steps in production planning.

Production Planning

The work of a production planner can be grouped into four parts as shown in **Figure 25-2.** These actions are:

- *Routing.* Determining the production path for each product going through the plant.
- *Scheduling.* Deciding when each production activity will take place and when it will be finished.
- *Dispatching.* Giving orders for completing the scheduled tasks.
- *Expediting (follow-up).* Ensuring that the work stays on schedule.

routing. The process of determining the best path through the manufacturing system for parts and products.

scheduling. The process of determining the timing of each phase of production.

dispatching. The process of issuing production orders to start the manufacturing process.

expediting. Following up to make sure work stays on schedule.

forecast. Management's estimates of the demand for products.

Routing and Scheduling

Routing is determined by the operations needed to make the product. If the part needs sawing, drilling, and sanding, it will be routed in that order to the saw, drill, and sander.

Routing is done using flow and operation process charts. You will recall that these charts were shown in Chapter 22. They outline the sequence (order) for making each part and product.

The purpose of *scheduling* is to have enough material and products to meet demand. See **Figure 25-3.** To meet this challenge planners must know how many products are needed. Using this information, they will set levels of production. (This is the amount of product that must be made in a certain time period.) The schedule can be set up for a day, week, or month.

Planners never guess at how many products are needed. This information is gathered from three sources. As shown in **Figure 25-4,** the sources include:

- **Forecasts of needs.** *Forecasts* are management's estimates of demand for products.
- **Customer orders.** These are sales orders from other enterprises such as stores.
- **Internal orders (shop or stock orders).** These are orders to produce parts or products that will be used by the enterprise itself.

Some companies produce products only when they have orders for them. Major industrial and military equipment are first sold. They are then manufactured. Aircraft, locomotives, and transit buses are also produced *after* being sold.

Other companies build to forecast. They first estimate their sales. Then, products are produced to the forecast. High-volume consumer goods are examples of items produced using this technique. Television sets, jeans, and toothpaste are manufactured to forecast.

Many manufacturing plants produce parts and products for internal use. Automotive engines, bumpers, and alternators are produced in separate plants. They are then shipped to an assembly plant. At the plant these and other products are made into automobiles.

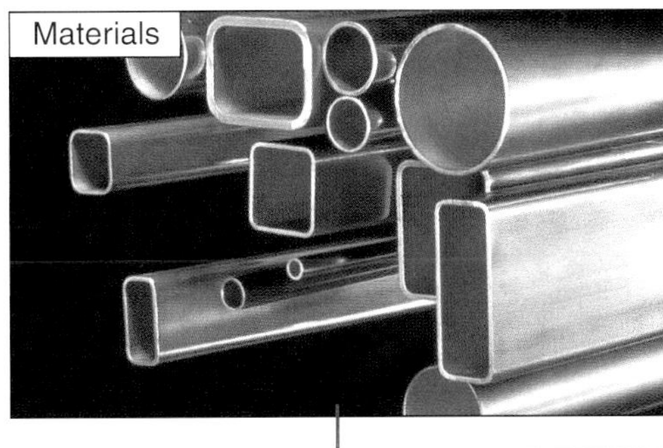

Figure 25-3. Production scheduling moves materials through the plant so that parts and products can be made for customers.

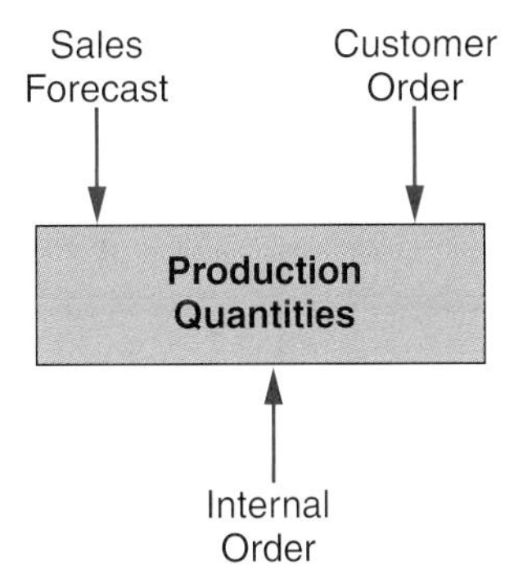

Figure 25-4. Production levels (amounts) are set from orders or forecasts.

Figure 25-5. Production scheduling insures efficient use of human, capital, and material resources. (Goodyear Tire and Rubber Co.)

All of these activities must be scheduled so parts will be ready when needed. See **Figure 25-5.** Quantities of products needed are written on a form like the one in **Figure 25-6.**

Production schedules must consider *lead time*. See **Figure 25-7.** This is the time needed to get a product or part made. Not all parts take the same time to make. Parts are needed at different times. All these things affect the production schedule. Production scheduling is very important in manufacturing.

lead time. **The time that must be allowed between a decision to build a product and actually building it.**

Producing Products

Once developed, the schedules are given to the production supervisors. These managers assign workers to complete specific tasks. They must also supervise and motivate the workers.

PRODUCTION SCHEDULE

DATE ______________________

Product Name and Number	A Projected Sales	B Desired End Inventory	C Total Required (A + B)	D Beginning Inventory	E Total to Make (C – D)	To Make			
						Week 1	Week 2	Week 3	Week 4

Figure 25-6. Sample production schedule form.

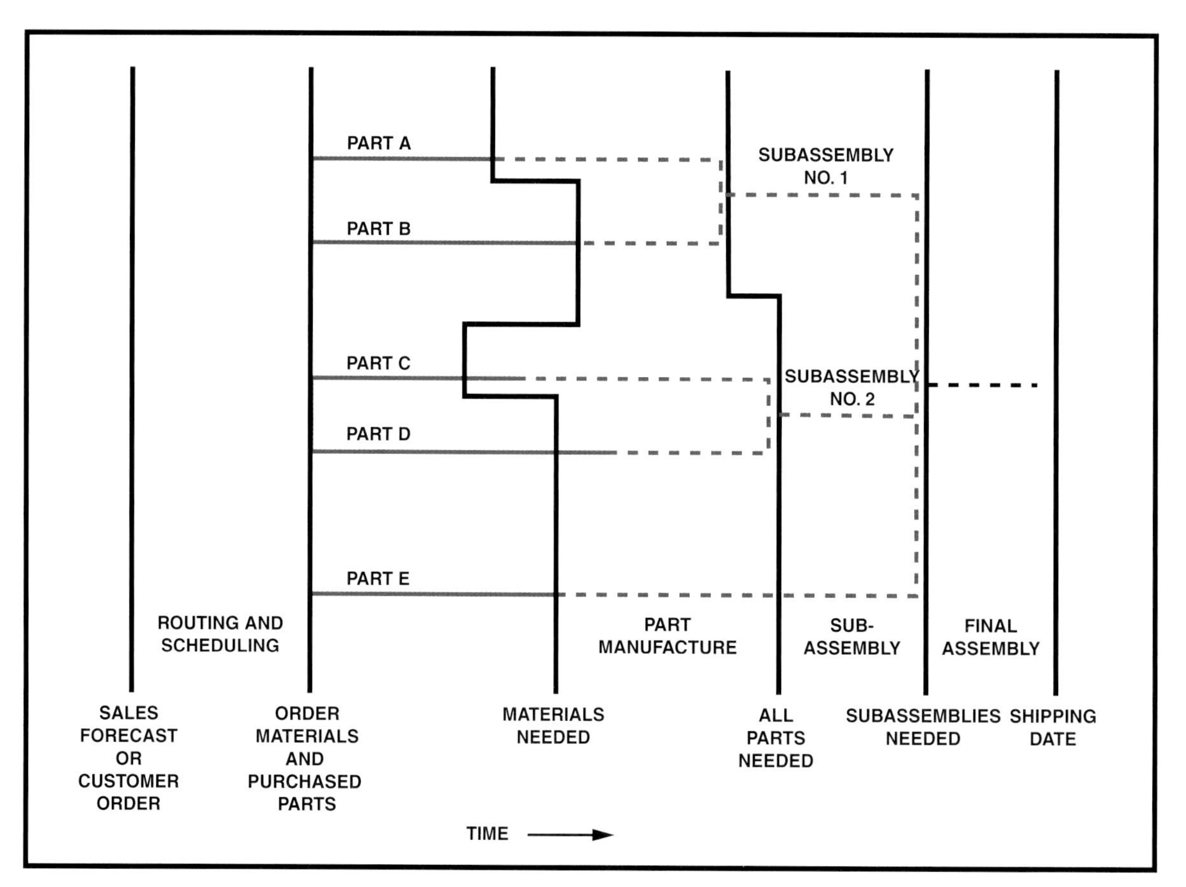

Figure 25-7. A lead time diagram shows major tasks to be done. Spaces between vertical lines represent periods of time. They could be days, weeks, or months. Such a diagram tells a manager when to order materials and assign tasks.

Line management must see that products are produced on time. This involves:

- Getting all parts manufactured
- Making subassemblies
- Making final assembly

Pilot Run

Often, a production system is tried out to see if it works. The system is tested before full manufacture starts. This test, called a *pilot run*, is designed to "debug" (correct) the system. The line runs for a short time while a few products are produced and evaluated. Design errors are caught and corrections can be made before high-volume manufacture starts.

pilot run. **A small-scale production test for a product, assembly line, or entire plant.**

The pilot run can turn up a number of problems. These could cause changes to be made in the:

- Product design
- Tooling
- Materials used
- Plant layout
- Material handling systems
- Type and sequence of operations

Full production starts after the pilot run. Various parts are put together into subassemblies, which are then assembled to make the final product. Refer to **Figure 25-8.**

Controlling Production

Once production is started, it must be checked often for problems. This activity is called control. Three basic types of data (information) are important for controlling production. These are:

- Product output data
- Quality control data
- Labor utilization (use) data

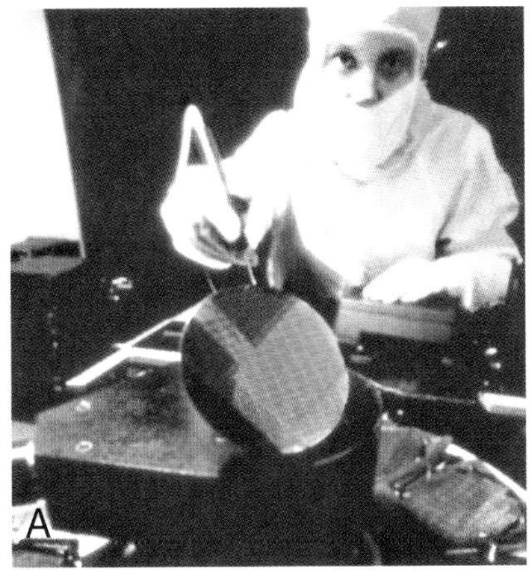

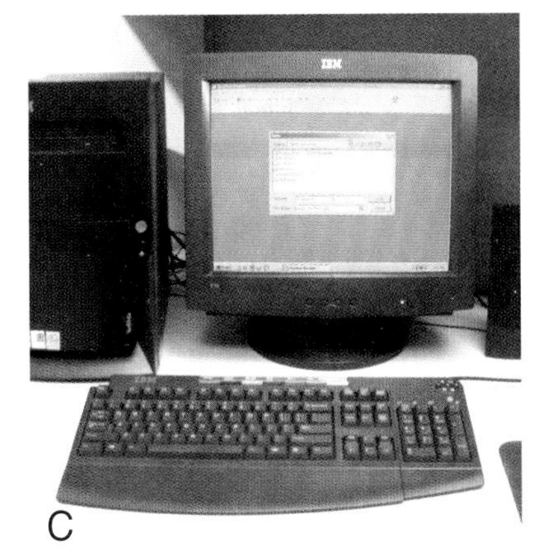

Figure 25-8. Product manufacture involves making parts, putting together subassemblies, and assembling final products. A—This worker is making computer chips. B—The chips will become part of an electronic circuit. C—The circuit is used in a computer.

PRODUCTION REPORT

DEPARTMENT ________________________ DATE ____________

Production or Part Description	Quantity Scheduled	Started	Scrap	Finished

Figure 25-9. A typical daily production report form.

Product Output Data

A production department schedules levels of product manufacture. Each plant, department, and worker is expected to complete certain tasks on time.

During manufacture, production data is collected. Various reports are prepared to tell whether production schedules are being met.

A production record may be kept for each worker. These records are often summarized on departmental production reports. See **Figure 25-9.** The plant manager will review each department report. Then she or he prepares a plant production report that usually goes to the corporate office.

Daily production reports show areas needing corrective action (change in plan). Overtime can be scheduled to keep from falling behind or workers may be hired. New equipment or tooling may be installed. Training and motivational (reason to work harder) programs can be used.

Production reports are not just a record of what happened. They are also a record for future action.

Quality Control Data

Success or failure in producing quality products is shown by quality control reports. These reports tell about three major activities:

- Receiving materials
- Processing materials
- Testing final products

Figure 25-10. This inspector is using a water test to check fuel tanks for leaks. (AC Rochester)

Inspecting materials

Since quality starts with good materials, inspectors must check incoming stock. See **Figure 25-10.** If the material fails to meet standards, the product will also fail. Each batch of material is inspected (compared to a set standard). The inspectors report all rejected material. See **Figure 25-11.**

Work-in-process inspection

As materials are processed they are inspected at various stages. See **Figure 25-12.** Work-in-process is divided into:

- **Rejects.** Items failing to meet standards. These items cannot be repaired.
- **Rework.** Items failing to meet standards. These items can be repaired.
- **Accepted.** Items that meet standards.

MATERIAL REJECTION FORM

QUANTITY ____________ DATE ______

DESCRIPTION ____________

SUPPLIER ____________

Reason for Rejection:

AUTHORIZED BY: ____________

Figure 25-11. This type of form can be used to report rejected materials and purchased parts.

The results of these inspections are reported to management. See **Figure 25-13.** The data will be used to decide what corrective action to take. The errors are first studied to see why they happened. The reject can be caused by:

- Material defects
- Machine error
- An operator making a mistake

Material defects can be corrected by better material inspection. Machines and tooling can be adjusted to reduce machine error. Workers may need more training or supervision.

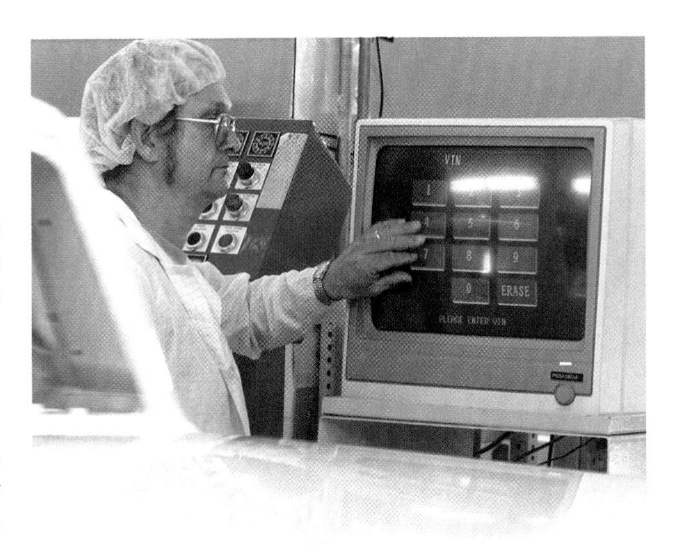

Figure 25-12. This inspector is checking the quality of the paint finish on this automobile. (Chrysler Corp.)

Product inspection

Final products must meet quality standards. See **Figure 25-14.** Some products may be visually (by eye) inspected to assess their overall appearance. Other products are inspected by machines to check their size and performance. These tests are called nondestructive because the products are not damaged during inspection.

Some products, however, undergo destructive testing. A sample product is destroyed as it is checked. It may be operated until it breaks or parts are cut apart to check welds and other features. Other products are dropped to check durability.

Products that pass inspection are so marked. An "OK" may be stamped on them. Or a sheet saying "Inspected by _____ _____" may be placed in a pocket. The customer is thus shown that the product met company standards.

Labor Data

Workers expect to be paid for their work; therefore, records for pay purposes are kept. The type of records maintained depends on the pay system. The two basic systems are:

INSPECTION REPORT OF WORK IN PROGRESS

DATE ______________________ INSPECTOR ______________________

Part No.	Passed	Wrong Size	Chipped	Improper Sanding	Scratched or Dented	Other

Figure 25-13. An inspection report such as this one is used for work in progress. Such reports will help managers decide what corrective actions need to be taken.

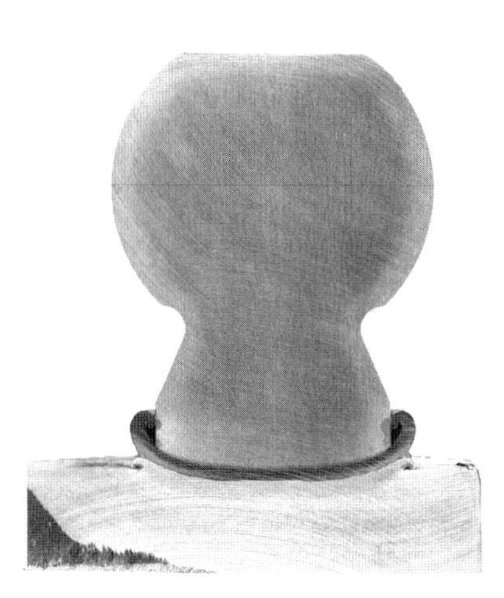

Figure 25-14. Finished goods are inspected in many ways. A—Visually. B—A machine guides a striping applicator on a car. The striping will be checked visually later. C—Destructive testing. (GM, Chrysler, Caterpillar)

- **Standard pay systems.** Workers are paid for time spent on the job.
- **Incentive pay systems.** Workers receive a base pay for time worked. The company may pay a bonus for production beyond a set number of units called a quota. Workers are encouraged to be more productive by providing additional pay for extra production.

Standard pay systems generally require the worker to keep a time card. A time clock is used to automatically record starting and quitting times.

Incentive systems use time cards and production records. The workers' starting and ending times are recorded. Also, the output of each worker is recorded. Any reject parts produced are subtracted from the output. The result is then used to calculate the worker's pay.

For example, a worker may work for eight hours. The base production rate may be 300 parts per day. The worker produced 325 parts but five of these were rejects. The worker receives pay for eight hours work and he or she also receives a bonus for producing 20 extra good parts.

Labor records are used to determine paychecks for each pay period. Hourly employees (paid by the hours worked) commonly receive a paycheck each week.

Summary

Products must be produced efficiently for a company to make money. This production must be scheduled, carried out, and controlled.

Production levels are set to match known or expected sales. They are based on orders or forecasts. People, machines, and materials are scheduled to meet production goals.

Parts, subassemblies, and products are manufactured. The output is controlled. The numbers produced are compared to the schedules.

Data is gathered to measure the success of the manufacturing activity. Product output, quality control, and labor utilization (use) information is kept. Reports are prepared. Corrective action is taken to keep the activities on schedule.

Key Words

All of the following words have been used in this chapter. Do you know their meaning?

dispatching
expediting
forecast
lead time
pilot run
production planners
routing
scheduling

Test Your Knowledge

Please do not write in this text. Place your answers on a separate sheet of paper.

1. What three steps are involved in the manufacture of products?
2. One who schedules production is concerned with three major tasks. Indicate which of the following are included:
   a. Deciding how many of a product to make.
   b. Calculating how much the product will cost.
   c. Indicating when the parts/products will be built.
   d. Telling what numbers of workers, equipment, and amounts of materials are needed to produce the product or parts.
   e. Determining how much workers should be paid.
3. _____ is determining the path of production for each part or product as it moves through the plant.
4. What are internal orders?
5. True or false? Lead time is the amount of time it takes to produce a part once it is ordered.
6. Indicate what changes can be brought about after a pilot run.
7. Quality of the product is the responsibility of those who (schedule, control) production.
8. List the defects or errors that may cause a part to be rejected.

Applying Your Knowledge

1. In a team, develop a plan for and produce five kites. Keep a daily production report form.
2. Read a "how-to" book or follow a set of instructions that teach you how to make something you have never made before.
3. Help a child learn to do something. What incentives are there for doing things besides earning money?

Technology Link

Agricultural Technology

Manufacturing is a unique and broad technology. In basic terms, it is the process that changes resources into products. Manufacturing affects many other technologies. Other technologies, such as *agricultural technology*, have an effect on manufacturing as well. The link between manufacturing and agricultural technology is a two-way street.

Have you ever sat down at the dinner table and thought about the food you are eating? At one time, to produce the food needed to feed the people of this country, many people had to be farmers. As manufacturing processes evolved, so did the machinery used on the farm. This meant that larger and more efficient farm equipment could be manufactured. Today, a farmer can farm larger fields and produce more crops. Manufacturing has also directly affected the processing of crops. Instead of purchasing fresh food daily or canning food for later use, the manufacturing and processing of most food is as automated as the production of an automobile or a computer.

What most people don't realize is the direct effect other technologies have on manufacturing. With the advancements in agricultural technology, we now have a healthier workforce. Not only do today's agricultural products provide the workforce with nutrition, but many agricultural products have disease preventive characteristics. This results in a healthier, more experienced, attentive, and efficient workforce.

Chapter 26 Maintaining Worker Safety

Objectives

After studying this chapter, you will be able to:

✓ Give reasons why safety on the job is important to yourself and a company.
✓ List and describe the five E's of safety.
✓ List four steps in proper job safety training.

Both the company and the employees suffer when accidents happen. Injured employees suffer physical pain and they may lose income. Their insurance benefits may not cover all their expenses.

Through accidents, companies can lose the services of valuable employees. They cannot use the employees' special skills and knowledge for a period of time. The company will also lose money.

Insurance rates are based on accident history. The more accidents a company has the higher their rates will be.

Safety Programs

Modern companies have well-designed safety programs. These programs start with the new worker and continue throughout the employee's working life. See **Figure 26-1.**

These safety programs contain five major phases, or what is called the five E's of safety. The phases are engineering, education, equipment, encouragement, and enforcement. See **Figure 26-2.**

Figure 26-1. The safety program continues through the work-life of the employee.

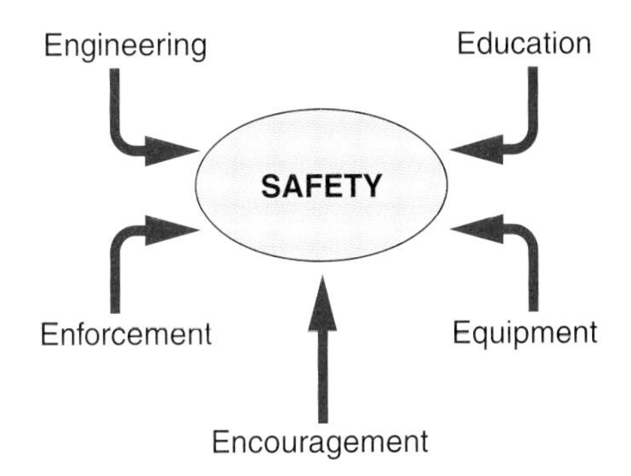

Figure 26-2. The five E's of safety.

Figure 26-3. Look at these two photos. Which job is most dangerous? Why? (American Iron and Steel Institute)

Figure 26-4. These high speed metal stamping presses are designed with safety in mind. Note the doors in front of the ram. Also, the controls are on the stand to the right. The operator is a safe distance away from the actual press operation.
(American Metal Stamping Assn.)

Safety Engineering

Some jobs are safer than others. The person with a desk job is relatively safe. A person operating a forging press is more likely to get hurt. See **Figure 26-3.**

Safety engineering designs processes and machines to be safer. Equipment can be engineered to be safely operated. See **Figure 26-4.** Hazards (sources of danger) can be designed out of a process. Guards can be placed over cutter heads and saw blades. Controls can be placed in convenient locations. Emergency stop switches can be located within easy reach.

safety engineering. A practice that designs processes and machines to be safer.

Tooling can also be designed to be safe. Small parts may be held by clamps. Parts may be guided safely into cutters. The worker's hands are kept away from tools and dies.

Workers may be removed from very dangerous operations. Mechanical devices may be engineered to do the job instead. Robots do many dangerous jobs. See **Figure 26-5.** They are often used to load punch presses, do welding, and apply finishes.

Safety Education

Workers must be educated about safety. They must be taught its importance through a structured program. Throughout the year, the company holds meetings to discuss these concepts.

A worker must receive careful safety instruction for his or her assigned job. This instruction includes the four steps presented in **Figure 26-6.**

The trainer never assumes a person knows how to do the job safely. Each worker should receive complete instructions on safe practices. They should also receive printed materials that summarize the training. These materials should present safety rules and procedures. See **Figure 26-7.**

Figure 26-5. Robots such as this one are capable of doing very dangerous operations. When robots are used, humans do not have to be exposed to dangerous environments. (Cincinnati Milacron)

Safety Equipment

During training, the worker should be taught about required *safety equipment.* The need for personal protection is important. Each job has its own equipment requirements. Special protection devices may be provided. Safety equipment protects the worker's sight, hearing, lungs, and skin. Refer to **Figure 26-8.** Note that the worker is wearing:

- Safety glasses
- Respirator
- Protective coat

You can also see that the equipment is attached to a dust collection system. Also the cutter is carefully guarded.

safety equipment. **Devices that protect a worker's sight, hearing, lungs, and skin.**

Encouragement

Working safely is a habit. It is developed over time. Therefore, workers must be encouraged to work safely.

Most large companies have an ongoing safety program. Posters remind workers to work safely. Company newsletters will have articles on safety. Signs will announce the number of days worked without an accident. Safe workers may win special recognition. Often, safety awards will be given to individuals or departments.

Employees should also be encouraged to report unsafe conditions. Alert workers can always see ways to improve operations. They should be encouraged to share their ideas.

Figure 26-6. The steps in educating workers in job safety.

Figure 26-7. Safety education programs include materials that reinforce safety. (Ball Corp.)

Figure 26-8. Proper safety equipment is very important. (Nilfisk of America)

Enforcement

Everyone tends to become careless over time. We start to take chances. We drive too fast or run yellow lights. The police will remind us that this is wrong. They give us a traffic ticket.

The same is true about safety. Workers often have jobs where they must do the same task over and over again. The job becomes routine. Boredom can creep in. With boredom comes carelessness.

Enforcement is making sure that workers follow safety rules. An alert supervisor will remind workers of careless acts. All workers must be required to work safely. Unsafe workers are a *hazard* (danger). They can hurt themselves and others.

The unsafe worker, like the unsafe driver, may need to be disciplined. She or he must be made to understand that unsafe practices will not be allowed.

enforcement. **Making sure that workers follow safety rules.**

hazard. **Sources of danger in a manufacturing plant.**

Summary

Safety is everybody's job. Employees must be provided a safe work environment. The operations must be engineered to be safe. Proper safety education and equipment must be provided. All workers must be encouraged to work safely. Finally, safety rules must be constantly enforced.

Key Words

All of the following words have been used in this chapter. Do you know their meaning?

enforcement
hazard
safety engineering
safety equipment

Test Your Knowledge

Please do not write in this text. Place your answers on a separate sheet of paper.

1. List and describe the five E's of safety.
2. What are the four steps in proper job safety training?

Applying Your Knowledge

1. Design a safety poster that warns of a safety hazard in the home, school, or shop.
2. Make a videotape of bicycle safety.
3. Design a safety device for a machine in your laboratory.
4. Use the Internet to discover the types of workplace hazards regulated by the Occupational Safety and Health Act (OSHA).
5. Write instructions for the safe handling of a tool, equipment, or substance.

Occupational Health and Safety Workers

Occupational health and safety specialists promote occupational health and safety within organizations by developing safer, healthier, and more efficient ways of working. They evaluate work environments and design programs to control and eliminate disease or injury caused by chemical, physical, biological agents, or ergonomic factors. They may conduct inspections and enforce adherence to laws, regulations, or employer policies governing worker health and safety. Occupational health and safety technicians collect data on work environments for analysis by the specialists. Technicians usually work under the supervision of specialists, and help implement and evaluate programs designed to limit risks to workers.

Occupational health and safety specialists and technicians identify hazardous conditions and practices. Sometimes, they develop methods to predict hazards from experience, historical data, and other information sources. They identify potential hazards in existing or future systems, equipment, products, facilities, or processes. After reviewing the causes or effects of hazards, they evaluate the probability and severity of accidents that may result. They develop and help enforce a plan to eliminate hazards. Specialists conduct training sessions for management, supervisors, and workers on health and safety practices and regulations, as necessary. They also check on the progress of the safety plan after its implementation. If improvements are not satisfactory, a new plan might be designed and put into practice.

Specialists inspect and test machinery and equipment, such as lifting devices, machine shields, or scaffolding, to ensure they meet appropriate safety regulations. They may check that personal protective equipment, such as masks, respirators, safety glasses, or safety helmets, is being used in workplaces according to regulations. Specialists also check that dangerous materials are stored correctly. They identify work areas for potential accident and health hazards. They implement appropriate control measures, such as adjustments to ventilation systems. Their investigations might involve talking with workers and observing their work, as well as inspecting elements in their work environment, such as lighting, tools, and equipment.

Chapter 27
Working with Labor Unions

Objectives

After studying this chapter, you will be able to:

✓ Explain the purpose of a labor union and tell why workers would want to belong to one.
✓ List the legal steps in forming a union.
✓ List the officers of a union and describe their duties.
✓ Explain the procedures for negotiation of a new contract between a union and a company.
✓ Define a contract and outline the elements in a typical contract.

union. A legal organization that represents workers' interests.

Workers often feel they have no way to assert their rights and to express their ideas to their bosses. Individual employees feel alone and powerless in dealing with management.

These feelings are the reason workers join together in labor unions. Workers want to get together to promote their rights. They form a union to be officially heard. See **Figure 27-1.** The *union* is a legal (has rights by law) organization that represents the workers' interests.

The National Labor Relations Act of 1935 (also called the Wagner Act) provided workers a way to be heard. It guaranteed the rights of employees:

* To organize and join a union of their choice.
* To bargain collectively (as one voice) with their employer.

Figure 27-1. Union organizers help workers establish a union. (United Steelworkers of America)

Organizing a Union

Organizing a union takes several steps. The actual activities will vary from union to union. They will also be different from company to company. The six most common organizing activities, **Figure 27-2,** are:

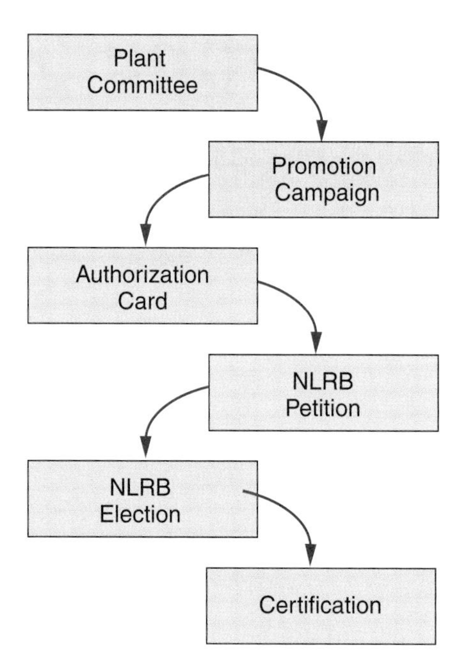

Figure 27-2. These are the most common steps in forming a union. Unions sometimes organize demonstrations to inform the public of an issue. (United Steelworkers of America)

Figure 27-3. Union organizers pass out leaflets promoting the union.

- **Forming a plant committee.** The union organizers meet with workers who want a union. The organizers work for a national union. They cannot come into the plant to organize the union. However, workers can promote it during nonworking time. The plant committee is a group of workers who want to form the union. They coordinate (bring order to) the organizing effort within the plant.
- **Developing a program to promote the union.** Not all workers are aware of the union activities. Some do not understand the issues. Others may not want a union. Much like political candidates looking for votes, union organizers try to convince workers that they will be better off if they vote for and join the union.

Promoting the union is done in a number of ways.

- Holding meetings to explain how they work.
- Passing out literature after work. See **Figure 27-3.**
- Setting up large group meetings and rallies.
- **Getting** *authorization cards* **signed.** The plant committee wants a card from each worker. See **Figure 27-4.** This card lets the union represent the employee. It gives your bargaining rights to the union. The National Labor Relations Act requires authorization cards from 30 percent of the workers. If these are obtained, the union asks management to accept the union. Generally, management will not recognize a union at this point.
- **Submitting a** *NLRB* **(National Labor Relations Board) petition.** This is a request for an election. The board makes certain that the union has met all legal requirements. If it has, a date for elections is set.
- **Holding a NLRB supervised election.** All workers who will be eligible to join the union can vote. However, they are not required to do so. If most workers choose the union, it wins.
- **Making the union the official bargaining agent for the workers.** This is called *certification*. Management must now recognize the union. The two must meet and set pay rates, hours, and working conditions.

authorization cards. Cards that are signed by union members giving a union the authority to negotiate for employees.

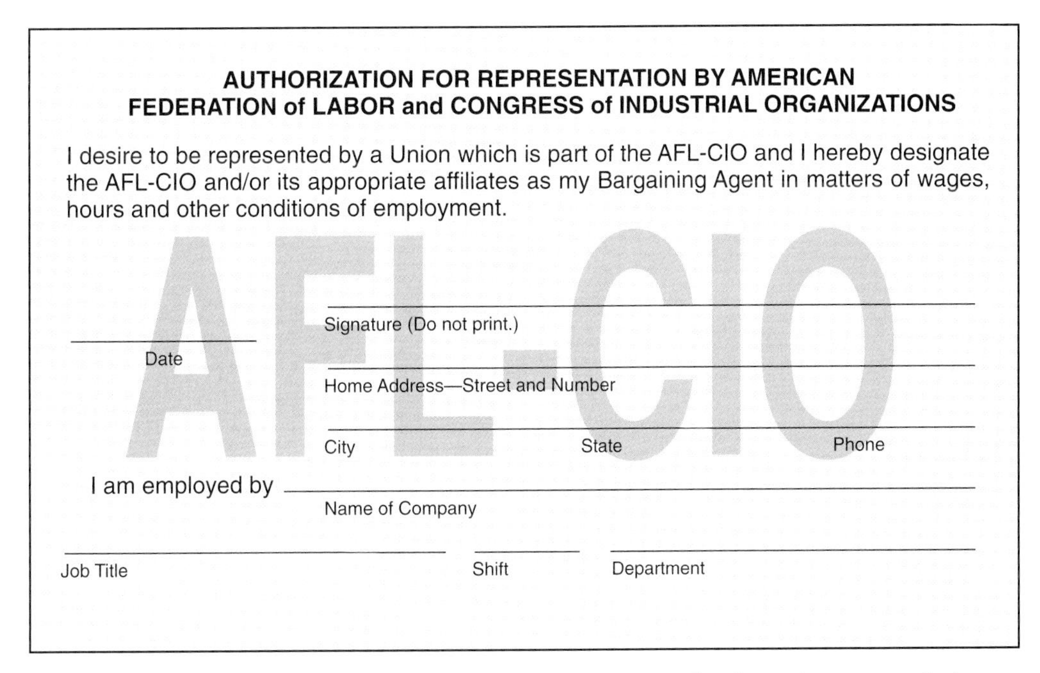
AUTHORIZATION FOR REPRESENTATION BY AMERICAN FEDERATION of LABOR and CONGRESS of INDUSTRIAL ORGANIZATIONS

I desire to be represented by a Union which is part of the AFL-CIO and I hereby designate the AFL-CIO and/or its appropriate affiliates as my Bargaining Agent in matters of wages, hours and other conditions of employment.

AFL-CIO

Date

Signature (Do not print.)

Home Address—Street and Number

City State Phone

I am employed by

Name of Company

Job Title Shift Department

Figure 27-4. A sample authorization card. If enough workers sign cards, the union may ask the company to recognize it as an "agent" for workers.

Organizing a Local

NLRB. The National Labor Relations Board, a federal agency that oversees labor relations between workers and employers.

certification. Making the union the official bargaining agent for the workers. Management must now recognize the union, and the two must meet and set pay rates, hours, and working conditions.

shop steward. A union officer who represents a group of workers.

Most national unions are made up of a number of "locals." These are local chapters (groups) of the national union. The national union's constitution and bylaws set the method of organizing locals.

Generally, three major officers run the local. These officers are elected by the membership. The officers and their duties are:

- President
  - Conducts all meetings of the local union
  - Appoints committees
  - Calls special elections as needed
- Vice president
  - Assists the president
  - Becomes president if the elected president leaves office
- Secretary-treasurer
  - Keeps minutes of all meetings
  - Receives dues
  - Keeps financial records
  - Pays union bills

Other important union officers are the *shop stewards*. Each steward represents a group of workers. The members elect the shop stewards. Often, one steward is elected from each major department in the plant. The shop stewards are the union officials closest to the workers. They represent the employees in contract discussions.

Figure 27-5. The contract is being negotiated. The union team is on one side of a table. Management representatives are on the other.

Figure 27-7. The labor agreement is submitted to the union (rank and file) members. They must approve it before it goes into force.
(United Steelworkers of America)

Figure 27-6. The contract is signed by both management and union officials.
(United Steelworkers of America)

Negotiating a Contract

The biggest job of the union is to get the workers a fair labor agreement. This is a legal document. It spells out what work must be done for what pay. Union officers and management work out the agreement together. Several meetings may be necessary. These meetings are called *collective bargaining*. See **Figure 27-5.** Each side makes *proposals* (tells what they want in the agreement). These are followed by each side making counterproposals. This is called bargaining. Each side is trying to change the proposal of the other side.

Next, the union and management will begin to work out their differences. It is a time of give and take. Each side will give up something but tries to get the other side to give up something, too. This is called *negotiation*. Finally, they reach an agreement. This is how pay rates, hours, and working conditions are set.

Both sides sign a contract agreement. See **Figure 27-6.** It is then submitted to the workers for approval. This is done at a meeting. The union bargaining team first explains the contract to its members. See **Figure 27-7.** Then the workers vote. A majority must support it before the contract is adopted.

collective bargaining. A negotiating process aimed at bringing about an agreement between the company and its workers (as a group).

proposals. A statement that tells the other side in a negotiation what one side wants in an agreement.

negotiation. A complaint that is made by employees who feel that management has broken the rules of a labor contract.

contract. Contains the rules that the workers and the managers must obey.

Contents of a Contract

A *contract* contains the rules that the workers and the managers must obey. See **Figure 27-8.** It is generally in force for three years; then a new contract must be made. A typical contract will outline:

- What employees are to be covered by the contract. (For example, all hourly production workers.)
- Wage rates for various jobs
- Hours of work for a normal day and overtime

- Vacation and holidays
- Policies for hiring and firing workers
- Work rules that describe the employees who can do each type of job
- Working conditions, including lighting, ventilation, and safety measures
- Grievance procedure (way to handle complaints from the workers)
- Expiration date (when the contract ends)

Figure 27-8. The union contract spells out what workers and management can and cannot do.

Grievances

The contract is like a rulebook. Different people may see its meaning differently. There may be arguments over the use of the contract. If employees feel that management acted against the contract, they can file a complaint, called a *grievance*. There are generally four steps in a grievance. It can be settled at any step. These steps, shown in **Figure 27-9**, are:

1. The employee and the shop steward discuss the complaint with the supervisor.
2. If the matter is not settled, the committee of stewards reviews the grievance. If it has merit (deserves to be heard), they present it to the plant manager or personnel director.
3. If still unsettled, a representative from the national union studies the grievance. It can be presented to top management.
4. If not resolved at this point, the grievance is submitted to *arbitration*. An outside person (arbitrator) hears both sides. He or she gives a ruling. This judgment is binding on both sides. They must live by it.

The grievance procedure is an orderly way to settle disputes. Both sides have to face the problem. It keeps small problems from piling up into a big problem.

grievance. **A complaint filed by an employee.**

arbitration. **The process where an outside person hears both sides of a dispute and issues a ruling that both sides must live with.**

strike. **A refusal by unionized workers to perform their jobs. Strikes usually occur when labor and management fail to agree on a contract.**

wildcat strike. **An illegal strike called during the life of a contract.**

Strikes

At times workers and management cannot settle their differences. The workers then *strike*. See **Figure 27-10.** This is an action that withholds labor from the company and applies economic pressure on management. Its goal is to force the company to agree with the workers viewpoint.

Most often a strike happens after a contract expires and a new contract cannot be agreed upon. This is a legal activity for the union and its members. A few strikes take place while a contract is in force. Usually these strikes go against a "no strike" clause that is in most contracts. When this happens, the strike is called a *wildcat strike* and it is illegal. The workers lose all their rights and can be fired.

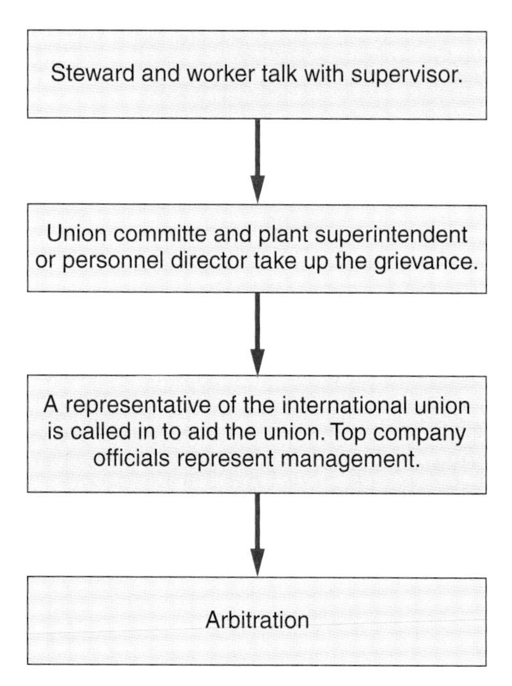

Figure 27-9. Steps in the grievance procedure. (AFL-CIO)

Figure 27-10. These workers are on strike and picketing a manufacturing plant.

In rare cases, the company may close until a contract dispute is settled. This action is called a *lock-out*. The workers of the company lose income. This may apply pressure on the workers to accept the contract offer.

lock-out. **A company closes business until contract is agreed upon.**

Summary

Through unions, workers can express their views as a group. They are organized through a legal procedure. First, plant committees are formed to promote the union. Authorization cards are signed. The union asks for the right to represent workers. A NLRB petition seeks an election. The workers vote for or against the union. If successful, the union is certified. It becomes the bargaining agent for the employees.

A local chapter of the union is organized and the members elect officers.

Once organized, the union, through its officers, negotiates with the company for a fair labor contract. Union officers and labor union officials work out the agreement (contract) that is then presented to the membership for approval.

The contract sets down conditions under which the workers will work. It sets pay rates, hours, and working conditions. It also contains a procedure to settle disputes. This system is called the grievance procedure.

Key Words

All of the following words have been used in this chapter. Do you know their meaning?

arbitration
authorization cards
certification
collective bargaining
contract
grievance
lock-out
negotiation
NLRB
proposals
shop steward
strike
union
wildcat strike

Test Your Knowledge

Please do not write in this text. Place your answers on a separate sheet of paper.

1. Following are the steps for organizing a union.
   Place the steps in the proper order.
   a. Develop a program to promote the union.
   b. Get workers to sign authorization cards.
   c. Form a plant committee.
   d. Hold a NLRB supervised election.
   e. Petition the NLRB for an election.
   f. Make the union the official bargaining agent for the workers.
2. A smaller union that is part of a large, national union is called a _____ _____.
3. *True or false?* A union will usually have three main officials: a president, a vice president, and a secretary-treasurer.
4. A _____ contains the rules that the workers and managers must obey.
5. Working conditions are (never, usually) a part of the contract.
6. List the major steps in settling a grievance.

Applying Your Knowledge

1. Invite an officer from a union local to discuss organizing and operating a union.
2. Invite a union officer or labor relations director from a company to discuss negotiating a contract.
3. In class, debate the reasons for and against joining a union.
4. Role-play the four steps in the grievance process, from an initial complaint through arbitration and resolution.

Unit 9 Marketing the Product

Chapter 28 Developing Marketing Plans

Objectives

After studying this chapter, you will be able to:

- ✓ Define the term marketing.
- ✓ List five major elements in a marketing plan.
- ✓ Define market research and list the data that is collected.
- ✓ Explain the difference between trade names and trademarks.
- ✓ Explain the method used in arriving at a factory price for a product.

Companies face a common challenge after products are designed and produced. They must sell the products to generate income and profit. This is not as easy as it sounds. There is usually more than one company producing the same or similar products. Each product vies for the customer's attention and money, **Figure 28-1.**

Products do not sell themselves; they must be marketed. See **Figure 28-2.** Marketing brings together those who make products with those who buy them. This two-way exchange is shown in **Figure 28-3.**

Figure 28-1. The marketplace offers customers many different products from which to choose. (UAW Solidarity)

Elements of a Marketing Plan

A complete marketing plan has five major elements:

- ***Product.*** A company must have the right product. It must meet the consumers' wants.
- ***Price.*** Customers must see value in the product. The price must be right.
- ***Promotion.*** The customer must be made aware of the product and its good features must be explained.

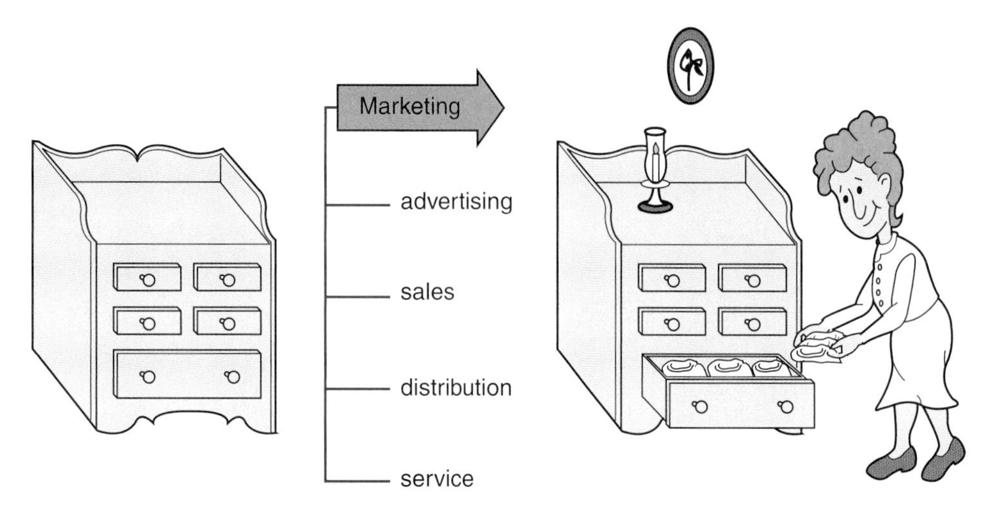

Figure 28-2. Marketing promotes, sells, distributes, and services products.

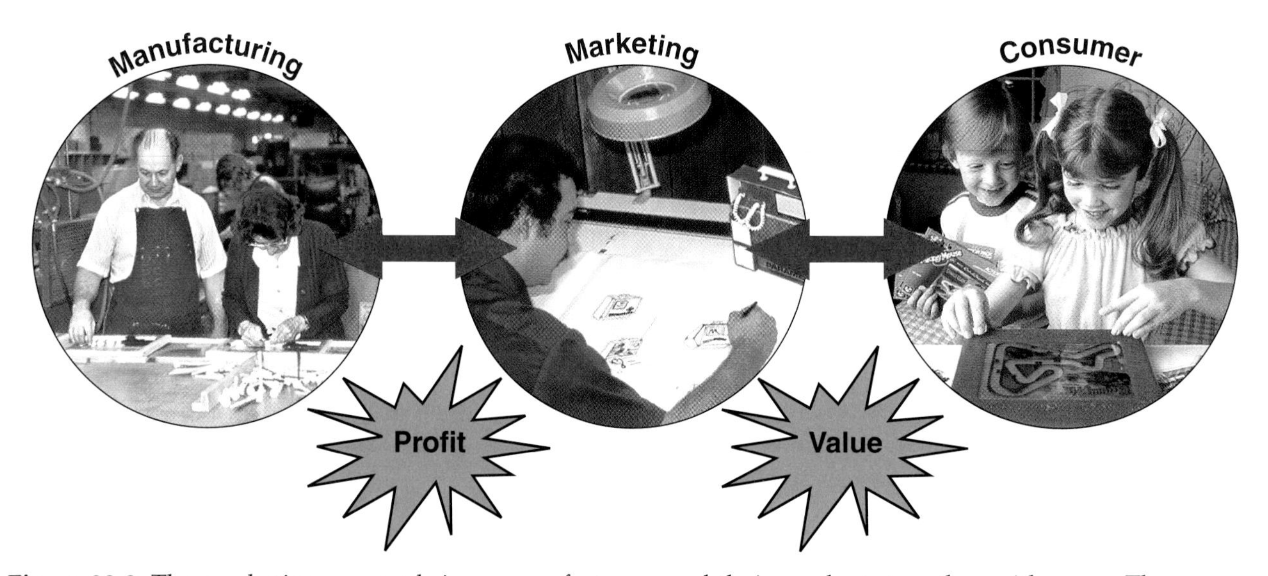

Figure 28-3. The marketing process brings manufacturers and their products together with users. The user receives value; the manufacturer receives a profit.

- *Distribution.* When a customer decides to buy, the product must be available. There must be a method of moving the product to its users.
- *Service.* Products may break down or fail. The company must have a way to repair or replace what it makes.

distribution. **The system needed to get products from the manufacturer to the consumer.**

marketing. **The area of managerial technology concerned with moving a product from the manufacturer to the customer by means of advertising and selling activities.**

Beginning a Marketing Plan

Marketing is based on market research. This area was introduced in Chapter 18. *Market research* gathers data about:

- Who will buy the product?
- What is the typical customer's background? (age, gender, education, income, etc.).

- Where do the customers live?
- How will the customer use this product?
- How much would people expect to pay for the product?
- Where is the customer expected to buy the product? (discount store, specialty store, hardware store, supermarket, etc.).
- What type of product service is expected?

market research. The process of gathering and analyzing information about customers' desires, competing products, and sales results.

The results of this market research provide a foundation for many marketing and product development activities. The information can be used in marketing for activities such as:

- Developing a trade name and trademark
- Selecting a marketing theme
- Pricing the product

Market research information can also be used in planning advertising, sales, distribution, and servicing activities. These areas will be discussed in later chapters.

An important part of marketing is the communication between a company and its customer. Early on, much of the promotion of a product was done through word-of-mouth communication.

The study of communication systems and how they can be used for marketing is big business. As communication systems developed, the number of ways a product can be promoted increased. To reach its customers, a company now may use a print ad, signage, radio ad, television spot, Internet ad, etc. Selecting the right communication system for promoting a product can help sales of a product.

Developing a Trade Name and Trademark

Companies identify their products in two ways. They use trade names and trademarks.

Trade Names

trade name. The legal name of a company; the name that identifies a company to its customers.

A *trade name* is the official name under which a company may do business. Through it the customer comes to recognize the manufacturer. The trade name

Figure 28-4. Some sample trade names on products.

may be registered and protected by law. No one else can use a company's registered trade name. See **Figure 28-4.**

Picking a trade name takes careful thought. The company must live with it for a long time. Generally, it is not wise to use the product name in the trade name. The company's product line may change. It may enlarge or be completely different.

A trade name using the product name may become out-of-date. Changing it can be expensive. Packages, signs, stationery, and thousands of other items must be changed. However, trade name changes cannot always be avoided. Sometimes the nature of the business changes. Technology may change so the busines must change with it. An early example of this type of change happened with Radio Corporation of America. The company was known by this name for many years. It then stopped making radios and it changed its name to the initials RCA. The new name was used by a company that owned a television network and a number of radio and telvision stations. It also manufactured records, tapes, television sets, and other electronic devices. Today, RCA no longer exists as a company. The trade name is used by another company for telvision sets and its television and radio networks have been sold to another company.

Trade names also change as businesses merge or are bought and sold. Chrysler Corporation is now part of DaimlerChrysler. The mergers of oil companies have seen the names of Mobil Oil and Exxon become Exxon Mobil. The purchase of Amoco by British Petroleum created BP Amoco.

The trade name should identify the company. It should provide an image of the enterprise.

Trademarks

trademark. **A name, symbol, or combination of name and symbol used to identify a product.**

A *trademark* identifies a product. It is the registered property of a company. Like the registered trade name, no one else can use a registered trademark. A trademark, by definition:

- Is a word, name, or symbol
- Is used by a single company
- Identifies goods made or sold by the company
- Separates products made by one company from those made by others. See **Figure 28-5.**

Trademarks are designed to stick in your mind. How many trademarks can you think of? Can you "see" several trademarks in your mind's eye?

Trademarks are carefully designed and selected using some basic rules. A trademark should be:

- Easy to see and recognize
- Timely (modern, up to date, in keeping with the times) See **Figure 28-6.**

- Easily used on advertisements and packages
- In good taste (not offensive, obscene, or negative to most people)
- Refer to **Figure 28-5.** Do you feel these trademarks follow the rules?

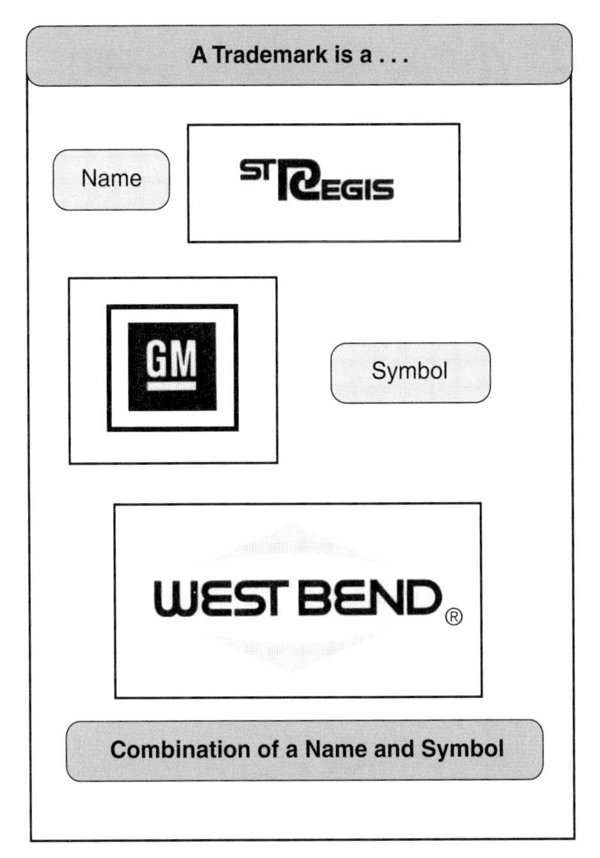

Figure 28-5. Trademarks are words, symbols, or a combination of the two.

Selecting a Marketing Theme

Trademarks are one important part of a marketing plan. A second part is a theme. The company develops a slogan that describes their activities or products. Trademarks are fairly stable and change very slowly. The new Sears trademark in **Figure 28-6** is only the fourth one used in over 100 years. On the other hand, marketing themes change as often as markets and customer demands change.

At one time Ford Motor Company used "A better idea" as its theme to suggest that engineering was most important. It reflected the emphasis of the company and the desires of its customers. But times change and foreign car manufacturers changed the marketing scene. Customers became more concerned with quality than they were about styling. People wanted better cars with higher quality. Ford responded by improving the quality of their cars and changing its theme to "At Ford, quality is Job 1."

Today, many companies have used themes that conveyed an image. See **Figure 28-7.** They encouraged people to think a certain way. Through their theme advertising they wanted people to believe that you "Care Enough to Send the Very Best" when you send a Hallmark card or that "You're in Good Hands" if you have Allstate insurance.

The thoughts that these and hundreds of other slogans developed caused people to take action. Good themes create positive thoughts and images about the company. Consumers will believe in the company and its products.

Figure 28-6. Sears, Roebuck and Co. changed its trademark to show a more modern image. The new trademark shows action and movement. It shows that the stores are progressive.

Pricing Products

The process of setting product prices is not simple. Pricing must consider three factors. See **Figure 28-8.** First, the cost of making and selling the product must be considered. Then, the price of similar products on the market must be part of the pricing decisions. Finally, the mind-set customers have about product cost must be factored in.

Figure 28-7. These are some typical company and product themes that have been used in advertising and promotional efforts.

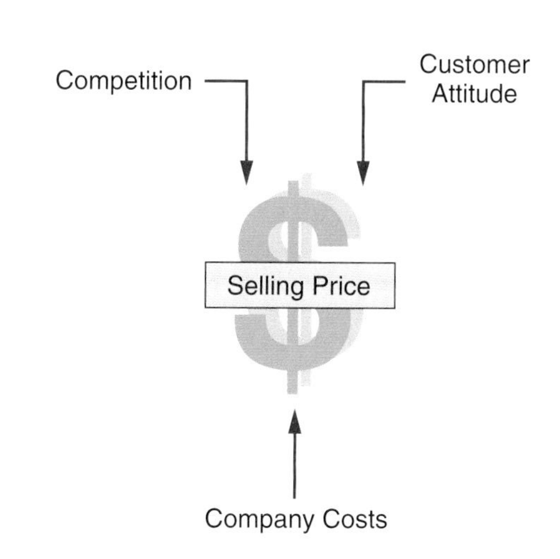

Figure 28-8. Factors affecting selling price.

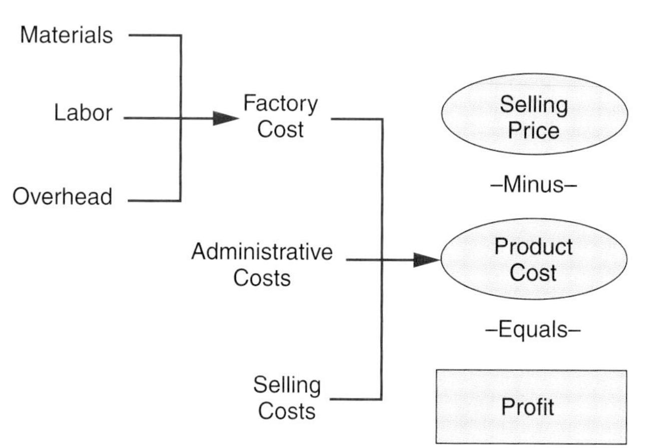

Figure 28-9. Elements in product cost.

Product Costs

The basic costs, **Figure 28-9,** that are considered in pricing a product are:

- *Factory cost.* The actual cost of producing the product.
- ***Labor costs.*** Wages and salaries earned by people producing the product.
- ***Material cost.*** Cost of material used in making the product.
- *Overhead.* Cost of equipment, utilities, and insurance.
- ***Administration cost.*** Cost of developing products and managing support functions.
- *Selling cost.* Cost of promoting, selling, and distributing.

factory cost. **A company's actual cost of producing a product.**

overhead. **The cost of equipment, utilities, and insurance needed to produce the product.**

selling cost. **The cost of promoting, selling, and distributing a product.**

These costs, plus some profit, set the minimum price the manufacturer wants for the product. This is the price the manufacturer must charge if the company is to make a profit and stay in business. Any price below this would produce a loss or poor profit. The product would soon be discontinued.

However, you and I must pay a much higher price. The manufacturer seldom sells products directly to us. They sell to wholesalers and retailers who also expect to make a profit. Their expenses and profit margins are added to the price of the product.

The final selling price is the sum of many levels of expense and profit. Manufacturers, transporters, distributors, and retailers all add their expenses and profit margins.

Competition

Manufacturers' products will have to compete. Customers compare competing products on several points. One is quality. Generally, higher quality products will sell for a higher price. Product function (how well it works) is also considered. Customers buy what best serves their needs. We are willing to pay more for a product that fits our needs.

The company's reputation is also a major factor in choosing a product. The image that we have of the company is important. We often prefer to buy products from a manufacturer we know. Finally, we react to price. If everything else is equal, the cheaper product sells best.

The company will have to decide the way it will face competition. Will it be on quality? Will function be most important? Will the company stress service and guarantees? Will price be the deciding factor?

This decision will affect selling price. In all cases, the customer must see the price as fair. The product must earn the selling price. It must deliver quality, function, or service, or it must compete on price.

Customers and Price

A company must "know" its customers. Their tastes, values, and income are important. A product must appeal to customers. It must be attractive to them.

Summary

All products are marketed. Early in the process an image is built. Company trade names and trademarks are selected. The product takes on an image and an identity.

A general marketing theme is then developed. The slogan ties all marketing efforts together. Advertising, selling systems, and packaging are designed to use the theme.

Finally, the product is priced. Costs are considered. Profit margins are added. The customer will decide if the price is fair. If so, the product will sell.

Key Words

All of the following words have been used in this chapter. Do you know their meaning?

distribution
factory cost
market research
marketing
overhead
selling cost
trademark
trade name

Test Your Knowledge

Please do not write in this text. Place your answers on a separate sheet of paper.

1. Advertising, selling, distributing, and servicing a product come under a company activity known as _____.

Matching questions: Match the definition (2 through 8) with the correct term (a through g).

2. _____ Making customers aware of products.
3. _____ Method of moving product to customers.
4. _____ Arrangement a company makes to have products repaired or replaced.
5. _____ Gathers data about customers' wants and expectations about new products.
6. _____ The official name of a company.
7. _____ Identifies goods made or sold by a company.
8. _____ A slogan describing company activities or products.
   a. Market research.
   b. Trademarks.
   c. Trade name.
   d. Promotion.
   e. Marketing theme.
   f. Distribution.
   g. Service.
9. _____ _____ is the actual cost of producing a product. It includes labor costs, material costs, and _____.
10. How does a manufacturing firm set the minimum price that it charges its customers for its products?

Applying Your Knowledge

1. Develop a trademark and advertising slogan for one of your school's athletic teams or service clubs.
2. Inform others in your school about marketing. Collect ten examples of company trade names, trademarks, and advertising themes and arrange them in a hallway display case. Include written information about the type of data gathered by market research.
3. Research the marketing of American products in other countries. Discuss these questions: What types of products are being sold? Do the people of these countries like the presence of American marketing or are they opposed to it? What would be the positive and negative effects of American products on other cultures?

Chapter 29 Promoting Products

Objectives

After studying this chapter, you will be able to:

✓ List and describe the two different types of advertisements.
✓ List and explain the functions of an advertisement that is effective.
✓ Give the three basic steps needed to create an advertisement.
✓ List the functions of a good package.
✓ Give the basic steps in designing a package.
✓ List and describe four types of plastic packages.
✓ Name five main considerations in selecting a package.
✓ Discuss the three important design considerations in package graphics.

People buy products and services almost daily and these purchases require making a choice. This is where product promotion enters the scene. It is designed and produced by companies to encourage sales. Promotion influences the customer to select one product over another. See **Figure 29-1.**

Product promotion can be done in many ways. Two very important methods are advertising and packaging. Both of these approaches are designed to attract attention to the product.

Figure 29-1. Consumers must constantly make product choices. (Richard Barella)

Advertising

Advertising is getting the attention of the public by using print and electronic messages. You read, view, or listen to the promotional message. The customers can decide to receive the message or not. They are in control of the communication.

advertising. A marketing function aimed at informing customers and persuading them to buy a product.

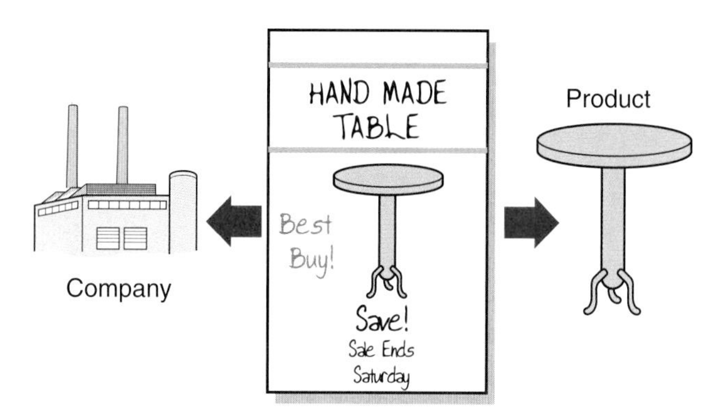

Figure 29-2. Advertising promotes either the product or the company.

Figure 29-3. A public service ad draws us to some action for the common good.
(The Advertising Council)

Types of Advertising

There are basically two types of advertisements. See **Figure 29-2.** Most ads are designed to promote either:

- The company or an idea
- A product

Company and idea advertising

Some advertisements try to get people to think a certain way. These advertisements promote the company image or an idea.

Many *public service advertisements* stress ideas. They want the public to act a certain way. Such advertisements are designed to protect the environment, improve health standards, or increase donations to a cause. See **Figure 29-3.** Typical idea ads promote forest fire prevention, use of seat belts, and giving to charity.

Other ads promote companies. They do not ask us to buy a specific product; instead they communicate an image of the company. "Get a Piece of the Rock" (Prudential Insurance) and "Quality Goes in Before the Name Goes On" (Zenith) are two examples. Both want you to respect the company.

Product advertising

Most advertising promotes specific products. Television and radio promote products on every program. Newspapers and magazines are full of product advertisements. Billboards describe all kinds of products. You cannot go through a day without seeing or hearing a product advertisement.

Functions of Advertisements

public service advertisements. **Advertisements that are designed to get the public to act a certain way, such as to protect the environment, improve health standards, or increase donations to a certain cause.**

Advertisements call us to act. They are saying, "buy this product," or "believe in this company."

These advertisements are fulfilling a function. To make us act, they take us through four steps. These, shown in **Figure 29-4,** are:

- The advertisement must attract our attention. We must want to read, see, or hear the message.
- It must inform us. We learn of the product's availability. We are told of its features and advantages.
- It must persuade us. We must want to use the product or support the idea.
- It must cause action. We must seek the product or behave differently.

Figure 29-4. An advertisement should attract attention, provide information, persuade customers, and cause action.

Figure 29-5. An artist is preparing an advertising layout. (Ohio Art Co.)

Creating Advertising

People with special training, **Figure 29-5,** create advertising. Often special companies are hired to do the job. These enterprises are called *advertising agencies*. Some companies have their own staff to develop their ads. They are called *in-house advertising departments.*

In either case, creating advertisements follows three basic steps. See **Figure 29-6.** First, the message is developed. The information that the customer is to receive is developed. Second, the presentation is designed. The layout for the advertisement is developed. Finally, the advertisement is produced. It can be printed or recorded for distribution to the target audience.

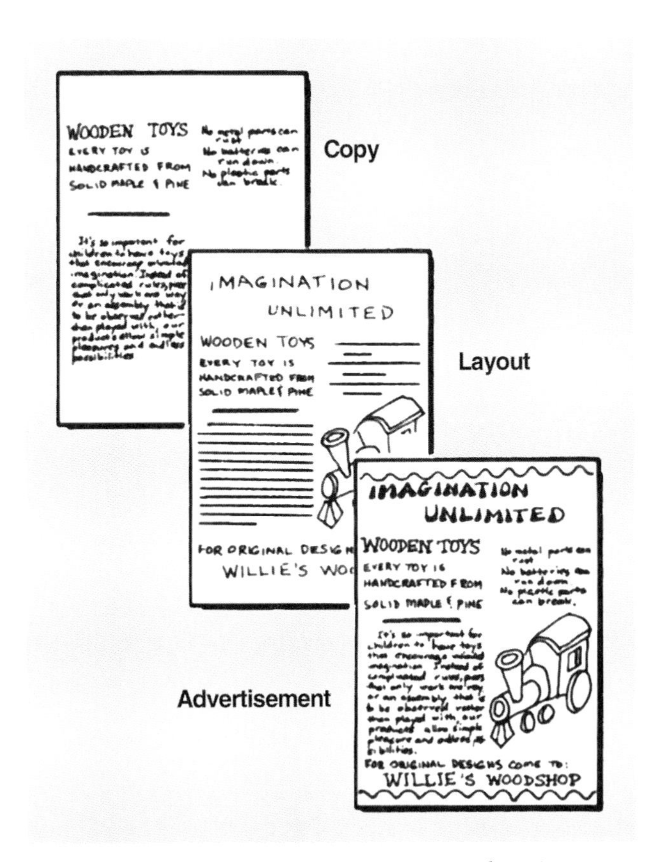

Figure 29-6. Steps in developing a print advertisement.

Developing the message

Advertisements must have a message. The message is developed from the theme. Developing a theme was discussed in Chapter 28. This theme provides the focus for the advertising campaign.

The message is then developed from the theme. Facts and ideas are chosen. The product or idea is described in words.

Messages can take two forms. Those meant to be used on radio or television are called *scripts*. Those used in print media (newspapers, magazines) are called *copy*.

advertising agencies. **Specialized service companies that have experience and talent in developing effective advertising materials.**

Rough

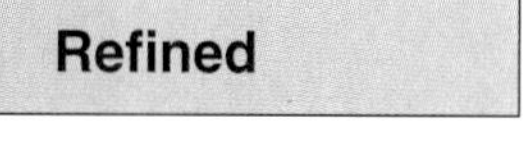

Refined

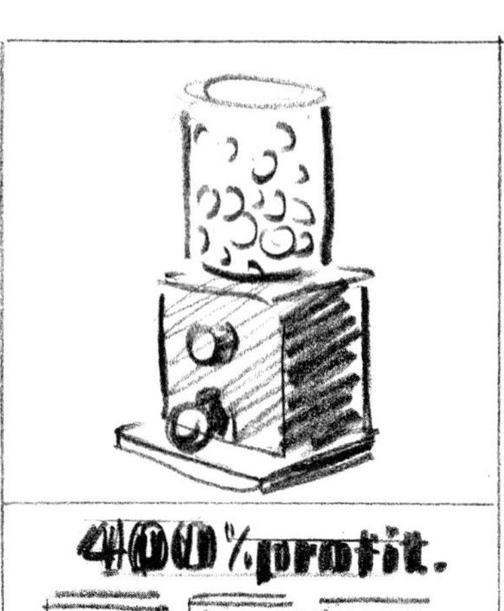

Comprehensive

Figure 29-7. Preparing a layout includes making rough, refined, and comprehensive layouts. (Manufacturing Forum)

Designing the presentation

A message is not enough. The ad must attract and hold attention. It must be presented well. The method used to do this is called the presentation. It is important to the success of an ad.

For print advertising, the presentation is called a *layout*. It is the way of arranging the information and pictures. Generally, layouts are produced in three steps. See **Figure 29-7.** First, the basic ideas for the advertisements are developed using rough layouts. These drawings are also called "thumbnail sketches." A number of these are developed to show various possible advertisements. Then, the best ideas are developed more fully. These ideas are presented with refined layouts. The final layout communicates exactly how the advertisement will look. This layout is called a *comprehensive* (almost complete) layout. The comprehensive is presented for approval so that the advertisement can be produced.

Television advertising uses a storyboard to show its presentation. The *storyboard* contains a sketch of each scene. It is used to guide the director. It shows the position of the actors. The background to be included is also shown.

script. **The written message of a radio or television advertisement. It is followed as the advertisement is produced.**

copy. **The written portion of an advertisement, whether for use in print or as a radio/TV commercial.**

layout. **The suggested arrangement of copy and illustrations for a print advertisement. A layout is submitted by an advertising agency for the company's approval.**

comprehensive. **Finished advertisements, ready for reproduction. Also referred to as mechanicals or camera-ready art.**

Producing the advertising

The creative ideas must be reproduced so customers get the message. Print and electronic media are often used. (Media means all the ways of communicating-newspapers, radio, and TV).

If the message will be in print, type will be set. An artist or photographer will provide illustrations. A layout artist brings together the type and illustrations to make a printed advertisement. See **Figure 29-8.** Then, the ad is printed.

If the media is radio or television, the advertising department may need actors. The art director or an advertising agency will arrange for a stage or location to produce the commercial (name for a radio or TV ad). The ad, **Figure 29-9,** comes into our homes.

Figure 29-8. The art and copy are changed into a finished advertisement. Left—Comprehensive layout shows how ad will look. Right—The ad is printed.

Packaging

A second way to promote a product is with its package. See **Figure 29-10.** A product's container probably has attracted you. It gave you information you thought useful. Perhaps you acted on the information and bought the product.

Functions of a Package

A *package* can serve three main functions. See **Figure 29-11.** It can protect and contain the product as it moves from the manufacturer to the customer. Some products can be damaged during shipment. Others are small and easily lost. Still other products can be damaged by moisture. Products like toothpaste, corn flakes, and orange juice need to be contained and protected. In fact, most products need some protection as they travel from the factory to you.

A package can also promote the product. It can be designed to attract people to the product. It can use color, graphics, and shapes to cause people to investigate the product inside.

Figure 29-9. The script for a television commercial can be recorded and become a radio advertisement. (Richard Barella)

Figure 29-10. Packages are a good way to promote a product. (James River Corp.)

Protect

Persuade

Inform

Figure 29-11. Packages attract attention. They also protect and promote the product and provide information. (Merisant)

storyboard. **A set of drawings that contains a sketch of each scene of a television commercial. It is used to guide the production of the commercial.**

package. **A device that protects, contains, promotes, and provides information to the customer about the product.**

Finally, it can provide the customer with information. It may contain health warnings, nutritional information, and assembly instruction. It can tell the consumer who made the product and how it is made.

Designing Packages

Package design follows three steps:

1. Package type is selected.
2. Package graphics (printing) are designed.
3. Package is printed.

Types of packages

Designers can choose from a large number of package types. Bottles, tubes, cans, cartons, bags, and trays are just a few of those available.

Bottles, cans, and tubes are often used. They can hold liquids, pastes, and granules (grains). These are popular for both consumer and industrial goods.

Two important materials for packaging are paperboard and plastic. Other materials are glass (jars and bottles) and metal (cans).

Paperboard packages

Paperboard (cardboard) can be formed into trays and boxes. The material is cut and scored (creased). Then it is bent into shape. Often, a window is cut into the box. The opening, often covered with plastic film, lets the customer see the product.

Paperboard packages take many different shapes. **Figure 29-12** shows four. Also shown is the layout for these packages.

Plastic packages

Plastic bottles, tubes, and jars make good containers, too. A number of other packages are made from plastic sheet and film. These, shown in **Figure 29-13,** include:

- **Blister pack.** A plastic blister (formed bubble) is shaped by thermoforming. (See Chapter 10 for a description of the thermoforming process.) The product is placed in the blister. The blister is then glued to a card. The card contains the package graphics (printing). Blister packs are used for hardware, toys, batteries, and other small products. Often the packages are displayed on a rack.
- **Skin pack.** The product is placed on a special paperboard sheet. Heated plastic film is drawn tightly around the product. (This process also uses thermoforming machines.) The film sticks to the coating of the paperboard. The paperboard sheet contains the package graphics. Skin pack is used for the same type of products as blisters.
- **Bags.** Plastic bags are used to hold a variety of products. The bags are usually made from long tubing. The tubing is unrolled to the right length. One end is sealed and cut. The product is then placed into the tube. The second end is sealed, forming the bag.
- **Shrink packaging.** Shrink packaging usually puts a plastic film over another package. The package is placed between two layers of special film. The sides are sealed. This produces a plastic envelope around the product. The film is then heated. The heat causes the film to shrink. Shrink packaging is used around products like cassette tapes, games, and computer discs. It provides a clear moisture- and dust-resistant cover. Usually, it keeps customers from opening boxes and packages in the store.

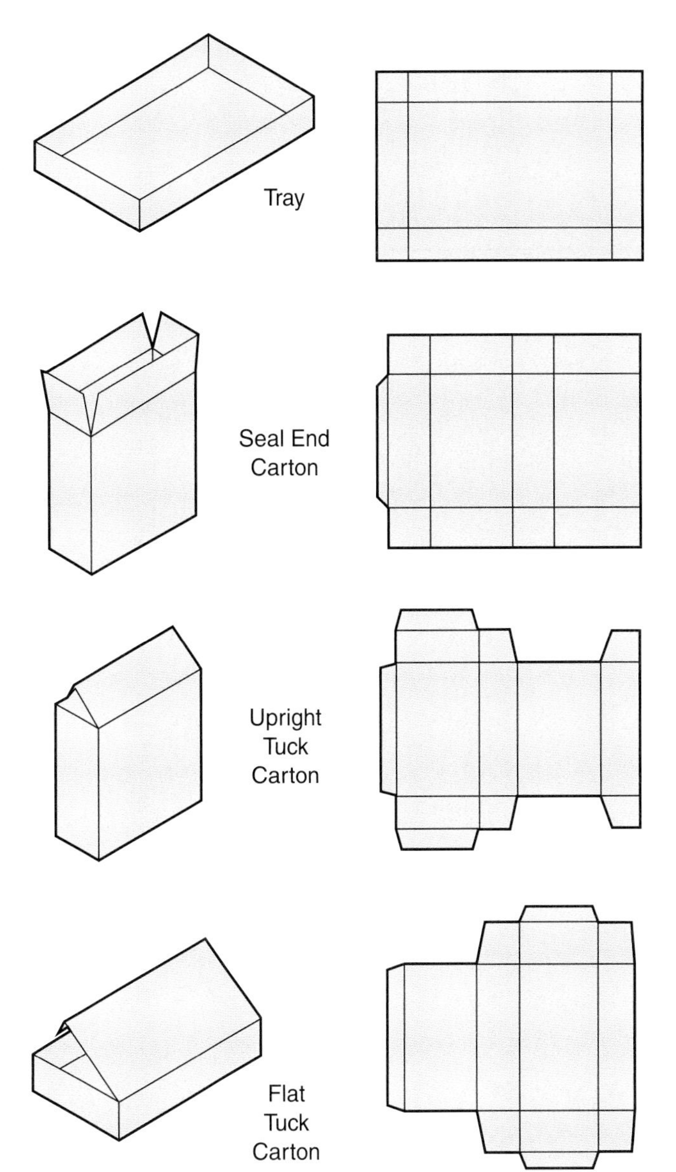

Figure 29-12. These are typical layouts for selected paperboard packages.

Selecting a Package

A package must be right for the product. See **Figure 29-14.** Five major factors are considered. A package is selected to meet the:

- Product requirements.
  - Shape. Form—gas, liquid, or solid.
  - Characteristics—fragile, sharp corners, poison, etc.
  - Number to be sold as a unit.

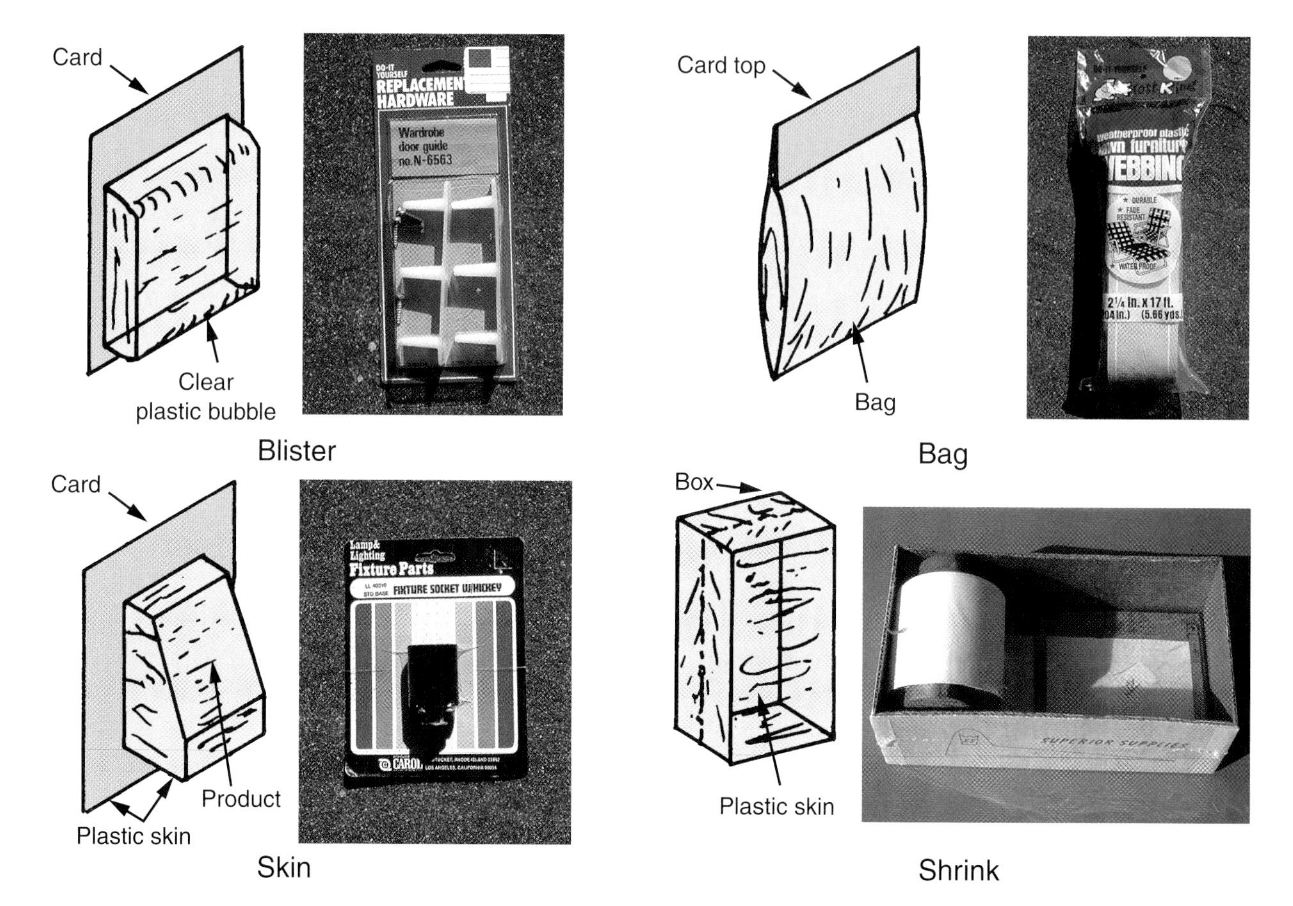

Figure 29-13. Some common plastic packages.

Academic Link

Ever wonder what type of planning went into the packages you see when you go to the store? Packaging science is the study of the use of materials, methods, and machinery to develop and produce packages that protect and preserve products, instruct the consumer, and help market the product. Environmental concerns are very important in packaging selection and design.

Scientific instruments and equipment are useful in package development and evaluation. Packaging science has a close relationship with the food science and graphics fields. Knowledge in biology, chemistry, physics, English, mathematics, graphics, statistics, microbiology are needed as well.

Packaging is a large, international industry and an extremely dynamic, rapidly growing field. On the basis of sales, it is the third largest industry in the United States. Virtually everything grown or manufactured is packaged in some fashion. The food industry is the largest user of packages, but nonfood packaging is essential also.

- Protection requirements.
  - Keep out moisture, dirt, grease, etc.
  - Prevent breakage.
  - Discourage theft.
- Legal requirements.
  - Information required by law.
  - Need for safety.
  - Weight and measurement requirements.
- Market requirements.
  - Sizes and shapes preferred by customers and retailers.
  - Quantities bought at one time.
  - Method of display (shelf, rack, etc.).
  - Ease of handling.
- Cost considerations.
  - Material cost.
  - Cost to manufacture the package and fill it.

Figure 29-14. The package must be appropriate for the product that it contains. (Graphic Arts Technical Foundation)

Preparing Package Graphics

Packages must be attractive. They also must provide information. Many of the advertising principles discussed earlier apply to packages. There are three important design considerations. The first is product identity. The package should show the trademark and brand names properly. Also, a description of the product should be clearly presented.

The second consideration is package graphics. The package should be designed with colors that are appropriate for the product. See **Figure 29-15.** The graphics (drawings, pictures, and type) should be pleasing to the customer. Also, all information required by regulations and needed by customers should be included.

The third consideration is customer attention and acceptance. The package should be attractive, both close-up and at a distance. It should meet customers' needs. It should be easy to handle, open, close, and recycle.

Color	Impression	Associated With
Black	Solemn	Death, mourning, darkness, emptiness
White	Purity	Cleanliness, winter, Mother's Day
Red	Official, close to the heart	Christmas, Valentine's Day, Fourth of July, danger, fire, warmth
Blue	Prize winning, cool and refreshing	Water, sky, cleanliness, clearness
Yellow	Cheerfulness, brightness	Sunlight, caution, daytime
Green	Comfort, natural	St. Patrick's Day, mature, growing things
Purple	Dignified, exclusive	Easter, reality, evening
Orange	Happy, glowing, friendly	Halloween, Thanksgiving, fall

Figure 29-15. Colors are important in packaging. They should support the product's image.

Producing Packages

Packages are a manufactured product. The actual cutting, forming, and assembling use manufacturing processes. These were presented in Chapters 10, 11, and 13. Graphics are printed using several methods. These include silkscreen and offset lithography. Typical printing processes are taught in communication classes.

Summary

Products must be promoted. Often advertisements and packages are used for this task. Both are carefully designed. The package, as well as the advertising, should attract customers' attention. Beyond that, they should provide information, persuade customers, and cause action.

Advertising should encourage customers to buy the product. The package should present the product attractively. Also, it should contain and protect the product.

Good advertisements and packages can improve sales. People can be encouraged to buy the product.

Key Words

All of the following words have been used in this chapter. Do you know their meaning?

advertising
advertising agencies
comprehensive
copy
layout
package
public service advertisements
script
storyboard

Test Your Knowledge

Please do not write in this text. Place your answers on a separate sheet of paper.

1. Indicate which of the following are the two main types of advertisement:
   a. Promotes company or an idea.
   b. Promotes a product.
   c. Promotes a person in the company.
   d. Promotes a process.
2. List and explain the four functions of advertisements.
3. The three steps of creating advertising are: developing the _____, _____ the presentation, and _____ the advertisement.
4. The _____ provides the focus for the advertising campaign.
5. A(n) _____ contains a sketch of each scene in a television commercial.
6. Name the three basic steps needed to design a package.
7. There are four types of plastic packaging. They are:
   a. _____
   b. _____
   c. _____
   d. _____
8. What are the five main considerations in selecting packages?

Applying Your Knowledge

1. Find an advertisement you feel is good and one you feel is bad. Analyze them based on how the person(s) depicted in the ad might feel.
2. Design a package for a baseball, pair of socks, or ten cookies.
3. Develop a point-of-purchase display for ballpoint pens.
4. With a partner, design a poster to promote the sale of a product to help someone with a physical disability.

Chapter 30 Selling and Distributing Products

Objectives

After studying this chapter, you will be able to:

✓ List and describe the three major channels of distribution used for consumer goods.
✓ Explain the two major types of sales.
✓ Describe the role of the sales manager in a company.
✓ List the steps in making a sale.

Products must move from the manufacturer to the consumer. They may follow any of several paths that are called *channels of distribution*. Consumer goods follow one of three main routes. These are shown in **Figure 30-1.**

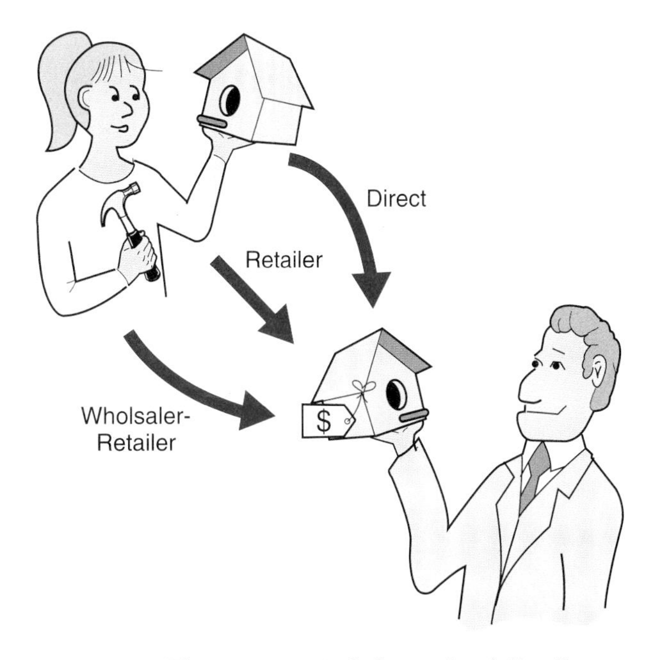

Figure 30-1. These are typical channels of distribution for consumer goods.

Channels of Distribution

The simplest path is called direct selling. This is where a manufacturer sells products directly to the customer. Many encyclopedias, cosmetics, and vacuum cleaners follow this channel. Mail order and catalog sales are also considered direct selling.

Some manufacturers sell directly to retailers who then sell the product to the customers. An *"authorized" dealer* or "franchised" dealer is this type of retailer. Such dealers are the only ones allowed to sell the product. Most new automobiles are sold this way.

Most consumer products follow a third route. They are first sold to wholesalers who resell the products to retail stores. See **Figure 30-2.** The retail stores make the final sale to the consumer.

channels of distribution. The paths that products follow from the manufacturer to the consumer. A product may follow any of several paths: retailer, direct, and wholesaler-retailer.

authorized dealer. A retailer who buys directly from the manufacturer and is the only person allowed to sell a product in a certain area.

Sales

Each step in the channel of distribution involves sales. During a sale the ownership of goods changes hands. Products move from warehouses to stores and then from the stores to customers. See **Figure 30-3.**

Types of Sales

There are two major types of sales. These, shown in **Figure 30-4,** are industrial sales and retail sales.

Figure 30-2. These regional wholesalers are buying products from a furniture manufacturer. (Hon Industries)

Industrial sales

Industrial sales involve several types of action. Raw materials may be sold to primary processors. Industrial materials are sold to secondary manufacturing companies. Finished products move to wholesalers and retailers. These are all examples of industrial sales.

Retail sales

Retail sales means selling to the final consumer. The customer pays for the products and he or she receives them immediately.

The Sales Force

Manufacturers may hire a special sales force. This includes *salespeople,* the people who do the selling to the customer. Sales managers who manage the sales effort are also hired. The sales effort takes three major steps. These are shown in **Figure 30-5.**

Warehouse

Store

Customer

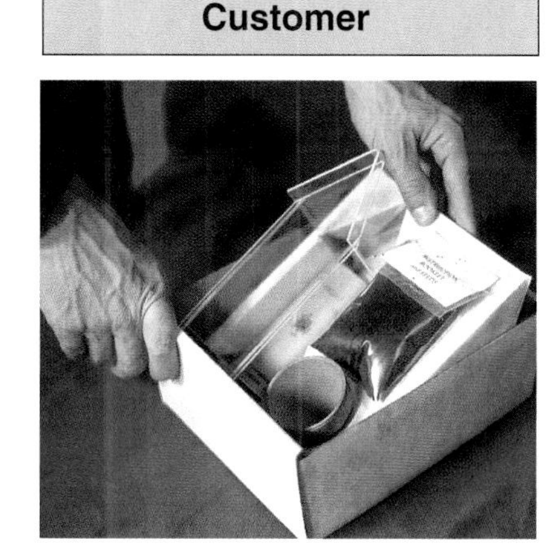

Figure 30-3. Sales move products from warehouses to stores to the customer. (Jack Klasey, Ohio Art Co.)

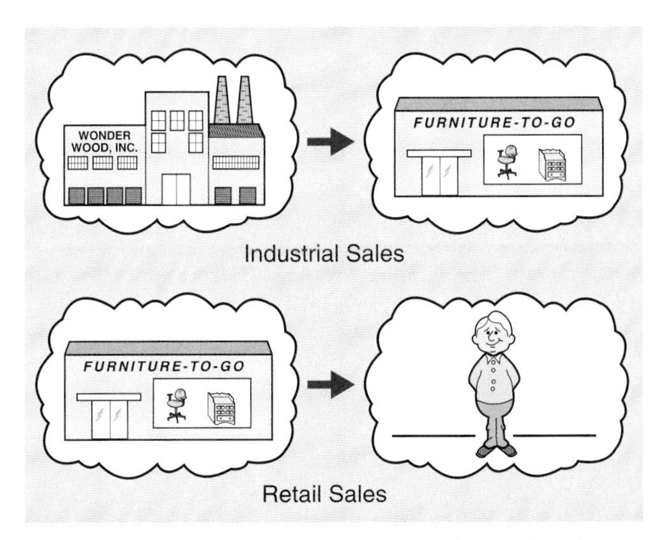

Figure 30-4. Sales are classified as industrial and retail.

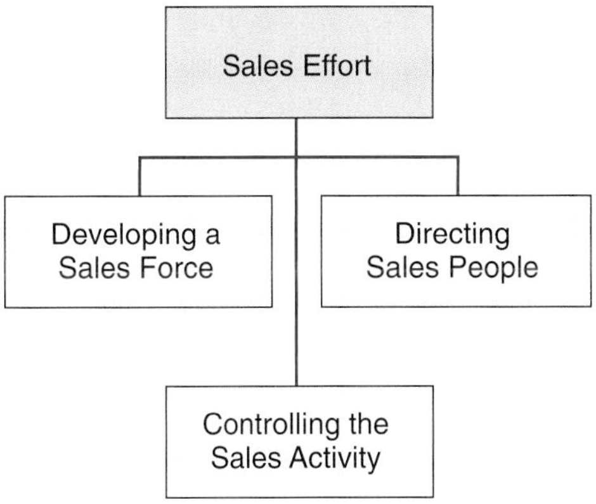

Sales Effort

Developing a Sales Force

Directing Sales People

Controlling the Sales Activity

Figure 30-5. There are three activities needed to develop a good sales program.

Developing a Sales Force

A sales force is developed in the same way as other workforces. The people are first recruited. Recruiters look for qualified people to fill the sales jobs. Human resource people place ads in newspapers and visit schools. Some people approach the company looking for sales jobs.

Applicants are screened to find the best qualified people. They complete applications and are interviewed. Previous sales experience is checked. The best applicants are hired.

Those hired receive special training in the art of selling. See **Figure 30-6.** The pay plan is also carefully outlined. Many salespeople work on salary plus commission. They receive a base salary. The *commission* is a reward for sales completed. They may get a percentage of their total sales.

Figure 30-6. These salespersons are receiving special training about the company's products. (AMP, Inc.)

Directing Salespeople

At least in large companies, regional or district sales managers direct the work of the sales force. They supervise and motivate the sales force. They assign sales people to areas called *territories* that may cover a city, several counties, an entire state, or several states.

To supervise means to give salespeople direction in their work. A manager must explain company policy, correct poor performance, and reward good performance.

Another part of supervision is to get people to do their best. This is called motivation. We all let down once in a while. It is why coaches take time during the game to give their teams a "pep talk." Sales managers must also find ways to encourage better sales effort from salespeople. They hold sales meetings to urge their people to work harder. They give out sales awards to those who sell

industrial sales. **The sales of goods to anyone but the final customer.**

retail sales. **Selling products to the final consumer, who receives the product immediately.**

salespeople. **People hired by a company to present products to the customer.**

commission. **A fee paid to the member of a sales force for the sales he or she completes.**

territory. An area that is assigned to a salesperson. A territory may cover a city, several counties, an entire state, or several states.

quota. The amount of product that a salesperson is expected to sell.

sales forecast. A company's estimate of expected sales, for a period of time.

more than their quota. (A *quota* is the amount each salesperson is expected to sell.) As new goals are discussed, the sales people are "fired up" to "get out and sell."

Controlling the Sales Effort

All managed activities need to be controlled. By now, you can see that selling is certainly a managed activity. Of the various goals, an important one is the *sales forecast*. This is a sales budget that estimates overall expected sales for each reporting period. A period may cover a week, month, or quarter (three months).

Sales forecasts are broken down by region, which are further divided. Each salesperson works toward a sales quota so the sales forecast is met. Various territories are expected to produce scheduled sales. Refer to **Figure 30-7.**

Failure to reach quotas will call for action. New motivation techniques may be required. Some salespeople might be replaced. Better training may be provided. All action is designed to increase sales.

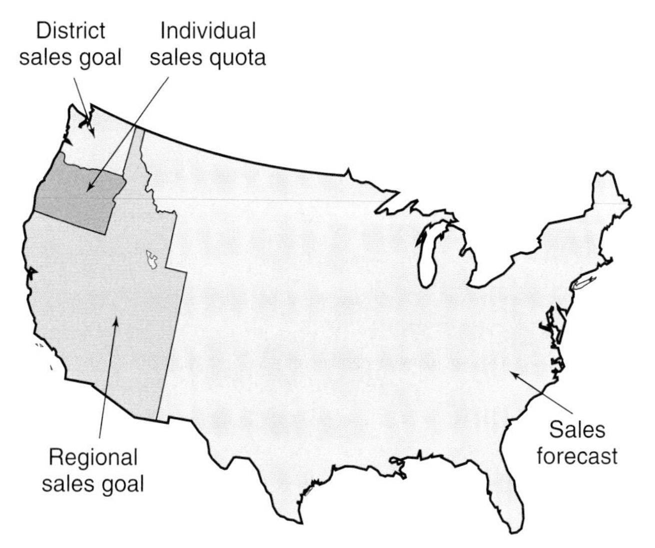

Figure 30-7. Types of sales goals.

The Art of Selling

Selling is an art that takes talent. People can develop their skill in this art. Good salespeople are skilled in:

- Approaching the customer.
- Presenting the product.
- Closing the sale.

Each sales call follows these steps. See **Figure 30-8.** Each step must be successful; otherwise, there will be no sale.

Approaching the Customer

The approach to a customer is critical. The salesperson must attract the buyer's attention and develop interest in the product.

The approach is based on two key principles. First, an appointment is usually arranged. The salesperson must be on time. Salespeople often wait for a customer; however the customer should never have to wait for a salesperson.

Secondly, the salesperson should not waste a customer's time. The product should be introduced in a few words. The first words will set the tone for the meeting. Most customers like a businesslike approach.

Presenting the Product

The product presentation should:

- Clearly explain the product.
- Describe its features and benefits.

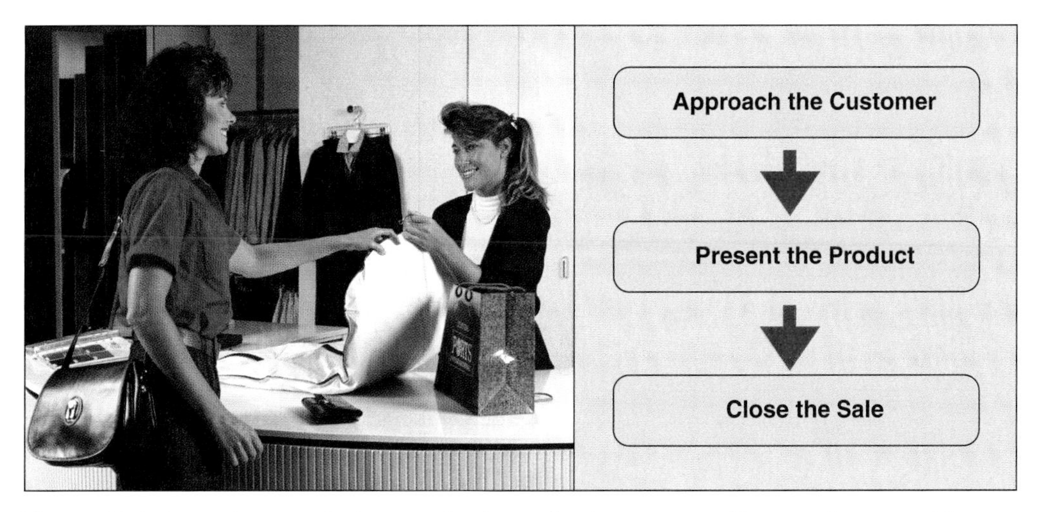

Figure 30-8. Steps in completing a sale. A good salesperson will try to improve his or her skills in selling.

- Answer all the customer's questions.
- Prepare the client for a "yes" buying decision.

The product explanation must be designed for the customer. One type of customer may buy a product to resell it. Another may be buying for his or her own use. Their reasons for wanting the product are different. The presentation must take this into consideration. Retailers want to make a profit from selling products. The customer wants to know how it will make a job easier or life more comfortable.

Closing the Sale

If the presentation is successful, the sale can be closed. When you buy a product in a store, the close is simple. You just pay for it and the clerk gives you the product. An industrial sale does not work this way. The salesperson takes the order, which is recorded on an order form. The order form, such as the one shown in **Figure 30-9,** will list:

- Item.
- Quantity ordered.
- Cost.
- Shipping instructions
- Billing instructions
- Other important information (shipping dates, discounts for early payment, etc.).

The order information goes to the factory where the information is entered on a computer. See **Figure 30-10.** Often, the order is filled from inventory, **Figure 30-11,** and the items are shipped to the customer. Some products are not built until they are ordered. Aircraft companies do not keep airliners in inventory. These and other products are designed and built for the customer. See **Figure 30-12.**

Order Form

Ball

Consumer Products Division
1509 South Macedonia Avenue, Muncie, Indiana 47302

Warehouse No.	Date of Shipment	Delivery Order No.	Ball Customer No.
P. O. No.	Ppd/Col	Name of Carrier	Type Sale / B/L No.
Date to Ship/Arrive	Date Ordered	Order Placed By:	Routing:
Special Instructions			

SHIP TO:	SOLD TO:
C/O	Street
Street	City / State / Zip
City / State	Terms

Cases	Wt. Per cs.	Product Code	Description	Cs. per Pallet	No. of Pallets	Cases	Wt. Per cs.	Product Code	Description	Cs. per Pallet		No. of Pallets
	6	60000	½ PT. CAN OR FRZ JAR	170					BAGS			
	8	61000	PT. REG. MASON JAR	208			4	51600	PINT 30 COUNT		REGULAR PAK	
	12	62000	QT. REG. MASON JAR	121			4	52400	1-½ PINT			
	20	64000	½ GAL. REG. MASON JAR	56			4	53200	QUART 20 COUNT			
	9	66000	PT. W/M CAN OR FRZ JAR	154			4	56400	2 QUART			
	12	12400	1-½ PT. W/M CAN OR FRZ JAR	121			4	57800	GALLON			
	13	67000	QUART WIDE MOUTH JAR	121			4	58600	2 GALLON			
	21	68000	½ GAL. WIDE MOUTH JAR	56			5	71600	PINT 50 COUNT		ECONOMY PAK	
	29	80800	REG. JELLY GLASS 6/12's	42			5	73200	QUART 40 COUNT			
	28	81000	DLX QC JELLY GLASS 6/12's	49								
	34	81200	DLX QC JELLY JAR 6/12's	45					BOXES			

Figure 30-9. An order form such as this may be used by a manufacturer's sales force. (Ball Corp.)

Figure 30-10. Many sales orders are entered in computer systems.

Figure 30-11. Most consumer products, like these automobile parts, are sold from inventory. The manufacturer keeps products in a warehouse. (AC Rochester)

An industrial sale requires time to complete. The product may have to be built after the order is placed. When the customer receives the order the salesperson is long gone. He or she is calling on other customers trying to make additional sales. Many times, salespeople never see the products they sell because they sell from catalogs or samples.

Figure 30-12. This airliner was built to fill an order. (Alaska Airlines)

Summary

Products move from manufacturers to customers through distribution channels. At each step in the channel products are bought and sold.

Salespeople create these sales. The sales force is developed, directed, and controlled. They are motivated to approach customers, present products, and close sales.

The success of a company is related to the success of the sales force. Products can be built and also have to be sold.

Key Words

All of the following words have been used in this chapter. Do you know their meaning?

authorized dealer
channels of distribution
commission
industrial sales
quota
retail sales
sales forecast
salespeople
territory

Test Your Knowledge

Please do not write in this text. Place your answers on a separate sheet of paper.

1. The simplest type of distribution is called _____ selling.
2. List two other types of distribution or selling.
3. If you are selling airplanes to an airline company you are engaged in (retail, industrial) sales.

Matching questions: Match the definition (4 through 10) with the correct term (a through g).

4. _____ Selling products to a wholesaler.
5. _____ Area to which a salesperson is assigned.
6. _____ An estimate of the future sales volume.
7. _____ Getting customer's attention.
8. _____ Explaining product and features.
9. _____ Getting customer to buy.
10. _____ Place to record an order.

a. Order form.
b. Closing the sale.
c. Territory.
d. Product presentation.
e. Sales approach.
f. Sales forecast.
g. Industrial sales.

Applying Your Knowledge

1. Develop a list of products that are sold through (a) direct, (b) retailer, and (c) wholesaler-retailer distribution channels.
2. Develop a sales presentation for decorative candies that will be sold to support a trip to Washington, DC.
3. Arrange to visit a company in your area. Talk to the person who heads the sales department. Ask him or her how they approach sales.
4. Design a sales receipt that could be used by a company that sells a wide range of school supplies.

Chapter 31
Maintaining and Servicing Products

Objectives

After studying this chapter, you will be able to:
- ✓ Tell the difference between durable and nondurable goods.
- ✓ Explain the difference between repair and maintenance of products.
- ✓ List and describe the steps in a product use cycle.
- ✓ List and explain the steps used to repair products.
- ✓ Explain the economics of replacement versus repair.

When a customer buys a product its life cycle (time it will be used) starts. The manufacturer's work is not done, however. The customer must know how to properly operate the product. The manufacturer provides directions for use and maintenance. Service information or facilities are also made available.

You will recall that earlier we divided products into two groups. These were nondurable or soft goods, and durable or hard goods. See **Figure 31-1.** *Nondurable goods* last less than three years under normal conditions. Clothing and food are nondurable goods as are pencils, paper, lightbulbs, and motor oil. *Durable goods* are designed to last over three years. Automobiles, bicycles, and refrigerators are durable goods.

nondurable goods. **Products that are made to last less than three years. Clothing, food, pencils, paper, lightbulbs, and motor oil are nondurable goods.**

durable goods. **Products made to last more than three years. Automobiles, bicycles, and refrigerators are durable goods.**

Figure 31-1. Two types of goods. Left—The food and the packages are nondurable goods. (James River Corp.) Right—The wood lathe is a durable product. (Delta International Machinery Co.)

maintenance. The tasks that are performed on a product to keep it in good working order.

Both types of goods require maintenance and service. Clothing must be washed. This is called *maintenance.* Buttons have to be replaced. In this case, the clothing is being serviced.

Cars, trucks, and buses also need maintenance. Engine oil must be changed while air and oil filters are replaced. Tires are rotated. Many automobiles need service, too. Worn parts are replaced or leaking radiator is repaired.

Most servicing and maintenance is done on durable goods because they are usually more expensive. We want them to last longer and it is usually cheaper to repair them than to replace them.

Product Use Cycle

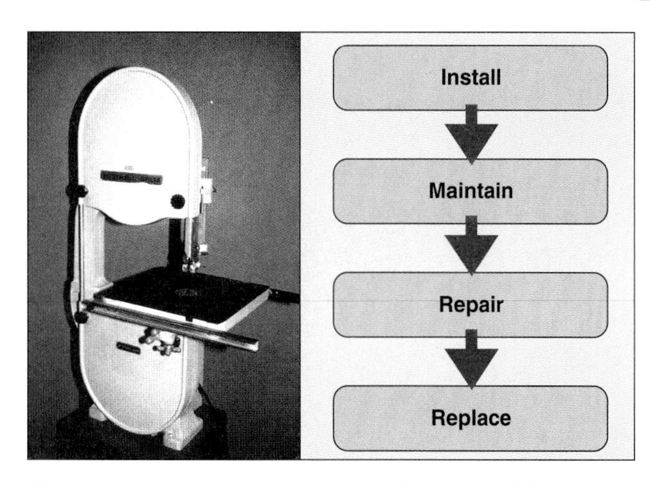

Figure 31-2. Steps in a product's life cycle. Most durable goods follow this cycle.

Products usually follow a common life cycle. See **Figure 31-2.** First the product is installed. When a product is installed, it is set up where it will be used. From time to time it is *maintained* to keep it in working order. After a time, the product is *repaired*: worn and broken parts are replaced. Finally, the product is taken out of service when it is not possible or economical to repair. It is *replaced.*

Installing Products

Some products are not ready to use when they leave the factory. They must be *installed.* See **Figure 31-3.** The product must be set in place and leveled. Often it is permanently located. This means the product is fastened down or it is so large that it cannot be moved.

maintained. When service has been performed on a product to keep it in good working order.

replaced. The product is taken out of service when it is not possible or economical to repair it, and a new product is purchased to do the old product's job.

installed. When a product is installed, it is set up in the place where it will be used. It is unpacked, hooked up to utilities, adjusted, and tested.

Many products must be connected to utilities. Installers must run water lines, electrical wiring, or natural gas lines. They attach drains to sewers. These activities are all part of installing products.

Product testing is performed after installation. See **Figure 31-4.** For example, a dishwasher is run through its cycle or a furnace is operated for a period of time. The installed product is then turned over to the customer.

Maintaining Products

Often, products require attention to keep them working. Durable products require the most maintenance. Automobiles have maintenance schedules provided by the manufacturer, **Figure 31-5,** as do many industrial machines. The product must be lubricated and the controls must be adjusted.

Nondurable products are also maintained. Dishes and clothing are washed, hiking boots are coated with silicone, shoes are polished, and rugs are vacuumed.

All maintenance is designed to make the product last longer. It reduces the amount of wear and breakage the product experiences.

Repairing Products

Products will not work forever, however. Parts go out of adjustment and sometimes they break. Surfaces wear thin or rust through and bearings wear out. The product stops working and needs to be *repaired*.

Repair is a set of actions that puts the product back into working order. The product is made as much like a new product as possible. There are three major steps in repairing a product, as shown in **Figure 31-6.**

Diagnosing

You cannot fix something unless you know what is wrong. Finding the defect is called *diagnosing*. This activity seeks the cause for the breakdown. Refer to **Figure 31-7.**

The owner or repair technician studies the defect. The cost of repair is estimated and, then, the owner must make a decision. Is the product worth fixing? If it is, the defect is repaired.

Correcting Defects

Most defects can be corrected. We can hire repair people to work on the product and sometimes we can repair it ourselves. Repair technicians often receive factory training where they learn how to repair the product. See **Figure 31-8.** They learn to use special tools and techniques developed by the manufacturer.

The repair people, then, put the product into working order. They order replacement parts for the product. See **Figure 31-9.** They install and adjust parts and, sometimes, replace entire subassemblies. See **Figure 31-10.**

Testing

All repair work needs to be tested to be sure that the repaired product meets operating standards. Testing serves as the "quality control" step for repair. Proper testing is important to make sure the product works.

Figure 31-3. This engine test equipment is being installed. After the components (parts) of the system are in place, they must be connected and tested. (Daimler Benz)

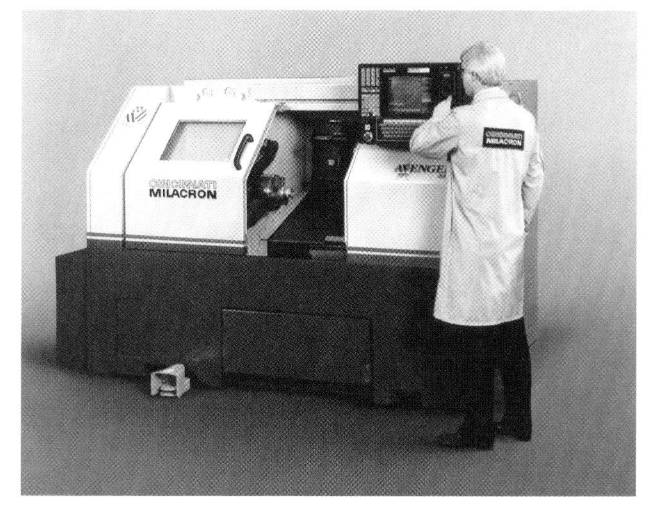

Figure 31-4. A technician is testing a new CNC turning center. (Cincinnati Milacron)

repaired. **Worn and broken parts are replaced to restore the product to proper working order.**

MAINTENANCE SERVICES

Schedule 2

I : Inspect and if necessary correct, clean, or replace
R: Replace or change A: Adjust

MAINTENANCE INTERVALS / MAINTENANCE ITEM	Number of months or miles (kilometers), whichever comes first											
	Months	5	10	15	20	25	30	35	40	45	50	55	60
	Miles × 1000	5	10	15	20	25	30	35	40	45	50	55	60
	(km × 1000)	(8)	(16)	(24)	(32)	(40)	(48)	(56)	(64)	(72)	(80)	(88)	(96)
Drive belts							I						I
Engine oil		R	R	R	R	R	R	R	R	R	R	R	R
Engine oil filter		R	R	R	R	R	R	R	R	R	R	R	R
Engine timing belt *1		Replace every 60,000 miles (96,000 km)											
Air cleaner element				I*3			R			I*3			R
Spark plugs							R						R
Cooling system							I						I
Engine coolant							R						R
Fuel filter													R
Fuel lines							I*2						I
Idle speed							A*3						A

*1 Replacement of the engine timing belt is required at every 60,000 miles (96,000 km). Failure to replace this belt may result in damage to the engine.
*2 This maintenance is recommended by Mazda. However, it is not necessary for emission warranty coverage or manufacturer recall liability.
*3 This maintenance is required in all states except California. However, we recommend that it also be performed on California vehicles.

Figure 31-5. This chart is a service schedule for an automobile. These services help keep the car in excellent working order. (Mazda Motors of America)

diagnosing. **Determining what is wrong with a product that needs repair.**

Academic Link

Repair and maintenance technicians study specifications from blueprints, sketches, or descriptions of parts to be replaced, and plan sequence of operations. The technicians must be knowledgeable of mechanics, shop mathematics, material properties, and layout machinery procedures.

Many industries have licensing or certification programs for their technicians. For example, many states and independent organizations offer licensing for electricians. They promote the quality of electrical service and repair through the voluntary testing and certification of electricians.

Replacing Products

Sometimes it costs too much to repair the product. In this case, the repair costs are higher than the product's value. Customers may decide they do not need the product that much. The cost of repair is greater than their need.

Other times a new product is cheaper to buy than the cost of repairing the old one. The original manufacturing system may have been highly automatic. Labor costs were, therefore, cheaper. Parts were purchased in large quantities and the product was efficiently manufactured. Its selling cost was reasonable.

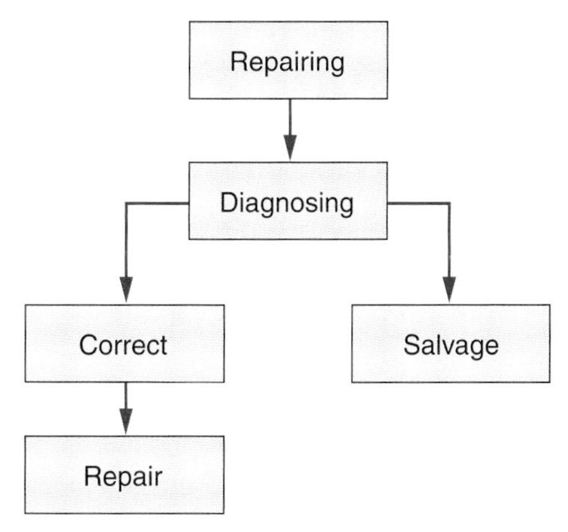

Figure 31-6. These are the basic steps in repairing.

Figure 31-7. These technicians are diagnosing a problem in a communication system. (AT&T)

Figure 31-8. A service person is receiving training on automotive suspension systems. (Arvin Industries)

Figure 31-9. This employee is filling orders for repair parts. (Goodyear Tire & Rubber Co.)

Figure 31-10. These trained technicians are repairing a communication switching system. (Northern Telecom)

Repair seldom enjoys these benefits. The products are repaired one at a time and replacement parts are ordered as they are needed. Often, a single worker repairs the devices, which makes repairs expensive.

Products not repaired must be recycled. Sometimes they are used for parts. Other times, the products are thrown away. Still other products are *recycled*. They are ground up, shredded, or melted down to become raw materials for primary processing. Many steel, aluminum, glass, and paper plants use scrap.

recycled. When a product cannot be repaired, the materials in that product are reprocessed for other use.

Summary

Many products need service and maintenance. Maintenance includes all the actions that keep products working, while servicing is repairing broken products.

Servicing involves determining the fault. The defect is diagnosed and corrective action is often taken. Parts and systems are replaced and operating adjustments are made. The repaired product is tested.

Eventually all products wear out. They can be disassembled for parts, placed in a dump, or recycled to make new products.

Key Words

All of the following words have been used in this chapter. Do you know their meaning?

diagnosing
durable goods
installed
maintained
maintenance
nondurable goods
recycled
repaired
replaced

Test Your Knowledge

Please do not write in this text. Place your answers on a separate sheet of paper.

1. Goods that last less than a year are called ______ goods.
2. Automobiles and other items that last longer than ______ years are examples of what we call goods.
3. Explain the difference between repair and maintenance.
4. Which of the following are steps in the repairing of products?
   a. See if it is working.
   b. Find the cause for the breakdown.
   c. Fix or replace the part.
   d. Check the product for proper working order.
   e. Determine the cost of the repair.
5. Explain why it is sometimes cheaper to replace than repair products.

Applying Your Knowledge

1. In a small group, write an owner's manual for a simple product such as a pencil sharpener.
2. Visit a repair service center (automobile, appliance, computer). Ask the manager to demonstrate the steps in servicing the products.
3. Tour a local recycling center or scrap yard to find out how certain products can become the raw materials for primary processing of other products.

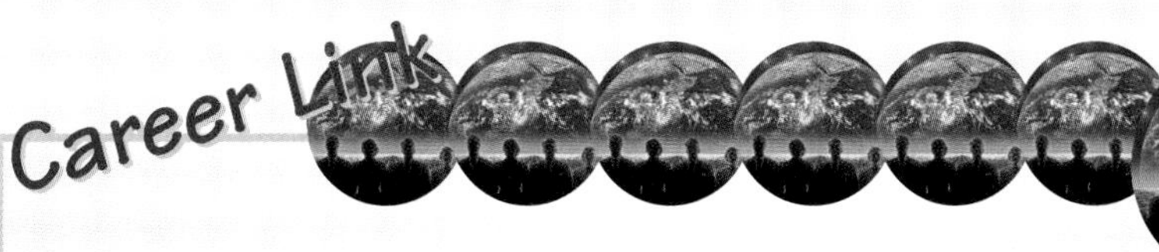

Service Technicians

Appliance service technicians keep home appliances working and help prevent unwanted breakdowns. Some technicians work specifically on small appliances, such as microwaves and vacuum cleaners. Others specialize in major appliances such as refrigerators, dishwashers, washers, and dryers.

Home appliance technicians inspect appliances visually and check for noises, excessive vibration, fluid leaks, or loose parts to determine why they fail to operate properly. Service manuals, troubleshooting guides, and experience are used to diagnose particularly difficult problems. Technicians disassemble the appliance to examine its internal parts for signs of wear or corrosion. They follow wiring diagrams and use testing devices such as ammeters, voltmeters, and wattmeters to check electrical systems for shorts and faulty connections.

After identifying problems, technicians replace or repair defective belts, motors, heating elements, switches, gears, or other items. They tighten, align, clean, and lubricate parts as necessary. Technicians use common hand tools, including screwdrivers, wrenches, files, and pliers, as well as soldering guns and special tools designed for particular appliances. When repairing appliances with electronic parts, they may replace circuit boards or other electronic components.

Technicians also answer customers' questions about the care and use of appliances. For example, they demonstrate how to load automatic washing machines, arrange dishes in dishwashers, or sharpen chain saws to maximize performance.

Technicians write up estimates of the cost of repairs for customers, keep records of parts used and hours worked, prepare bills, and collect payments. Self-employed technicians also deal with the original appliance manufacturers to receive monetary claims for work performed on appliances still under warranty.

Unit 10 Performing Financial Activities

Chapter 32 Maintaining Financial Records

Objectives

After studying this chapter, you will be able to:

✓ Name three major types of financial records kept by manufacturing companies.
✓ Tell why companies should keep financial records.
✓ List and describe the major types of budgets.
✓ Tell the difference between a company's assets and liabilities.

All companies must manage the use of their money. To do this, they must monitor (review) their financial results. This activity requires financial records and keeping financial records is often called *accounting*.

accounting. The practice and process of keeping financial records.

Many companies keep three types of financial records. These are:

* Budgets
* General accounts
* Cost accounts

These records are designed to control the use of money. They help managers measure the financial health of the company.

Budgets

Budgets were briefly introduced in Chapter 17. You may recall that they are plans written in terms of money. *Budgets* are plans that forecast income and expenses.

budget. Plan that a business uses to forecast income and expenses.

Money managers depend upon budgets because they are the basis for measuring financial success. The company's financial performance is compared to them.

Sales and Production Forecasts

Manufacturing companies use several types of budgets. Budgets are based on two major types of information, **Figure 32-1:**

Figure 32-1. Production schedules and sales forecasts provide a basis for budgets. (Ford Motor Company)

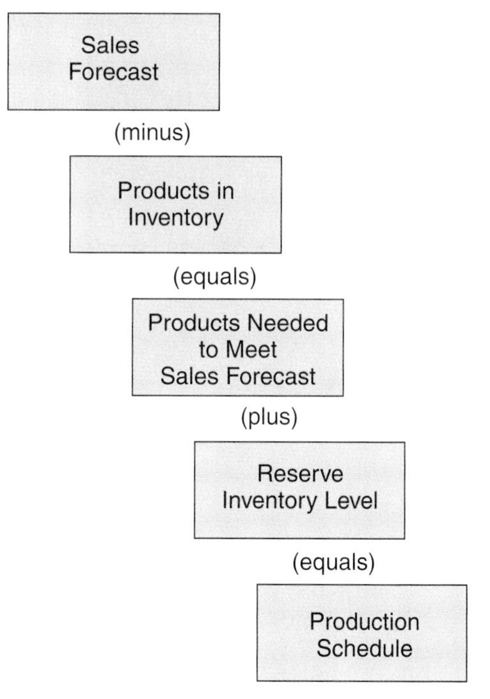

Figure 32-2. Note the relationship between sales forecasts and production schedules.

- **Sales forecast.** Anticipated sales for a period of time. Sales of each product and model are included.
- **Production schedule.** A record of products that are to be built in a given period. The schedules consider the sales forecast and the number of products in inventory.

The relationship between these two documents is shown in **Figure 32-2.** The sales forecast estimates the total need for products for a set period of time (week, month, year). Usually, some products are part of the finished goods inventory already in the warehouse. These products are subtracted from the forecast to provide the minimum production level. This many products would exactly meet the sales forecast. However, companies want some reserve inventory on hand. These products can cover production delays or meet the demands of extra sales. This reserve inventory is added to the minimum production level to reach a production schedule. The schedule will meet the sales forecast and will maintain a reserve inventory.

Types of Budgets

It costs money to produce and market products. These costs are predicted on some basic budgets. These budgets, as shown in **Figure 32-3,** are:

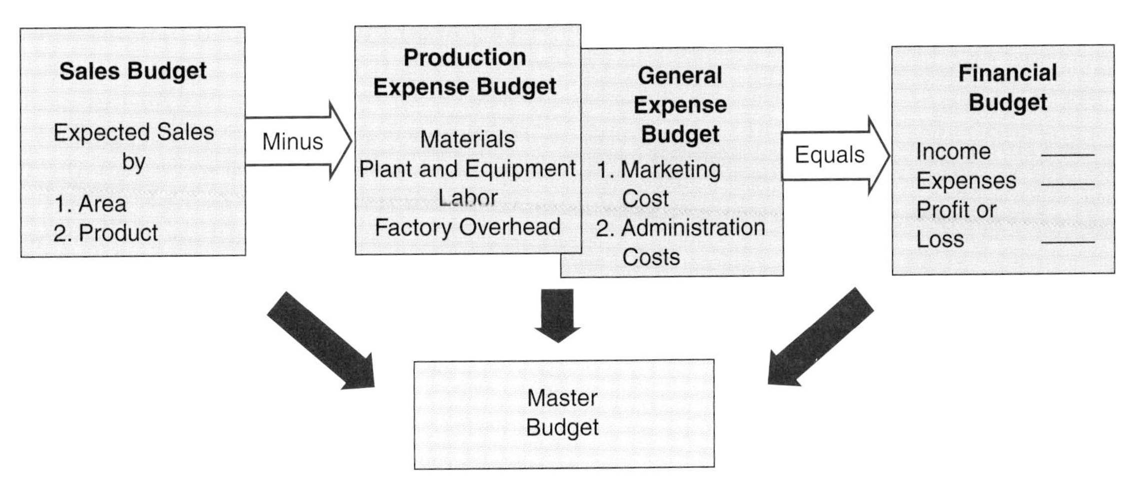

Figure 32-3. These are major types of budgets.

CDE INCORPORATED SALES BUDGET

	TOTALS		SCHOOL SALES		HOME SALES	
	UNITS	DOLLARS	UNITS	DOLLARS	UNITS	DOLLARS
PRODUCT A @ $2.00						
WEEK 1	15	$ 30.00	9	$ 18.00	6	$ 12.00
2	30	60.00	20	40.00	10	20.00
3	23	46.00	14	28.00	9	18.00
4	22	44.00	13	26.00	9	18.00
TOTAL	90	$180.00	56	$112.00	34	$ 68.00
PRODUCT B @ $1.50						
WEEK 1	20	$ 30.00	9	$ 13.50	11	$ 16.50
2	24	36.00	11	16.50	13	19.50
3	22	33.00	9	13.50	13	19.50
4	18	27.00	8	12.00	10	15.00
TOTAL	84	$126.00	37	$ 55.50	47	$ 70.50

Figure 32-4. A sample sales budget. It estimates sales of a company by the week.

- **Sales budget.** A projection of income from sales. See **Figure 32-4.** It estimates sales by product and region.
- **Production expense budget.** An estimate of production costs. This budget is based on the production schedule. See **Figure 32-5.** The production budget is often divided into four sub-budgets. These are:
  - **Labor budget**—Estimated cost of human resources needed to meet the production schedule.
  - **Materials budget**—Estimated cost of materials needed to build scheduled products.
  - **Equipment budget**—Estimated cost of new equipment needed.
  - **Factory overhead budget**—Estimated cost of utilities, maintenance, and supervision salaries.
- **General expense budget.** An estimate of the cost of activities that support production. These activities include administration (corporate management) and marketing (advertising and sales).

CDE Incorporated Production Schedule (Budget)			
Product A			
Week	Sales (1) Forecast	Scheduled (2) Inventory	Production (3) Required
I (April 1-5)	15	20	35
II (April 8-12)	30	20	30
III (April 15-19)	23	10	13
IV (April 22-26)	22	00	12
Product B			
I (April 1-5)	20	20	40
II (April 8-12)	24	20	24
III (April 15-19)	22	10	12
IV (April 22-26)	18	00	08

Figure 32-5. A typical production schedule. It estimates levels of production to meet predicted sales.

- **Financial budget.** An estimate of income and expenses. It can be presented on a daily, weekly, or monthly basis. This budget tells a manager if income will meet expenses at each point in time. The need to borrow money at some point is shown on this budget. It also shows when money will be available to pay debts.
- **Master budget.** A summary of all budgets. It provides an overview of financial activity.

General Accounting

Budgets are a prediction of financial activity. General accounting records this activity. It involves two major tasks:

- Recording financial transactions (dealings)
- Summarizing financial activities

Recording Transactions

Most accounting systems are based on a ledger or journal, **Figure 32-6,** where various financial actions are recorded. See **Figure 32-7.** Each action is described and the date it happened is entered in the ledger. The amount of the transaction is entered. Money going out, or expenses, is entered in the debit column. Money coming in, or *income*, is entered in the credit column. If the company is making a *profit*, the income will exceed debits.

income. The money a company receives, primarily from sales of products or services.

profit. The amount of money left over after all the expenses of a business have been paid.

Summarizing Financial Activities

Ledgers help accountants record financial actions. Other managers want a more general report such as balance sheets and income statements.

COST LEDGER

ITEM: ____________ CODE NO. ____________

Date	Description	Debit	Credit

Figure 32-6. A general ledger form. It keeps track of only one kind of expense or income.

Balance sheets

A balance sheet "balances" assets with liabilities. It lists all the things the company owns, for example, materials, equipment, and knowledge. These items, called *assets*, include:

- Current assets
  - Cash and securities (stocks and bonds)
  - Accounts receivable (money owed by others to the company)
  - Inventories (materials and goods in plants and warehouses)
- Other assets
- Property, plant, and equipment
  - Land, buildings, and equipment owned by the company

Figure 32-7. These employees are entering financial data into a computer system. (Union Pacific)

assets. **The things that a company owns, such as cash, property, and equipment.**

liabilities. **The money that the company owes to other companies and to individuals.**

shareholders' equity. **The actual value of the shareholders' ownership.**

Balance sheets also list what the company owes which is called *liabilities*. They include:

- Current liabilities
  - Accounts payable (money owed to other companies)
  - Salaries and fringe benefits for employees
  - Long-term debts due
  - Interest and taxes due
- Long-term debt (loan and bonds outstanding)
- Deferred taxes (taxes due at a later date)

The assets should exceed liabilities because a successful company should own more than it owes. The excess is shown as *shareholders' equity*. This is the actual value of the shareholders' ownership (equity). A recent balance sheet is shown in **Figure 32-8.**

Income statement

The second report is an income statement or profit and loss statement. See **Figure 32-9.** It shows the financial success for the year. It lists net (total) sales. The cost of the products sold is then subtracted to provide the *gross profit*. Next, the general expenses are subtracted to determine the *operating profit*. Finally, other income and expenses are entered. The resulting figure is *"profit before taxes"*. Taxes are subtracted giving the *net income*, which is the actual profit that can be used for:

- Enlarging the company (retained earnings)
- Paying shareholders a dividend

CONSOLIDATED BALANCE SHEET
December 31, 20XX and 20YY

ASSETS	20XX	20YY
Current Assets:		
Cash and marketable securities	**$ 209,030,000**	$ 203,350,000
Receivables	**201,610,000**	163,380,000
Inventories	**248,230,000**	243,620,000
Prepaid expenses	**10,210,000**	11,410,000
Total current assets	**669,080,000**	621,760,000
Investments in Partially Owned Companies	**45,550,000**	41,600,000
Receivables and Investments, Related-Party	**39,050,000**	36,880,000
Other Assets	**168,790,000**	171,440,000
Property and Equipment	**348,940,000**	350,480,000
	$1,271,410,000	$1,222,160,000

LIABILITIES and SHAREHOLDERS' EQUITY	20XX	20YY
Current Liabilities:		
Notes payable	**$ 13,530,000**	$ 123,890,000
Accounts payable	**43,740,000**	34,170,000
Income taxes	**22,060,000**	28,610,000
Accrued liabilities	**50,040,000**	48,540,000
Total current liabilities	**129,370,000**	235,210,000
Long-Term Debt	**367,640,000**	372,540,000
Deferred Income Taxes	**45,600,000**	25,020,000
Shareholders' Equity	**728,800,000**	589,390,000
	$1,271,410,000	$1,222,160,000

The accompanying notes are an integral part of the consolidated financial statements.

Figure 32-8. A balance sheet from a large company.

Cost Accounting

cost accounting. A system of charging expenses to specific categories.

profit center. A unit chosen for cost accounting. A profit center may be a single product, a department, an entire plant, or a group of plants.

The third financial record system is called *cost accounting*. It charges each transaction to a product line or plant. Each product or plant is called *profit center*. See **Figure 32-10.**

The income for each product or plant is recorded. Also, each expense item is charged to them and corporate office's expenses are divided among the profit centers.

Cost accounting is a valuable tool. It pinpoints plants and products that are doing well and the ones that need attention.

Cost accounting helps managers to decide which products to discontinue (stop making). Plants that need improvement or closing are also identified.

CONSOLIDATED STATEMENT OF INCOME
for the years ended December 31, 20XX, 20YY and 20ZZ

	20XX	20YY	20ZZ
Net sales	**$1,059,450,000**	$855,740,000	$876,530,000
Cost of sales	**687,850,000**	557,110,000	571,400,000
Gross profit	**371,600,000**	298,630,000	305,130,000
Selling, general and administrative expenses	**183,810,000**	150,440,000	139,910,000
Operating profit	**187,790,000**	148,190,000	165,220,000
Other expense (income), net:			
Interest expense	**40,840,000**	40,210,000	39,870,000
Other income, net	**(29,370,000)**	(27,070,000)	(25,390,000)
	11,470,000	13,140,000	14,480,000
Income before income taxes and extraordinary income	**176,320,000**	135,050,000	150,740,000
Income taxes	**69,760,000**	57,410,000	62,420,000
Income before extraordinary income ($1.50 per share in 1982)	**106,560,000**	77,640,000	88,320,000
Extraordinary income from retirement of debentures	—	14,510,000	—
Net income	**$ 106,560,000**	$ 92,150,000	$ 88,320,000
Earnings per share	**$1.93**	$1.78	$1.73

The accompanying notes are an integral part of the consolidated financial statements.

Figure 32-9. A typical income statement.

Figure 32-10. Cost accounting maintains a separate ledger for each profit center.

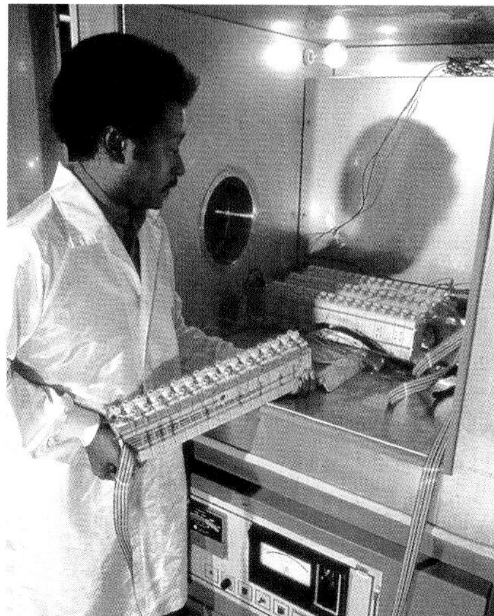

Figure 32-11. Financial records help managers make key decisions about product development, engineering, and production activities. (General Motors Corp.)

Summary

Managers control the company's money through a process called budgeting. In this process, income and expenses are predicted. Each transaction is entered into ledgers. They are usually kept on computer systems.

Each product and plant must make money. Cost accounting keeps track of their progress. Income and expenses for each profit center are recorded. The records are a decision-making tool.

Good financial management helps managers make important decisions. See **Figure 32-11.** It helps them ensure that the company makes a profit. This is, after all, the key to staying in business.

Key Words

All of the following words have been used in this chapter. Do you know their meaning?

accounting
assets
budget
cost accounting
income
liabilities
profit
profit center
shareholders' equity

Test Your Knowledge

Please do not write in this text. Place your answers on a separate sheet of paper.

1. What are three major types of financial records kept by manufacturing companies?
2. List and describe the major types of budgets.
3. All the things a company has are called _____ ; what the company owes is called _____.
4. List the uses for profits.

Applying Your Knowledge

1. Keep a general account of how much you spend (debits) and how much you earn (credits) for one month. Think about whether you are managing your finances well. Create a monthly personal budget based on this analysis.
2. List your assets and liabilities. Using the figures you have, calculate your net worth (assets minus liabilities).
3. Obtain the annual report of a company from someone you know who works for the company or from the library. Identify the company's earnings and expenditures for a given year.

Purchaser

A purchaser may also be called a purchasing manager, buyer, and purchasing agent. They search to obtain the highest quality merchandise at the lowest possible purchase cost. In general, they buy goods and services for their company or organization. They determine which commodities or services are best, choose the suppliers of the product or service, negotiate the lowest price, and award contracts that ensure that the correct amount of the product or service is received at the appropriate time. In order to accomplish these tasks successfully, they study sales records and inventory levels of current stock, identify foreign and domestic suppliers, and keep abreast of changes affecting both the supply of and demand for needed products and materials.

Purchasers evaluate suppliers based upon price, quality, service support, availability, reliability, and selection. To assist them in their search, they review catalogs, industry and company publications, directories, and trade journals. Much of this information is now available on the Internet. They research the reputation and history of the suppliers and may advertise anticipated purchase actions in order to solicit bids. At meetings, trade shows, conferences, and suppliers' plants and distribution centers, they examine products and services, assess a supplier's production and distribution capabilities, and discuss other technical and business considerations that influence the purchasing decision. Once all the necessary information on suppliers is gathered, orders are placed and contracts are awarded to those suppliers who meet the purchasers' needs. Contracts are often for several years and may stipulate the price or a narrow range of prices, allowing purchasers to reorder as necessary. Other specific job duties and responsibilities vary by employer and by the type of commodities or services to be purchased.

Computers continue to have a major effect on the jobs of purchasers. In manufacturing and service industries, computers handle most of the routine tasks, enabling purchasing workers to concentrate mainly on the analytical and qualitative aspects of the job. Computers are used to obtain instant and accurate product and price listings, track inventory levels, process orders, and help determine when to make purchases. Computers also maintain lists of bids and offers, record the history of supplier performance, and issue purchase orders.

Chapter 33 Closing an Enterprise

Objectives

After studying this chapter, you will be able to:

- ✓ Describe the process of dissolving a company.
- ✓ Name the two types of bankruptcy and explain how they differ.
- ✓ List the steps a company must follow in going through dissolution.

A company is a living entity that develops, manufactures, and sells products to meet customer's needs. When successful, the company makes a profit. However, this does not always happen. Sometimes competition or other factors keep the company from being profitable. In rare cases, the owners simply want to stop doing business whether the company is profitable or not.

Each year thousands of companies outlive their usefulness. See **Figure 33-1.** This leads to closing the company, which is called *dissolution*. During this process, the company that was built is taken apart. People that were hired leave. Plant, equipment, materials, and finished goods are sold. Patents and product plans are sold. The company that was once a sum of its parts is now broken into pieces.

Figure 33-1. For a variety of reasons, thousands of companies close their doors each year.

There are two types of dissolution. These, as shown in **Figure 33-2,** are:

- Voluntary dissolution
- Involuntary dissolution

dissolution. The process of closing down a business.

Voluntary Dissolution

voluntary dissolution. When the owners of a business close the company because they want to do so.

The owners may close the company because they want to do so. They may enter into *voluntary dissolution*. They file a form to dissolve the corporation. It is called a "Certificate of Dissolution" or "Articles of Dissolution."

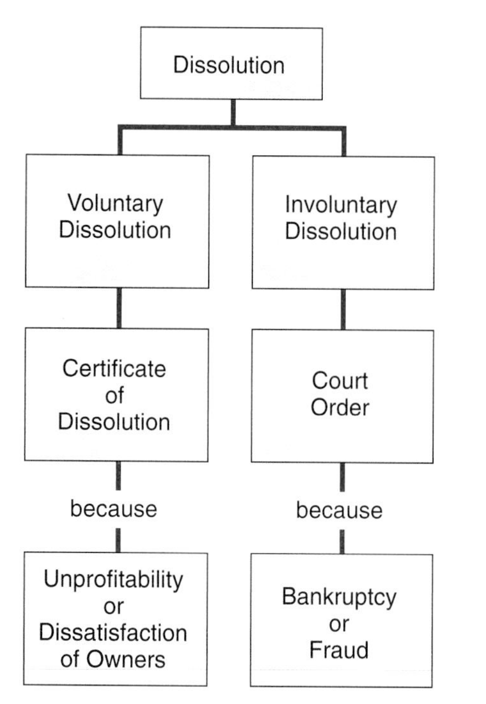

Figure 33-2. Types of company dissolution actions.

Voluntary dissolution may be caused by unprofitability. The company may be losing money and the owners close the company before it fails. Dissatisfaction with the company is another reason. The owners may feel they have better uses for their money. They can close the business and sell the assets so their money can be invested elsewhere.

Involuntary Dissolution

Some companies are forced to dissolve by a court order. The assets are sold by court appointed agents to raise money. Three main reasons that may cause *involuntary dissolution* are:

- Inability to pay debts (called *bankruptcy*)
- Dishonest financial activity (fraud and other criminal acts)
- Loss of a charter (state refuses to renew the charter, usually because of illegal activity)

involuntary dissolution. When a company is forced to dissolve. The assets are sold to raise money. Three reasons may cause involuntary dissolution: bankruptcy, dishonest financial activity, and loss of state charter.

bankruptcy. A state in which a company cannot pay its debts; the company is dissolved or reorganized.

creditor. Individuals, companies, and financial institutions to whom the company owes money.

Steps in Dissolution

Dissolving a company is an orderly activity that includes four major tasks. These are:

1. Filing the proper legal documents.
2. Closing operations.
3. Selling the corporate assets.
4. Distributing the company's money.

Filing Legal Documents

Two legal documents can start dissolution. The company officers can file a Certificate of Dissolution that has been approved by the stockholders. It tells everyone that the owners are closing the company and the corporation is starting voluntary dissolution.

The company may receive a court order to close the company. This action is usually started by a lawsuit. *Creditors* (people who are owed money) demand payment that the company cannot meet. The corporation files for bankruptcy and starts involuntary dissolution.

Another type of bankruptcy does not dissolve the company but allows the company to *reorganize*. This action is called Chapter 11 Bankruptcy. In this process, the company's debt is restructured (changed). Operations are improved and the company is reborn.

Closing Operations

During dissolution the company closes its operations. All sales activities stop and deliveries of products cease. Employees lose their jobs. The busy corporation becomes an empty shell, see **Figure 33-3.**

Figure 33-3. A busy plant becomes an empty place during dissolution.

Selling Assets

The company still owns property, which is called the assets. Most companies have four types of assets. These, as shown in **Figure 33-4,** are:

- Plant and equipment
- Material and work-in-process
- Finished goods
- Knowledge (this includes product plans, process information, etc.)

These assets must be sold to convert them into cash. Some items may be sold in private sales. Another company may buy the equipment. The real estate (buildings and land) may be sold separately. Materials may be sold back to the suppliers.

Often, assets are sold at a public sale. Advertisements, such as those shown in **Figure 33-5,** are placed in newspapers to announce the auction. Buyers come to bid on the assets.

Figure 33-4. When a company dissolves, it sells its assets. Common types of assets are shown. A—Plant and equipment. B—Materials and work-in-process. C—Finished products. D—Information.

Figure 33-5. These advertisements for auctions of the assets of companies were placed in newspapers.

Distributing Cash

The final step in dissolution is paying debts. Often there is not enough cash to meet all debts. The debts are then paid in a specific order that is set by law. See **Figure 33-6.** Taxes are paid first. If money is left, then legal costs are paid next. After taxes and legal fees are paid, the employees receive payment for wages and salaries. Any remaining money must pay creditors such as suppliers and utilities. If there is still money left, people holding loans and bonds receive their payment. Finally, the owners (stockholders) receive any money that is left. In involuntary dissolution, the owners seldom receive any money.

Figure 33-6. When a company closes and there is not enough money to pay everyone, the debts are paid off in a certain order.

Summary

The stockholders are the greatest risk takers. They bought into the company to make money but they can also lose their investment (money). This is the basic element of capitalism—our form of business ownership. People invest their money to provide the capital needed to buy the plant and equipment. They also pay early development costs for the product. The investors are "betting on a winner." If they are right, they make money and receive dividends. Also, the stock becomes more valuable.

Likewise, they can lose money. A wrong decision can lead to bankruptcy where the company is dissolved. The assets are sold to raise cash and as many bills as possible are paid. The company dies.

Key Words

All of the following words have been used in this chapter. Do you know their meaning?

bankruptcy
creditor
dissolution
involuntary dissolution
voluntary dissolution

Test Your Knowledge

Please do not write in this text. Place your answers on a separate sheet of paper.

1. What are the two types of dissolution?
2. Give the causes for each type of dissolution.
3. List the steps in dissolving a company.
4. One of the legal documents that starts dissolution is called a _____ of Dissolution.
5. Companies going into bankruptcy have four types of assets that are sold. Indicate which of the following are assets:
   a. Goodwill.
   b. Orders from customers.
   c. Workforce.
   d. Plant and equipment.
   e. Material and work-in-process.
   f. Finished goods.
   g. Knowledge.
   h. All of the above.
   i. None of the above.

Applying Your Knowledge

1. Obtain an announcement for an auction sale related to closing a business. Determine who caused the closing and if it was voluntary or involuntary. (Note: most ads of this type will tell who ordered the sale to be held.)
2. Go to the local library and look for music cassettes or CDs whose theme is the dissolution or dividing of businesses (farms, factories, corporations) and the effect on the people who used to work there. Listen to the music and write about what you heard.
3. Invite a stockbroker to speak to your class about investments and risk taking. Prepare questions for the speaker beforehand.

Unit 11
Automating
Manufacturing
Systems

Chapter 34 Automation in Manufacturing

Objectives

After studying this chapter, you will be able to:

✓ Give a brief history of automation.
✓ Name a device that has revolutionized factory automation.
✓ List five reasons for employing automation.
✓ Name at least three components of CIM.
✓ Explain what CAD and CAM are.

Throughout the history of manufacturing, people have sought more efficient ways to make products. These efforts can be divided into two tasks:

- Designing products for efficient manufacture
- Making products more efficiently

Although divided, these tasks are not independent. Often the work of one directly impacts the other. For example, Eli Whitney's work in producing muskets with interchangeable parts required design changes. Musket parts were designed so that all like parts were exactly alike so they could be easily manufactured. The parts were made so that any like part would fit any musket. This enabled the muskets to be assembled more efficiently. Thus, the first task impacted the second.

This interdependence must be addressed if a company is to compete in the world market. See **Figure 34-1.** Key to this focus is automatic or automated manufacture, which is often called automation. *Automation* is the application of control to an apparatus, a process, or a system by means of mechanical or electronic devices. Automation uses computer and automatic controls to replace direct human control.

automation. Replacement of human control for a machine, process, or system, with control by mechanical or electronic devices.

Development of Automated Manufacturing

It is difficult to trace automated manufacture to any specific event or invention. Ideas for automatic manufacturing machines have many roots.

Efficient Design Efficient Manufacture World-Class Product

Figure 34-1. Manufacturers must develop efficient designs and manufacturing processes to produce world-class products. (American Petroleum Institute, Reynolds Metals Co., Deere and Co.)

Figure 34-2. This is the Cutty Sark, an early sailing ship that brought tea from India.

Early efforts to accomplish automation used strictly mechanical controls. One of the earliest attempts in factory automation came around 1800. At that time there was a growing need for blocks (pulleys) to raise and lower sails, move heavy cargo, and position cannons on ships. See **Figure 34-2.** A typical warship of that age required over 1400 blocks. At that time, each block was individually made by artisans working on lathes, drills, and other machines. In England, Henry Maudslay undertook the task to produce blocks using an automated production line. After six years of hard work, the world saw its first large-scale, *mass production,* manufacturing line. However, mass production did not extend to other industries for another 50 years. Furthermore, its adoption was not in England, but in America. Such a movement, marked by revolutionary changes in production, is called an industrial revolution. This industrial revolution ushered the Western world into what became known as the industrial age.

Even earlier roots of factory automation trace back to mechanized toys. They were developed to entertain members of high society. One such toy, a wooden model of a pigeon designed by Archytas of Tarentum, dates back to about 350 B.C.

Soon after 350 B.C., Hero of Alexandria began to experiment with pneumatics and hydraulics. He developed a number of water-driven mechanisms including automatic temple doors and singing mechanical birds. His work was applied to many mechanical toys, music boxes, and musical clocks developed in Europe during the eighteenth century. These devices had mechanisms that used *disks*. The disks had ridges or holes that encoded a *program* (detailed operating instructions) that guided the apparatus. The program code held on the disk caused the item to chime or move according to a set plan.

The idea of programmed cards or disks was adapted in 1804 by the French loom maker, J. M. Jacquard. His looms used a series of punched cards to program the machine to automatically produce a set pattern in fabrics and carpets. In 1822, Babbage proposed building a machine called the Difference Engine to automatically calculate mathematical tables. The Difference Engine was only partially completed when Babbage conceived the idea of another, more sophisticated machine called an Analytical Engine. Later, Herman Hollerith applied this principle to an automatic data-recording and accounting machine using punched cards. This invention was first used for the 1890 United States census. It reduced the time needed to tabulate the results of the census by 50%. His punched cards became the forerunner of the computer card, which was used to enter data in early computers.

In the twentieth century, the idea of a *punched-tape* program was tried on manufacturing machines. These machines, called *numerical control (NC) machines,* became the backbone for the second industrial revolution—the automation age.

The arrival of modern factory automation came with the recent development of the computer. This useful machine is the dominant feature of the computer revolution, which impacts our lives every day. Its advent brought us out of the automation age and moved us into the information age. See **Figure 34-3.**

Figure 34-3. Computers can be used to control machines and processes. They can also process written, graphic, and numerical data. (American Iron and Steel Institute, PPG Industries)

Academic Link

Computer science is the study of the design of computers and their processes. It is a young discipline that is evolving rapidly from its beginnings. In its most general form, computer science is concerned with the understanding of information transfer and transformation. Particular interest is placed on making processes efficient and providing them with some form of intelligence.

Computer science includes theoretical studies, experimental methods, and engineering design all in one discipline. This differs from other physical sciences that separate the understanding and advancement of the science from the applications of the science in fields of engineering design and implementation. In computer science there is a natural intermingling of the theoretical concepts of computability and mathematical efficiency with the modern practical advancements in electronics, which continue to stimulate advances in the discipline. It is this close interaction of the theoretical and design aspects of the field that binds them together into a single discipline and makes computer science a rewarding area of study.

Figure 34-4. This is an example of an early computer. (John W. Mauchly Papers; Rare Book and Manuscript Library, University of Pennsylvania)

This period in history began in 1945 with the development of the first electronic computer. It was called the ENIAC (Electrical Numerical Integrator and Calculator). See **Figure 34-4.** From this dramatic start, came the mainframe computer, the minicomputer, and the microcomputer that operate today in almost every facet of the public and private community.

Components of an Automated System

All efforts to automate manufacturing are guided by a few basic goals. These include:

- Reducing manufacturing costs
- Increasing product quality and consistency
- Enhancing manufacturing flexibility
- Improving marketability of products
- Reducing the reaction time to changes in market demands

These goals have caused most manufacturing companies to adopt labor- and material-saving practices, programs, and systems. A number of these are

- * Artificial Intelligence (AI)
- * Computer-Aided Design (CAD) and Computer-Aided Manufacturing (CAM)
- * Flexible Manufacturing System (FMS)
- * Process Automation
- * Just-In Time (JIT)
- * Manufacturing Resource Planning II (MRP II)
- * Materials Requirement Planning (MRP)
- * Basic Machine Controls

Figure 34-5. Computer use in manufacturing extends from relatively simple to highly complex tasks. (AT&T Co.)

listed in **Figure 34-5.** The complexity of the systems increases as they move upward on the chart. The simplest are mechanical, numerical, and computer machine controls. These range from simple tracer lathes to complex, computer-controlled machining centers. The most complex is *artificial intelligence* in which computers are programmed to process data and make decisions. This is a radical departure from the normal computer tasks of data processing.

All of the various components presented in **Figure 34-5** are used to make up a package called *computer-integrated manufacture (CIM)*. CIM is defined by the National Research Council as all activities from the recognition of a need for a product; through the conception, design, and development of the product; and on through production, marketing, and support of the product in use.

computer-integrated manufacture (CIM). An approach in which all steps in producing a product, from recognition of need for it through manufacture and marketing, are integrated into a single, dynamic computer-controlled system.

All of these activities use written, numeric, or graphic data. This data is integrated into a working system by computers. The result is diverse operations integrated into a single, dynamic system.

CIM is an ultimate goal of many manufacturing enterprises. It is only beginning to be implemented into manufacturing companies today. The two main components of CIM are computer-aided design (CAD) and computer-aided manufacturing (CAM). See **Figure 34-6.**

computer-aided design (CAD). A computer-based system used to create, modify, and communicate a plan or product design.

CAD is a computer-based system used for creating, modifying, and communicating a plan or product design. Activities performed by CAD, as shown in **Figure 34-7,** include:

- **Engineering design.** This is the preparation and documentation of various product design ideas and solutions.
- **Design analysis.** This is the evaluation of product designs using computer-simulated tests, such as for stress, heat transfer, and deflection.
- **Design presentation.** This is the preparation of engineering drawings that are used to communicate approved designs to manufacturing personnel.

Some references say CAD stands for computer-aided drafting. Some say that CAD is strictly a drafting tool. They refer to a system that also has design and analyzing capabilities as computer-aided drafting and design (CADD).

CAD
Engineering Design Design Analysis Design Presentation (Engineering Drawings)
CAM
Computer Process Control Computer Numerical Control Robotic Control Adaptive Control Manufacturing Support Production Scheduling Inventory Control Material Requirement Planning Numerical Control Tapes

Figure 34-6. CAD and CAM are important parts of computer-integrated manufacture.

Engineering Design

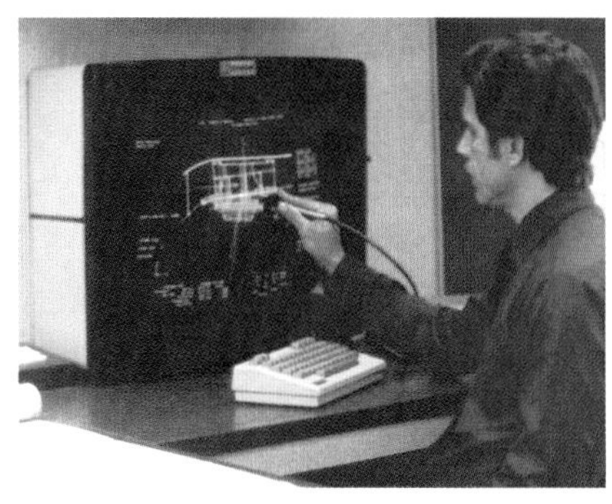

Design Analysis

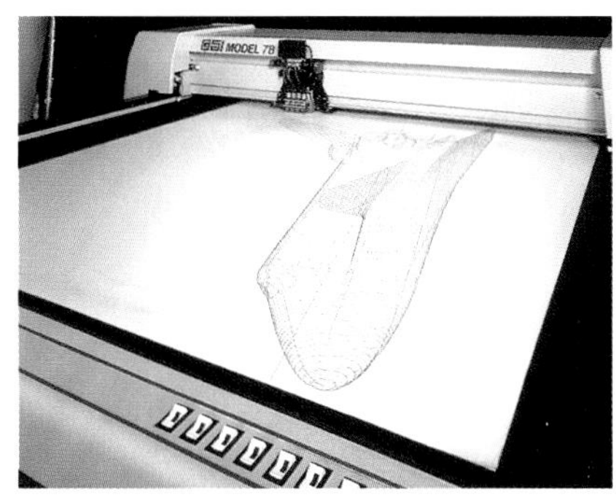

Design Presentation

Figure 34-7. Computer-aided design is used for engineering design, design analysis, and design presentation activities. (International Business Machines Corp., Ford Motor Co., Gerber-Scientific Instrument Co.)

More recent practice accepts CAD as standing for all drafting, design, and analyzing functions.

CAM is the use of computer technology in the management and control of manufacturing operations. CAM operation is concerned with one or both of the following:

- **Computer process control.** This is the process of using mechanical and electronic means (NC machines and robots, for example) to change the shape, size, or composition of materials.
- **Manufacturing support.** This is the processing of data including such things as machine status, part counts, and processing variables.

CAM uses a computer to assist in monitoring and controlling either or both components of manufacturing operations. See **Figure 34-8.** Computers may be used to direct machines through a series of steps to process materials. Also, computers may gather and process data about the effectiveness of an operation.

Much more may be said about CAD and CAM operations. They will be the focus of the next two chapters of this book.

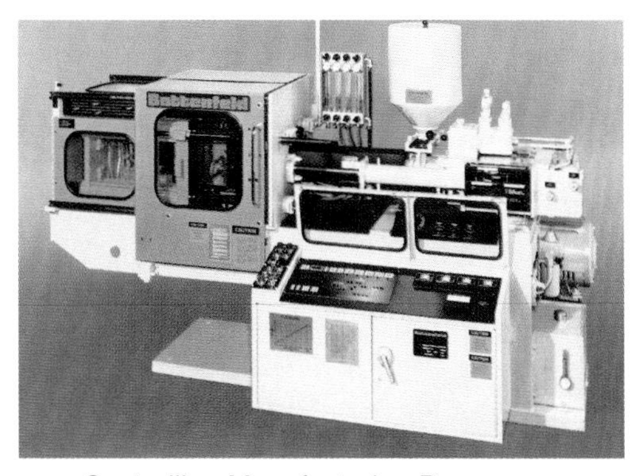

Controlling Manufacturing Processes

Providing Manufacturing Support

Figure 34-8. Computer-aided manufacturing involves controlling operation and processing manufacturing data. (Battenfeld Inc., AT&T Co.)

computer-aided manufacture (CAM). **A system that uses computer technology in the management and control of manufacturing operations.**

Summary

Manufacturing companies are rapidly automating their operations in an effort to remain competitive in the world market. Computers are being used to make product design and manufacturing more efficient.

Two major systems are widely used in industry today. They are CAD and CAM. Unified into a total system, they are a part of CIM.

Key Words

All of the following words have been used in this chapter. Do you know their meanings?

automation
computer-aided design (CAD)
computer-aided manufacture (CAM)
computer-integrated manufacture (CIM)

Test Your Knowledge

Please do not write in this text. Place your answers on a separate sheet.

1. *True or False?* Automation evolved as a result of a desire for more efficient ways to manufacture products.
2. Name three roots of automation.
3. What modern device revolutionized factory automation?
4. *True or False?* One reason manufacturers choose to automate is to increase product quality and consistency.
5. List three components of CIM.
6. Which of the following tasks do CAD systems help designers and engineers complete?
   a. Engineering design.
   b. Design analysis.
   c. Design presentation.
   d. All of the above.
7. What are the two major functions of a CAM system?

Applying Your Knowledge

1. Prepare a bulletin board that shows the major tasks of CAD and CAM systems.
2. Visit a local manufacturing company to observe CAD and/or CAM systems in operation. Write a brief report about what you observed.

Assembler and Fabricator

Assemblers and fabricators produce a wide range of finished goods from manufactured parts or subassemblies. They produce intricate manufactured products, such as aircraft, automobile engines, computers, and electrical and electronic components.

Assemblers may work on subassemblies or the final assembly of an array of finished products or components. For example, electrical and electronic equipment assemblers put together or modify missile control systems, radio or test equipment, computers, machine-tool numerical controls, radar, or sonar, and prototypes of these and other products. Electromechanical equipment assemblers prepare and test equipment or devices such as appliances, dynamometers, or ejection-seat mechanisms. Coil winders, tapers, and finishers wind wire coil used in resistors, transformers, generators, and electric motors. Engine and other machine assemblers construct, assemble, or rebuild engines and turbines, and office, agricultural, construction, oilfield, rolling mill, textile, woodworking, paper, and food wrapping machinery. Aircraft structure, surfaces, rigging, and systems assemblers put together and install parts of airplanes, space vehicles, or missiles, such as wings or landing gear. Structural metal fabricators and fitters align and fit structural metal parts according to detailed specifications prior to welding or riveting.

Assemblers and fabricators involved in product development read and interpret engineering specifications from text, drawings, and computer-aided drafting systems. They also may use a variety of tools and precision measuring instruments. Some experienced assemblers work with engineers and technicians, assembling prototypes or test products.

Chapter 35
Computers and Product Design

Objectives

After studying this chapter, you will be able to:

✓ List five types of computer hardware.
✓ Describe what software is.
✓ Name various applications of CAD.
✓ Explain the benefits of CAD.
✓ Describe the functions of engineering design and analysis.

Traditionally, products have been designed by people working on drawing boards. See **Figure 35-1.** They use items like T-squares, triangles, and engineer's scales to develop a complete set of drawings for a product. In the realm of manufacturing, this set might include:

- Detailed engineering drawings showing the size and shape of the various parts.
- Assembly drawings showing how various parts fit together.
- Systems drawings showing electrical, mechanical, pneumatic, and hydraulic systems of the product.
- Tool and fixture drawings for special purpose devices used in manufacturing the product.

Figure 35-1. Until recently, product drawings were prepared by drafters working at drawing tables.

Today, traditional board drafting is becoming obsolete. Many companies are turning to Computer-Aided Design (CAD) systems to improve the accuracy of their drawings. See **Figure 35-2.** Also, they are finding that CAD reduces the time it takes to prepare and/or modify drawings. In addition, they are using CAD because it allows the user to test and analyze a product design at the computer terminal.

Figure 35-2. CAD systems allow drafters and engineers to prepare drawings more quickly and accurately.

hardware. The physical components of a computer system.

computer. A device that includes a central processing unit, input/output unit, and a memory.

Components of a CAD System

A CAD system is made up of several components. Some of these are shown in **Figure 35-3.** Components of CAD consist of computer hardware and computer software. *Hardware* is the physical equipment of the computer system. It is comprised of:

- *Computer.* This is an electronic machine with internal parts that include a central processing unit, an input/output unit, and a memory. The *central processing unit (CPU)* controls and sequences the operations in the computer. The *memory* is used by the CPU for storing data. The *input/output unit* communicates with the peripheral devices (*external* input and output devices).
- *Video display terminal.* This is a peripheral device. It is referred to as a monitor. It is sometimes called a VDT or a CRT. It is a video screen onto which the computer displays data including graphics (drawings, graphs, charts, etc.) for the operator's use.
- *Input devices.* These, too, are peripheral devices. They are used to enter data and directions into the computer. Input devices of a CAD system commonly include a keyboard, a digitizing tablet, and a mouse or other tracking device.

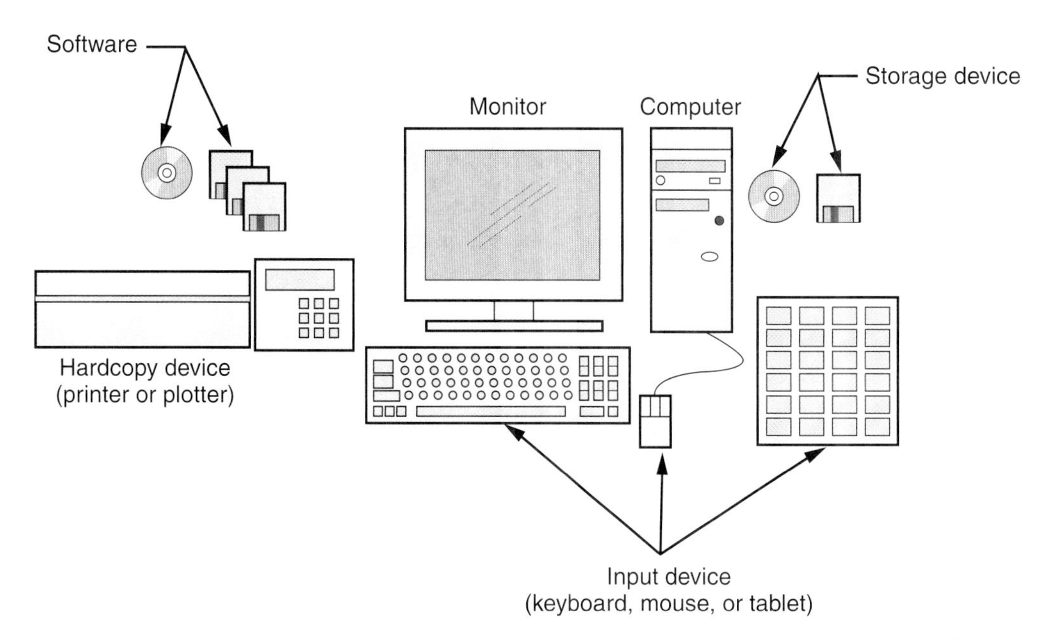

Figure 35-3. A variety of parts makes up a CAD system. Note that most drawings in this book were produced on a CAD system.

- *Hard copy devices*. These are also peripheral devices. They may be plotters, film recorders, or printers. They are devices that are used to obtain hard copies (paper printouts) of the information processed by the computer. The output can be text, numbers, and/or graphics. Hard copy devices are output devices (as are video display terminals).
- *External storage devices*. These are CDs, disks, or tapes used for storing data.

Software contains the set of instructions used by the computer. It directs general computer tasks like data processing and system operation. In a CAD system, it allows the computer to handle graphic data. It also allows drafting applications to be performed. Now let us take a look at applications of CAD.

Applications of CAD

CAD software allows for different applications of the system. Generally, these applications can be divided into three groups: mechanical, electrical/electronic, and architectural engineering and construction. See **Figure 35-4.**

Mechanical applications of CAD involve the design and layout of parts and products. They are represented by 2- and 2 1/2-dimensional drawings and 3-dimensional models. A *2-dimensional drawing* is used to define the shape of an object using one or more profiles. A *2 1/2-dimensional drawing* is used when the side of an object does not need to be profiled. A *3-dimensional model* is used to give a pictorial view of an object. See **Figure 34-5.** A model aids evaluation of the size, shape, and performance of a product. Models are essential when the part is to be manufactured by multiaxis, computer-controlled machines.

Electrical/electronic applications of CAD involve the design and layout of components and wiring in electrical and electronic circuits. The circuit designer is concerned with the location of components and the interconnections between them.

Architectural engineering and construction applications of CAD involve the design and layout of buildings and their related systems and/or surroundings. In this application, architectural, mechanical, and electrical drawings are produced. These are in the form of plans, elevations, and section views. Plans

central processing unit (CPU). An integrated circuit that controls and sequences operations of a computer.

memory. Component inside a computer used by the CPU to store data.

input/output unit. Component of the computer that communicates with external input and output devices.

video display terminal. A television-like screen upon which the computer displays information.

input devices. External components, such as a keyboard or a digitizing tablet, that are used to enter data and directions into a computer.

hard copy devices. Equipment that can be connected to a computer and used to obtain a paper or film copy of information processed by the computer.

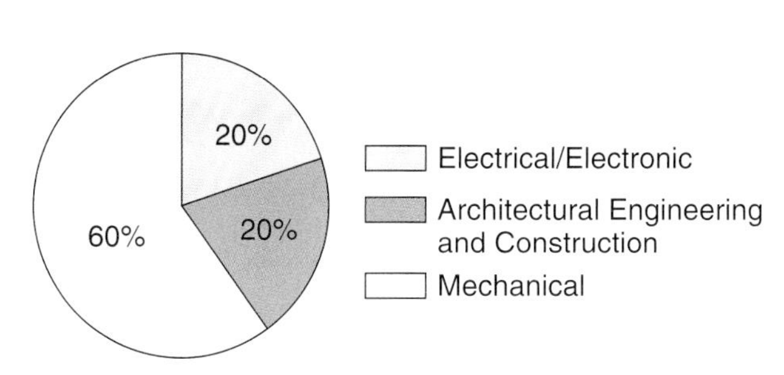

Figure 35-4. This chart shows the percentage of the total for each of the three major applications of CAD.

Figure 35-5. CAD systems can show parts in three dimensions. (Thomas Short, Anthony Dudek)

external storage devices. **Disks or tapes used to store data outside a computer.**

software. **The set of instructions used by a computer. Different software is used for various tasks, such as word processing, filing, operation of machines, or planning.**

show overall layout; elevations show side views; section views show cross sections.

CAD serves other uses as well. It is used in applications of civil drafting of land terrain, road systems, and utilities. In the applications mentioned, it is often used, not only for design and layout, but also for such things as preparing bills of materials and generating production schedules and cost estimates.

Benefits and Functions of CAD

CAD is used for a variety of beneficial reasons. It is quicker than traditional drafting. Standard symbols on file can be used, and repetitive objects can be quickly reproduced. It produces better quality drawings. It facilitates drawing modification. It allows drawing manipulation and product analysis. For example, a three-dimensional model can be changed into a three-view, orthographic drawing in seconds. This task could take hours or days using traditional drafting methods.

Three basic functions of CAD were briefly introduced in Chapter 34. These were:

* Engineering design
* Design analysis
* Design presentation

These functions, **Figure 35-6,** are important and interrelated. All contribute to an effective and efficient product design.

Engineering Design

Engineering design involves developing drawings from product sketches. See **Figure 35-7.** Traditionally, products were designed with little attention to their manufacture. Appearance was considered the primary challenge for

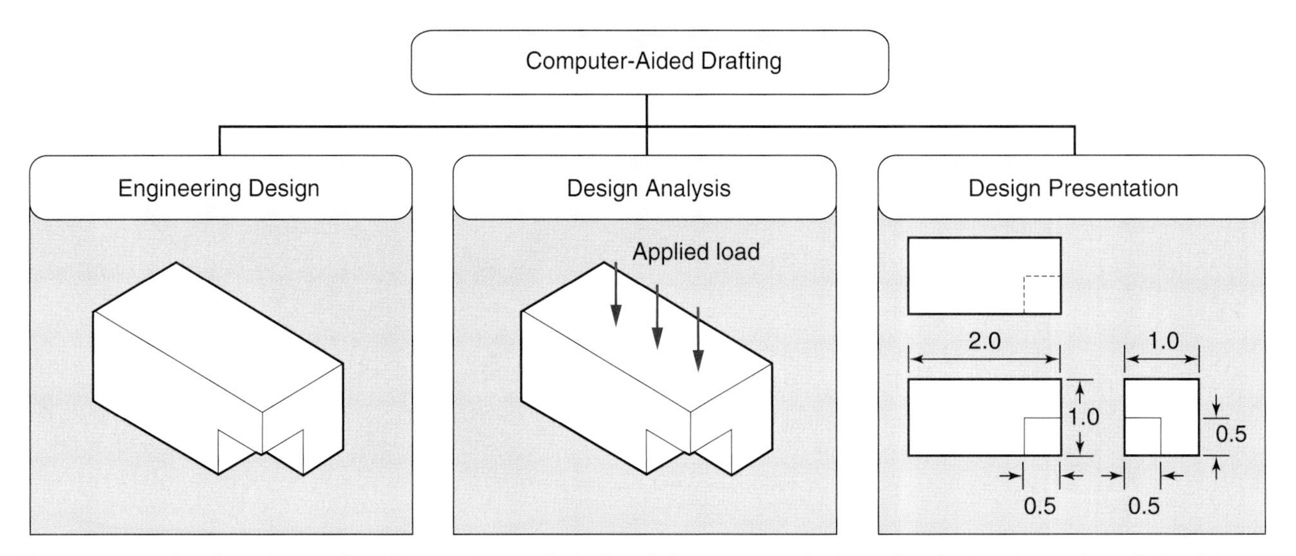

Figure 35-6. The functions of CAD are pictured. A sketch becomes a design; the design is analyzed (in this case, for reaction to external forces); a multiview is presented showing shape and dimensions of the design.

designers. However, this has changed because of today's competitive environment. A new approach called *design for manufacture (DFM), design for assembly (DFA),* or *design for manufacture and assembly (DFMA)* is being used.

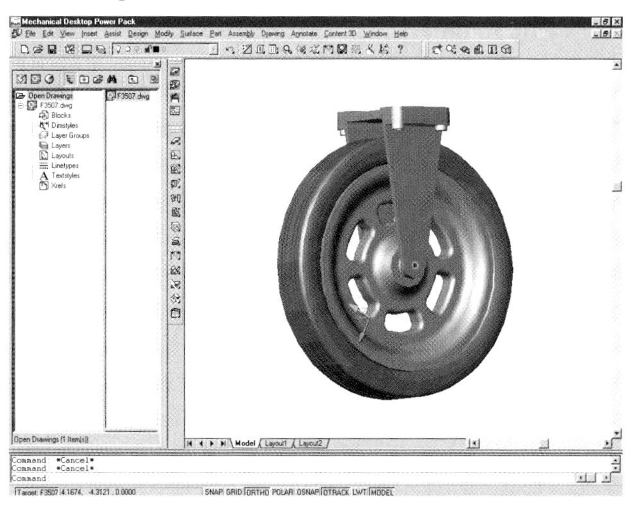

Figure 35-7. CAD systems make product design and engineering easier. (Thomas Short, Anthony Dudek)

This new approach strives to design an appealing, functional product that can be easily manufactured and assembled. The results are dramatic. One company redesigned the sight used for the guns it made. It reduced the part manufacture time by 71%, the number of parts by 75%, and the sight assembly time by 85%. Another company used DFA to reduce the assembly time of a printer it manufactured from 30 minutes to 3 minutes. Still another company redesigned its cash register so that it could be assembled in less than 2 minutes. Such accomplishments translate to cheaper, more competitive products for the customer and more profit for the company.

DFM includes carefully defining the appearance and the shape of a product. All parts are designed to be as simple as possible. Unnecessary parts are eliminated, and, wherever possible, small parts are combined with larger parts.

The design effort refines the styling of a product. Product shapes and lines are simplified. Curves that are hard to manufacture are removed wherever possible. Then, drawings for each part and a finished assembly are produced.

Developing CAD drawings

CAD makes these tasks much simpler. Generally, designers working with CAD start by producing a wireframe model of the object being designed. The wireframe model is used to produce a surface and solid models. See **Figure 35-8.**

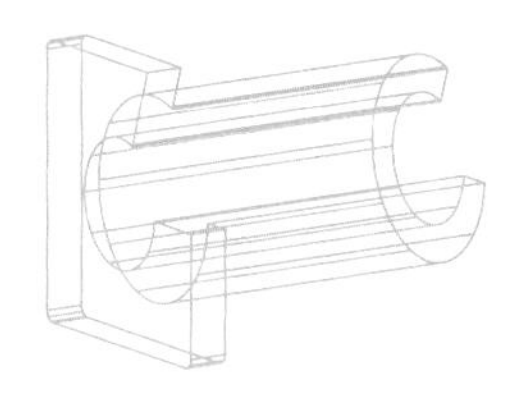

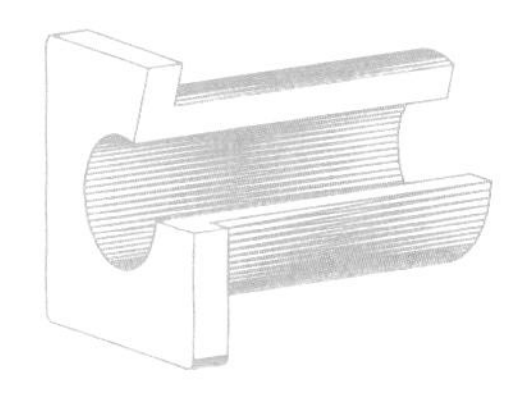

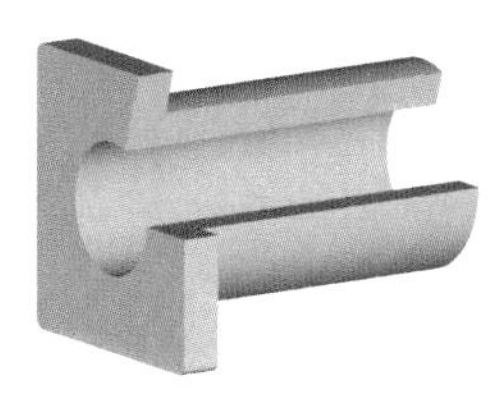

Figure 35-8. These different models were produced with a CAD system.

A *wireframe model* is a series of lines, circles, and arcs that together form an outline of an object. This type of model is a series of overlapping lines. A wireframe model is not the best way to see an object. The model fails to adequately show the relationship between surfaces and edges. The lines of the wireframe often produce a confusing outline.

A CAD system can quickly change the wireframe model into a surface or a solid model. A *surface model* is a representation whereby an object is defined only by its surface. It is created by putting a "cover" around a wireframe model. An empty can could be properly defined by a *wireframe model.* A series of lines, arcs, and circles that together form the outline of an object.

surface model. A computer model in which an object is defined only by its surface.

solid model. A computer representation that takes into account both the surface and the interior substance of an object.

perspective model. A lifelike view of an object that is shown as the eye would see it naturally.

surface model. Such a model could represent a solid object such as a brick, but to the computer, it would be hollow. A surface model better shows how an object will look when manufacture of it is complete.

A *solid model* is a true representation of an object whereby its surface *and* its interior substance is defined. These are sometimes referred to as *rendered models.* A computer treats a solid model as though it were a truly solid object. Solid models enable cross sections of a product to be viewed. They also are helpful for product and design analysis. Many CAD systems allow different colors and engineering materials to be assigned to each part of the product.

The surface and solid models are very important in making styling decisions. They allow the designer to "see" the object before it is built. In fact, most CAD systems are able to rotate an object through a series of views. See **Figure 35-9.** This feature lets designers view the product at any angle. It allows them to decide if the styling is pleasing from all vantage points.

A lifelike view can be produced by having the CAD system change the solid model into a *perspective model.* This type is like a surface model shown in perspective as the human eye would see it. See **Figure 35-10.**

Finally, as mentioned before, a CAD system can take a model of an object and show it in three views. These are the traditional views used in orthographic projection—top, front, and right side. See **Figure 35-11.**

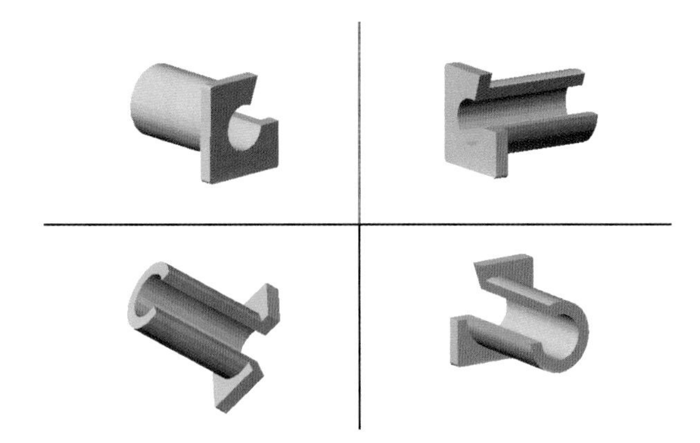

Figure 35-9. These drawings show a part rotated through a series of views.

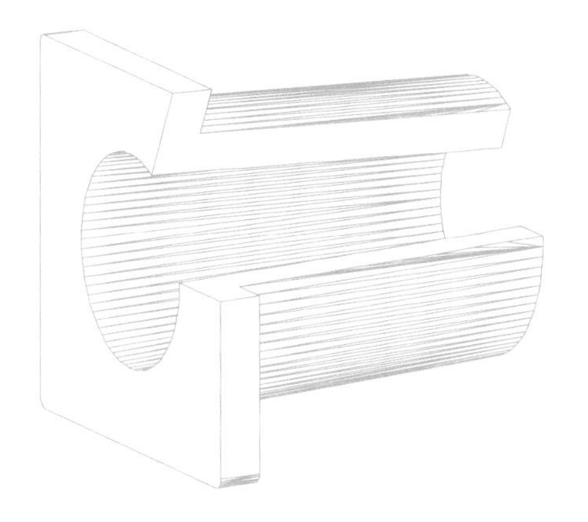

Figure 35-10. The perspective presentation shows the surface model as the human eye would see it.

Design Analysis

Almost every new product must be subjected to design analysis. See **Figure 35-12.** This review determines if the product will function as intended. Typically, the analysis may evaluate the performance of the product under normal and extreme operating conditions. This would involve evaluating the physical properties of the product. Its ability to withstand stress from stationary or moving forces, for example, might be analyzed. As previously mentioned, many CAD systems allow the user to apply engineering materials to the product. They then can simulate the operating conditions and evaluate the product design. Various loads may be applied on the design and simulate operating demands.

Also, advanced CAD systems can simulate manufacturing operations. From this, numerical control codes can be developed to control machines.

Design Presentation

Engineering drawings document a designer's product ideas. The drawings become a vital tool for many company activities. Their use extends far beyond the realm of research and development. Tooling designers, industrial and manufacturing

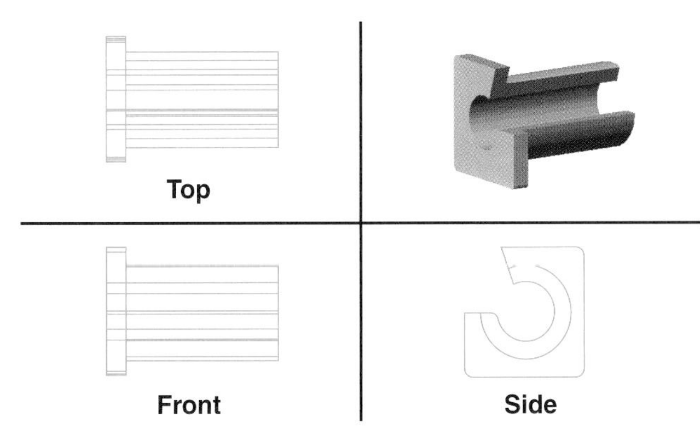

Figure 35-11. These drawings show a multiview of the product shown in the upper-right corner.

engineers, production planners, package designers, machine setup people and operators, and production supervisors are but a sample of the people that need accurate drawings.

Many projections suggest that drawings on paper will become obsolete. It is expected that data will move from the CAD system directly to manufacturing machines and inspection stations. The product specifications will be stored in computer memories. The information will be readily available to anyone in the company who has access to a computer terminal. The costly activities of printing, distributing, and storing drawings will be eliminated. Also, the need to retrieve and destroy out-of-date drawings will be removed.

However, such a time is well off in the future for all but the most advanced companies. Most manufacturing enterprises still need typical engineering drawings. See **Figure 35-13.** These are quickly produced by CAD systems. The information in the computer's memory is used by plotters, film recorders, and printers. They produce up-to-date, accurate drawings and pictures.

Figure 35-12. These engineers are using a CAD system to conduct a design analysis of a product.

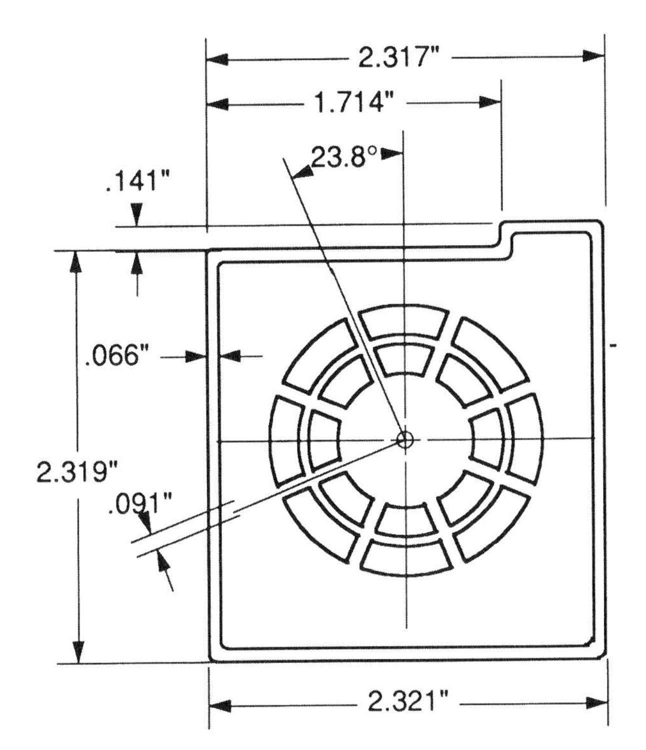

Figure 35-13. This is a simple drawing produced on a CAD system.

Summary

CAD is an increasingly important system to most contemporary manufacturing enterprises. It allows a company to speed up its engineering design, design analysis, and design presentation activities. The systems promise to become part of a larger, more effective, computer-integrated manufacturing system.

Key Words

All of the following words have been used in this chapter. Do you know their meanings?

central processing unit (CPU)
computer
external storage devices
hard copy devices
hardware
input devices
input/output unit
memory
perspective model
software
solid model
surface model
video display terminal
wireframe model

Test Your Knowledge

Please do not write in this text. Place your answers on a separate sheet.

1. Study Figure 35-2. Which components of a CAD system do you see in the photo?
2. List and describe the three major applications of CAD in industry.
3. Describe the concept of DFM.
4. As listed in the chapter, the three functions served by a CAD system are _____, _____, and _____.
5. Which of the following is a type of model used by CAD for product design?
   a. Wireframe.
   b. Solid.
   c. Perspective.
   d. All of the above.
6. *True or False?* CAD is limited in that it is unable to simulate operating conditions for product evaluation.

Applying Your Knowledge

1. Use a simple CAD or graphic software package to make a drawing of the part shown in **AYK 35-1.**
2. Select and analyze a product that you use often. On a form like that in **AYK 35-2,** list ways you could improve its DFM and DFA.
3. Visit a company to see how they use CAD in their operations.
   a. Prior to your visit, research CAD operation in the school library or resource center.
   b. Prior to your visit, prepare a list of questions to ask about the company's CAD operation.
   c. After your visit, list ways you could use a CAD system in this class.

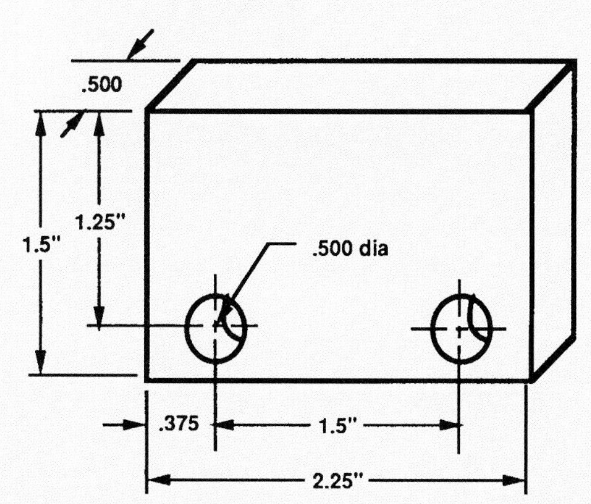

AYK 35-1. This part is easily drawn on a CAD system.

PRODUCT: ____________________

SUGGESTED IMPROVEMENTS FOR:	
MANUFACTURE	**ASSEMBLY**
1.	1.
2.	2.
3.	3.
4.	4.
5.	5.

AYK 35-2. A suggested form is shown.

Chapter 36
Computer Systems in Product Manufacturing

Objectives

After studying this chapter, you will be able to:

✓ Name and describe three kinds of automated machines and systems.
✓ Give examples of classes of NC machines.
✓ Explain the importance of NC programming coordinates.
✓ List and describe four systems that have advanced NC.
✓ Cite applications where robots are useful.
✓ Explain what end effectors are.
✓ Name and describe four management information systems.

Recent history has seen companies dramatically increase their use of computers to help improve manufacturing efficiencies. You have been studying manufacturing throughout this book. You have probably become aware that manufacturing is not a simple process. It involves two major and often complex activities. These, as shown in **Figure 36-1,** are:

- **Material processing.** This is the physical act of changing the form of material to make it more valuable to customers. This involves making industrial materials out of natural materials and finished products out of industrial materials.
- **Management.** This is the action taken to plan, organize, actuate, and control an organization's activities. Management is apparent in research and development, production, marketing, employee relations, and financial affairs.

Computers are used in all of these areas to help promote efficiency. Specifically, computer-aided manufacture shows great promise in helping to improve a company's ability to make products that are competitive in a world market. It addresses areas of manufacturing that move a concept from engineering drawings to finished product, as shown in **Figure 36-2.** CAM may be integrated with computer-aided design to produce a truly computer-supported manufacturing system. This integration, as discussed in Chapter 34, is computer-integrated manufacture.

MANUFACTURING

=

Figure 36-1. Manufacturing involvews both management and material processing activities.

There are a number of ways to view the elements of CAM. For this discussion, we will divide it into two major areas:

- Machine and process control
- Management information systems

Machine and Process Control

The ability to work with metals was developed nearly 6000 years ago. The processes used were, at first, simple and crude. Later, machines were invented to help the artisan produce more and better products. However, it was not until after the mid 1700s that automation really began to be applied to product manufacturing. This movement was hastened by the invention of some basic machines including the turret lathe and the automatic screw machine.

The most significant breakthrough in machine automation came with the development of numerical control. This event took place in the early 1950s. It made way for more complex, automated machines and process control systems. From it evolved computer-directed numerical control, industrial robots, and flexible manufacturing systems. See **Figure 36-3.**

Numerical Control

Numerical control (NC) is the programmable control of manufacturing machines as directed by a code of numbers, letters, and other symbols. This code makes up the program of instructions. It tells the machine controls what to do. The controls then activate the machine drive mechanisms (motors, gear trains, etc.) in order to produce a specific part. Whenever the part changes, the numerical code must also change.

NC machines have three basic parts. See **Figure 36-4.** These are:

- *Machine tool.* This is a power-driven machine used for the shaping, cutting, turning, boring, drilling, grinding, or polishing of solids, especially metals.
- *Control unit.* This is a unit that receives the machining program, possibly through a tape reader or a computer interface, and control devices. It sends signals to and receives signals from the machine tool. It affects machine tool operation.

numerical control (NC). A method of programming manufacturing machines to be controlled by a code consisting of numbers, letters, or other symbols.

- *Program.* This is a detailed set of instructions. The program directs machine tool operation via the control unit.

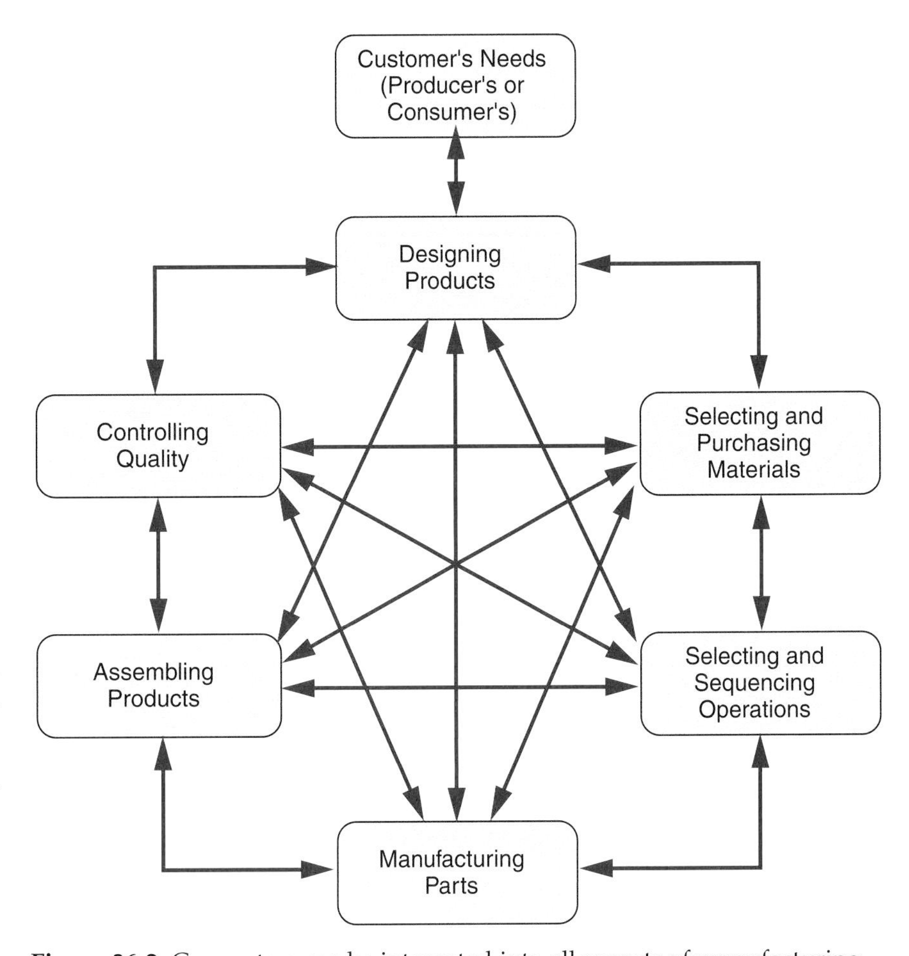

Figure 36-2. Computers can be integrated into all aspects of manufacturing.

Classes of NC Machines

NC has been applied to a number of machines. Commonly included are metal- and wood-cutting machines, tube- and rod-bending machines, assembly machines, bending and shearing presses, and welding and flame-cutting apparatus. NC machines may be separated into three classes based on their mode of tool (cutting element) positioning. See **Figure 36-5.** These are:

- *Point-to-point machines.* PTP machines position the tool above a point on a workpiece while it is clear of it. The machine then performs a specific operation. After completion, the tool is moved to the next point and the operation is repeated.

  An NC drill press is a good example of a PTP machine. In operation, the workpiece is first clamped on the machine table. The table is then moved until the workpiece is positioned under the drill bit at the location for the first hole. The table stops, and the drill is fed downward producing the hole. The drill is then retracted, and the workpiece is positioned for the next succeeding holes.
- *Straight cut machines.* Straight cut machines make straight-line cuts through a workpiece. Only cuts parallel to a major machine axis (discussion of which follows) are possible. Straight cut machines will produce parts with rectangular configurations. They cannot produce angular cuts. Straight cut machines can perform point-to-point operations.
- *Contouring machines.* Contouring machines are the most complex and expensive NC machine. They can

Figure 36-3. Industrial robots evolved from NC machines. These robots are fitting the windows into an automobile. (Ford Motor Company)

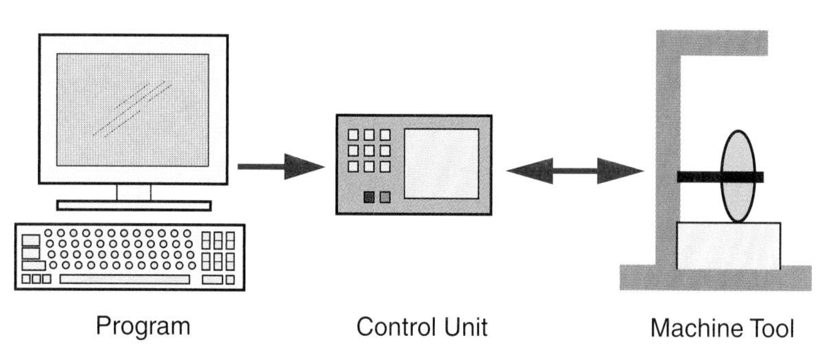

Figure 36-4. NC machines have three basic parts.

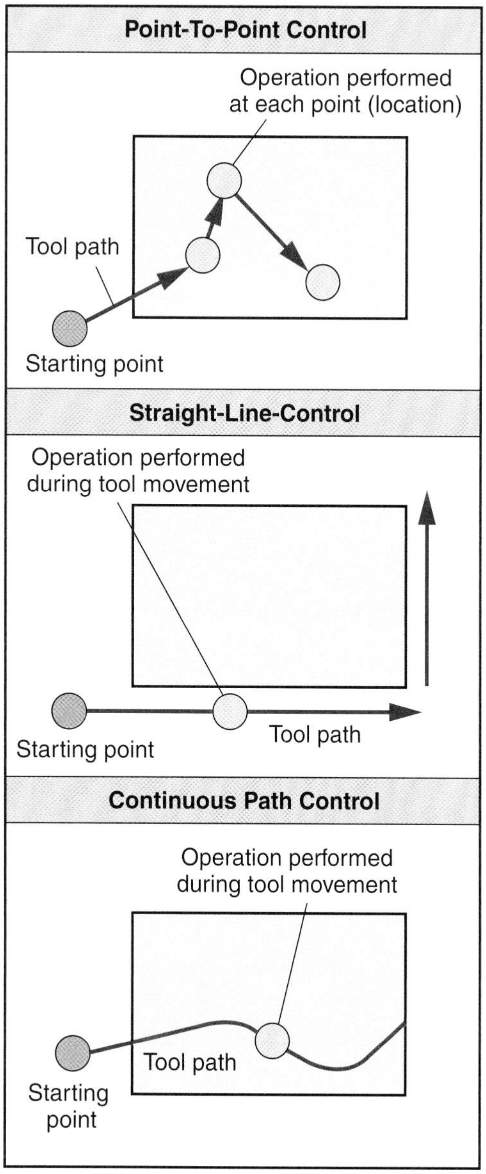

Point-To-Point Control
Operation performed at each point (location)
Tool path
Starting point

Straight-Line-Control
Operation performed during tool movement
Starting point
Tool path

Continuous Path Control
Operation performed during tool movement
Tool path
Starting point

Figure 36-5. There are three classes of NC machine.

move on any or all machine axes at the same time. This allows the tool to follow curved paths. Contouring machines can also be used to do point-to-point and straight-line work.

NC Programming Coordinates

Location, size, and shape data for all NC machines are developed around a coordinate system. This system specifies machine tool motions using X, Y, and Z rectangular coordinates. See **Figure 36-6.** The X axis is usually horizontal. It runs parallel to the movement direction of what is called the machine *table.* The Z axis runs parallel to the axis of what is called the machine *spindle.* The Y axis is at right angles to both the X and Z axes. It runs parallel to the movement direction of what is called the *traverse crossfeed* or *saddle.*

Coordinates make it possible to define the shape of a part. They are assigned to a part design relative to a chosen point of reference. This point, called the origin, is where all axes intersect. It is the zero point. Distances are measured from the origin and are assigned X, Y, and Z coordinate values. One direction from the origin is measured in positive units and the other in negative units. For example, a point located at position (1,2,–4) is one positive unit on the X axis, two positive units on the Y, and four negative units on the Z axis. This point can be communicated to the machine controls in numerical code.

NC Programming

Early NC programs were commonly found on punched tape. The holes in the tape enable the input of data to the machine control unit. The data is relayed through pulses of electricity caused by the holes. The control unit commands different tasks with different pulses.

Data was a number value or an instruction. Number values may be coordinate locations or speed and feed values. Instructions include axis commands, speed, feed, tool change, coolant start/stop, and machine start/stop.

The program is developed by a process planner and a machine programmer. Drawings are analyzed for proper machine X, Y, and Z coordinates. Materials for processing are considered. The proper machine and cutting tools are selected. The speeds and feeds for each operation are figured. Coolant requirements are determined. Based on these variables, a program is then written, developed and checked.

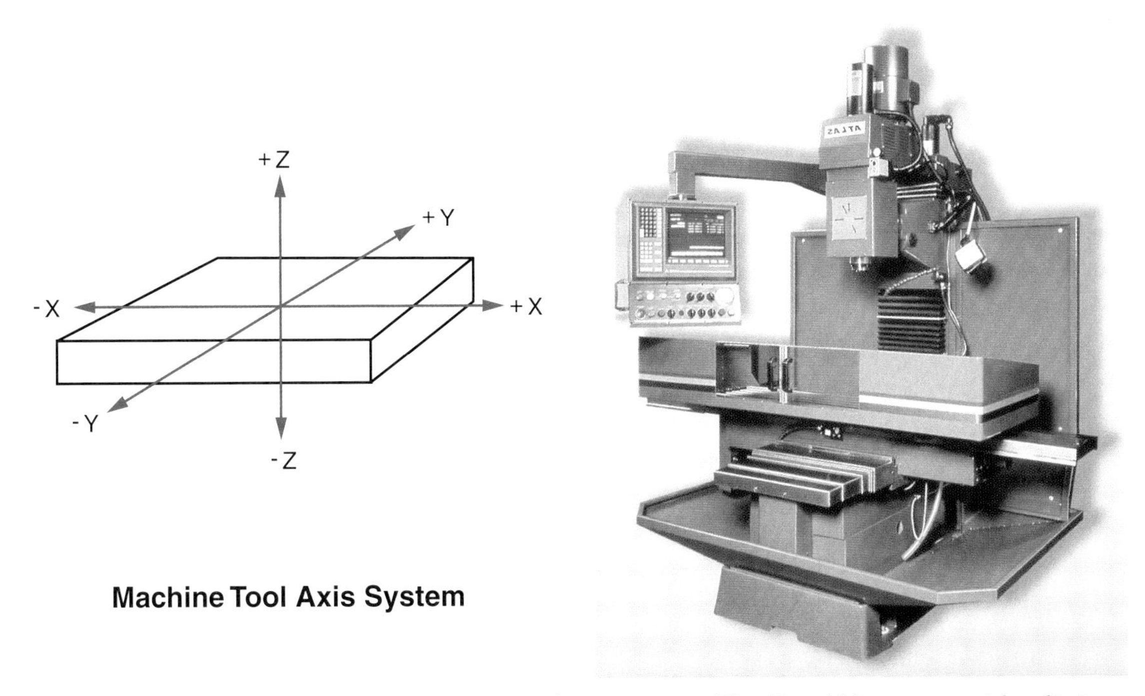

Figure 36-6. Left—A standard three-axis coordinate system. The X and Y axes are said to lie in the same plane with the Z axis passing through it. Right—The machine pictured has three primary machine axes. The Z axis is reserved for the spindle regardless of whether it is horizontal or vertical. (Atlas Corp.)

Advances in NC

NC revolutionized material processing and assembly. Since its development, it has not remained static. Additional systems have advanced NC. These are:

- *Direct numerical control (DNC).* DNC was the first extension of basic NC technology. This type of system was developed in the 1960s. It uses a master computer to control a number of machine tools. See **Figure 36-7.** The older tape reader of NC is replaced with a computer and its memory. The DNC computer provides operating instruction for each machine tool in its system. It communicates specific directions to each one as needed. DNC systems have a computer, which has a memory for storing the NC program, a communication link (wiring), and a number of machine tools.

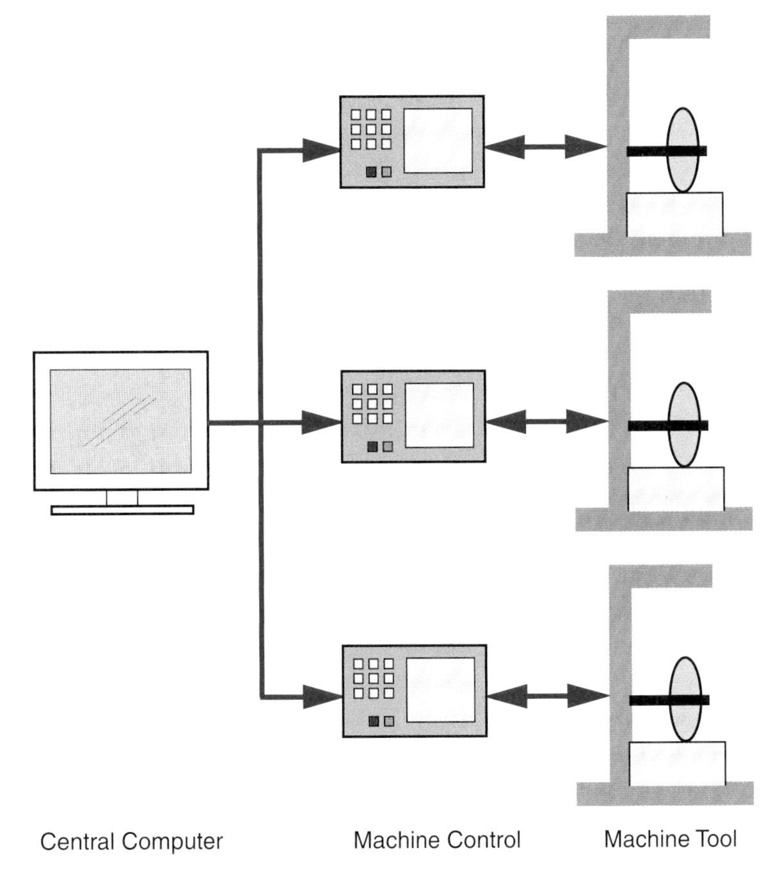

Figure 36-7. A schematic of a DNC system is shown here.

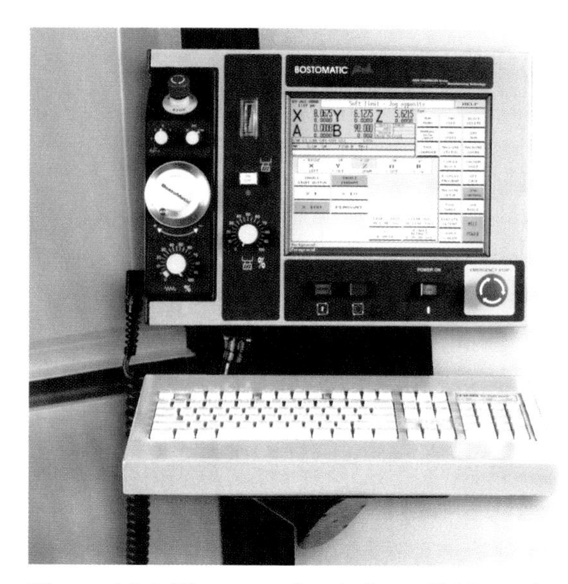

Figure 36-8. The control unit for a CNC machine is shown. (Bostonmatic)

Figure 36-9. This machining center is processing a large tractor part. Notice the tool cassette that holds tools for various programmed operations. (Caterpillar Tractor Co.)

- *Computer numerical control (CNC).* The development of fast and inexpensive minicomputers has allowed a new type of computer-directed NC system to be introduced. This system, called CNC, looks very much like the original tape-driven NC machine. The main difference, as with DNC, is that a computer replaces the NC tape reader. See **Figure 36-8.** Unlike DNC, each machine tool is controlled by its own computer. This feature allows for more flexibility and, in most cases, is cheaper to install.
- *Machining centers.* These take NC one step further. They are very complex, manufacturing machines. See **Figure 36-9.** These machines are computer controlled. They have five basic features. First, they can perform several different types of operations. They may drill, bore, ream, tap, mill, and face a part. Second, they can change tools automatically. Third, they can position a workpiece so that operations can be performed on several sides. Fourth, they have two tables so that a part can be machined on one while a new part is positioned on the other. Fifth, they are numerically controlled and run without a machine operator.
- *Adaptive control.* NC machines removed the operator from positioning work and from operating machines. The NC program took the operator's place. However, these programs have limitations. Machine speed and feed settings are written into the NC programs. These settings cannot be adjusted for specific operating conditions and problems. For example, as a cutting tool starts to become dull, it cuts slower. The tool feed needs to be reduced, but the program cannot sense this need. In another case, a drill bit tries to wander as it starts to cut a hole. It also grabs as it finishes the cut. At the start and finish, the speed and feed of the bit should be reduced to compensate for these problems and to lessen drill breakage. To address this type of problem, adaptive control was developed. Adaptive control automatically adjusts a machine tool so that it adapts to operating conditions. It measures machine variables and then adjusts feed and/or speed rates. Machine variables measured may be spindle torque,

point-to-point (PTP) machines. **Numerical control machines that position a cutting tool above one point on a workpiece, carry out an operation, then move the tool to the point where the next operation is to be carried out.**

straight cut machines. **Equipment that makes straight-line cuts through a workpiece. Straight cut machines can produce only rectangular parts.**

contouring machines. **The most complex and expensive NC machine. They can move on any or all machine axes at the same time.**

direct numerical control (DNC). **A system using a computer to communicate directions to each machine tool, as needed, to perform a sequence of operations.**

horsepower, vibration, or cutting temperature. Adaptive control increases tool life, machine productivity, and product quality.

Industrial Robots

Automation in manufacturing often requires automatic material handling. Parts must move from one machine to another. There are a number of automatic material handling systems. One of these systems is the computer-directed *automatic storage and retrieval system (AS/RS).* This system will automatically locate and retrieve parts and materials stored in special warehouse racks. AS/RS will also store parts and materials. Another material handling system uses *automatic guided vehicles (AGVs)* as shown in **Figure 36-10.** These driverless vehicles will move parts to and from machines. A computer commands the AGV to move to a specific location. It then follows guideways in the floor to that location.

One very important material-handling device is the *industrial robot*. This device is defined as a programmable, *multi*functional manipulator designed to move material, parts, or tools for the performance of a *variety* of tasks. A key distinction of a true robot is that it may be reprogrammed.

Figure 36-10. Pictured is an AGV in a manufacturing setting.

Figure 36-11. These industrial robots are used for spray painting a truck body. (FANUC Robotics)

Applications of Robots

Robots are used for a number of reasons. They can perform routine tasks, for example, loading and unloading parts into machines. They can perform repetitive tasks, for example, painting and welding. See **Figure 36-11.** Robots can also perform operations that would be unsafe for humans. As an example, they are used for loading and unloading materials into punch presses. They are ideal for handling hot, corrosive, or toxic materials, which are also considered unsafe for humans.

Fixed and Variable Sequence Robots

Robots perform their tasks through the specific direction of a control unit. This unit may be designed for fixed or variable sequences. A *fixed sequence robot* is a simple robot. It has limited versatility. It often employs a *pick-and-place* action such as that used to load or unload a machine. The robot moves to a position. It picks up a workpiece. It moves to a second position. It then places the workpiece in a die or other machine part. Later, the robot picks up the machined part and places it in a tray or on a conveyor. This sequence of picking up a part in one location and placing it in another gives the robotic action its name, i.e., pick-and-place.

computer numerical control (CNC). Control of an individual machine by a computer that replaced an NC paper tape reader.

machining centers. A type of separating machine that uses computer control and can perform many different operations.

adaptive control. A system of automated machine control that adjusts tool feed rates to match changes in operating conditions.

industrial robot. **A programmable manipulator capable of a variety of functions in a manufacturing setting.**

fixed sequence robot. **A simple robot that performs a simple sequence of events, over and over.**

variable sequence robot. **A robot that can be programmed to do one task, then easily reprogrammed to do a different one.**

Variable sequence robots are more complex and versatile. They are programmed to complete a specific task. This may be a pick-and-place task. It may be welding a bead. It may be spray painting a surface. When the task no longer needs doing, the robot is reprogrammed to complete a different type of task.

Programming a Robot

A robot may be programmed several ways. It is often done with a special control device called a *teach pendant.* This device uses a computer to walk the robot through the actions it will perform. The robot's memory stores the sequence of actions. It may then repeat or play back these actions in exactly the way it was taught (programmed).

A robot may be programmed in a way similar to the way NC machines are. That is, a computer is used to encode a program that is used to direct the robot's actions.

Robotic Movement

Robots, **Figure 36-12,** move in specific ways. They may be classified according to how they move. The groups can be defined by a coordinate system that best describes this movement. These include:

- **Rectangular or Cartesian coordinate system.** The robot moves on three linear axes, X, Y, and Z. A series of slides allow it to move in a straight line. The outline of the robot's possible movements forms a rectangular box.
- **Cylindrical coordinate system.** The robot has slide action for its vertical and horizontal movement. It will pivot on a center post creating a cylindrical motion. The outline of the robot's possible movements forms a cylinder.
- **Spherical or polar coordinate system.** The robot will pivot on its vertical and horizontal axes. The robot arm will extend and retract. The outline of the robot's possible movements forms a shape described by part of a sphere.
- **Revolute, jointed, or articulated coordinate system.** The robot will pivot on all three axes. This action most closely resembles the arm and shoulder movement of a human. Therefore, this type of robot can perform the most complex operations.

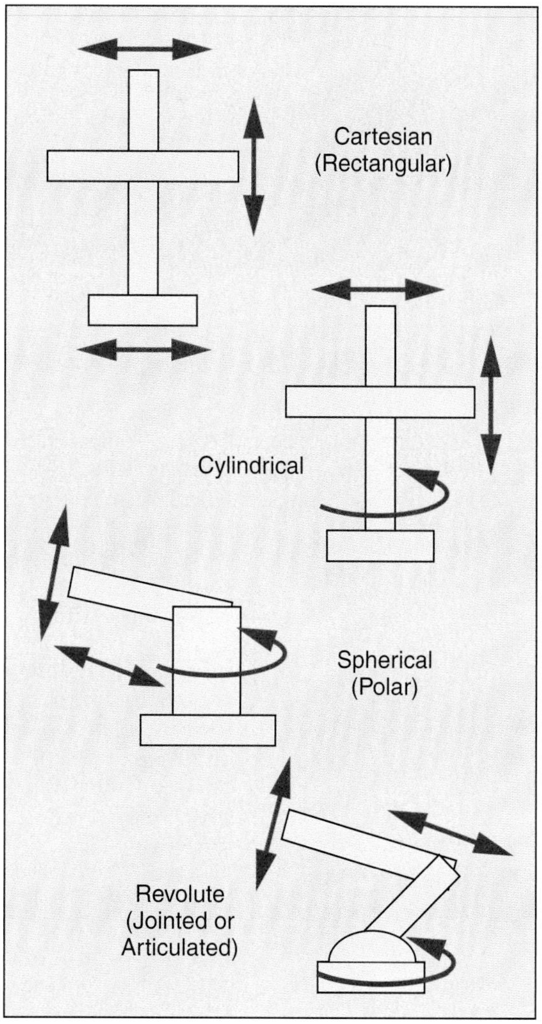

Figure 36-12. There are four types of robot movement.

End Effectors

Robotic *end effectors* are the tools at the end of a robot's arm that do the actual work. Robots may be classified into groups according to the type of end effectors they use. The most frequently used fall under two broad categories: *grippers* and *special-purpose.* Grippers, as the name implies, are used for gripping an object. Types of grippers include grasping, hooking, scooping, as well as adhesive, vacuum, and magnetic. Special-purpose end effectors are used for welding, painting, and sanding. They are used for drilling, grinding, polishing, deburring, and nut and screw driving. They are used in inspection operations for measuring as well.

Figure 36-13. This self-monitoring, high-speed FMS will machine both automotive and truck water pumps. (Ford Motor Company)

Flexible Manufacturing Systems

A system that integrates computer-controlled machines and material handling equipment with manufacturing control (production control, quality control, etc.) systems is called flexible manufacturing system (FMS). See **Figure 36-13.** FMS may include NC machines and robots connected by automatic material handling devices.

FMS allows companies to make products in small lots, yet with the cost advantages of continuous manufacturing. Computers control machine operations and adaptive control, material movement, inspection activities, and other segments of the operation. If the process fails to maintain quality standards, internal sensors and controls will shut it down and call for help.

end effectors. **Tools at the end of a robot's arm, used to perform work.**

Closely associated with FMS is *group technology (GT).* This practice was developed a number of years ago. It groups products and parts that have similar manufacturing requirements together so that they may be manufactured in small lots. GT allows the departments or factories making the parts to concentrate on producing quality items with a minimum of equipment and required skills.

Many flexible manufacturing systems depend on manufacturing cells. See **Figure 36-14.** A *manufacturing cell* is a small group of machines connected by a material handling system. Often, the individual elements of the cell are computer-controlled to increase flexibility and efficiency.

manufacturing cell. **A small group of machines connected by a material handling system. Often, the individual elements of the cell are computer-controlled to increase flexibility and efficiency.**

Management Information Systems

Computers were originally designed to process data. Therefore, it is not surprising that they are used for this type of work by manufacturing companies. They are used to maintain financial records, process sales orders and reports, and prepare invoices and shipping orders.

In product manufacture, computers are playing an ever-increasing, management support role. Computerized systems of information processing help

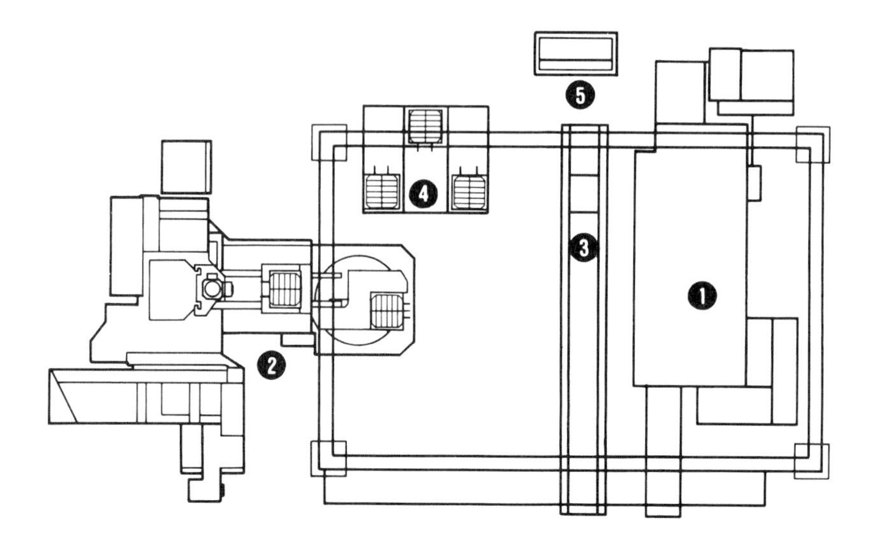

1. CNC Turning Center
2. CNC Machining Center
3. Gantry-Mounted CNC 5-Axis Robot
4. 3-Position Load/Unload Station
5. Controller

Figure 36-14. The illustration above is a schematic of the manufacturing cell shown below. The manufacturing cell shown has a turning center and a machining center with a gantry robot loader. (Cincinnati Milacron)

achieve peak operational efficiency. These are *management information systems.* Four important systems are:

- Computer-aided process planning
- Material requirements planning
- Manufacturing resource planning II
- Just-in-time manufacturing

management information systems. **Computerized systems of information processing that help a company achieve peak operational efficiency.**

Computer-Aided Process Planning

As you learned earlier, process planning involves determining production methods, then selecting and sequencing operations. Expanding on this, it also involves:

- Determining the operations to be used to manufacture a product.
- Selecting the equipment that will be used for each operation.
- Determining the tooling requirements for each machine operation.
- Developing a logical sequence for the operations.

Computer-aided process planning (CAPP) allows the methods engineers to complete these tasks efficiently. It enables them to see the total operation as an integrated system. It coordinates all tasks associated with the manufacturing process.

CAPP requires a large, computer database. It must have stored information about material properties; machine characteristics and capacities; tooling geometry; processing effects on material surface finish, mechanical properties, and dimensions; etc. This data is used to determine the optimum method of manufacturing and assembling the product.

CAPP is used to generate production schedules, flow charts, routing sheets, and a number of other process planning documents. It allows for easy modification of these and other similar planning tools.

computer-aided process planning (CAPP). **An integrated system that coordinates all tasks associated with the manufacturing process.**

materials requirement planning (MRP). **A system that allows schedulers to place priorities on materials and products to be produced.**

manufacturing resource planning II (MRP II). **A complex, computer-controlled system that can be used to control most aspects of production from scheduling to recording output.**

just-in-time (JIT) manufacturing. **A system that reduces inventories of raw materials and product components to the absolute minimum. Materials and parts are scheduled to arrive at the production line just when needed for use.**

Material Requirements Planning

Material requirements planning (MRP) is part of a complete, manufacturing resource planning system. It allows schedulers to place priorities on material and parts to be produced. The priority is based on production schedules or customer orders.

Products are divided into parts based on material requirement priority. Required material is obtained from the bill of materials stored in the computer file. The computer then schedules delivery of material resources (raw materials, industrial goods, tooling, etc.) needed to produce these parts. It also maintains work-in-progress and inventory records.

Manufacturing Resource Planning II

Manufacturing resource planning II (MRP II) is a complex, computer-based system that goes well beyond MRP. MRP II not only performs the functions of MRP, it produces production schedules, monitors production performance, and records manufacturing outputs.

Just-In-Time Manufacturing

Just-in-time (JIT) manufacturing is a computer-based monitoring system. It was developed in Japan. The JIT philosophy stresses the effective use of resources. Its focus is on eliminating tasks, such as transportation, inspection, and storage, which are thought to be wasteful. Raw material is scheduled to

Academic Link

Manufacturing is covered with a number of types of communication sources. These sources include sketches, product drawings, bill of materials, flow process charts, etc. The understanding of these sources is essential.

A communication tool in a just-in-time production system is called kamban. Being a very important tool for just-in-time production, kamban has become synonymous with the JIT production system. Kamban, meaning label or signboard, is used as a communication tool in JIT systems. For example, a kamban is attached to each box of parts as they go to the assembly line. A worker from the following process goes to collect parts from the previous process leaving a kamban signifying the delivery of a given quantity of specific parts. Having all the parts funneled to the line and used as required, the same kamban is returned back to serve as both a record of work done and an order for new parts. Thus, kamban coordinates the inflow of parts and components to the assembly line, minimizing the processes.

arrive *just in time* to be manufactured into parts and products. Parts are scheduled to be produced *just in time* to be assembled into products. Products are scheduled to be produced *just in time* to meet customer delivery orders. At all stages, inventory levels are reduced to the absolute minimum. This saves time and money, which translates into lower product costs and greater company profit.

Summary

Computers have become an indispensable part of manufacturing. They are used to control machines and processes and to provide management information. Manufacturing has been enhanced through NC, DNC, and CNC machines, machining centers, and adaptive control. It has been further enhanced through industrial robots, automated material handling, and FMS.

Management information systems include CAPP, MRP, and MRP II. JIT manufacturing is also a management information system. The integration of all of these systems is called CAM. When CAM and CAD are integrated the result is CIM. CIM is becoming the manufacturing method of the future.

Key Words

All of the following words have been used in this chapter. Do you know their meanings?

adaptive control
computer numerical control (CNC)
computer-aided process planning (CAPP)
contouring machines
direct numerical control (DNC)
end effectors
fixed sequence robot
industrial robot
just-in-time (JIT) manufacturing
machining centers
management information systems
manufacturing cell
manufacturing resource planning II (MRP II)
materials requirement planning (MRP)
numerical control (NC)
point-to-point (PTP) machines
straight cut machines
variable sequence robot

Test Your Knowledge

Please do not write in this text. Place your answers on a separate sheet.

1. Two major elements of CAM are _____ and _____.
2. Which of the following is a kind of automated machine or system?
   a. An NC machine.
   b. An industrial robot.
   c. FMS.
   d. All of the above.
3. *True or False?* A PTP machine can move on three machine axes at the same time.
4. With regard to NC, for what reason are rectangular coordinates used?
5. _____ automatically adjusts a machine tool so that it adapts to operating conditions.
6. *True or False?* Robots are useful in applications where human safety is a concern.
7. End effectors may be divided into the broad categories of _____ and _____.
8. Name four important management information systems.
9. The focus of JIT manufacturing is:
   a. Development of a logical sequence of manufacturing operation.
   b. Prioritizing part production based on customer needs.
   c. Eliminating tasks thought to be wasteful.
   d. All of the above.

Applying Your Knowledge

Note: Be sure to follow accepted safety practices when working with tools. Your instructor will provide safety instructions.

1. Draw X, Y, and Z axes and plot the following coordinates: (1,3,2); (–2,4,2); and (3,–2,2).
2. Visit a local manufacturing company that uses computers to control machines or provide manufacturing data. Prepare a bulletin board to communicate the information you gained from the trip.
3. Produce a simple part using a CNC machine.
4. A manufacturing cell is a series of computer-controlled machines connected by automatic material handling devices. Design a manufacturing cell for a simple product or for the one you are manufacturing in the enterprise activity.
   a. Look at the schematic of the manufacturing cell shown in Figure 36-13.
   b. Determine what machines will be needed in order to manufacture the product you have chosen.
   c. Secure drafting or sketching paper with a grid on it from your instructor. Locate the machines on the paper and sketch them. Try to locate the machines in the best arrangement for an efficient flow from one to the next. (Note: If you wish, use simple rectangles, squares, and circles to indicate the machines and the robot.)
   d. Label all machines and robots.

Computer Support Specialist

Computers have become an integral part of everyday life. We use them for a variety of reasons at home, in the workplace, and at school. Almost every computer user has encountered a problem, whether it is the disaster of a crashing hard drive or the annoyance of a forgotten password. The increased use of computers has created a high demand for specialists to provide advice to users, as well as day-to-day administration, maintenance, and support of computer systems and networks.

Computer support specialists provide technical assistance, support, and advice to customers and other users. These troubleshooters interpret problems and provide technical support for hardware, software, and systems. They answer phone calls, analyze problems using automated diagnostic programs, and resolve recurrent difficulties. Support specialists may work either within a company that uses computer systems or directly for a computer hardware or software vendor. Increasingly, these specialists work for help-desk or support services firms, where they provide computer support on a contract basis to clients.

Support specialists are troubleshooters, providing valuable assistance to their organization's computer users. Because many nontechnical employees are not computer experts, they often run into computer problems they cannot resolve on their own. Support specialists install, modify, clean, and repair computer hardware and software. They also may work on monitors, keyboards, printers, and mice.

Support specialists answer phone calls from their organizations' computer users and may run automatic diagnostic programs to resolve problems. They also may write training manuals and train computer users how to properly use the new computer hardware and software. In addition, they oversee the daily performance of their company's computer systems and evaluate software programs for usefulness.

Due to the wide range of skills required, there are a multitude of ways workers can become a computer support specialist. While there is no universally accepted way to prepare for a job as a support specialist, many employers prefer to hire persons with some formal college education. A bachelor's degree in computer science or information systems is a prerequisite for some jobs. However, other jobs may require only a computer-related associate degree.

Unit 12 Manufacturing and Society

Chapter 38
Manufacturing and the Future

Objectives

After studying this chapter, you will be able to:

- ✓ Discuss the future trend of management.
- ✓ Describe the changing image of "American industry."
- ✓ Discuss the emerging "thinking" computer.
- ✓ Explain why continuing education will be so important in the upcoming years.

Benjamin Franklin said all people could be certain of death and taxes. The future adds a third certainty—change. Throughout history, change has been one thing that all people had to deal with. Sometimes change comes about rapidly. In just one lifetime, animal power was replaced with gasoline engine power. Custom handcrafting was replaced with automation and mass production. See **Figure 38-1.** In spite of the rapid pace, people could grow with the change. It was still moderate enough.

Figure 38-1. In just one lifetime, the automobile has changed from the Quadricycle (left), Henry Ford's first automobile, to a modern, mass-produced automobile (right). (Ford Motor Company)

Things are different now and will continue to be because change is accelerating. It is moving from *rapid* to *radical.* Change is coming so fast now many people neither understand it nor accept it. Cynics suggest that if you understand something, it is already obsolete. This is partly true for much of the technology around us. A look at the radical advances in computer hardware and software supports this view.

We do not have to be victims of change. We can adapt and grow with it if we look to the future. You have probably heard the saying, "This is the first day of the rest of your life." It suggests that you must keep looking to the future. That is where you are going to live for the rest of your life. It is nice to have memories of the fun times you have had. However, memories give few clues about the world in which you will be living and working. What kind of world will it be? What new technologies will emerge? What will be the role of manufacturing in this world? What will be the role of managers and workers in industry? How will products be designed and produced? See **Figure 38-2.** From what materials will they be made?

These and hundreds of other questions could be asked about manufacturing and technology in the future. No one has the answers; however some people, called *futurists*, try to predict possible future conditions. They use historical perspectives and current trends to suggest future events.

futurist. A person who tries to predict possible future conditions on the basis of historical perspective and current trends.

There are a number of areas of manufacturing and technology that will be different in the future. For this discussion, predictions about the following will be explored:

- Expanding technologies
- Management values and entrepreneurship
- Materials
- Manufacturing methods and processes
- Products and the consumer
- Jobs and the worker

Expanding Technologies

Technology and science, by their very nature, are constantly expanding. The next several decades will see major advances in a number of fields of study.

Figure 38-2. The future will bring many new manufactured products. Left—Solar-powered automobile. (General Motors Corp.) Right—Radically different designed aircraft, like this B-2 Spirit. (Northrop Grumman)

Advances in physics will allow computers to increase in speed. Presently, they are slowed by the speed of electrons and switching gates. Use of the new fiber-optic type computer will be one way of overcoming this barrier. The other will be superconducting materials, which will offer virtually no resistance to electron flow.

New computer memories will be developed that will be based on the human brain function. The human brain can locate data, retrieve it, and process information faster than any known computer. New silicon memory systems that can see, hear, feel, and remember will appear on the market.

Figure 38-3. Biotechnology can be used to develop stronger, faster-growing, and more disease-resistant plants.

Other advances in physics will expand telecommunication technologies. Fiber optics will continue to grow as a preferred channel for data and voice communications. Increasingly complex satellites will be present around the globe. Telecommunications will be within the financial reach of almost everyone.

Biotechnology, an offshoot of molecular biology, will also become a central technology. See **Figure 38-3.** It focuses on using organisms as part of a technological system. Biotechnology is already being used to develop disease resistant plants and natural pesticides. Among other things, it is being used in industrial processes and in oil spill cleanup.

Figure 38-4. New technologies will help people diagnose, treat, and cure illness.

Technology will continue to impact medicine and health. See **Figure 38-4.** People will live longer because of new drugs and treatment. New artificial body parts (prostheses) will be developed and organ transplants will become more common.

Technological advancements will allow alternate energy sources to provide new sources of energy. These will help reduce the demands for petroleum, natural gas, and coal. See **Figure 38-5.**

Management Values and Entrepreneurship

Business experts are saying that many of the problems with American industry can be traced to top management. They have lacked confidence in their workers and suppliers. They have placed great value on short-term goals in the interest of favorable quarterly dividends. They have placed little value on long-term growth. Investments in product development and plant and equipment upgrades have been neglected. This problem is being addressed and a promising new type of business is developing. It is known as entrepreneurship.

The future should be bright for a new breed of owner/managers—the daring entrepreneurs of the future. There will be venture capital (initial start-up money) to finance new ideas and companies. These new entrepreneurs will have high levels

Fossil Fuel

Alternate Energy

Figure 38-5. Right—Technological advancements will help reduce the demands for fossil fuels such as petroleum. Left—People will use more alternate energy sources such as wind and solar energy.

Figure 38-6. Entrepreneurship and participatory management will bring the manufacturing team (workers, technical people, and managers) closer together.

of self-confidence and be willing to take risks. Likely, they will not have lived through a depression, which causes people to be financially cautious. The new business titans will not be driven solely by financial success. They will take financial risks for the challenge of pushing the enterprise system to the limit.

These people will organize and develop a number of small companies. Many of these companies will be part of the new information age. They will push new ways to communicate, apply computer systems, and conduct commerce. They will develop new products and services quickly. They will abandon older, less-profitable products just as quickly. These smaller companies will generate the majority of new jobs in the future. The large manufacturing giants of today will be present but not grow very rapidly. Production of many goods that require high degrees of manual labor will move to countries with large populations and low wages.

This entrepreneurship will be coupled with new management styles. See **Figure 38-6.** Most successful companies will practice participatory management. There will be fewer layers of management between the owners and the workers. The workers will have significant input to managerial decisions. They will be expected to accept higher levels of responsibility in their work for the success of the company.

Materials

Many materials are becoming scarce. The supply of copper, lead, and mercury could be exhausted during the twenty-first century. Likewise, major fuels, such as petroleum and natural gas, are being depleted. New materials and energy sources will be developed as replacements. A wave of the future will be engineered materials.

Historically, products and structures were engineered to fit a material. For example, homebuilders chose the size of rafters in terms of the span and the expected dead load (weight of roof and snow). Longer rafters had to have a larger cross section. They were made from 2 × 6s instead of 2 × 4s. Likewise, larger metal parts were made of thicker steel to provide more strength. The characteristics of the material dictated many design solutions.

The future will see material engineered to meet product demands. See **Figure 38-7.** Flexible glass that bends without breaking, metals that stretch, and plastics that are as tough as steel have been developed. Ceramics, a favorite material of primitive humans, are finding their way into such products as space shuttle heat shields and tooth implants. Solid wood construction members will continue to be replaced with materials engineered for wood chips and flakes, **Figure 38-8.**

Figure 38-7. Laboratory research will develop new manufacturing materials. (PPG Industries)

Figure 38-8. Engineered materials will play a larger role in construction.

Plastics and composites (especially, ceramic and plastics reinforced with synthetic or carbon fibers) will be the new materials of choice. They will replace metals in many products. These new materials have some exceptional qualities, one of which is tolerance of high temperature. Plastics have been developed that can withstand 600°F (316°C). Newer composite materials can endure temperatures as high as is found in most types of combustion chambers.

Products will be designed to use less material. Currently, Americans use about 10 tons of nonrenewable materials per person per year. This is causing a strain on the world's natural material resources. It must be and will be reduced as products will be made lighter and more obsolete goods are recycled.

New and improved technologies will supplant existing technologies and reduce the amount of material required. Communication is already a case in point. The Transatlantic Cable used 150,000 tons of material. A communication satellite uses only 500 lb. of materials, and also provides better communication service.

Manufacturing Methods and Processes

American industry dominated the world for nearly a century. The country successfully mechanized manufacturing, agriculture, mining, and forestry. The

Figure 38-9. Industrial robots will be increasingly important to manufacturing companies. (DaimlerChrysler)

picture of "American industry" used to conjure up images of workers doing repetitive tasks in dark and dingy factories. Thoughts of people looking bored or wiping their brow were brought to mind. This has all changed. Industry is gaining a new image because of the new wave of change. Mechanization has been largely enhanced by automation. The image of the future is of a gleaming workplace with highly skilled technicians and state-of-the-art machines.

Almost everyone who describes manufacturing in the future tells of a greater use of computers, automated machines, and industrial robots. See **Figure 38-9.** On the horizon are "thinking" computers. The name assigned to this branch of computer science is ***artificial intelligence (AI)***. A branch of AI is called *expert systems (ES)*. Both of these systems can perform human-like activities. They have the ability to reason. They can see, hear, and feel. They can interpret audio and visual data. Like people, AI uses the senses to gather information. The computer then processes the sensory information and acts upon it. ES features the ability of a computer to function on "learned" rather than programmed operations and data. These systems will be used in manufacturing to monitor processes, check quality, and direct machines and material handling devices (robots, AGVs, etc.).

artificial intelligence (AI). **Computer-based systems that have the ability to learn from information gathered through sensory inputs.**

CIM will become common. Products will be conceptually designed, manufactured, and marketed on the computer well before actual production begins. Every step of manufacturing will be interconnected by computer systems. Designing, engineering, manufacturing, quality control, purchasing, financing, and marketing will be integrated into one system.

In another arena, manufacturing will live in better harmony with the environment. In some cases, waste from one process will be the raw material of another. Materials will be recycled after they serve their useful life. Discarded metal cans could become the base materials of automobiles. When worn out, these cars could become the structural steel of high-rise buildings. This type of system would create an industrial ecosystem much like the biological ecosystem of the earth. There would be a constant recycling of materials. This type of system is becoming essential as the demands of our rapidly growing global population accelerate.

Manufacturing will move beyond its traditional bounds. Experiments of manufacturing in space are now being performed. A leading manufacturer of agricultural equipment is a major participant in space research. It is using the low gravity environment of space to conduct research on the microstructure of iron. This environment allows for a better study of what happens when molten iron solidifies. The results of the research may help the company improve its iron-casting techniques. Another experiment involves growing organic crystals from organic solvents.

As a result of these and other experiments, some manufacturing processes will move into space. See **Figure 38-10.** This environment will allow for better process control. Weightless and pollution-free environments will improve many processes. Recently, perfectly round spheres were manufactured in space. These beads have become essential for calibrating delicate laboratory measuring instruments.

Manufacturing in space will thrive. One estimate suggests that by the twenty-first century, several billion dollars worth of pharmaceuticals and electronic materials will be manufactured in space. Earthbound manufacturing will not go away, however. It will still flourish.

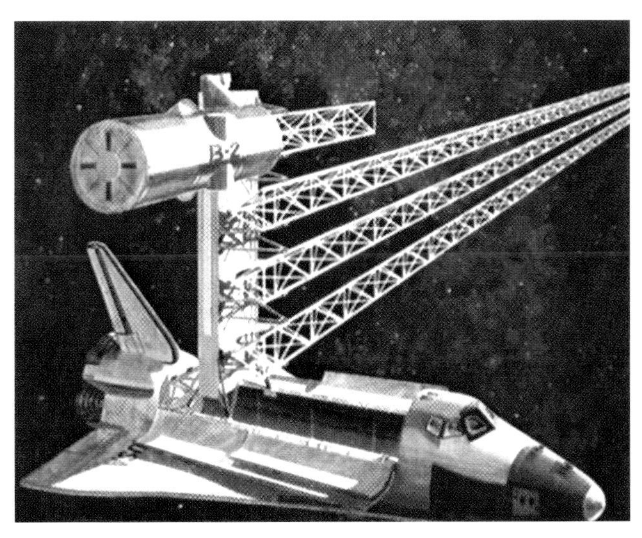

Figure 38-10 This is an artist's idea of what a "space factory" might look like. (NASA)

Products and the Consumer

The products of the future will be world-class. They will be high quality and low priced. Quality will become a central issue, because the average consumer will accept only quality products. Those companies producing this type of product will remain competitive. They will significantly lower costs by reducing scrap and the need for repairs. Today, these costs can run as high as 30% of production costs.

Also, these products will be global. The parts will be manufactured in many countries. Manufacturing will move around the globe as economic conditions change. Labor intensive (requiring large amounts of human labor) products will be made in densely populated, poorer nations. High-tech products will be the domain of developed countries. Manufacturing efficiency will be the key to economic survival of every country.

Products will be a paradox between simple and complex. They will continue to be technologically more complex; however, they will be simpler to fabricate and assemble.

Jobs and the Worker

The percentage of the total workforce involved in the actual manufacture of products continues to decline as a percentage of the total workforce. In contrast, employment in service jobs is increasing. In 1990, 15.3% of the workforce was in manufacturing. However, by 2000 only 12.7% worked in manufacturing. It is estimated by 2010 only 11.4% of workers will have manufacturing jobs. In 2000, service-producing workers accounted for 71.1% of the workforce, up from 67.4% in 1990. By 2010 it is estimated 74.4% of the workers will be employed in service area. Many of these service-producing jobs will be in computer and information-processing fields. Some of these computer-related jobs will be directly involved in manufacturing. These manufacturing jobs will be in areas that design, produce, and market products, but will not be involved in the physical manufacture of the products.

Automation will continue to cut the ranks of the blue-collar worker. Jobs for machine operators, which are traditionally well paying work that requires limited skills, will continue to shrink. The "steel-collar" worker-robots and other automated machines will replace them. In addition, low-skill and semi-skilled

Figure 38-11. Jobs requiring creativity and imagination will always be in demand.
(Ford Motor Company)

jobs will move off-shore where cheaper labor is available. The blue-collar worker will have two choices: accept lower paying jobs or upgrade education and skills.

The jobs with a future will involve technology and science. Both of these fields are rapidly changing. Job-related knowledge will quickly become obsolete. The half-life of engineers (the point at which one-half of their knowledge is obsolete) is three to ten years and shrinking. Technicians and technologists face the same problem. This will dictate the need of employees to continually involve themselves in training programs and other educational activities.

Nearly all jobs will require at least a high school education. The majority of new jobs will be white-collar positions. They will be mostly managerial, technical, or scientific in nature. People possessing computer skills will be in demand. The best-paying jobs will require creativity, decision-making abilities, and human relations skills. See **Figure 38-11.** Almost universally, they will require post-high school training. There will be few meaningful jobs for high school dropouts.

Coupled with the change in job demands will be a continued loss of membership in labor unions. Traditional strongholds of organized labor, the factory worker and miner, will weaken. The new scientists, engineers, and technicians will not readily join unions. This will create a new work environment. The "us" and "them" (labor and management) attitude will disappear.

Employer Expectations

Many general qualities are needed to be successful in the workplace. Behavior required for professional success and advancement includes the following:

- **Cooperation.** An employee must cooperate with supervisors, other employees, and customers.
- **Dependability.** A dependable employee is timely, completes all assignments, and sets realistic goals for completing projects. A dependable employee is trusted by others.
- **Work ethic.** Good employees put an honest effort into their work.
- **Respectful.** In order to be respected, employees must show respect for others, the company, and themselves.

Today's workplace emphasizes *equality*—that is, the idea that all employees are to be treated alike. *Harassment* (an offensive and unwelcome action against another person) and *discrimination* (treating someone differently due to a personal characteristic such as age, sex, or race) are not tolerated. These negative behaviors often result in termination of employment.

Summary

New technologies in the fields of physics and molecular biology are emerging. Superconducting materials, humanoid robots, and genetic manipulation will be in the forefront of technology. The future should bring a change in management values. Entrepreneurs will start up small companies. They will be dynamic and profitable. Larger companies will be present but will stagnate. High-quality engineered materials will be developed.

Manufacturing methods and processes will change. The image of a dirty factory will become one of a gleaming workplace run by high-tech machines and people. AI and CIM will be common. Goods will be recycled. Products will be manufactured in space. Manufacturing will change globally. Products will be better quality.

Jobs will become more service oriented and less manufacturing oriented. These new technologies will demand highly trained people. There will be no meaningful work for the unskilled or semiskilled worker or for the high school dropout. Summing up, the future in manufacturing, as always, promises change; it promises to be interesting.

Key Words

All of the following words have been used in this chapter. Do you know their meanings?

artificial intelligence (AI)
futurist

Test Your Knowledge

Please do not write in this text. Place your answers on a separate sheet.

1. *True or False?* Change in the future will occur at the same pace as it occurred in the past.
2. Biotechnology, an offshoot of molecular biology, is being used _____.
   a. To develop disease resistant plants and natural pesticides
   b. In some industrial processes
   c. In oil spill cleanup
   d. All of the above.
3. Small companies, formed by a new breed of owner/managers called _____, will generate the majority of new manufacturing jobs in the future.
4. *True or False?* The workplace of the future will have highly skilled technicians and state-of-the-art equipment.
5. Some computers of the future will be able to_____.
   a. Reason.
   b. Learn.
   c. See, hear, and feel.
   d. All of the above.
6. Explain why continuing education beyond high school will be so important in the upcoming years.

Applying Your Knowledge

1. Develop a bulletin board showing changes seen in your life and in the lives of the nearest two generations before you.
2. Select a product. List the materials in it that have been developed in the last 50 years.
3. Examine a product of your own choice.
   a. Identify the materials of which it is made and list them on a sheet of paper.
   b. Research the history of the materials using information found in the resource center or school library.
   c. Indicate on the sheet which materials in the product have been developed in the last 50 years.
   d. Report your findings to the class.

Market and Survey Researchers

Market researchers, or marketing research analysts, are concerned with the potential sales of a product or service. They analyze past sales numbers to predict future sales. They gather data on competitors and analyze prices, sales, and methods of marketing and distribution. Market researchers devise methods and procedures for obtaining the data needed, and they often design telephone, personal, or mail interview surveys to assess consumer preferences. Trained interviewers, under the market research analyst's direction, usually conduct the surveys.

After compiling the data, market research analysts evaluate it and make recommendations to their client or employer based upon their findings. They provide a company's management with information needed to make decisions on the promotion, distribution, design, and pricing of products or services. The information can be used to determine the advisability of adding new lines of merchandise, opening new branches, or otherwise diversifying the company's operations. Market researchers may conduct opinion research to determine public attitudes on various issues, which may help political or business leaders and others assess public support for their electoral prospects or advertising policies.

Survey researchers design and conduct surveys. They use surveys to collect information that is used for research, making fiscal or policy decisions, and measuring policy effectiveness, for example. As with market researchers, survey researchers may use a variety of mediums to conduct surveys, such as the Internet, personal or telephone interviews, or mail questionnaires. They may also supervise interviewers who conduct surveys in person or over the telephone.

Survey researchers design surveys in many different formats, depending upon the scope of research and method of collection. For example, interview surveys are common because they can increase survey participation rates. Survey researchers may consult with economists, statisticians, market researchers, or other data users in order to design surveys. They may also present survey results to clients so they can use the information to plan future business decisions.

Chapter 39 You and Manufacturing

Objectives

After studying this chapter, you will be able to:

- ✓ Explain the difference between leadership and control.
- ✓ Explain three principles shared by effective leaders.
- ✓ List seven elements of an effective plan.
- ✓ List five general considerations to help you choose the right job.
- ✓ Discuss details you should consider before making a purchase.
- ✓ Apply manufacturing lessons to everyday life skills.

Much of what has been learned about manufacturing is concerned with group efforts—groups of managers directing groups of operations and groups of employees. It is relevant at this point to discuss how manufacturing relates to you—the individual. See **Figure 39-1.** You may wonder, "What role can I play in manufacturing? What lessons related to manufacturing can I apply to my own life?" This chapter attempts to answer these questions.

Leadership

Leadership is a vital part of manufacturing. It is equally vital in other aspects of your life. *Leadership* is the leading or guiding of others. A key to good leadership is having the ability to inspire all kinds of people to work together.

leadership. **The ability to inspire others to work together.**

Some people equate leadership with giving orders or directing people and demanding results. This is actually *control.* Leadership and control are not the same thing. Prison guards exercise control while they strongly direct and monitor the actions of people. Leadership, on the other hand, connotes a more positive, guiding approach at reaching desired results.

Effective leaders employ the ideals of participatory management. Striving to maintain an open atmosphere, they involve all group members in the decision-making process. They value input from every person. See **Figure 39-2.**

Effective leaders are good at getting things done. Effective leaders share common principles in regard to leadership that help them reach their goals. These are:

- A "team" attitude
- A shared vision
- A plan of action

Figure 39-1. Individuals serve manufacturing in leadership, general employment, and consumer roles.

Figure 39-2. Effective leadership employs the ideas and skills of every individual.

Team Attitude

Leaders know that group success depends on developing an effective team. Effective teams have some common characteristics.

First, the members have *mutual respect* for each other. Differing opinions are respected and welcomed. A good team will have differences and even heated arguments. However, in the end, the members will remain friendly and cooperative.

Second, an effective team is made up of people who have *specific knowledge and talents.* Each person brings different ideas and skills to a group. When unified, the overall effectiveness of the team is greater than the sum of the ideas and talents contributed by the team *members.* This phenomenon is called *synergism.* It is a state wherein the whole effect is greater than the sum of the individual effects.

synergism. A state in which the whole effect is greater than the sum of the individual effects.

Third, an effective team has a *designated leader.* See **Figure 39-3.** This leader should be enthusiastic and have a positive attitude. He or she should be the "cheerleader" for the group. The leader should encourage and reinforce the success of each group member.

Shared Vision

Leaders know that group success depends on a shared vision. The leader and the team members must want the same results. Our nation's space program provides a good example of this. See **Figure 39-4.** President Kennedy said that an American would walk on the moon within a decade. It happened because thousands of people were taken up by the challenge. They shared the vision of an American walking on the moon and were willing to strive for this vision. They worked long hours—gave up hobbies, vacations, and other pursuits. Such dedication helped Neil Armstrong become the first person to walk on the moon.

Getting people to share a vision is not always easy. The key is for them to see personal benefit in pursuing the vision. Leaders sometimes make appeals to the workers' sense of pride or loyalty in trying to obtain their support.

Have you ever watched an outstanding school band? The band members share a vision of putting on a good show for the audience. The members take pride in their music and receive satisfaction from the applause. They are loyal to the same school and want to promote school spirit through music. Finally, all the members benefit from the experience and friends they gain while playing in the band.

Figure 39-3. Effective groups have appointed or selected leaders.

Figure 39-4. The American space program depends on thousands of people sharing the same vision.

Plan of Action

Few things of value get done without a plan of action. Columbus *planned* to sail around the world. Rembrandt *planned* to paint great pictures. George Washington *planned* to win the Revolutionary War. Kennedy *planned* to have America place a man on the moon in a decade. Lee Iacocca *planned* to set a new course for the Chrysler Corporation. Executives from General Motors *planned* to start Saturn, a new automobile manufacturer, from the ground up.

Leaders also operate from a plan by helping individuals and groups set goals and a course of action. This is best done by working with the group and guiding them instead of issuing a predetermined plan. Guiding people is always easier when they understand and accept the overall goal.

An effective plan:

- Sets goals and deadlines for an overall job.
- Establishes and organizes specific tasks.
- Challenges members to accept portions of the job.
- Completes the job within a given time frame.
- Monitors progress and compares it to an approved schedule.
- Determines corrective actions to be taken when problems arise.
- Rewards members with praise, money, or other incentives for completing tasks and overall job.

A good leader helps people to become team members, establish a vision, and develop a plan of action. With these elements in place, the group is better positioned to succeed at the job before them.

Figure 39-5. Manufacturing careers can be challenging and rewarding. (AMP Inc.)

Employment

Working for a living can be challenging and rewarding. See **Figure 39-5.** Everyone has some basic needs that can be fulfilled by working. We need to have money to pay our bills and buy goods and services. We need to have money saved to give us security and to meet future financial needs.

However, money does not fulfill all of our wants and needs. People generally want to be needed. They want to be important in the eyes of others. Most of us feel better if we belong to a group, contribute to its success, and are recognized for our contributions. Working is one way a person can do this—being a contributing family member is another. Spouses can contribute to the family. People can contribute both time and money to religious groups, service clubs, charity organizations, and governmental agencies.

Choosing the Right Job

The arena of manufacturing is a major source of employment. Manufacturing jobs use a diverse range of human talents. See **Figure 39-6.** Some jobs require high levels of scientific and mathematical knowledge and skill. Other jobs require artistic abilities. Technical knowledge and skills are required of engineers, technicians, maintenance workers, and machine operators, to name a few. Managerial skills are needed by all workers; however, they are particularly important for supervisors and higher-level managers. Human relations or "people" skills are needed by almost everyone in a company. No job is isolated from other people because all employees must work together and get along.

Like other careers, manufacturing careers should be carefully selected by the prospective worker. A major share of life will be spent working. Therefore, it is unwise to settle for just any job. It is unfortunate to have a job that fails to provide satisfaction.

Selecting a job that will suit personal goals is not an easy task for most people. It involves research. It involves careful consideration. To be satisfied, you must choose a direction based on what you know about yourself and what you have learned from your research. Consider the following and weigh the importance of each before choosing your direction:

- Market demand
- Job demands
- Personal traits, interests, and abilities
- Educational requirements
- Lifestyle

Market Demand

It is a good idea to look into the market demand of the jobs that are of interest to you. Find out how hard it would be to find employment in them before obtaining special training or additional education. There may already be a glut of people in the field of your choice or there may be simply no need for a certain specialty. Unless you are focused on a particular field, if the demand is not there, consider an alternative job. It would be a shame to spend time pursuing additional training if you could not find a job in the area in which you trained.

If, in spite of lacking job opportunities, you still strongly favor a certain job, by all means, pursue it. You may have to search harder to find a position. Yet, the rewards of working in a job that you really like can make the effort worthwhile.

Job Demands

Consider the demands of the job. Manufacturing jobs can be analyzed by what they require of the worker. Most jobs require employees to work with a variety of people, machines, materials, and data. See **Figure 39-7.** However, assembly people work largely with machines and tools. Accounting people process large quantities of data. Salespeople meet other people constantly. Labor jobs can be physically demanding and require good physical condition.

Some jobs require continuing education on and off the job. Other jobs are stressful. Some have fixed work hours while others involve long hours or frequent overnight travel. If the demands of a job do not appeal to you, consider something else.

Personal Traits, Interests, and Abilities

A manufacturing job should be matched to personal traits, interests, and abilities. People who like active work, as opposed to desk work, may select a job in plant maintenance. People who like mechanical things might select jobs in engineering design or product manufacture. People with artistic ability might seek jobs in advertising, product development, and package design. Friendly, outgoing individuals might want to pursue jobs in sales or public relations. Analytical people, who like detail, may be attracted to research, accounting, finance, and quality control positions.

Opportunities exist in manufacturing for skilled workers. These people can work as machinists, assemblers, tool-and-die makers, and cabinetmakers. They can also work as metalworking, plasticworking, and NC machine operators. High school courses in technology, mathematics, and science, and a mechanical aptitude are helpful for people who would like to pursue careers in these occupations.

Educational Requirements

Almost all jobs have educational requirements. Most jobs require a high school diploma. Other jobs might require a master's or even a doctorate degree. Personal abilities (mechanical, analytical, artistic, etc.) are usually not enough

Scientific

Managerial

Artistic

Human Relations

Technical

Figure 39-6. Manufacturing requires a diversity of talents. (Ford Motor Co.)

Figure 39-7. Manufacturing attracts people because of the balance between working with people, data, and things.

to get by on for most jobs. They need to be enhanced through furthering education.

If you have a particular area of interest, consider the education that is required. Consider how much time and effort you are willing to spend in its pursuit. If you have a particular job goal, it is best to go after it. If hard work is required to obtain it, stay determined. You may get discouraged at times, but the satisfaction you will receive in accomplishing your goal will make all of your efforts worthwhile.

The amount of education you receive can determine at what level of a particular field you may work. For example, different jobs that

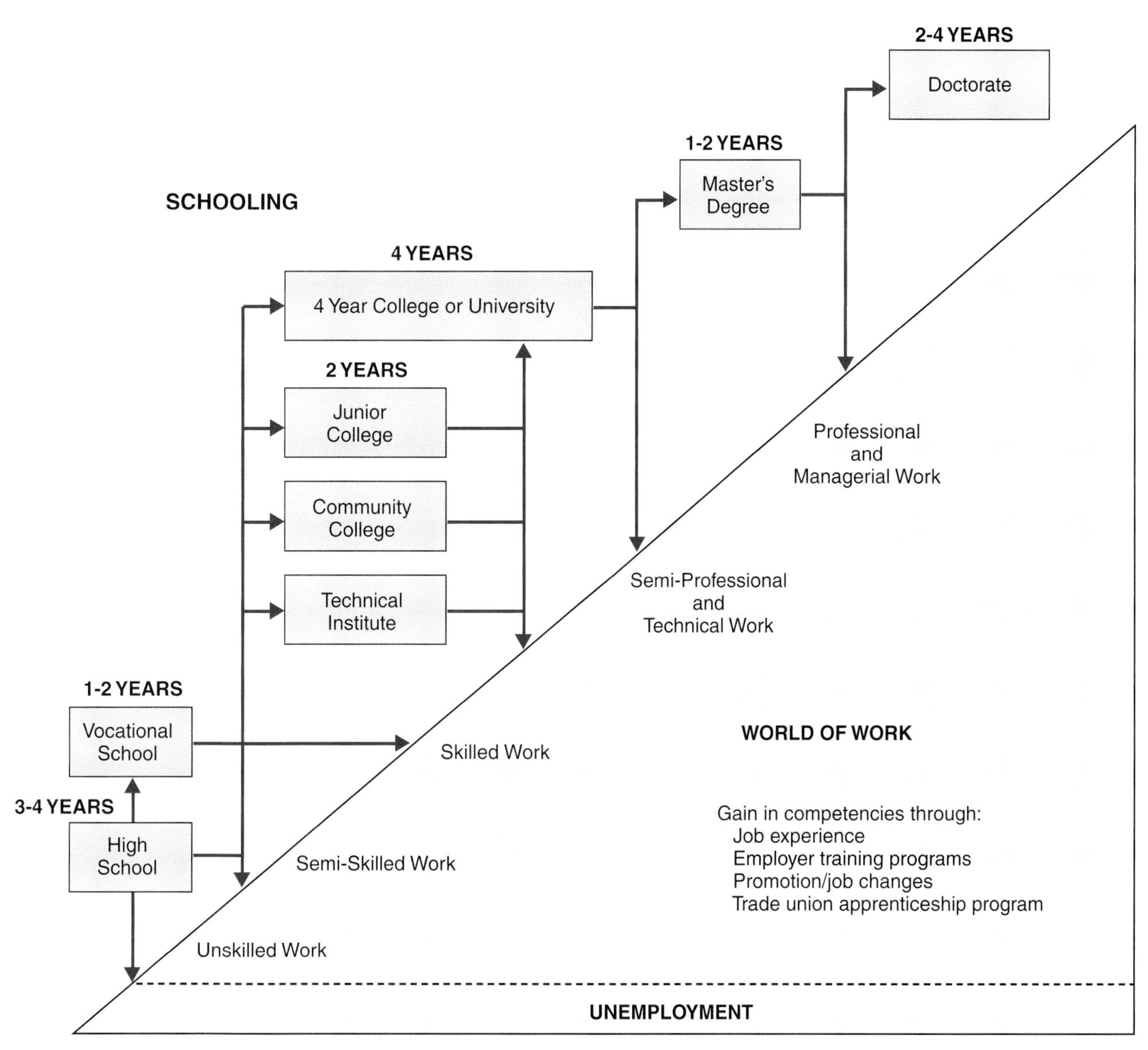

Figure 39-8. This chart relates the type and amount of training available to the type and level of employment in the world of work.

involve mechanical skills require different levels of education. Machine operators generally need a high school education. Robot repair personnel may need an associate (two-year) degree. Mechanical design engineers usually have a bachelor's (four-year) degree. The person who wants to manage the company's mechanical division, in many cases, needs a master's degree.

The chart in **Figure 39-8** shows a relationship between jobs and their educational requirements. The diagram can be summarized by these points:

- You can generally enter the working world at a higher level of responsibility if you have completed a higher level of training.
- There are many kinds of schools, and you can move from one to another.
- You can move from one school to the working world and return to schooling at the *next* level. You can also return to the same level of schooling in another field of study.

- You can enter the working world at any level that you have prepared to enter. You can then advance by taking advantage of opportunities to improve your skills on the job. Usually, however, there are limits to how far you can advance by this method.

Lifestyle

Employment has a direct effect on a person's lifestyle. Choose a job that will fit yours. Determine how much money you need to support your lifestyle. Determine where you would like to live. How will you spend your leisure time? Do you like city life or rural life? Are you an outdoor-type person? If so, do you like summer or winter activities? What kind of climate do you like? Also, determine if the type of lifestyle you want will permit a job, which requires overnight business travel or long hours.

Being Successful on the Job

Suppose that you have done your homework and you have landed the job of your dreams. You may wonder what you can do now to be as successful on the job as you have been in your job search effort. Key to being successful on the job lies in having a positive attitude. It will affect your productivity and the productivity of others. Attitude is closely related to ability to get along with people.

Getting along with others requires more than getting people to like you. You need to know how to handle certain "sticky" situations. Human relations skills are vital. You should know how to deal with the unfair supervisor, the harassing coworker, or the office "gossip." You have to understand yourself and others. Being smart about getting along with others does not mean being insincere or a phony. The person who understands human relations tries sincerely to develop friendly, honest, and open relationships with all coworkers.

The successful employee tries not to irritate others. Poor grooming habits are offensive and will likely irritate other employees. Cleanliness and proper appearance is important. It will affect your attitude. If you are clean and properly dressed, chances are you will feel good and others will notice your positive attitude. You will likely be more productive. Your coworkers will have more respect for you. Your supervisor will perceive you more favorably. Always dress for success.

There are many other habits that can irritate people. Speaking or yelling in a harsh, whining, or loud voice is one. Bragging is another. The list can be quite long. Ask a friend or relative to be honest with you about annoying habits you might have.

To help you be successful on the job, try to build a good relationship with your supervisor. There are many ways to do this. Here are a few:

- Make your supervisor look good.
- Do not take a problem to someone higher in the management chain without your supervisor's knowledge.
- Do not fall into a "buddy-buddy" relationship with your boss.
- Be honest. Also, do not try to bluff your way around a question you cannot answer.
- Accept responsibility for your mistakes. Do not blame others.

Figure 39-9. Manufacturing companies produce a spectrum of products ranging from relatively simple (left) to futuristic and complex (right). (Lockheed Martin)

There are a number of other factors that will contribute to your being a success on the job. It is important that you maintain good health. Eat right and get enough sleep. Come to work well rested and prepared to meet the challenges of the day. Avoid absenteeism and tardiness. Exercise self-control. Follow company standards and policies. Avoid unethical practices and keep your personal standards high. Be a good producer—maintain high productivity on the job and do quality work.

The Individual as a Consumer

As consumers, each individual influences manufacturing. See **Figure 39-9.** Manufacturing companies react to consumer demands because they want to make products that are needed or wanted.

As you select products, you should keep several factors in mind. One is need. Each customer should ask, "Do I really need this product, or is it a passing fancy?" Another is value. A product should be worth what it costs. Closely related to this is price. First, can you afford the product? Second, is it competitively priced? Shop around to be sure you are getting the best price.

Other factors to consider before making a purchase are:

- **Function.** "Will the product do what I want or expect it to do?"
- **Appearance.** "Do I like the looks of the product?"
- **Operation.** "How hard is it to use?"
- **Durability.** "How long will it last? How hard or expensive would it be to maintain or repair?"

People should be concerned about the environment. Each customer should ask if the product can be recycled or disposed of safely. The best products for future generations will be those that can be recycled. They can become the raw material of new products.

Other Impacts on Everyday Life

One underlying theme of this chapter is that studying manufacturing is helpful. It has provided you with valuable information that can be used in everyday situations. For example, through studying unions, you have learned how to set up an agenda and run a meeting using proper parliamentary procedure. You have learned that a meeting should have a purpose, be short and to the point, and be operated under a set of rules. Through the study of manufacturing you have also been learning about:

- Managing your activities
- Financial affairs
- Communications skills

Managing Your Activities

You have learned about management activities as you have studied manufacturing. You have learned that among the many things managers do are setting goals, determining courses of action, and establishing priorities. You will do the same thing as you approach a job. Often, you will do this subconsciously. In any event, if you approach your activities as a manager would, you will find them easier to complete. You will also be contributing to your personal satisfaction and success.

Furthermore, a lot of *stress* is caused by trying to reach unclear goals through poorly defined routes. Properly managing your activities will be less stressful for you. You will feel less pressured because you will know where you are going and how you plan to get there.

Financial Affairs

In your study of manufacturing, you have covered the area of financial affairs. What you have learned about general ledgers, credits, and debits can be applied to your personal life. The knowledge you have acquired will aid you in maintaining a personal budget and a checking account.

Let us look at a checking account. See **Figure 39-10.** You deposit money into it periodically. These are *credits* or *deposits.* Then, you write checks to buy things and pay bills. These are *debits.* Each time you deposit money, the amount is added to the account balance. Each check is subtracted from the balance. Keep track of your balance so that you will know how much money is available to spend.

By studying how companies raise money, you have indirectly learned about borrowing money. Companies use debt financing to obtain operating capital. They sell bonds or obtain bank loans. Likewise, you could also obtain a loan. You would follow the same basic procedure that a company does. That is:

1. Determine how much money is needed.
2. List personal assets (value of property, bank accounts, etc.)
3. Determine the amount of money available each month to repay the loan.
4. Select a bank or other financial institution (savings bank, loan company, insurance company, etc.) to approach for the loan.

Joseph Edward
1930 Clinton Avenue
Washington, IL 61571

Date_________________________ 20____

CHECKS AND OTHER ITEMS ARE RECEIVED FOR DEPOSIT SUBJECT TO THE TERMS AND CONDITIONS OF THIS BANK'S COLLECTION AGREEMENT

CURRENCY		
COIN		
CHECKS		
TOTAL FROM OTHER SIDE		
TOTAL		
LESS CASH RECEIVED		
Total Deposit		

89-1216
915

DEPOSIT TICKET

PLEASE ITEMIZE ADDITIONAL CHECKS ON REVERSE SIDE

Sunnyland
Bank and Trust
Sunnyland, Illinois 61571

195 212 7 ıı|

Joseph Edward
1930 Clinton Avenue
Washington, IL 61571

SAMPLE

1966

______________20______

89-1216
915

PAY TO THE ORDER OF___________________________________ $__________

____DOLLARS

Sunnyland
Bank and Trust
Sunnyland, Illinois 61571

FOR______________________ ______________________

⑆:0915 1216:⑆ 195 212 7 ıı|

PLEASE BE SURE TO **DEDUCT** ANY PER CHECK CHARGES OR SERVICE CHARGES THAT MAY APPLY TO YOUR ACCOUNT

CHECKING	DATE	CHECK ISSUED TO OR DESCRIPTION OF DEPOSIT	(–) AMOUNT OF CHECK		✓ T	(–) CHECK FEE (IF ANY)	(+) AMOUNT OF DEPOSIT		BALANCE	

Figure 39-10. Keeping a checking account means filling out deposit slips (top), writing checks (middle), and keeping track of your account balance (bottom).

5. Complete the loan application.
6. Receive the money.
7. Pay back the loan in installments over a period of years.

Remember that the loan company charges interest for the loan. This means you will be paying back more money than you borrowed.

Communications Skills

Finally, throughout your study of manufacturing, you have had to develop reports, make presentations, design advertisements, and maintain records. You have been developing communications and other skills. Take a closer look. In doing these assignments, you have had to:

- Determine what information an audience needs.
- Gather data to develop a message or a report.
- Organize material using an outline.
- Prepare and edit a written report.
- Prepare graphics and merge them into the report.
- Present a message or a report to an audience.

Perhaps you have not been aware, but you have been acquiring some basic skills that will be of benefit to you throughout your life.

Summary

Knowledge gained through learning about manufacturing can impact a person's life. Leadership opportunities are present in manufacturing and most aspects of life. The best approach to leadership is similar to a participatory management. Manufacturing provides many opportunities for employment. However, a thorough study of personal desires should be conducted in order to determine the right career path to follow.

Study of manufacturing can be related to the individual in other ways. Consumers will determine the products a company makes. Before buying, a consumer should consider whether or not the product fits certain criteria.

Finally, through the study of manufacturing, you have learned of matters relating to successful management, financial affairs, and communications. These matters can be applied to everyday life. From them, you have learned about effectively managing the activities that you do, keeping checking accounts, obtaining loans, and preparing and presenting reports.

Key Words

All of the following words have been used in this chapter. Do you know their meanings?

leadership
synergism

Test Your Knowledge

Please do not write in this text. Place your answers on a separate sheet.

1. A key to good _____ is having the ability to inspire all kinds of people to work together.
2. *True or False?* Effective leadership is accomplished through giving orders, directing people, and demanding results.
3. Effective leaders have a:
   a. "Team" attitude.
   b. Shared vision.
   c. Plan of action.
   d. All of the above.
4. *True or False?* Effective team members do not express opinions when they may cause friction in a group.
5. *True or False?* Working for a living can be rewarding.
6. *True or False?* Human relations skills are not so important to machine operators because they work with machines all day.
7. What are five general things to consider that will help you choose the right job for you?
8. Indicate which of the following statements is most correct.
   a. Buy a product right away as long as you need it and can afford it.
   b. One way of getting the best price on a product is to shop around.
   c. The consumer has little effect on determining what products a company makes.
   d. None of the above.
9. Lessons learned by studying manufacturing can be applied to everyday life in areas of:
   a. Managing activities.
   b. Financial affairs.
   c. Communications skills.
   d. All of the above.
10. A lot of _____ is caused by trying to reach unclear goals through poorly defined routes.
11. *True or False?* When you borrow money from a lending institution, the amount of money you pay back is equal to the amount you borrow.
12. What basic skills have you acquired in studying manufacturing? Explain how they will be of value to you.

Applying Your Knowledge

1. On a chart similar to that of **AYK 39-1,** list what you feel are characteristics of a good leader. Then, for each characteristic, list three state or national figures that possess the quality.
2. Explain synergism in terms of a choir or a football team.

3. Design a bulletin board display that would communicate the attributes of a good leader.
4. Prepare an agenda for a meeting. List steps you would use in conducting the meeting by parliamentary procedure.
5. Select five manufacturing jobs you know about.
   a. Develop a chart like that in **AYK 39-5.**
   b. Rate each job in terms of its requirements for working with people, data, things (machines and materials). Use a scale of 1 to 5 (with 5 being the highest score) to do the rating.
6. Select a job you might be interested in having. Analyze it in terms of its market demand, job demands, appeal to your personal traits, interest, abilities, educational requirement, and the lifestyle it would afford.

Leadership characteristic	State or national figure possessing this quality
1.	1. 2. 3.
1.	1. 2. 3.
3.	1. 2. 3.
4.	1. 2. 3.
etc.	1. 2. 3.

AYK 39-1. A suggested chart is shown.

MANUFACTURING JOB	RATING		
	PEOPLE	DATA	THINGS

AYK 39-5. A suggested chart is shown.

Appendix

This appendix contains a number of product ideas. They may be used as product designs for class projects. They may also be used as examples as class products are researched, designed, marketed, manufactured, and distributed. This appendix also contains a sample of a Material Safety Data Sheet (MSDS). It is helpful for training employees on how to use the chemicals and how to respond properly should an accident occur.

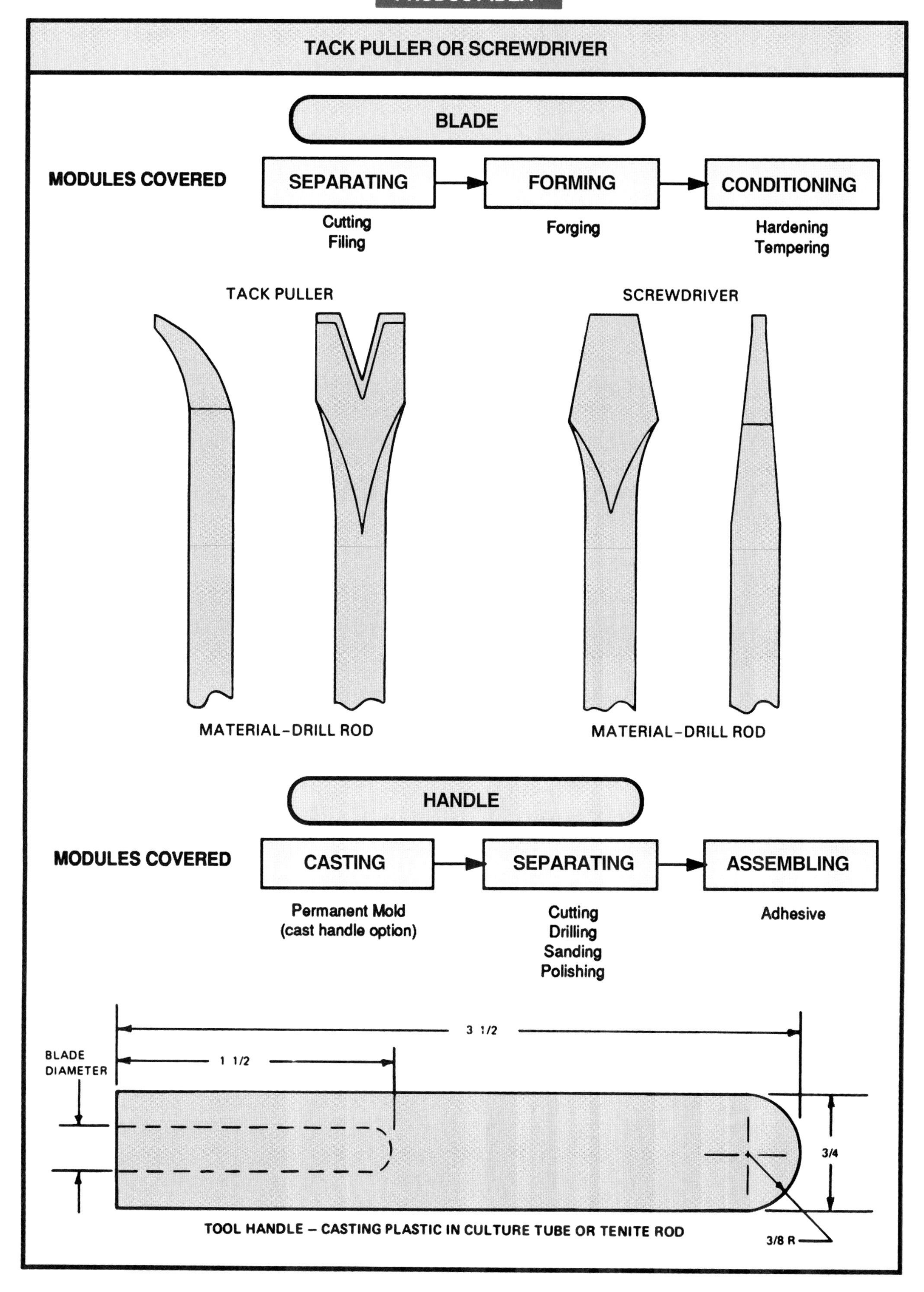

PRODUCT IDEA
TACK PULLER OR SCREWDRIVER
BLADE
MODULES COVERED
SEPARATING
FORMING
CONDITIONING
Cutting
Filing
Forging
Hardening
Tempering
TACK PULLER
SCREWDRIVER
MATERIAL-DRILL ROD
MATERIAL-DRILL ROD
HANDLE
MODULES COVERED
CASTING
SEPARATING
ASSEMBLING
Permanent Mold
(cast handle option)
Cutting
Drilling
Sanding
Polishing
Adhesive
3 1/2
1 1/2
BLADE
DIAMETER
3/4
3/8 R
TOOL HANDLE – CASTING PLASTIC IN CULTURE TUBE OR TENITE ROD

TACK PULLER OR SCREWDRIVER

BILL OF MATERIALS

Qty.	Part Name	Size T	x W	x L	Material
1	Blade	3/16 dia or 1/4 dia		5	Drill rod
1	Handle	3/4 dia		3 1/2	Tenite rod

OPERATION PROCESS CHART

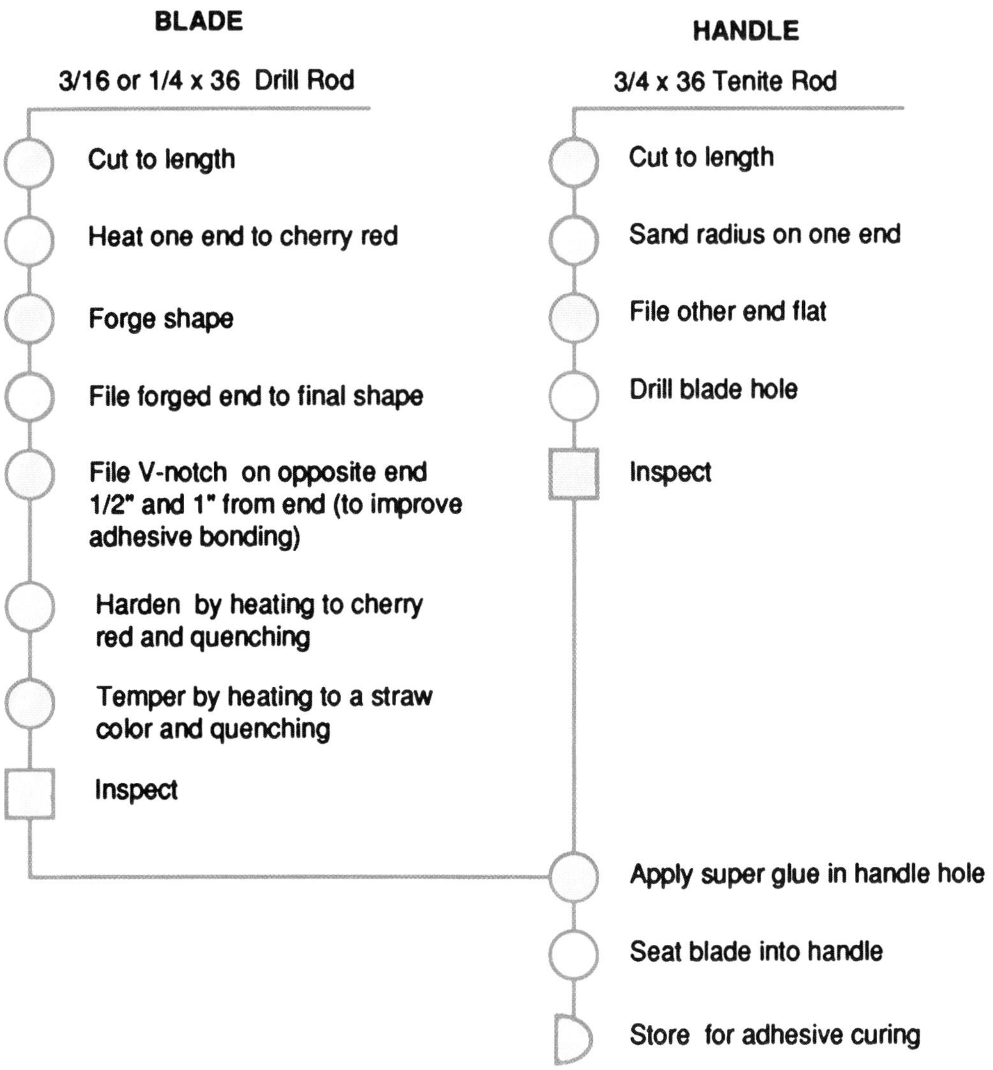

BLADE

3/16 or 1/4 x 36 Drill Rod

- Cut to length
- Heat one end to cherry red
- Forge shape
- File forged end to final shape
- File V-notch on opposite end 1/2" and 1" from end (to improve adhesive bonding)
- Harden by heating to cherry red and quenching
- Temper by heating to a straw color and quenching
- Inspect

HANDLE

3/4 x 36 Tenite Rod

- Cut to length
- Sand radius on one end
- File other end flat
- Drill blade hole
- Inspect

ASSEMBLY

- Apply super glue in handle hole
- Seat blade into handle
- Store for adhesive curing

PRODUCT IDEA

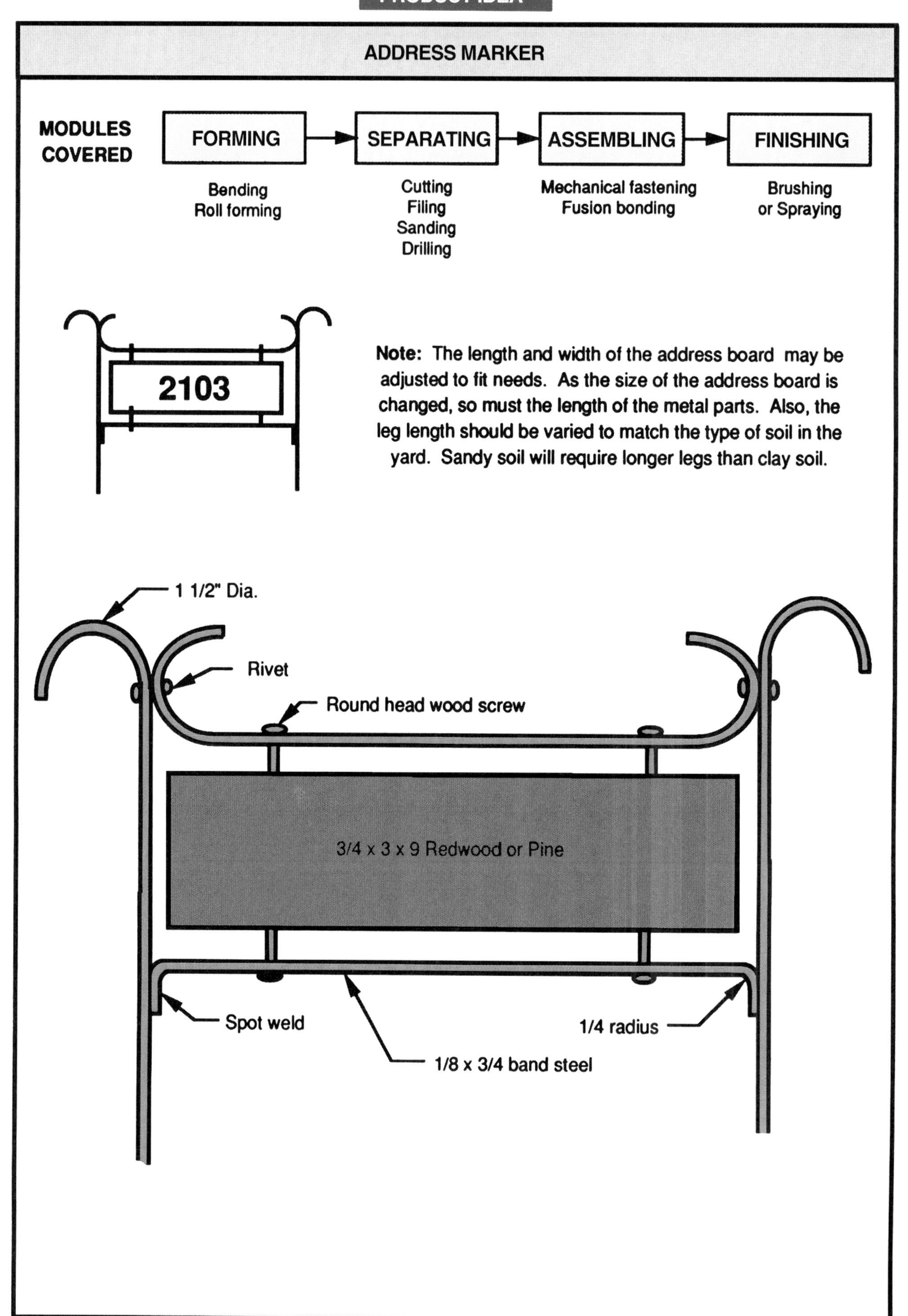

ADDRESS MARKER

BILL OF MATERIALS

Qty.	Part Name	Size T	x W	x L	Material
1	Board	3/4	3	9	Redwood or pine
2	Legs	1/8	3/4	18	Steel
1	Top bar	1/8	3/4	13	Steel
1	Bottom bar	1/8	3/4	11	Steel

OPERATION PROCESS CHART

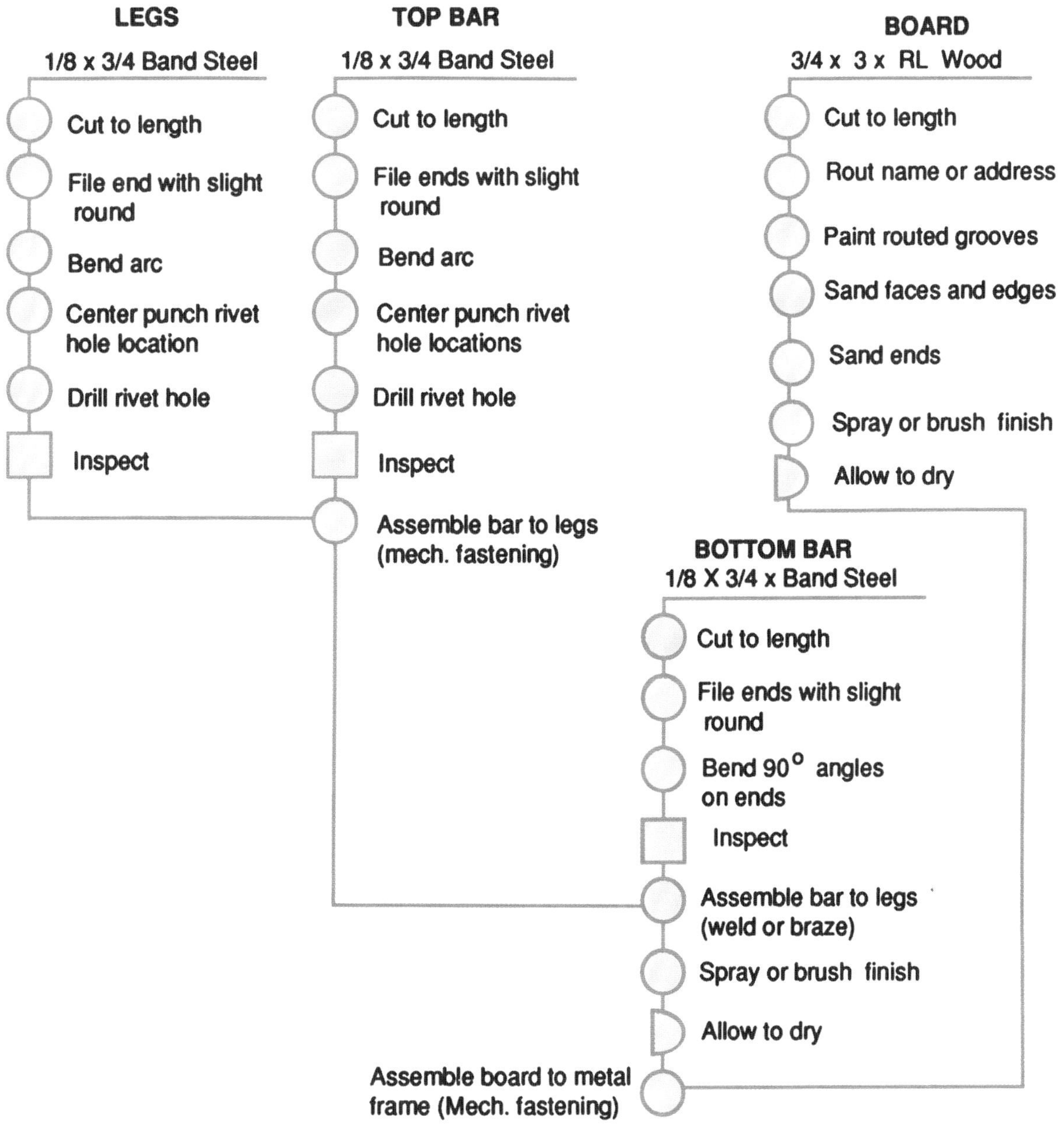

LEGS
1/8 x 3/4 Band Steel
- Cut to length
- File end with slight round
- Bend arc
- Center punch rivet hole location
- Drill rivet hole
- Inspect

TOP BAR
1/8 x 3/4 Band Steel
- Cut to length
- File ends with slight round
- Bend arc
- Center punch rivet hole locations
- Drill rivet hole
- Inspect
- Assemble bar to legs (mech. fastening)

BOARD
3/4 x 3 x RL Wood
- Cut to length
- Rout name or address
- Paint routed grooves
- Sand faces and edges
- Sand ends
- Spray or brush finish
- Allow to dry

BOTTOM BAR
1/8 X 3/4 x Band Steel
- Cut to length
- File ends with slight round
- Bend 90° angles on ends
- Inspect
- Assemble bar to legs (weld or braze)
- Spray or brush finish
- Allow to dry

Assemble board to metal frame (Mech. fastening)

PRODUCT IDEA

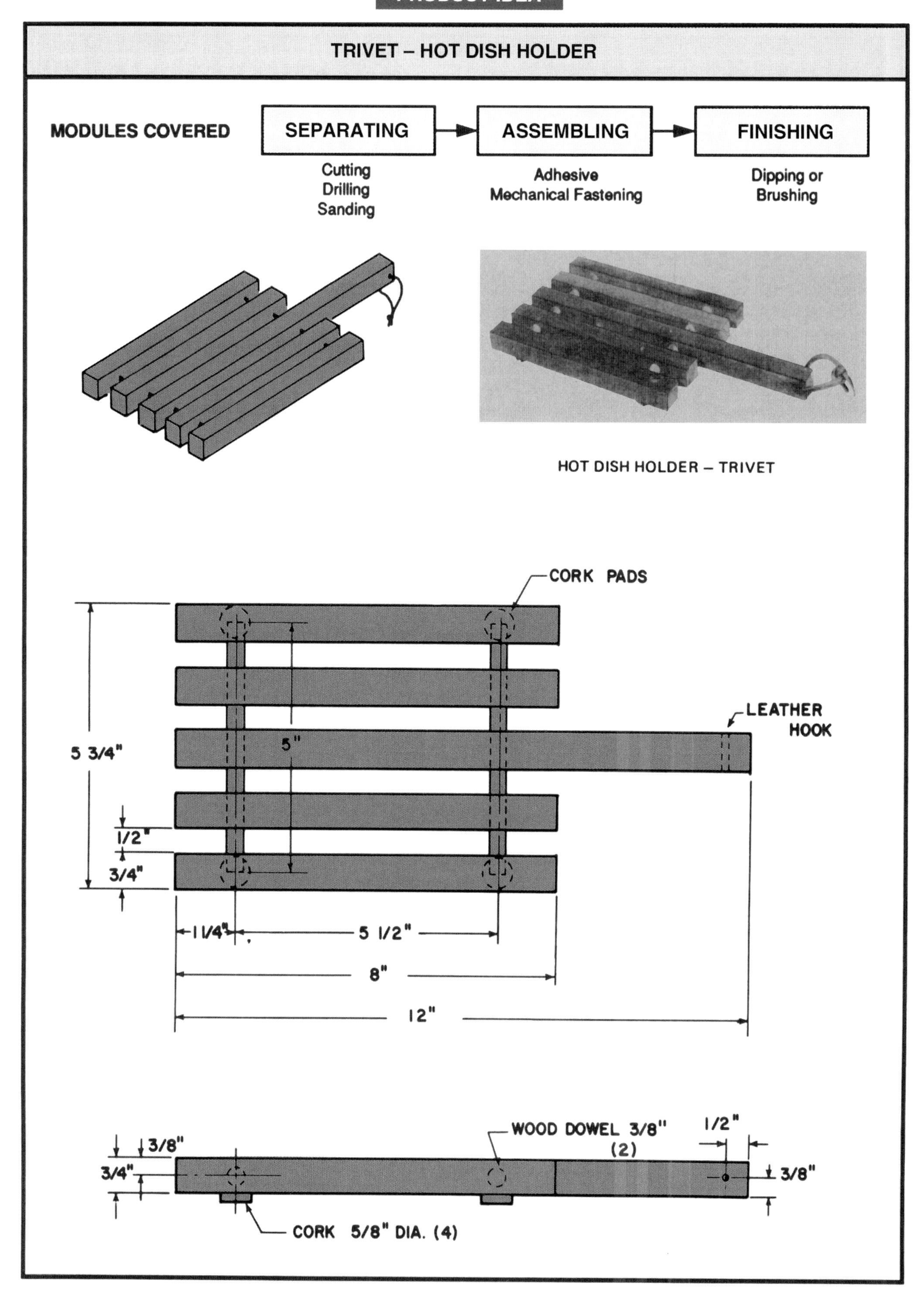

TRIVET – HOT DISH HOLDER

BILLS OF MATERIALS

Qty.	Part Name	Size			Material
		T	x W	x L	
4	Short slats	3/4	3/4	8	Maple, walnut, or poplar
1	Long slat	3/4	3/4	12	Maple, walnut, or poplar
2	Separator rods	3/8		5	Birch dowel
4	Cork feet	5/8 dia			Sheet cork
1	Leather hook			6	Boot lace

OPERATION PROCESS CHART

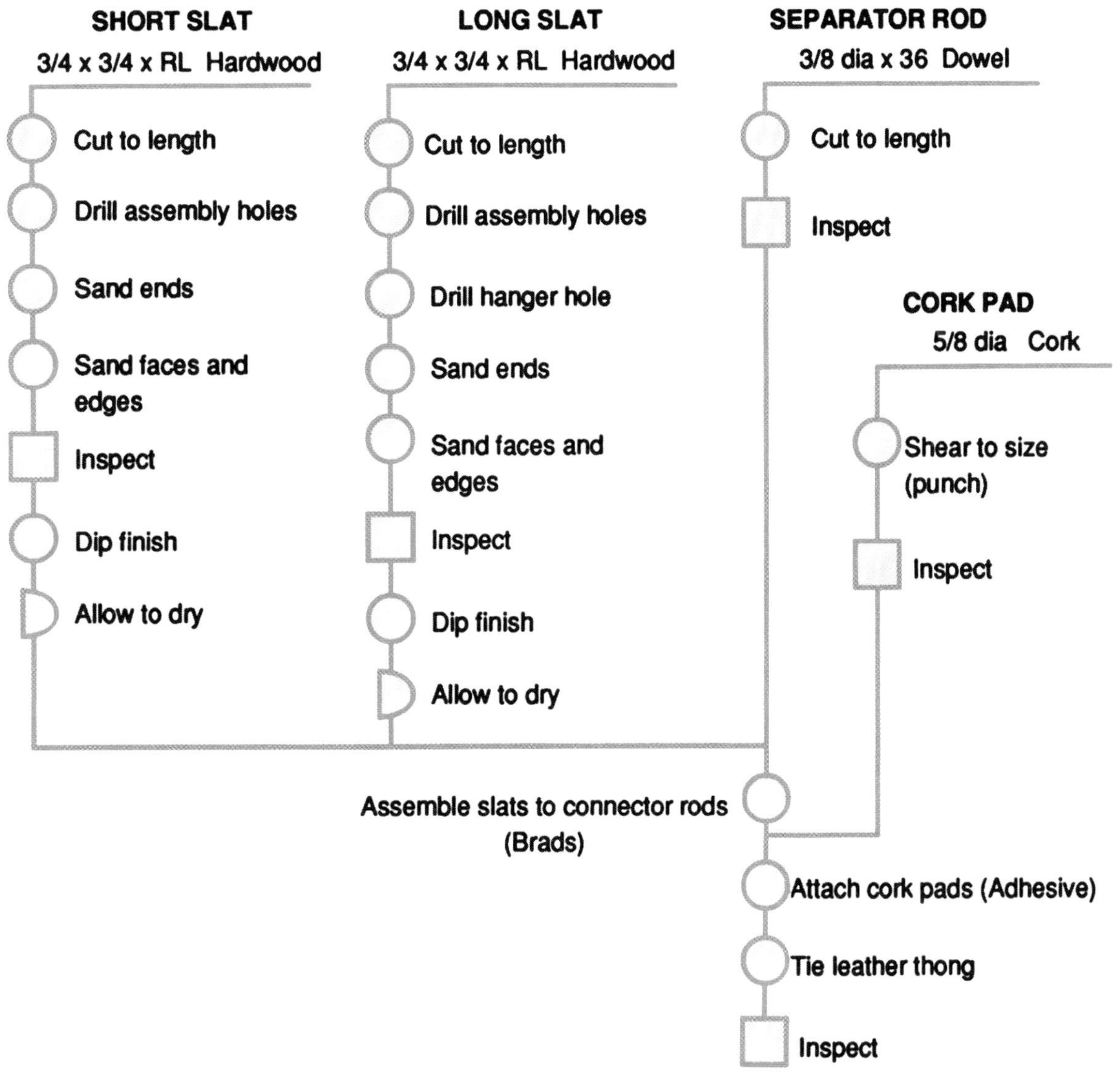

SHORT SLAT
3/4 x 3/4 x RL Hardwood

- Cut to length
- Drill assembly holes
- Sand ends
- Sand faces and edges
- Inspect
- Dip finish
- Allow to dry

LONG SLAT
3/4 x 3/4 x RL Hardwood

- Cut to length
- Drill assembly holes
- Drill hanger hole
- Sand ends
- Sand faces and edges
- Inspect
- Dip finish
- Allow to dry

SEPARATOR ROD
3/8 dia x 36 Dowel

- Cut to length
- Inspect

CORK PAD
5/8 dia Cork

- Shear to size (punch)
- Inspect

Assemble slats to connector rods (Brads)

Attach cork pads (Adhesive)

Tie leather thong

Inspect

PRODUCT IDEA

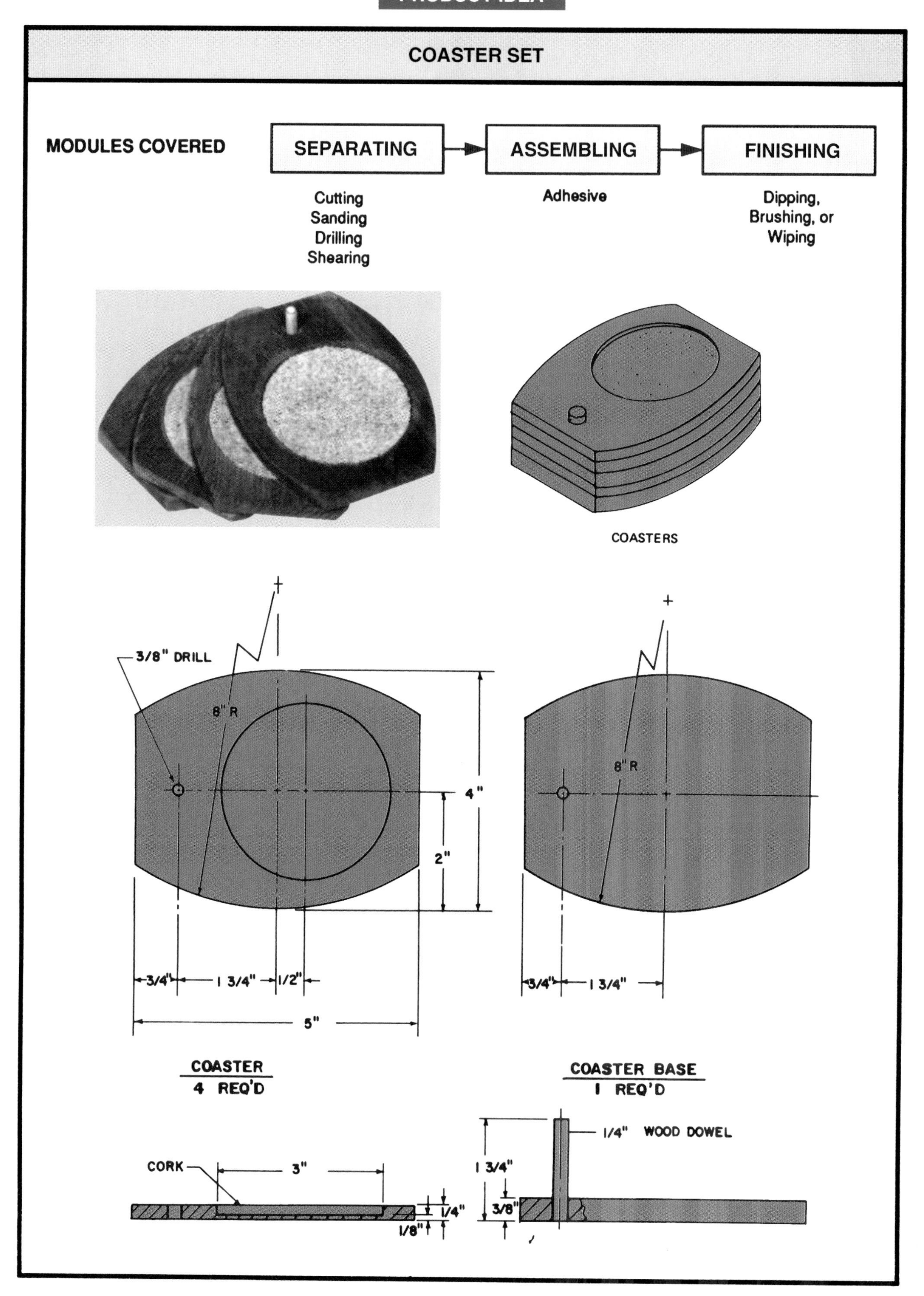

COASTER SET

BILL OF MATERIALS

Qty.	Part Name	Size T	x W	x L	Material
4	Coasters	1/4	4	5	Walnut, maple, or redwood
1	Base	3/8	4	5	Walnut, maple, or redwood
1	Post	1/4 dia		1 3/4	Birch dowel

OPERATION PROCESS CHART

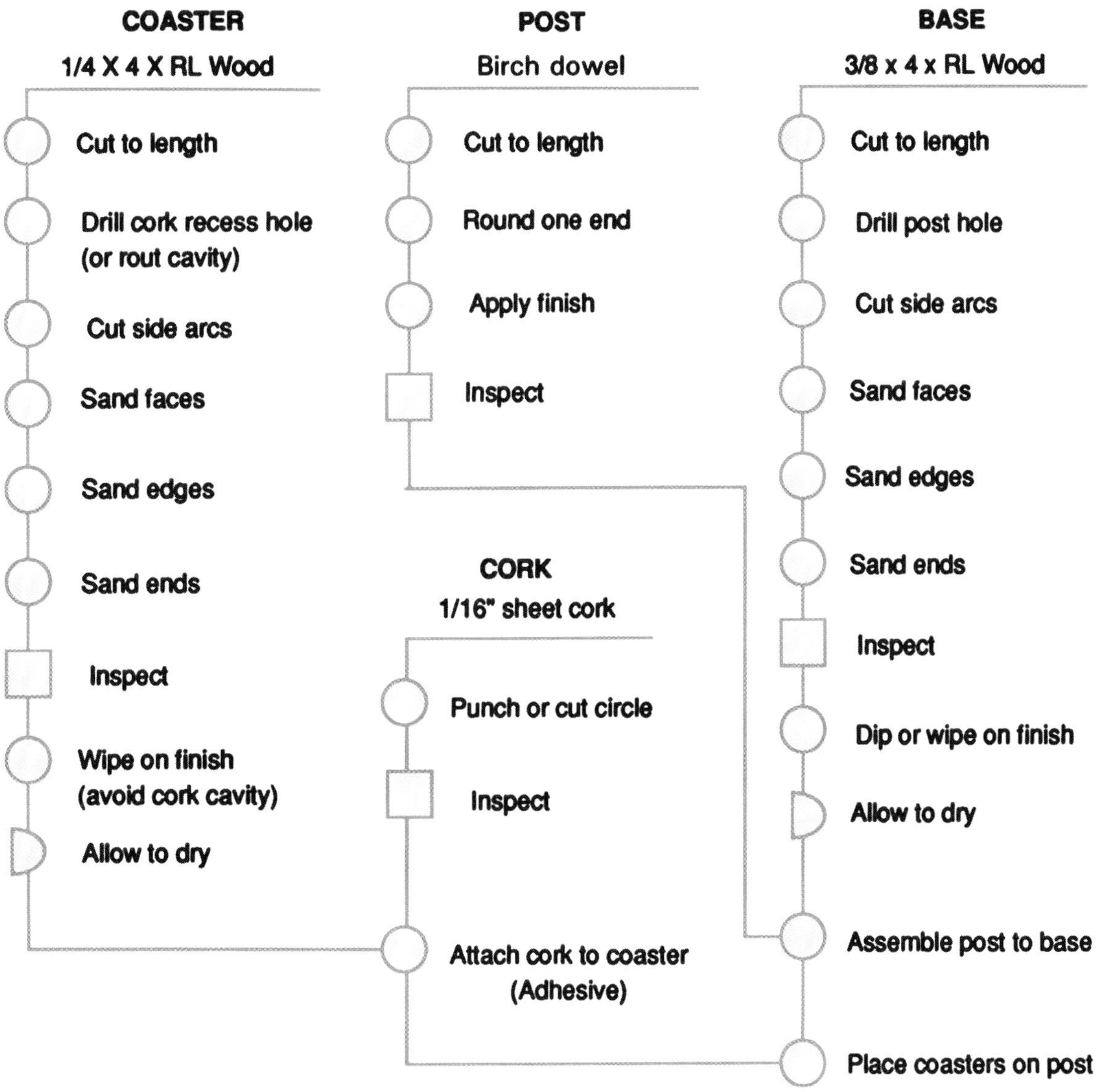

COASTER
1/4 X 4 X RL Wood

- Cut to length
- Drill cork recess hole (or rout cavity)
- Cut side arcs
- Sand faces
- Sand edges
- Sand ends
- Inspect
- Wipe on finish (avoid cork cavity)
- Allow to dry

POST
Birch dowel

- Cut to length
- Round one end
- Apply finish
- Inspect

BASE
3/8 x 4 x RL Wood

- Cut to length
- Drill post hole
- Cut side arcs
- Sand faces
- Sand edges
- Sand ends
- Inspect
- Dip or wipe on finish
- Allow to dry

CORK
1/16" sheet cork

- Punch or cut circle
- Inspect

Attach cork to coaster (Adhesive)

Assemble post to base

Place coasters on post

PRODUCT IDEA

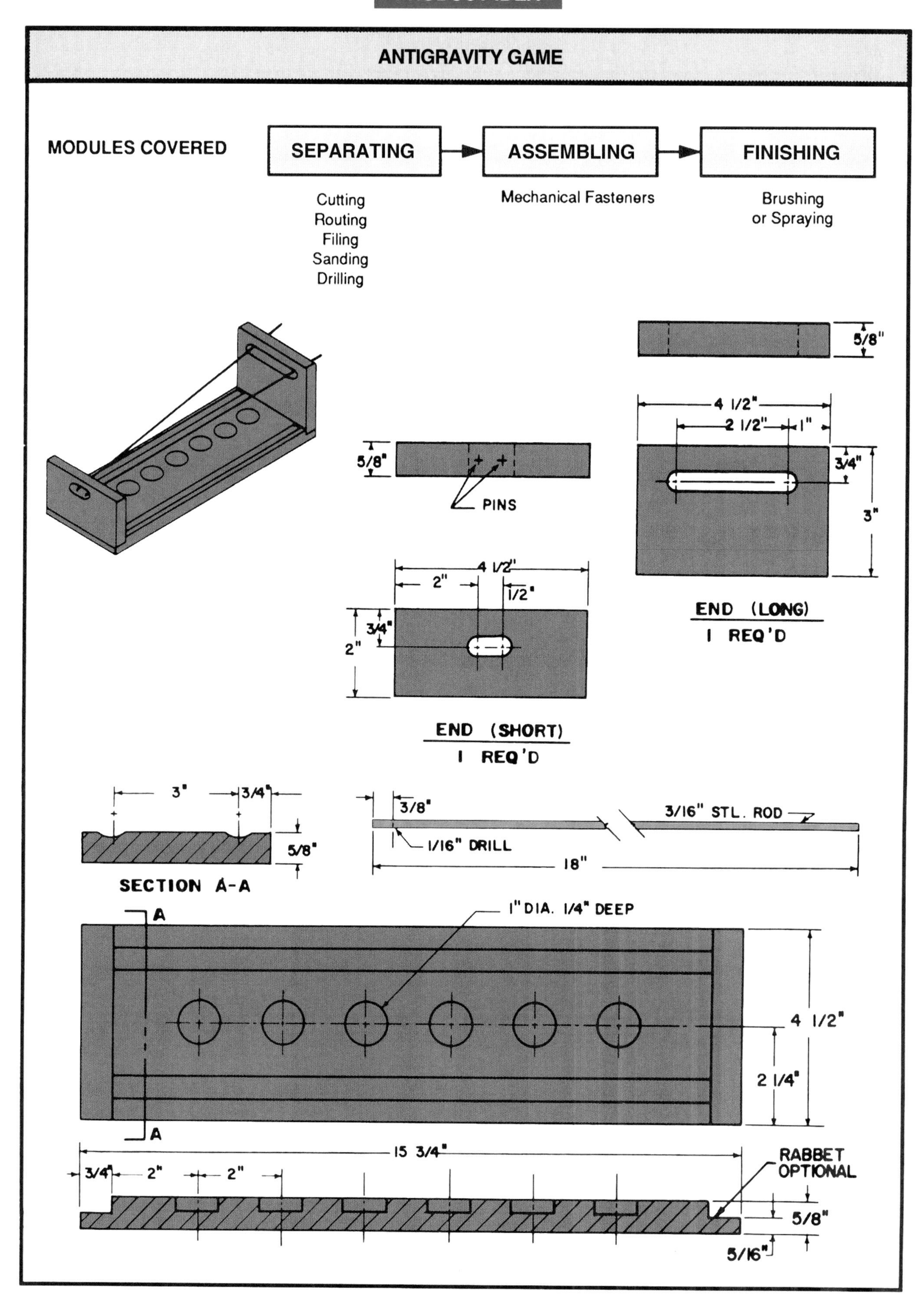

ANTIGRAVITY GAME

BILLS OF MATERIALS

Qty.	Part Name	Size T	x W	x L	Material
1	Base	5/8	4 1/2	15 3/4	Redwood or pine
1	Small end	5/8	4 1/2	2	Redwood or pine
1	Large end	5/8	4 1/2	3	Redwood or pine
2	Rod	1/4		18	Welding rod
1	Ball	1 dia			Reject ball bearing

OPERATION PROCESS CHART

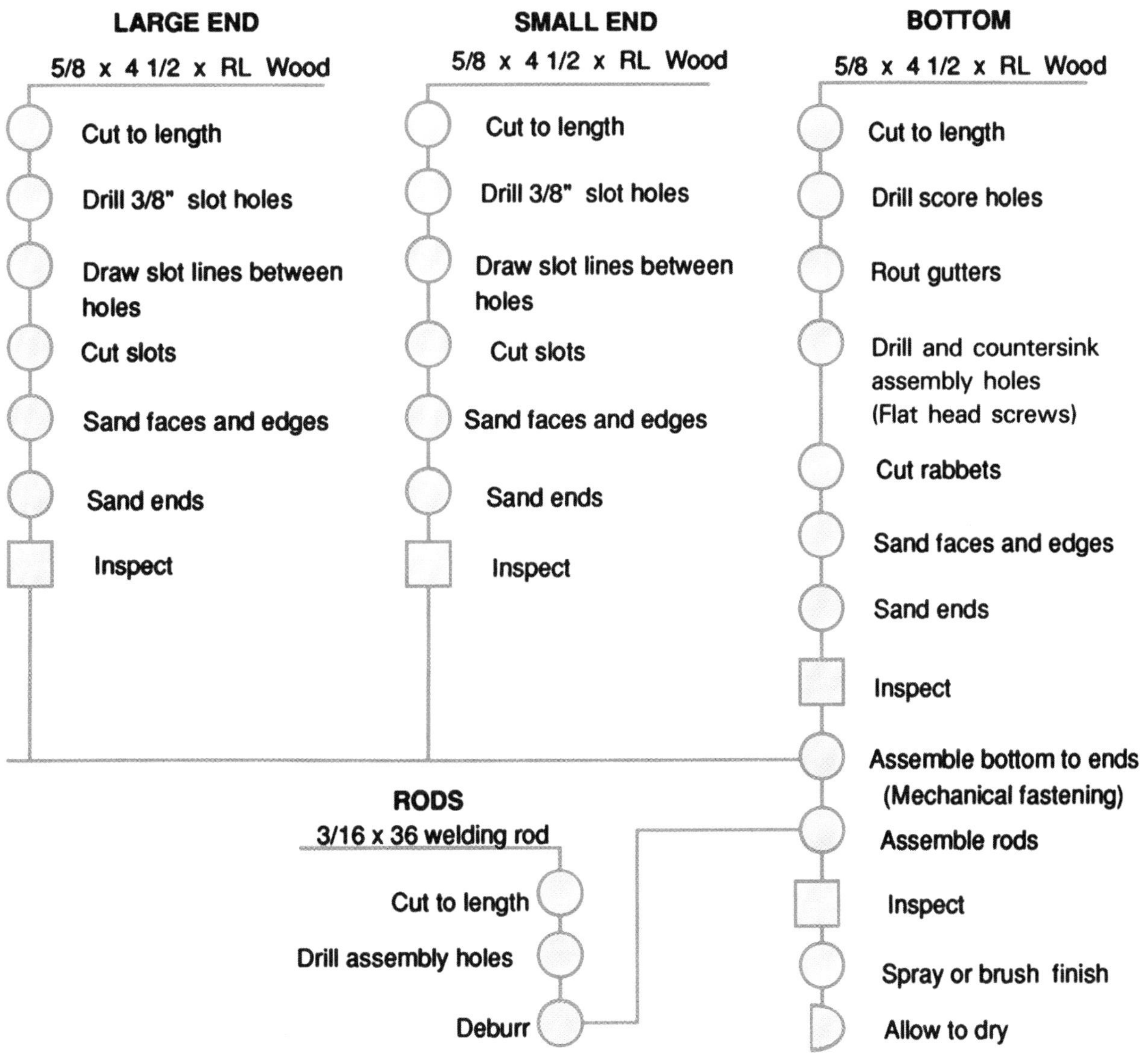

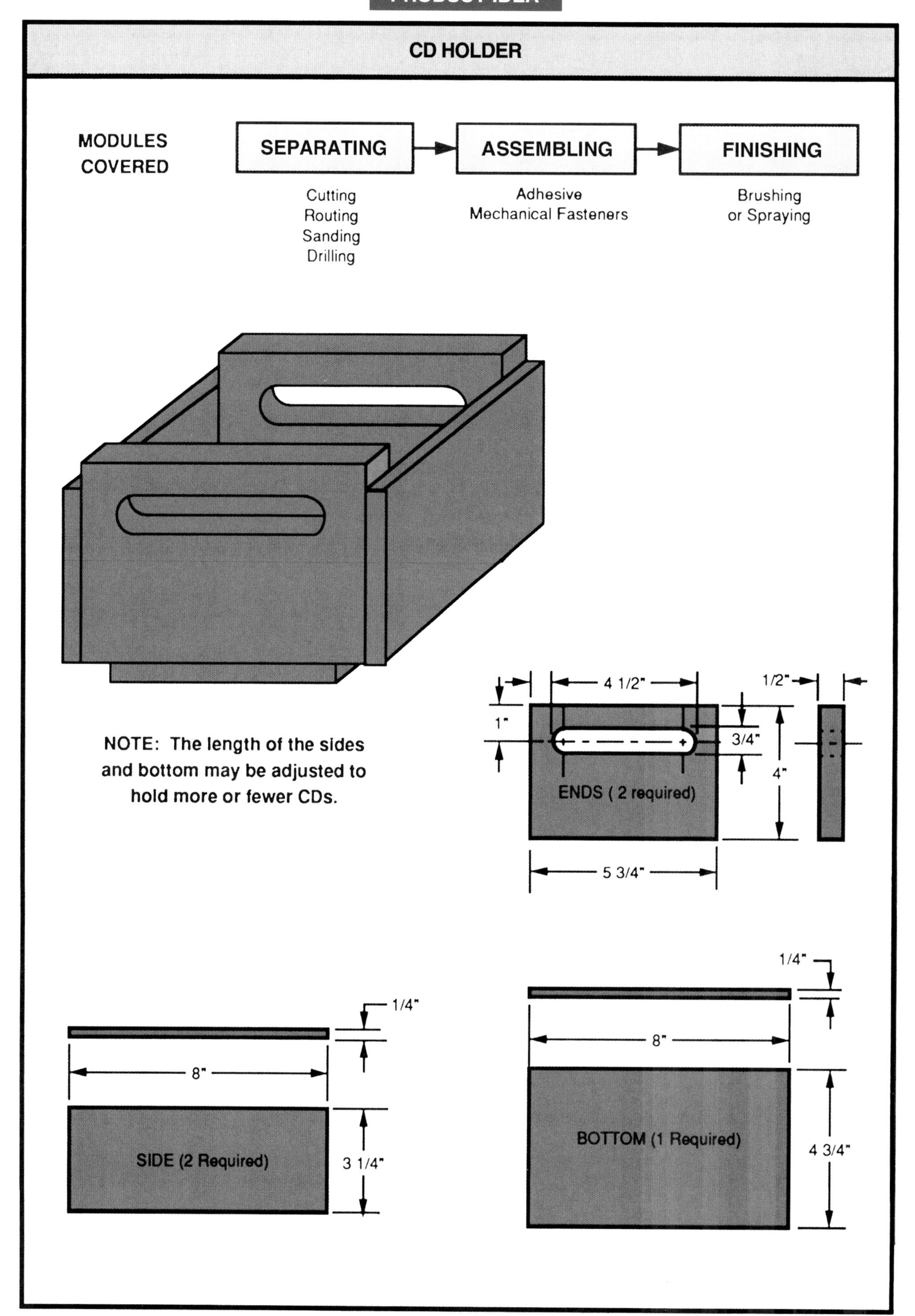
PRODUCT IDEA
CD HOLDER
MODULES COVERED
SEPARATING
Cutting
Routing
Sanding
Drilling
ASSEMBLING
Adhesive
Mechanical Fasteners
FINISHING
Brushing
or Spraying
NOTE: The length of the sides and bottom may be adjusted to hold more or fewer CDs.
4 1/2"
1/2"
1"
3/4"
4"
ENDS (2 required)
5 3/4"
1/4"
8"
SIDE (2 Required)
3 1/4"
1/4"
8"
BOTTOM (1 Required)
4 3/4"

CD HOLDER

BILL OF MATERIALS

Qty.	Part Name	Size T	x W	x L	Material
2	Ends	1/2	4	5 3/4	Maple, walnut, or pine
2	Sides	1/4	3 1/4	8	Maple, walnut, or pine
1	Bottom	1/4	4 3/4	8	Maple, walnut, or pine

OPERATION PROCESS CHART

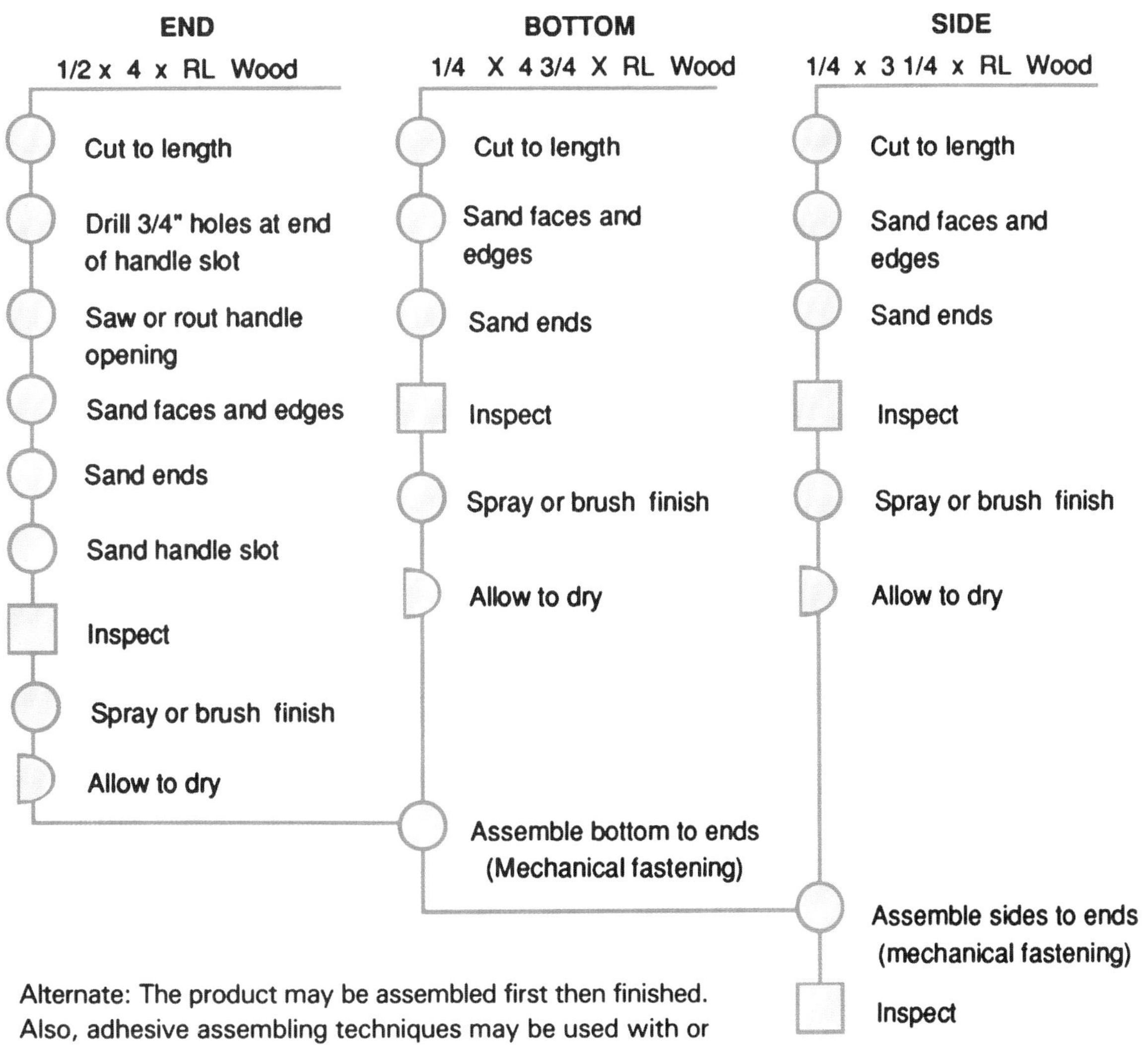

END
1/2 x 4 x RL Wood

- Cut to length
- Drill 3/4" holes at end of handle slot
- Saw or rout handle opening
- Sand faces and edges
- Sand ends
- Sand handle slot
- Inspect
- Spray or brush finish
- Allow to dry

BOTTOM
1/4 X 4 3/4 X RL Wood

- Cut to length
- Sand faces and edges
- Sand ends
- Inspect
- Spray or brush finish
- Allow to dry
- Assemble bottom to ends (Mechanical fastening)

SIDE
1/4 x 3 1/4 x RL Wood

- Cut to length
- Sand faces and edges
- Sand ends
- Inspect
- Spray or brush finish
- Allow to dry
- Assemble sides to ends (mechanical fastening)
- Inspect

Alternate: The product may be assembled first then finished. Also, adhesive assembling techniques may be used with or in place of the mechanical fasteners.

PRODUCT IDEA

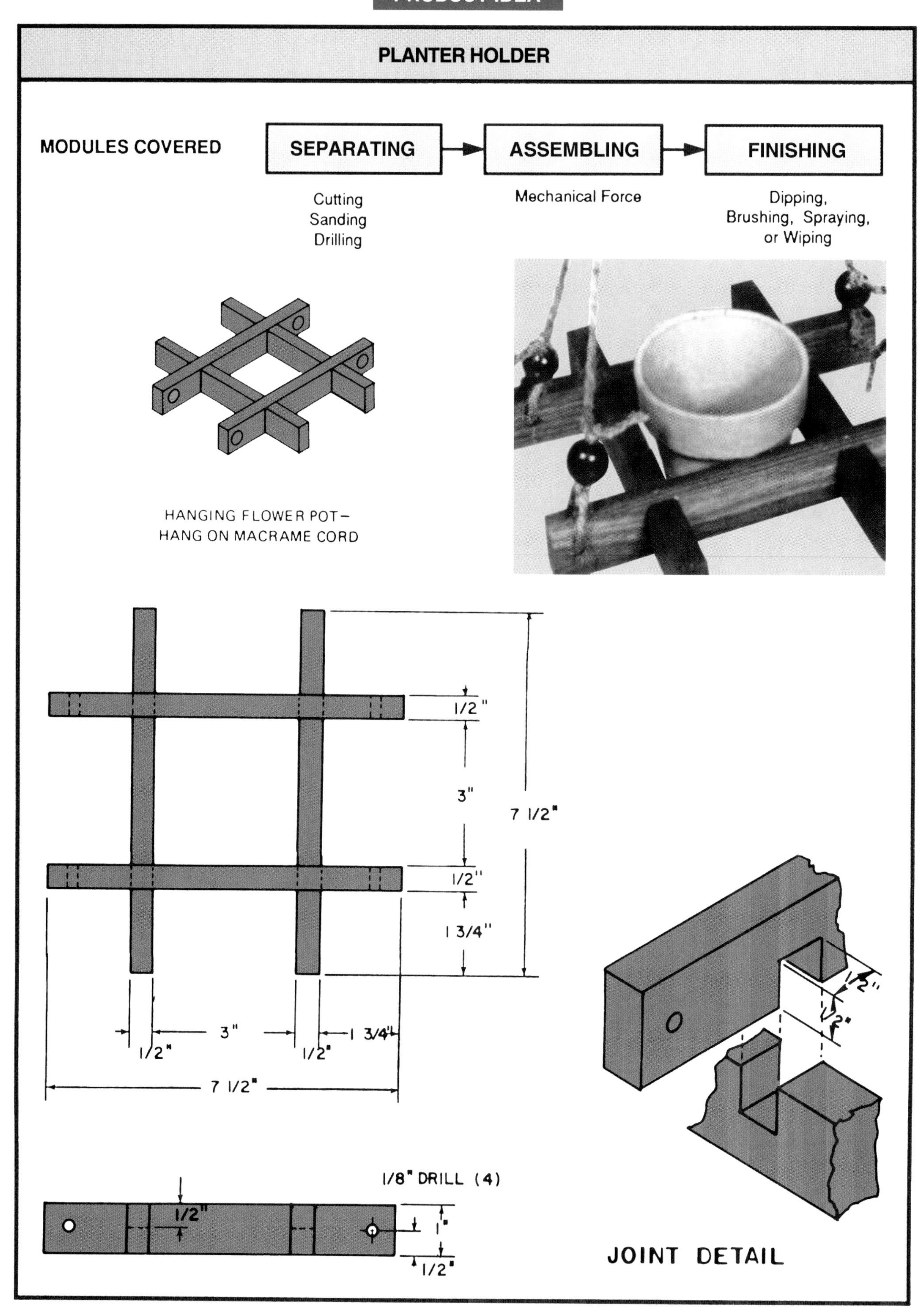

PLANT HOLDER

BILL OF MATERIALS

Qty.	Part Name	Size T	x W	x L	Material
4	Slats	1/2	1	5	Walnut, maple, or redwood
2	Cords			24 - 28	Hemp cord
4	Beads	1/2 dia			Colored wooden beads

OPERATION PROCESS CHART

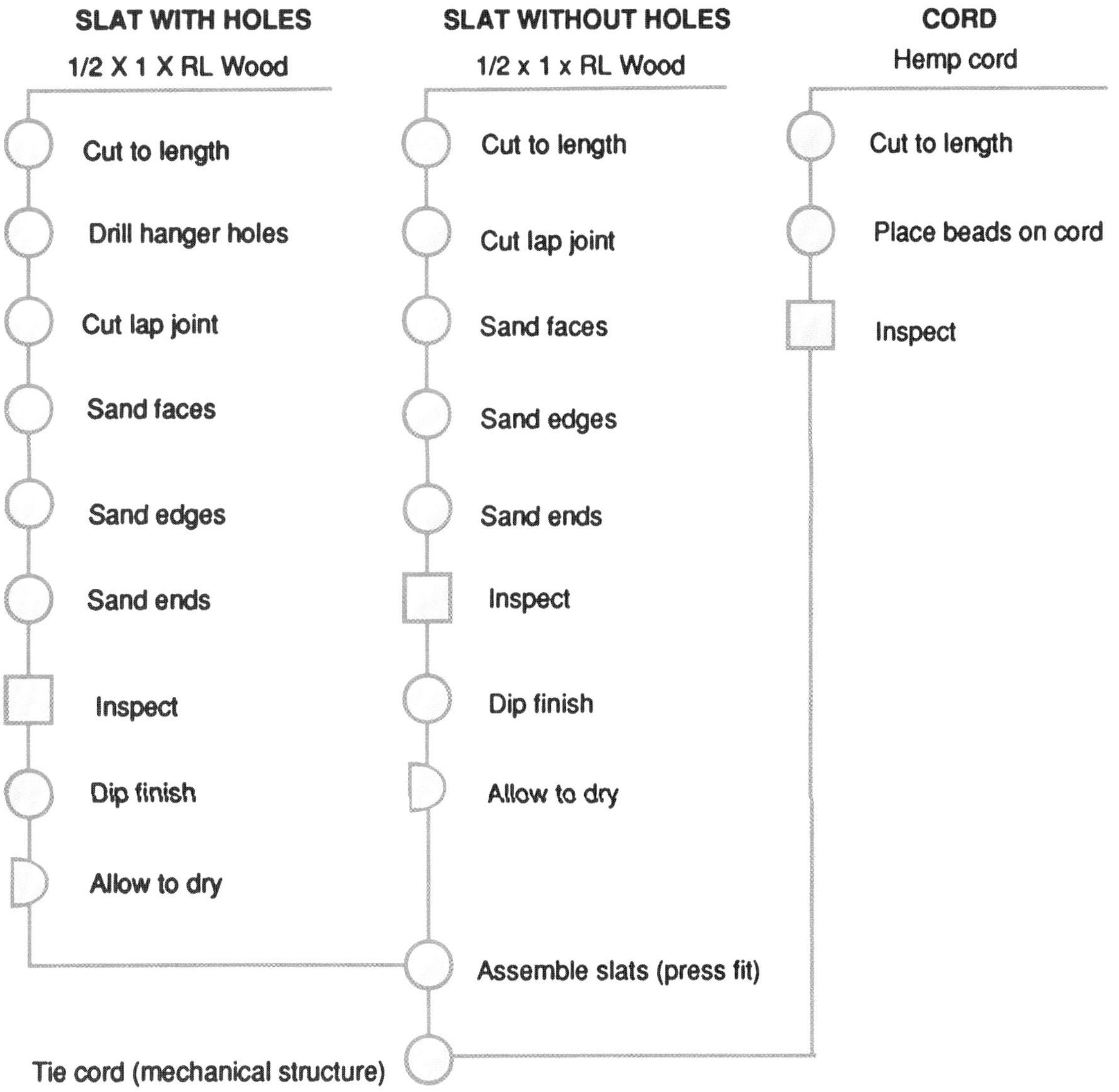

SLAT WITH HOLES
1/2 X 1 X RL Wood

- Cut to length
- Drill hanger holes
- Cut lap joint
- Sand faces
- Sand edges
- Sand ends
- Inspect
- Dip finish
- Allow to dry

SLAT WITHOUT HOLES
1/2 x 1 x RL Wood

- Cut to length
- Cut lap joint
- Sand faces
- Sand edges
- Sand ends
- Inspect
- Dip finish
- Allow to dry

CORD
Hemp cord

- Cut to length
- Place beads on cord
- Inspect

Assemble slats (press fit)

Tie cord (mechanical structure)

Material Safety Data Sheet

Preparation/Revision Date: 12/16/XX

ACME Chemical Company

24 hour Emergency Phone: Chemtrec: 1-800-424-9300
Outside United States: 1-202-483-7616

Trade Name/Syn: **DICHLOROMETHANE * METHYLENE DICHLORIDE**
Chem Name/Syn: **METHYLENE CHLORIDE**
CAS Number: **75-09-2**
Formula: **Ch_2Cl_2**

NFPA Rating:

Health 4
Flammability 1
Reactivity 1

Statement of Hazard:

Possible cancer hazard. May cause cancer based on animal data. Harmful if swallowed or inhaled. Vapor irritating. May cause eye injury and/or skin irritation. May cause damage to liver, kidneys, blood, and central nervous system.

Effects of Overexposure-Toxicity-Route of Entry:

Toxic by ingestion and inhalation. Irritating on contact with skin, eyes, or mucous membranes. May cause eye injury. Inhalation of high vapor concentrations causes dizziness, nausea, headache, narcosis, irregular heartbeats, coma, and death. If vomiting occurs, methylene chloride can be aspirated into the lungs, which can cause chemical pneumonia and systemic effects. Medical conditions aggravated by exposure: heart, kidney, and liver conditions. Routes of entry: inhalation, ingestion.

Hazardous Decomposition Products:

HCL, phosgene, chlorine

Will Hazardous Polymerization Occur?

Will not occur under normal conditions.

Is the Product Stable?

Product is normally stable.

Conditions to Avoid:

Contact with open flame, welding arcs, and hot surfaces.

Spill Procedures, Disposal Requirements/Methods:

Evacuate the area of all unnecessary personnel. Wear suitable protective equipment. Eliminate any ignition sources until the area is determined to be free from explosion or fire hazard. Contain the release with a suitable absorbent. Place in a suitable container for disposal. Dispose of in accordance with all applicable federal, state, and local regulations.

Ventilation:

Local exhaust: Recommended
Mechanical (Gen): Recommended
Special: NA
Other: None

Respiratory Protection:

NIOSH/MSHA air supplied respirator.

Protective Gloves:

Viton, PVA, or equivalent to prevent skin contact.

Other Protective Equipment:

Safety glasses with side shields must be worn at all times; eyewash; fume hood.

Part of a Material Safety Data Sheet for a hazardous solvent, providing information on ventilation and personal protective equipment. Always check the MSDS before working with any chemical.

Glossary

A

Abrasive cleaning: rubbing a surface with sandpaper or a similar material to smooth it and remove grease or other contaminants.

Abrasive machining: a process that uses an abrasive, in wheel or sheet form, to remove material from the workpiece.

Accounting: the practice and process of keeping financial records.

Accounts payable: money owed by a company to its suppliers (for material) and its employees (for labor).

Accounts receivable: money owed to the company by customers who have purchased products or services.

Acoustical properties: how a material absorbs, transmits, or reflects sound waves.

Acoustical reflectivity: a measure of how well or poorly sound waves bounce off a material.

Acoustical transmission: a measure of a material's ability to transmit sound waves.

Actuating: assigning tasks to specific employees and encouraging them to complete the assigned work.

Adaptation: changing a product by improving it. For example, the word processor is an adaptation of the manual typewriter.

Adaptive control: a system of automated machine control that adjusts tool feed rates to match changes in operating conditions.

Adhesive bonds: a means of joining materials that makes use of the stickiness or tackiness of the bonding substance.

Administrative budgets: estimates of costs for such activities as industrial relations, research and development, and financial affairs.

Advertising agencies: special companies that are hired to create advertisements and marketing messages.

Advertising: a marketing function aimed at informing customers and persuading them to buy a product.

Advertising agency: a specialized service company that has experience and talent in developing effective advertising materials.

AFL-CIO: the American Federation of Labor/Congress of Industrial Organizations, the only major labor federation in the United States.

Agency shop: a system under which nonunion workers pay union dues and receive all benefits gained by the union, but do not actually have to be union members.

Agricultural period: the period in human history when primitive people began to grow crops for food instead of hunting and gathering food from the wild.

Air drying: reducing wood moisture by allowing air to move around the lumber and carry away excess moisture.

Alloy: a mixture of two or more metals or a metal and another inorganic material that yields certain qualities. For example, brass is an alloy of copper and zinc.

Amorphous structure: an internal arrangement of atoms in a material that does not follow any specific pattern.

Annealing: a heat-treating process that relieves internal stresses in a part and softens any work hardening.

Annual report: a complete description of a company's financial activities during the preceding year; usually accompanied by other information about the company and its products.

Anodizing: a process that oxidizes the surface of a part, giving a layer of oxide that will resist corrosion. Anodizing is often used on aluminum and magnesium parts.

Applied research: investigations aimed at getting specific results; applied research is aimed at solving problems or developing products.

Apprenticeship: training that combines on-the-job and classroom instruction. The on-the-job portion is provided by a skilled worker. A special teacher provides the classroom portion.

Apprenticeship training: a lengthy process, involving both classroom training and work experience, for highly skilled positions.

Arbitration: the process where an outside person hears both sides of a dispute and issues a ruling that both sides must live with.

Arbitrator: in a labor dispute, a person who acts as a judge and rules in favor of either the company or the union.

Arc welding: a means of bonding metal, using heat developed by an electric arc to melt and fuse the materials.

Area: the measure of a surface in which the length is multiplied by the width. Area is used to measure materials and the floor space of buildings.

Articles of incorporation: the application for a corporate charter, containing information on the activities of the proposed corporation and the name of persons involved in it.

Artificial intelligence (AI): computer-based systems that have the ability to learn from information gathered through sensory inputs.

Assembling: a manufacturing process in which the final product is put together from separate parts.

Assembly: a group of parts that make up one product. For example, a pencil is an assembly of wood, graphite, metal, rubber, glue, and paint.

Assembly drawing: engineering drawing that provides information to show assemblers how parts of a product fit together.

Assets: the things that a company owns, such as cash, property, and equipment.

Atom: the basic unit in the study of the microstructure of materials.

Atomic closeness: in cohesive bonding, atoms of each part must have the same atomic closeness (be as close together) across the joint as they do within the materials themselves.

Automation: replacement of human control for a machine, process, or system, with control by mechanical or electronic devices.

Authorization cards: cards that are signed by union members giving a union the authority to negotiate for employees.

Authorized dealer: a retailer who buys directly from the manufacturer and is the only person allowed to sell a product in a certain area.

B

Balance sheet: a financial document that gives a picture of the company's financial situation on a specific date.

Bankruptcy: a state in which a company cannot pay its debts; the company is dissolved or reorganized.

Bargain: to negotiate on behalf of a union or company.

Basic oxygen process: a steelmaking process that reduces the carbon content of iron by injecting pure oxygen into molten iron.

Basic research: seeking after knowledge without being concerned about a specific product or process as a result.

Batch processing: the intermittent manufacture of raw materials in groups called batches.

Bench rule: a ruler for measuring small parts, marked in 1/8" increments.

Bending: one of the forming forces used to form material; produced by presses.

Bending forces: used to form sheet metal, tubing, and bar stock into new shapes.

Bid: a price from a vendor that will remain in effect for a stated period of time.

Biennially: occurring every other year.

Bill of materials: a list of all the parts and hardware items needed to make one product.

Biotechnology: the use of technological systems to grow genetic resources or employ living organisms to reach specific goals (such as the use of enzymes to clean up oil spills).

Blow molding: a process that uses air pressure to form hollow glass or plastic objects.

Board of directors: a group of people elected by the stockholders to represent their interests. The board sets company policy.

Bonding agent: a substance used to fasten parts together.

Bonding: A—permanently fastening parts together using heat, pressure, and/or a bonding agent. B—an attractive force that holds atoms and ions together.

Bonds: certificates of indebtedness issued by a corporation as a means of borrowing money.

Branched layout: a form of plant layout used for assembly operations. Subassemblies are produced operate on branches, then fed to the main stem for use in putting together the final product.

Brand name: a name used to identify a product and differentiate it from competing products.

Brazing: a method of joining metals that involves heating them and melting a filler material that will act as an adhesive.

Break-even chart: a graphic representation of income and expenses, showing the point at which sales income for a product will meet all the expenses involved in producing it.

Brittleness: a measure of a material's lack of resistance to force. A brittle material fractures easily.

Brushing: a cleaning method that uses a wire-bristled brush to dislodge and remove contaminants and dirt from a surface.

Budget: plan that a business uses to forecast income and expenses.

Business plan: a document or program that describes all aspects of a projected new business.

Bylaws: detailed information that outlines how a company will be run. The stockholders have the power to develop and change the bylaws.

C

Calipers: a tool that is used to measure two parallel surfaces. The distance between the caliper legs is measured with a rule. There are inside and outside calipers.

Cant: the square center of a log that is cut into lumber on a gang saw in a sawmill.

Capital resources: a company's plant and equipment.

Capital: the plant and equipment used to produce products, the permanent physical resources of a company.

Carbon steels: the most common forms of steel, widely used in manufacturing products such as automobiles, machinery, and appliances.

Career ladder: the concept that if you start at an entry-level job you can advance to other jobs after you gain experience.

Case hardening: a hardening process that produces a layer of hard metal on a soft core. Heated parts are placed in a carbon-rich substance. The parts absorb the carbon into a thin layer. The interior of the parts is not changed.

Cash flow: the money handling activities of a company, consisting of income and expenses.

Casting: a process in which an industrial material is first made into a liquid. The liquid material is poured or forced into a prepared mold. The material is allowed to solidify. The solid material is then removed from the mold.

Caustic cleaning: a surface preparation method employing strongly alkaline solutions.

Cavity: the void in a casting or forming mold where the part will take shape.

Central processing unit (CPU): an integrated circuit that controls and sequences operations of a computer.

Centrifugal casting: a forming process in which molten metal is placed in a spinning mold to form a hollow product.

Ceramics: a range of materials that have a crystalline structure, are inorganic (never living), and can be either metallic or nonmetallic. Ceramic materials are generally stable and are not greatly affected by heat, weather, or chemicals. They have high melting points and are stiff, brittle, and rigid.

Certificate of representation: recognition awarded by the National Labor Relations Board, allowing a union to serve as the bargaining agent for a group of workers.

Certification: making the union the official bargaining agent for the workers. Management must recognize the union, and the two must meet and set pay rates, hours, and working conditions.

Channels of distribution: the paths that products follow from the manufacturer to the consumer. A product may follow any of several paths: retailer, direct, and wholesaler-retailer.

Charter: a certificate, issued by a state, that gives a corporation legal standing and permits it to operate.

Chemical cleaning: using liquids or vapors to remove dirt and grease. Chemical cleaning is a basic part of many finishing processes.

Chemical conditioning: a method of changing the internal properties of a material through chemical reaction.

Chemical machining: a process that uses powerful chemicals to etch (eat away) excess material from the workpiece.

Chemical preparation: use of solvents or steam and detergents to remove grease and other contaminants from a surface prior to finishing.

Chemical processes: methods of processing raw materials that use chemical reactions. For example, plastics are formed by chemical reactions.

Chemical properties: characteristics of a material that determine how it will react chemically.

Chromate conversion coating: a finishing process that provides a clear coating on cadmium- and zinc-plated steel parts.

Chuck: a spinning mold used to shape metal in metal spinning. Also, the part of a lathe that holds and spins the workpiece.

Circular saw: a cutting device with a rotating blade. The workpiece may be fed into the blade, or the blade moved across the work.

Clamping devices: machines that hold or support the workpiece or a cutting element.

Clear-cutting: a method for harvesting trees that cuts down all trees regardless of age or species.

Coke: a clean-burning high-carbon fuel made from coal. Coke is coal with the impurities burnt out of it.

Cohesive bonds: a means of joining materials that uses the same forces that hold the molecules of the substance together.

Cold bonding: a means of joining ductile metals, such as aluminum or copper, by applying very high pressure in a small area.

Cold forming: forming of material at temperatures below the recrystallization point.

Collective bargaining: a negotiating process aimed at bringing about an agreement between the company and its workers (as a group).

Combination set: a set of measuring tools that has a 0° and 90° head, a protractor head, and a center-finding head. All the heads will fit on a single rule but are used separately.

Commission: a fee paid to the member of a sales force for the sales he or she completes.

Committeeman: another term for a union steward.

Communication: transforming information into messages that can be transmitted (moved) from the source to a receiver.

Communication technology: the use of technological means to convey information and ideas.

Communicators: employees that are becoming information handlers. They do more planning than actual production work.

Company: a group of people who work together to produce a product or service for a profit.

Company or idea advertising: print media ads or radio/TV commercials designed to promote a company's image or a specific idea.

Competitive analysis: companies carefully study the products of their competitors. From this study a company can determine the need to improve its own products.

Competitive price: a price at which a product can be produced cheaply enough to sell at a price similar to like products on the market.

Composites: a combination of two or more materials with properties the materials do not have by themselves.

Compound: a material consisting of atoms of two or more different elements in definite proportions.

Compounded: mixed together; used to describe blending of liquid or finely divided solid materials in preparation for molding or casting.

Comprehensives: finished advertisements, ready for reproduction. Also referred to as *mechanicals* or *camera-ready art*.

Compression: a forming force that squeezes the material into the desired shape.

Compression strength: the ability to resist squeezing forces.

Computer: a device that includes a central processing unit, input/output unit, and a memory.

Computer numerical control (CNC): control of an individual machine by a computer that replaced an NC paper tape reader.

Computer-aided design (CAD): a computer-based system used to create, modify, and communicate a plan or product design.

Computer-aided manufacture (CAM): a system that uses computer technology in the management and control of manufacturing operations.

Computer-aided process planning (CAPP): an integrated system that coordinates all tasks associated with the manufacturing process.

Computer-integrated manufacture (CIM): an approach in which all steps in producing a product, from recognition of need for it through manufacture and marketing, are integrated into a single, dynamic computer-controlled system.

Conciliator: a person whose sole duty in a negotiation is to try to clear up problems and help the parties reach agreement.

Conditioning: changing the properties of a material by mechanical, thermal, or chemical means.

Conductors: materials that permit a ready flow of electrical current.

Construction: using manufactured goods and industrial materials to build structures on a site.

Consumer products: manufactured goods that are ready to use; no further processing is needed to use the product. Consumer products are bought from retail stores, dealers, or catalogs.

Continuous manufacturing: a type of manufacturing that produces products in a steady flow. Materials go in and finished products come out at a steady rate.

Contouring machines: equipment that can move a tool, under numerical control, in any or all machine axes at the same time.

Constitution: the document governing operation of a union or other organization.

Construction technology: the use of technological systems to build houses, stores, industrial plants, bridges, and other structures.

Contouring machines: the most complex and expensive NC machine. They can move on any or all machine axes at the same time.

Contract: contains the rules that the workers and the managers must obey.

Control: the financial affairs area of a company that maintains financial records and develops and monitors budgets.

Controlling: comparing actual results to goals and the plans developed for reaching them.

Convention: a periodic national meeting held to develop union governing policies.

Conversion finish: a process that chemically alters the surface of a material into a protective layer.

Converted finishes: changing the surface of a material by chemical action. Converted finishes add no material as a coating. The surface molecules are changed to make a protective skin or layer. Chemicals in the air and water will not damage the material.

Cope: the top half of a mold.

Copy: the written portion of an advertisement, whether for use in print or as a radio/TV commercial.

Copy platform: the theme or basic objectives of an advertising campaign.

Corporate charter: a document issued by the state that legally authorizes the company to do business in the state.

Corporation: a legally created business unit. It is an "artificial being" in the eyes of the law.

Cost accounting: a system of charging expenses to specific categories.

Counterboring: a machining operation that produces two straight, round holes on the same axis in a workpiece. The upper hole is larger, to allow the head of a fastener to be set below the surface.

Countersinking: a machining operation that is similar to counterboring, but with a beveled upper hole. Countersinking is done to sink the head of a flathead fastener below the surface.

Court order: a legal action used to dissolve a business by forcing it into bankruptcy.

Covalent bonding: sharing of electrons by elements to form a stable structure.

Creditors: individuals, companies, and financial institutions to whom the company owes money.

Credits: in accounting terms, money coming into a company.

Crystalline structures: those composed of boxlike units, called crystals, arranged in a lattice form.

Curtain coating: a finishing process where the surface of the material is flooded with finishing material. The excess runs off and is collected. Curtain coating is often used to coat flat parts or sheet stock.

Custom manufacturing: used to make small numbers of products. The company produces them to a customer's specifications. Custom manufacture is the most expensive type of manufacturing; often used to make large products.

Cutting elements: tools or blades that are made of material harder than the material to be cut.

Cutting motion: relative movement between the workpiece and the tool that causes material to be removed.

D

Daily production reports: an important source of manufacturing data that allows comparison of actual production to scheduled activity.

Debits: in accounting terms, money being paid out by a company.

Debt financing: raising money for a company by borrowing from a bank or selling bonds.

Deferred taxes: taxes due on money earned by a company that have been legally postponed until a later date.

Delegates: persons elected by fellow union members to represent them at a national convention.

Density: a measure of the weight of a certain volume of material. The volume of the sample objects must be constant. Density is given as pounds per cubic inch (lb/in^3), pounds per cubic foot (lb/ft^3), or grams per liter (g/L).

Department head: a person who runs one department in a plant, such as accounting or assembly.

Depth gage: a tool used to measure the depth of a hole, groove, or other feature.

Depth of cut: the amount of material that a tool separates from the workpiece in one stroke or revolution.

Descriptive knowledge: the area of human knowledge that includes language and mathematics. It is the basis for communication.

Deskilling: a method used to control labor costs by making a job simpler.

Detail design: activities that give a product its final appearance and functional characteristics.

Detail drawing: a drawing that gives the exact size of a part as well as the size and location of all features. These features may include holes, notches, curves, and tapers.

Development: the process of turning applied research results into marketable products or workable processes.

Diagnosing: determining what is wrong with a product that needs repair.

Diameter: the length of a straight line that passes through the center of a circle and touches the periphery.

Die casting: a forming process in which melted metal is forced into a steel mold, or *die.* The mold is water-cooled and ejects the solidified part.

Dies: shaping devices which material is squeezed between or formed over to achieve a new shape. There are three types of dies: open dies, die sets, and shaped dies. Dies can also have blades designed to cut special shapes such as curves, circles, or whole outlines.

Die sets: a pair of dies that have shapes machined or engraved on their faces. The shape on one half of the die fits the shape on the other half.

Dimension: the desired size (length, width, etc.), as shown on an engineering drawing.

Dimensions: the size measurements for a part.

Dip coating: a finish obtained by submerging the material in a container of coating material.

Dipping: a finishing method in which the material is dipped into a vat of molten metal.

Direct labor: supervisory and processing labor that is directly involved in producing the product.

Direct materials: the materials that actually become part of a product.

Direct numerical control (DNC): a system using a computer to communicate directions to each machine tool, as needed, to perform a sequence of operations.

Directing: assigning employees to jobs and encouraging them to complete their work efficiently.

Directional drilling: a technique that allows wells to be drilled at an angle or along a curve to reach oil or gas.

Dispatching: the process of issuing production orders to start the manufacturing process.

Dispersed or filler composites: materials consisting of a skeleton filled with another material. Fiberglass is such a composite.

Dissolution: the process of closing down a business.

Dissolved: broken up into tiny particles and mixed with a liquid.

Distribution: the system needed to get products from the manufacturer to the consumer.

Dividend: share of profits that is paid to a company's stockholders.

Dividers: a tool used to measure the distance between two points. The points of the divider are placed on the two points to be measured. The divider is placed next to a rule and the distance is read off.

Division of labor: the task of making a product is divided into small jobs. Workers are trained to do one job. Also, the development of people who could practice a handicraft instead of growing food in early societies.

Draft: angled sides on a pattern that allow it to be removed from a mold.

Drag: the bottom half of a mold.

Draw bench: a forming machine with a shaped one-piece die and a drive mechanism. Draw benches are used to stretch a bar or sheet over a die, and to pull a wire or bar through a hole in a die.

Drawing: a forming process that involves stretching metal into the desired shape.

Drawing forces: forces that pull and stretch material into the desired shape.

Drift mining: a method used to reach a mineral vein that comes to the surface at some point. A tunnel is dug into the vein, which is serviced by rail cars that move along the drift shaft.

Drilling: a machining operation that produces a straight, round hole in a workpiece.

Drop forging: shaping of material between two dies using the force of a falling (drop) hammer.

Drop hammers: a forging machine that raises the upper die and then allows it to fall on the material to shape it. Gravity pulls the upper die onto the lower die.

Drying: a process that removes moisture from a material. Drying can happen naturally or can be aided by applying heat.

Ductility: a property that allows a material to undergo plastic deformation without rupturing.

Due date: the date when an invoice must be paid. Usually, 30 days from the invoice date.

Durable goods: products made to last more than three years. Automobiles, bicycles, and refrigerators are durable goods.

E

Edge: the second largest surface of a part.

Ejected: this occurs when a solidified part is pushed out of a mold.

Elasticity: a material's ability to return to its original shape after being deformed by an applied force.

Elastic stage: during this stage a material under stress will stretch. When the stress is removed the material will return to its original size and shape. No deformation has occurred.

Elastomer: an adhesive, often called contact cement, that has low strength. It is useful for attaching plastic laminates to panels.

Electrical and magnetic properties: the measure of a material's reaction to electrical current and external electromagnetic forces. The principal measures of these properties are electrical conductivity, electrical resistivity, and magnetic permeability.

Electrical discharge grinding (EDG): a process that uses a spark as a cutting tool between a rotating disc and the workpiece.

Electrical discharge machining (EDM): a process that uses a spark (electrical discharge) to erode a small chip from the workpiece. EDM is widely used to make cavities for forging and stamping dies.

Electrical discharge sawing (EDS): a process that uses a spark as a cutting tool between a moving knife edge band and the workpiece.

Electrical discharge wire cutting (EDWC): a process that uses a spark as a cutting tool between a taut traveling wire and the workpiece.

Electrical properties: characteristics of a material that determine how well or poorly it will carry electrical current.

Electrocoating: a finishing process that uses unlike charges to attract finishing materials to the parts.

Electromagnetic forming: a process used to change the shape of a material through use of a very strong magnetic field.

Electroplating: the use of an electric current to deposit a thin, uniform metal layer on the surface of a base material.

Electrostatic spraying: a finishing process that uses electrically charged paint and parts to reduce overspray and ensure good coverage.

Elements: the "pure" substances that are the basic building blocks for all materials on earth.

Employee relations: the department responsible for employee hiring, training, and related activities.

Employment: placing qualified people in jobs.

Enamel: a varnish to which a pigment has been added to produce a colored coating.

End: the smallest surface of a part.

End effectors: tools at the end of a robot's arm, used to perform work.

Energy: the capacity for doing work. Energy is a necessary input in all technological systems. The energy may be electrical, chemical, mechanical, thermal (heat), or nuclear.

Enforcement: making sure that workers follow safety rules.

Engineering change order: a document describing a change in a product and how it is to be made.

Engineering data: information related to the product and how it is manufactured.

Engineering drawings: drawings that show the size and shape of parts and products.

Engineering materials: solid materials that will hold their shapes without outside support.

Engineering testing: careful checking of product quality and function.

Enterprise: a business; a unit of economic organization or activity.

Entrepreneurs: persons who organize, manage, and assume the risks of a business enterprise.

Equipment: the machines that process materials. They are used to produce industrial, military, and consumer products.

Equity financing: raising money for a company by selling shares of ownership.

Evolving technologies: areas of technology that are undergoing continuing change.

Exhaustible resource: a resource whose supply is finite; it can be used up and no more of the resource will be available.

Expediting: following up to make sure work stays on schedule.

Expel: to take away membership in an organization.

Expendable mold: a mold that is destroyed after one use.

Expenses: money that a company spends to purchase goods and services.

Explosive forming: a forming process that uses explosive charges to force material against a die.

External storage devices: disks or tapes used to store data outside a computer.

Extracting: the process of removing a finished part from a mold.

Extractive technology: the use of technological systems to recover natural resources from the earth and seas.

Extrusion: a material is forced through a shaped hole in a one-piece die. The material flows through the die taking on the shape of the opening. Extrusion can produce shapes in metals, plastics, and ceramics.

F

Face: the largest surface of a part.

Face turning: a machining process in which the cutter moves across the face, or end, of a rotating workpiece.

Factories: special buildings where people use machines to manufacture products.

Factory burden (overhead) budget: a means of assigning portions of plant costs, such as utilities and maintenance, to each product produced.

Factory cost: a company's actual cost of producing a product.

Federations: organizations made up of groups that have banded together for a common purpose.

Feed motion: in machining, the movement that brings new material in contact with the cutting element.

Feedback: information that a system uses to control itself. By measuring an output, the system gains information needed to take corrective action.

Ferrous metals: alloys with varying percentages of iron and carbon.

Fiber composites: materials consisting of fibers that are bonded together.

Fiberboard: a synthetic composite made from wood fibers. The fibers are held in place by the natural glue in the fibers called lignin.

Filler: the substance that provides the bulk for a composite material. The filler can be fibers, flakes, sheets, or particles that are the base for a composite.

Final assembly: the process of combining subassemblies to form the completed product.

Finance: the process of planning to acquire money for the company and determining the use of available funds.

Finances: the money used to purchase the resources needed to develop and engineer new products, buy material, pay for labor, and purchase machines.

Financial affairs: the area of managerial technology concerned with raising money and maintaining financial records.

Financial budget: a document prepared to predict income and expenses for a specific period.

Financial data: an estimate of manufacturing costs, selling price, and expected profit from a new product.

Finishing: a surface treatment that protects or decorates a material. Finishing involves three steps: selecting a finishing material, preparing the surface to accept the finish, and applying the finish.

Fired: term used to describe someone whose employment has been terminated for violating company rules or other cause.

Firing: a thermal conditioning process used on ceramics. Firing melts the glassy part of the ceramic. Upon cooling, the product will be particles held together by the glassy material.

Five-step employment process: a system used by many companies to find and hire new workers.

Fixed costs: those costs, such as factory overhead, that remain constant even if production levels change.

Fixed path: devices that move product or materials from one fixed point to another in a plant.

Fixed path systems: material-handling systems that use specific routes from point to point. Conveyors and piping are examples of fixed path systems.

Fixed sequence robot: a simple robot that performs a simple sequence of events, over and over.

Flame cutting: a process in which a mixture of gases is burned to melt a path between the workpiece and the excess material.

Flask: the container that encloses the sand in an expendable mold.

Flexible manufacturing: a system of computer controlled manufacturing that permits production of small quantities without increasing costs.

Flow bonding: a method of joining materials that uses a filler metal that melts onto a heated base metal. When cool, the filler metal acts as an adhesive.

Flow coating: a process that floods the surface with a finishing material. The excess is allowed to drip off the material.

Flow diagram: a graphic method of monitoring movement of parts through the plant.

Flow process chart: a graphic means of showing all the steps and processes a single part goes through as it is manufactured.

Flow process chart: a chart that shows the step-by-step arrangement of tasks a single part goes through as it is manufactured.

Forecast: management's estimates of the demand for products.

Forest products: all products that are made from wood.

Forging: shaping of heated metal by forcefully closing die halves on it, using either a drop hammer or a hydraulic ram.

Form utility: changing the form of a material to make it more useful.

Forming: a process that changes the shape and size of a material by a combination of force and a shaped form (such as a die or forming rolls).

Fracture point: the point at which a material under stress breaks into two or more parts.

Free enterprise system: an economic system in which privately owned, for-profit enterprises compete for business.

Fringe benefits: things a company pays for beyond basic wage or salary, such as health and medical insurance programs.

Function: the ability of a product to do a job.

Fusion bonding: a bonding process that uses the same material as the base metal to create the weld. Thick parts require the use of a filler rod. This rod is made of the same material as the base metal. It provides more metal to produce a strong weld. The heat is provided by burning gases or by electric sparks.

Futurist: a person who tries to predict possible future conditions on the basis of historical perspective and current trends.

G

Galvanizing: the process of coating steel with zinc so that it resists corrosion.

Gas welding: a means of bonding metal, using the heat of an oxygen/acetylene flame to melt arid fuse the materials.

Gating: the path into the cavity that the liquid material follows.

General expense budgets: estimates of costs for company expenses that are not directly related to producing products.

General ledger: a central record of all the financial transactions made by a company.

General overhead: corporate costs for product development, marketing, and administrative activities.

Goals: the end to which effort is directed by a system, company, or individual.

Goodwill: the value of the company's name and reputation, as well as trademarks and patents it owns.

Gravity mold: a mold that is filled by pouring the liquid material into the top. No additional pressure is used.

Greensand casting: a casting process that uses moist sand packed around a pattern.

Greensand casting molds: a type of expendable mold, formed from a special sand and used for metal casting.

Grievance: a complaint that is made by employees who feel that management has broken the rules of a labor contract.

Grievance procedure: a means of settling disputes between workers and the company, as outlined in the labor contract.

H

Hammer: a machine that delivers a sharp blow with a movable upper head to forge metal.

Handicraft: a skill at making one type of product. Some people in early societies became carpenters, cobblers, and blacksmiths instead of having to grow food.

Hard copy devices: equipment that can be connected to a computer and used to obtain a paper or film copy of information processed by the computer.

Hardening: a process used to make a material more resistant to denting, scratching, or other damage.

Hardness: the ability of a material to resist scratching or denting.

Hardware: the physical components of a computer system.

Hardwoods: woods from broad-leafed deciduous trees, such as oak.

Harvesting: a method that collects a growing resource.

Hazard: sources of danger in a manufacturing plant.

Heat treating: the thermal conditioning of metals using a process of heating and cooling solid metal to produce certain mechanical properties. Heat treating includes the three major groups: hardening, tempering, and annealing.

High alloy steels: special steels alloyed with molybdenum, nickel, tungsten, or other elements; tool steel is an example.

High technology: the useful application of recent discoveries in science and other fields into new products.

High-energy rate (HER) process: a forming process that uses high-energy sources to provide forming force. These include: explosive materials and rapidly changing electromagnetic fields. These processes are used in special applications.

Horizontal milling machine: a machine that cuts a flat surface on a part held in a vise.

Hot forming: forming of material that is heated above the recrystallization point.

Hot melts: a type of finish that is made liquid by heating, then forms a coating as it cools to a solid state.

Human resource development (HRD): another name for employee relations.

Humanities: the study of areas of philosophy that relate to human concerns. Studying the humanities helps people develop personal priorities.

Hunting and fishing period: a period in history when humans lived off the land. They hunted animals and fish, and harvested wild berries, fruits, herbs, and roots.

I

Ideation: a process designers use to move from ideas in their mind to ideas on paper.

Imitation: a common product development technique. A company will produce a product much like those of other companies, letting someone else identify and build the market.

Impulse sealing: a bonding process used to seal plastic films. It employs heat generated by a pulse of electricity through a wire.

Incentive pay: payments made for completing work above set standards.

Income: the money a company receives, primarily from sales of products or services.

Income statement: a financial document that gives a picture of the company's financial situation over a period of time.

Indirect labor: labor that serves and supports the production activities.

Indirect materials: the supplies and other materials used in manufacturing the product.

Induction training: basic information on the company and its policies, usually required for all new workers.

Industrialization: the process of moving from an agricultural to an industrial economy.

Industrialized society: one in which people use machines to manufacture large quantities of products from materials.

Industrial materials: materials ready for secondary processing into manufactured products. Also known as *standard stock.*

Industrial products: materials and equipment used by companies to produce products. Industrial materials need further processing before becoming useful products.

Industrial relations: the area of managerial technology concerned with the human aspects of the manufacturing enterprise, such as personnel and labor relations.

Industrial robot: a programmable manipulator capable of a variety of functions in a manufacturing setting.

Industrial sales: the sales of goods to anyone but the final customer.

Industrial system: one designed to transform resources into outputs for a large number of people.

Industry: this term can be described as: all economic activity; or as the productive activities-communication, transportation, manufacturing, and construction; or as a group of companies that compete against one another, such as the publishing industry.

Information workers: professional workers that use great amounts of information.

Injection molding: a process similar to die casting, but usually used for plastics rather than metals.

Innovation: a product development technique where a totally new product is developed. The videocassette recorder, CD player, and the microchip are examples of innovations.

Inorganic coatings: coatings made up of metals or ceramic materials.

Inorganic materials: those that are nonliving; ores and earth elements.

Input devices: external components, such as a keyboard or a digitizing tablet, that are used to enter data and directions into a computer.

Input/output unit: component of the computer that communicates with external input and output devices.

Inputs: the resources needed to make a product. These inputs may be grouped into seven main classes: natural resources, human resources, capital, knowledge, finances, time, and energy.

Inspection: a process that ensures only quality products leave the plant. Inspection includes checking: materials entering the plant, purchased parts, work-in-progress, and finished products.

Inspection orders: forms used to schedule quality control checks.

Inspection tag: a card attached to a defective part, telling what is wrong and what should be done with it.

Installed: when a product is installed it is set up in the place where it will be used. It is unpacked, hooked up to utilities, adjusted, and tested.

Interference fits: a means of assembly that makes use of mechanical force to hold parts together.

Intermittent manufacturing: a system in which parts are produced in groups or job *lots,* in contrast to continuous manufacturing.

International union: a union with locals in both the United States and Canada.

Intrapreneurs: employees of an existing business who are encouraged to apply entrepreneurial skills within that business.

Inventory: the value of materials, work-in-process, and finished goods owned by a company.

Inventory control: controlling the amount of material on hand.

Inventory turn: the time it takes for a store's supply of a product to "sell out" and be reordered.

Invoice: a bill from a vendor for payment due on materials sold to a company.

Involuntary dissolution: this occurs when a company is forced to dissolve. The assets are sold to raise money. Three reasons may cause involuntary dissolution: bankruptcy, dishonest financial activity, or loss of state charter.

Ion: an atom with a positive or negative charge.

Ionic bonding: the bond formed between atoms by the attraction of opposite electrical charges.

J

Job analysis: the first step of a wage and salary study, involving an assessment of the job and the skills and knowledge required.

Job description: a written document describing the duties and needed skills involved in a specific position.

Job-lot manufacture: intermittent manufacture that produces finished products.

Job pricing: the process of attaching a wage rate to each job or job classification.

Job rating: the grouping of jobs (by skills required and other considerations) to establish rates of pay.

Job requirements: what each job demands of the employee.

Job security: the permanence and stability of employment with a company.

Joint: the point where two parts of an assembly come together and are fastened.

Just-in-time (JIT) manufacturing: a system that reduces inventories of raw materials and product components to the absolute minimum. Materials and parts are scheduled to arrive at the production line just when needed for use.

K

Kiln dried: lumber with a moisture content of 6% to 12%. Green lumber is placed in a large oven called a kiln. Air circulation, heat, and humidity are carefully controlled as the lumber is dried.

L

Labor agreement: a contract between a union and a company that sets wage rates, job conditions, and other matters affecting workers.

Labor budget: an estimate amount of labor needed for production, allowing a forecast of labor costs.

Labor relations: activities involved in solving problems that arise between workers and management.

Labor union: an organization that gives workers a united voice when dealing with management.

Lacquer: a material containing a polymer coating and a solvent. A lacquer dries as the solvent evaporates.

Laid off: term used to describe workers whose services are not required, temporarily or permanently, due to lack of business.

Laminate composites: materials composed of layers of wood.

Laminations: heavy timbers produced from layers of veneer or lumber. The timber is held together with synthetic adhesives.

Laser: a device that generates an intense light beam that can be used for cutting. The word laser stands for light amplification by stimulated emission of radiation. A laser amplifies light. This intense light beam produces heat when it strikes a surface. This heat will cut a workpiece.

Laser beam machining (LBM): a cutting method that uses a beam of focused light to melt a path in the material and separate the excess from the workpiece.

Layout: the suggested arrangement of copy and illustrations for a print advertisement. A layout is submitted by an advertising agency for the company's approval.

Lay out: the process of measuring and marking a part so that it can be made. The markings show where every feature should be on the part.

Lead time: the time that must be allowed between a decision to build a product and actually building it.

Leadership: the ability to inspire others to work together.

Length: the largest dimension of a part.

Levels of authority: responsibility and decision-making paths within a company.

Liabilities: the money that the company owes to other companies and to individuals.

Light reflection: a measurement of how well or poorly light waves bounce off a material.

Light transmission: a measurement of how well or poorly light waves pass through a material.

Linear: movement in a straight line; often used in reference to a machining operation.

Linear measure: the length of a material in units. Used to determine the size of a product or how much is being sold (such as pipe or rope).

Line organization: a form of business organization in which each position reports only to the manager immediately above it.

Line production routing sheet: a form that summarizes basic information about each production operation.

Loan: money borrowed by a company from a bank or other lender for a specific term at a stated rate of interest.

Local union: a labor union organized at the single plant or community level.

Lock-out: a company closes business until contract is agreed upon.

Long-term debt: money owed to creditors that is to be repaid over a period longer than one year.

Lumber: pieces of wood (a natural composite) that have been made to a certain size for construction and other uses.

M

Machine screw: threaded fasteners that are used to assemble metal parts. They have round, flat, or oval heads and a shank of uniform diameter and threads along its full length.

Machine tools: machines that are used to build other machines.

Machining: changing size and shape of a part by removing excess material.

Machining centers: a type of separating machine that uses computer control and can perform many different operations.

Machining operations: separating processes in which excess material is removed in the form of chips.

Machinist's rule: a ruler that has scales (markings) that will measure down to 1/64".

Macrostructure: the way in which different substances unite to form complex materials called composites.

Magnetic conductivity: the ability of a material to conduct magnetic lines of force

Magnetic permeability: a measurement of a material's ability to become magnetized.

Magnetic properties: characteristics of a material that govern its reaction to a magnetic field.

Maintained: when service has been performed on a product to keep it in good working order.

Maintenance: the tasks that are performed on a product to keep it in good working order.

Management: the group who supervises and brings together the inputs of people, machines, materials, finances, and people to produce and sell products.

Management information systems: computerized systems of information processing that help a company achieve peak operational efficiency.

Management processes: processes that are used to assure the efficient and appropriate use of resources.

Management structure: the way a company is organized, how responsibility flows and how decision making is divided.

Managerial technology: the use of systems and procedures to insure that transformation actions are efficient and appropriate.

Managers: company officials who make hiring, firing, and discipline decisions.

Manufacturing: the process of changing resources into more useful products.

Manufacturing cell: a small group of machines connected by a material handling system. Often, the individual elements of the cell are computer controlled to increase flexibility and efficiency.

Manufacturing engineer: a professional person who organizes manufacturing operations.

Manufacturing engineering: an activity devoted to making the physical layout of the plant, the manufacturing processes, and the production equipment as efficient as possible.

Manufacturing resource planning II (MRPII): a complex, computer-controlled system that can be used to control most aspects of production from scheduling to recording output.

Manufacturing technology: the use of technological systems to transform materials into products at a central location (manufacturing plant).

Manufacturing (variable) costs: those costs, such as labor and material expenses, that vary as production increases or decreases.

Market approach: a means of selecting products to manufacture based on the needs and desires of customers.

Marketing: the area of managerial technology concerned with moving a product from the manufacturer to the customer by means of advertising and selling activities.

Marketing budgets: estimates of costs involved in promoting, selling, and distributing the company's products.

Marketing data: information on potential customers and the competitive environment.

Marketing mix: a term used to describe the various activities carried out under a marketing plan.

Marketing system information: a component of market research that focuses on the effectiveness of advertising, packaging, sales approaches, and channels of distribution.

Market information: a component of market research that focuses on identifying the segment of the market most likely to purchase a product.

Market profile: a description of the type of customer a company is trying to reach with a new product.

Market research: the process of gathering and analyzing information about customers' desires, competing products, and sales results.

Master budget: a summary of all the specialized estimates (such as sales, production, and general expense budgets) prepared for different company activities. It provides an overall picture for top managers.

Mated dies: a set of dies with a raised shape on one die and a matching cavity in the other. Material is squeezed between the two to take the desired shape.

Material handling: the methods used to move material around a manufacturing plant.

Material handling engineering: the design and installation of systems to move material from place to place in a plant.

Material processing: changing the size, shape, and looks of materials to fit human needs. Changing the form of materials takes three steps: obtaining natural resources, producing industrial materials, and making finished products.

Material processing technology: another way of describing *transformation technology.*

Materials budget: lists of needed production materials and supplies, used as a basis for estimating costs.

Materials requirement planning (MRP): a system that allows schedulers to place priorities on materials and products to be produced.

Matrix: the agent that holds the filler together in a composite. It acts as the bonding agent for the filler.

Maximum profitability point: the price at which a company makes the greatest possible profit from a product.

Measurement: the process of describing a part's size using a standard for comparison. This allows the part to be duplicated from the measurements only.

Mechanical cleaning: cleaning processes that use abrasives, wire brushes, or metal shot to remove dirt and roughness.

Mechanical conditioning: changes to the internal structure of a material as a result of manufacturing processes. The changes are not always suitable; the material may have to be conditioned further to restore it to the desired state.

Mechanical conditioning: using physical force to modify the internal structure of a material.

Mechanical fastening: a method of holding parts together temporarily or permanently with mechanical fasteners or force.

Mechanical preparation: use of an abrasive or similar material to prepare a surface for finishing.

Mechanical processes: methods that use mechanical force to change the resource, such as cutting or crushing.

Mechanical properties: characteristics that govern how the material will react to a force or load, such as compression, tension, shear, and torsion.

Media cleaning: surface preparation methods that propel, vibrate, or tumble some form of abrasive against the surface of the material to be cleaned.

Mediator: a person who can recommend a solution to labor and management negotiators. They are not required to accept it.

Melted: heated until a liquid or semi-liquid state is attained.

Memory: component inside a computer used by the CPU to store data.

Mentafacturing. products are first made in the mind. This has been called "making in the human mind."

Metallic bonding: a combination of attracting and repelling forces that produces a very rigid structure.

Metallic coatings: finishes that consist of metal particles forming a surface layer on the base material.

Metallic materials: inorganic, crystalline substances with a wide range of physical and mechanical properties.

Metallizing: application of a very thin finish of metal particles by means of spraying or vaporizing in a vacuum chamber.

Metal planer: a machine that removes excess material by moving the workpiece back and forth against a stationary cutting tool.

Metals: an engineering material with a crystalline, inorganic structure. Usually processed using heat or using forming techniques.

Metal shaper: a machine that removes excess material by moving the cutting tool back and forth over a stationary workpiece.

Metal spinning: a forming method in which a rotating disc of soft metal is deformed by pressure of a tool.

Methods engineering: the task of determining how a product will be made.

Micrometer: a very precise measurement tool used for linear dimensions.

Microstructure: the way molecules and crystals are arranged in a material.

Military products: the materials and machines used for the defense of a country. These goods are special products made for the use of a country's military forces.

Mining: the process of digging resources out of the earth.

Mock-up: models designed to simulate the appearance of a new product.

Models: a three-dimensional representation of a product. The two major types of models are *mock-ups* and *prototypes*.

Moisture content: the amount of water trapped within a material's structure. It is expressed as a percentage of the dry weight of the material.

Mold: a container to hold a liquid material until it solidifies. A correctly shaped cavity is built into the mold. The two types of molds are *expendable* and *permanent*.

Molding: a process that heats plastic to a liquid or softened state, forces the plastic into a cavity of the desired shape, and then allows the plastic to cool and set into its new shape.

Molecular structures: a structure made up of polymeric chains held together with covalent bonds.

Monomer: a chemical compound that can unite with other monomers to form a polymer.

Move orders: forms used to schedule material movement in a plant.

Multiple point tools: an arrangement of two or more single-point tools to form a cutting device.

N

National Labor Relations Act of 1935: law that guarantees workers right to form or join unions and to bargain with an employer.

National union: a countrywide organization made up of local unions.

Natural polymers: those that occur in a usable form in nature, such as wood, wool, or natural (latex) rubber.

Negotiation: discussions held to work out a labor contract between an employer and the workers.

Negotiators: union and company representatives who meet to seek compromise on basic issues and work out a labor agreement.

Net income: a company's total income minus its expenses.

Net profit: what remains after subtracting all expenses from total income; also called *net income.*

Net worth: the value of a company's assets minus its liabilities.

NLRB: the National Labor Relations Board, a federal agency that oversees labor relations between workers and employers.

Nondurable goods: products that are made to last less than three years. Clothing, food, pencils, paper, light bulbs, and motor oil are nondurable goods.

Nonengineering materials: gases and liquids that must be confined in a container to hold a particular shape.

Nonferrous metals: metals that do not have iron as their principal ingredient. Copper, aluminum, gold, and silver are all nonferrous metals.

Nontraditional machining: machining processes that do not use a "tool" as we use the word. They use heat, light, chemical action, or electrical sparks to produce a cut.

Normalizing: a thermal conditioning process that relieves stress in steel and develops a fine, uniform grain structure.

Numerical control (NQ): a method of programming manufacturing machines to be controlled by a code consisting of numbers, letters, or other symbols.

O

On-the-job-training: a method of teaching skills at the workstation, often under the guidance of an experienced worker.

One-piece die: a type of die used to give shape to a material. Used in thermoforming and metal spinning processes. Also called *molds.*

Open die: the simplest type of die. An open die is two flat, hard plates. One half of the die does not move. The other half moves to hit (hammer) or put pressure (squeeze) on the material between the dies.

Open-pit mining: a type of mining used when the resource is close to the surface.

Open shop: arrangement under which workers are free to join or not join a union, as they choose.

Operation analysis charts: charts used by engineers to study individual manufacturing operations.

Operation process chart: a combination of all the flow process charts for parts making up a product. This chart shows where the parts come together to form assemblies, and finally, the product itself.

Operations: the processes that shape and assemble a product.

Operation sheet: a form used to record the sequence of operations needed to produce a product.

Optical properties: characteristics that describe how a material reacts to light waves. The two main optical properties are *opacity* and *color.*

Ores: earth or rock from which metal can be commercially extracted.

Organic coatings: finishing products derived from living or once-living sources. Natural gums or resins from trees are widely used.

Organic materials: those whose origins can be traced back to living things.

Organized labor: the general term for workers who have organized into unions.

Organizing: developing a structure to reach goals that have been established.

Orthographic projection: a method of presenting a product in drawings by showing the top, front, and right side views.

Outputs: the results of a process. Outputs may be desired (such as a product) or undesired (such as pollution, noise, waste, etc.).

Oxide conversion: a process that provides a finish that resists corrosion and abrasion.

Overhead: the cost of equipment, utilities, and insurance needed to produce the product.

Overspray: finishing material from a spray gun that misses the product or does not adhere to the surface. It is wasted and pollutes the air.

P

Package: a device that protects, contains, promotes, and provides information to the customer about the product.

Packaging: the process of designing a package for a product. The term is also used to refer to the package itself.

Paint: a coating that changes from a liquid to a solid by means of a polymerization reaction. (This is a linking of molecules into strong chains.) Many paints have a coloring agent added.

Parison: a heated tube of plastic or glass material that is placed in a mold for blow molding.

Parliamentary procedures: recognized rules and practices for running any type of meeting.

Participatory management: a system in which, workers and managers participate equally in making business decisions.

Particle composites: materials consisting of particles or flakes that are bonded together.

Particleboard: a synthetic wood composite made from chips, shavings, or flakes held together with a synthetic glue. The most common types are: standard particleboard, waferboard, flakeboard, and oriented strand board.

Parting compound: a powder that makes removing the pattern from an expendable mold easier.

Partnership: a form of business ownership involving two or more persons. Usually, partners will have equal shares of ownership of the business.

Pattern: a pattern is a device that is the exact shape of the finished part. Patterns are used to make expendable molds.

Payment: a transfer of funds, usually in the form of a check, to pay a debt (such as an invoice for material)

Payment terms: a statement on a vendor's invoice telling when payment is expected; sometimes, a prompt payment discount is offered.

Payroll data: information on hours worked by each employee, used to prepare the payroll.

Periodic table of elements: a listing of the chemical elements arranged by their properties and levels of chemical activity.

Permanent fasteners: joining devices designed to remain in place. To be removed, they must be destroyed.

Permanent molds: molds that can be used again and again to cast or mold a part. They will produce many parts before they wear out. Permanent molds are more expensive than expendable molds. There are two types of permanent molds: *gravity* and *pressure*.

Personal goals: what each person wants out of life.

Perspective model: a lifelike view of an object that is shown as the eye would see it naturally.

Phosphate conversion coating: a finishing process used as a prepaint coating on steel.

Physical properties: physical characteristics of a material, such as density, size and shape, structure, and surface texture.

Pickling: a cleaning method that uses acid to prepare the surface of a metal.

Pilot run: a small-scale production test for a product, assembly line, or entire plant.

Planning: setting goals and the major course of action to reach them.

Plant layout: in a manufacturing operation, the placement of machines and other equipment to promote the efficient flow of materials and people.

Plant manager: an individual in charge of an entire production facility. This person manages all activities at a single plant.

Plasma spraying: a finishing process that uses a gun to vaporize metal or ceramic materials. Hot gases carry the particles to the workpiece. This process deposits a thin, even coating on metal, plastic, and ceramic parts.

Plaster investment molds: expendable molds formed by pouring plaster around a wax pattern. Once the plaster hardens, heat is applied to melt and remove the wax pattern.

Plasticity: a material's ability to flow into a new shape when pressure is applied.

Plastics: another term for synthetic (manufactured) polymers.

Plastic stage: during this stage a material starts to yield to stress that is applied. The material will be permanently deformed into a new shape.

Plywood: a synthetic wood composite made of sheets of veneer. The grain of the core and the face veneers run in the same direction. Except in thin plywood, the crossbands are at right angles to the face veneers. Other plywood has cores made of particleboard or solid lumber.

Point of recrystallization: the lowest temperature that a material can be formed without causing internal stresses. At this point, the material will form easily.

Point-to-point (PTP) machines: numerical control machines that position a cutting tool above one point on a workpiece, carry out an operation, then move the tool to the point where the next operation is to be carried out.

Political power: the ability of an organization to gain the attention and cooperation of elected officials because of the large number of voters it represents.

Polymer: natural and synthetic compounds made of molecules that contain carbon. A single molecule is called a mer or monomer (meaning single mer). Polymers are made by combining monomers into a chain-like structure.

Polymeric materials: organic, noncrystalline substances with long, chain-like molecular structures.

Polymerization: linking together of chain-like molecules of a plastic, changing it from a liquid to a solid.

Porosity: the relationship of open space to solid space in a material. A porous material will be lighter than a nonporous material. Porous materials contain air within themselves and make good heat insulators.

Precision measurement: a very accurate type of measurement. Readings are given to greater than .001" accuracy.

Preliminary design activities: preparation of renderings, mock-ups, and other methods of refining the appearance and function of a product under development.

President: a full-time manager hired by the board of directors. The president is the top manager in a company.

Press: a machine that shapes or cuts material by applying steady pressure.

Press fit: a method of fastening that relies on the friction between closely fitting parts to hold them together.

Press forging: shaping of material between two dies using the force of a hydraulic ram.

Pressure bonding: a method of joining materials that uses both heat and pressure.

Pressure mold: a type of mold that has the liquid material forced into it. This type of mold must be built to withstand the force of a clamping system and the pressure from the material as it is forced into the mold.

Prevailing wages: the rates of pay generally established for jobs in a geographic area.

Primary processes: processes that change raw materials into a usable form for further manufacture. Changing logs into lumber is an example.

Primary processing: the first step in transforming raw materials into products. Example: Converting trees into lumber at a sawmill.

Private technological system: one designed to meet the needs of an individual.

Process control: using computers to control the operation of manufacturing processes.

Processes: the actions that are performed to raw materials to turn them into usable products.

Process layout: a type of plant layout where machines are grouped by the process they perform.

Product: the output of any manufacturing system. Products can be made for consumers, industry, or for military purposes.

Product advertising: print media ads or radio/TV commercials designed to sell a product, rather than promote a company.

Product-centered activities: managed areas of activity in manufacturing that directly move a product idea from the designer's mind to the market. These areas consist of research and development, production, and marketing.

Product developers: people who are responsible for seeking new product ideas compatible with the company that they work for.

Product engineering: the process of preparing a design for production. The design is specified and tested for operation and safety.

Product information: a component of market research that focuses on customers' reactions to existing products and desires for new products.

Production: the area of managerial technology concerned with using machines and tools to transform resources into products.

Production approach: a means of selecting products to manufacture based on capabilities of the factory and its equipment.

Production expense budget: an estimate of the material, labor, and overhead costs needed to carry out manufacturing activities.

Production overhead: all production costs that are not direct labor or direct materials.

Production planners: people who schedule and organize the manufacturing system to make sure that products are produced on time and at the lowest cost.

Production planning and control: an activity devoted to scheduling and coordinating the manufacturing of products.

Production schedule: a plan for making most efficient use of equipment and workers to manufacture products.

Production tooling: special devices that increase the speed, accuracy, and safety of production processes.

Productivity: a measure of the output per unit of labor.

Product layout: a manufacturing arrangement in which equipment is arranged in the sequence of operations needed to produce the product.

Product profile: a list of the general requirements that a new product must meet.

Profit: the amount of money left over after all the expenses of a business have been paid.

Profit and loss statement: another name for the document known as an *income statement;* often referred to as a "P & L."

Profit center: a unit chosen for cost accounting. A profit center may be a single product, a department, an entire plant, or a group of plants.

Profit-centered: companies that are formed to make money for the owners.

Profit sharing: a system that allows employees to benefit from efficient operation by sharing in the company's profits.

Prompt payment discount: payment terms allowing a company to take a small discount (usually 2 percent) off the amount of the bill, if it is paid in 10 days.

Properties: traits or qualities that are specific to a material.

Proposals: a statement that tells the other side in a negotiation what one side wants in an agreement.

Proprietor: the owner of a business.

Proprietorship: a business enterprise owned by one person.

Prototype: a working model. Prototypes are used to check the operation of the final product.

Protractor: a tool used to check angles other than 90° or 45°.

Public relations: a program that a company uses to inform people of its contributions to community welfare.

Public service advertisements: advertisements that are designed to get the public to act a certain way, such as to protect the environment, improve health standards, or increase donations to a certain cause.

Purchase order: a firm order for materials at a specific price.

Purchasing: the practice of buying the materials needed to manufacture products; it brings the suppliers and users together.

Purchasing system: a method used to purchase materials, following a regular sequence of activities.

Q

Quality circle: groups of workers that meet to discuss ways to improve the company and its operations.

Quality control: the process of ensuring that a product meets specifications and standards.

Quota: the amount of product that a salesperson is expected to sell.

Quote: a statement by a vendor of the most recent price for an item. The price quoted, however, is subject to change.

R

Rake: an angle that slopes away from the cutting edge of a tool. It reduces the force needed to produce a chip.

Raw materials: natural resources found on or in the earth or seas. Manufacturing starts with raw materials.

Reaming: a machining operation that uses a rotating cutter to shape or enlarge a hole.

Receiving report: a confirmation that the material ordered from a vendor has been received by the buyer.

Reciprocate: move back and forth. Some types of machining are done with a reciprocating motion of either the tool or the workpiece.

Recruitment: the process of attracting potential workers to a company.

Recycled: when a product cannot be repaired, the materials in that product are reprocessed for other use.

Refined sketch: a sketch that shows shape and size of a product idea. It gives a fairly accurate view of the designer's ideas.

Relief angle: the clearance behind the cutting parts of the tool. Relief angles also keep most of the tool from rubbing on the workpiece.

Rendering: a colored pictorial sketch used to show the final appearance of a product design.

Renewable resource: biological materials that can be grown to replace the materials we use. Trees and food plants are examples.

Repaired: worn and broken parts are replaced to restore the product to proper working order.

Replaced: the product is taken out of service when it is not possible or economical to repair it, and a new product is purchased to do the old product's job.

Research: the systematic search for unknown facts or principles.

Research and development: the area of managerial technology concerned with designing and specifying products.

Resistance welding: an assembly process that uses heat and pressure. This process is based on a material's ability to resist the flow of electric current. Resistance welding uses resistance to melt the material. The material is then squeezed and held to form the weld.

Resistors: materials through which electrical current will not readily flow; insulators.

Resource control: making sure that resources are being fully utilized, and not sitting idle or being wasted.

Resource flow: the flow of materials and people inside a plant.

Retail sales: selling products to the final consumer, who receives the product immediately.

Retained earnings: the portion of profits that a company uses to expand its operations or to develop new products.

Riddle: the sifting of sand over a pattern when an expendable mold is prepared.

Roll forming: a forming process that uses smooth rolls to give a curve to a straight piece without changing its cross section. For example, roll forming will produce a curved I-beam, but the "I" shape of the beam will not change.

Rolling: a forming process that uses a rotating applied force to change the thickness of a piece of steel or other material. Also, a finishing method in which a coating is applied by a roller.

Rolling machine: a device that changes the thickness of a material by squeezing it between rotating rolls.

Rolls: a type of forming device. They can be either smooth or formed.

Rotate: to turn about a central axis, or revolve. Either the workpiece or the tool rotates in many types of machining operations.

Rough sketch: preliminary drawing made by a designer to show a design idea.

Routing: the process of determining the best path through the manufacturing system for parts and products.

S

Safety engineering: a practice that designs processes and machines to be safer.

Safety equipment: devices that protect a worker's sight, hearing, lungs, and skin.

Sales: the exchange of products for money.

Sales budget: an estimate of sales for a specific period of time.

Sales call record: a form detailing the contacts made with customers by a salesperson.

Sales income: the money a company receives for its products when they are sold.

Sales forecast: a company's estimate of expected sales, for a period of time.

Sales order: a form that records the details of each product sale.

Salespeople: people hired by a company to present products to the customer.

Scheduling: the process of determining the timing of each phase of production.

Schematics: systems drawings used to convey information on electrical, hydraulic, pneumatic, or mechanical systems.

Science: the study of the laws and principles that govern the physical universe.

Scrap: waste material created by a manufacturing process, such as sawdust from a sawmill or steel chips from a machine shop. Every effort should be made to keep scrap to a minimum.

Script: the written message of a radio or television advertisement. It is followed as the advertisement is produced.

Seam weld: a resistance welding process that uses rolling electrodes. The process produces a continuous weld.

Seaming: a means of joining sheet metal and similar materials by folding and interlocking the edges.

Seasoning: the process of reducing the moisture content of lumber through drying in order to make it more stable.

Secondary processing: manufacturing methods that change standard stock into finished products.

Seed tree cutting: all trees in an area, except for four or five large ones, are cut. The large trees reseed the area.

Selective cutting: a harvesting technique in that mature trees are selected and cut. Younger trees are left standing.

Selling cost: the cost of promoting, selling, and distributing a product.

Semiconductors: materials that will conduct electricity under certain conditions.

Semipermanent fasteners: devices designed to remain in place, but capable of being removed without being seriously damaged. A wood screw is an example.

Sensors: devices that change heat, light, sound, or other types of energy into electrical signals that can be used by a computer or other control systems.

Separating: processes that remove excess material to change the size, shape, or surface of a part. There are two groups of separating processes: machining and shearing.

Severance pay: a special allowance often paid to workers who lose their jobs.

Shaft mining: mining method used for deeply buried mineral deposits. A main shaft and an airshaft are dug down to the level of the deposit. The material is mined by digging horizontal tunnels from the vertical shaft.

Shaped die: a one-piece die over which material is formed.

Shaped rolls: rotating forming devices with grooves machined into them.

Shaping device: a machine or tool used to determine the final shape of a part or product.

Shareholders' equity: the actual value of the shareholders' ownership.

Shear strength: the ability to resist opposing forces that would fracture the material along grain lines.

Shearing: a process that uses force applied to opposed edges to fracture excess material away from the workpiece.

Shearing machines: devices that cut material by fracturing it between opposing tool edges.

Sheet metal screw: a fastening device with threads that extend the full length of its shank.

Shell molds: molds formed as thin resin and sand shells on heated metal patterns.

Shop steward: a union officer who represents a group of workers.

Shot peening: a process that sprays tiny steel balls against metal parts. The balls pit the surface of the metal and increase its fatigue-resistance.

Shrink fit: a fastening method in which one part is heated to make it expand. After the parts are assembled, the heated piece cools and shrinks, resulting in a tight fit.

Single point tools: simple tools with only one cutting edge.

Single proprietorship: a business owned by one person. Many small businesses are single proprietorships.

Sinter: heating a material so that it becomes a solid mass without melting the material.

Size: how large or small an object is, described by a measurement standard.

Slip: fine particles of clay suspended in liquid for ease in molding ceramic products.

Slip casting: a process in which a solution of clay, particles suspended in water is poured into a mold. The water evaporates, leaving the clay behind. Either solid or hollow objects can be slip cast.

Slope mining: a method used for a shallow mineral deposit. A sloping tunnel is dug down to the deposit. The minerals are often carried to the surface on a moving platform called a conveyor.

Slurry: ground up coal that is mixed with water for pipeline transport.

Slush casting: a process in which a thin-walled hollow casting is formed from molten metal or powdered plastic.

Smooth rolls: rotating forming devices used to squeeze material to the desired thickness.

Societal goals: the meeting of the broad needs and wants of human society.

Software: the set of instructions used by a computer. Different software is used for various tasks, such as word processing, filing, operation of machines, or planning.

Softwoods: woods from evergreen conifers with needlelike leaves, such as pine or spruce.

Soldering: a process similar to brazing, but generally performed at temperatures below 600°F.

Solid model: a computer representation that takes into account both the surface and the interior substance of an object.

Solidifying: the hardening of a liquid or solid material in a mold.

Solvent bonding: a joining process for plastics that uses a solvent to soften the edges of the pieces to be bonded.

Solvent evaporation: the means by which many organic coatings change from a liquid to a solid state.

Specification sheets: written summaries of the characteristics of a product.

Specifying: to determine the size, material, and quality requirements for a product.

Spot welding: a heat and pressure bonding technique that uses electrical resistance to melt the material. The material is then squeezed and held to form the weld. Spot welding occurs at one "spot," hence the name.

Spraying: a finishing method in which paint or similar material is vaporized and deposited on the surface of the material.

Square: a measuring tool used to check right (90°) angles.

Squeezing forces: these forces compress a material to change its shape or thickness.

Staff employees: persons with specialized responsibilities who advise the company's operating managers.

Stamping: drawing operations that combine shearing and drawing.

Standard measurement: a type of measurement where the dimensions are held to the nearest fraction of an inch ($^1/_4$, $^1/_8$, etc.).

Standard stock: material output by primary processing operations, available in standard size units or standard formulations.

Stations: cavities in a single die. Each cavity performs one step of the forming to be done on a part. The material is moved from one station to the next. The part leaves the die completely formed.

Steam cleaning: the use of steam and detergents to remove grease and oily deposits from the surface of metal.

Stock: ownership rights to the company, bought by people investing in the company.

Stockholders: those who hold shares of stock, representing partial ownership of a company.

Stockholder's equity: the actual value of the investors' share of the company. Also called the company's *net worth.*

Storyboard: a set of drawings that contains a sketch of each scene of a television commercial. It is used to guide the production of the commercial.

Straight cut machines: equipment that makes straight-line cuts through a workpiece. Straight cut machines can produce only rectangular parts.

Straight turning: machining done by the cutting action of a single point tool to produce a uniform diameter on a rotating workpiece.

Strength: the ability of a material to resist an applied force.

Stretch forming: a draw bench pulls a material beyond its elastic limit and then pushes a die into the material. Commonly used in the aircraft industry.

Strike: a refusal by unionized workers to perform their jobs. Strikes usually occur when labor and management fail to agree on a contract.

Struck: this occurs when the mold is filled with sand. Before the mold can be turned over, the sand is leveled off, or struck.

Subassembly: a component, assembled from separate parts, that in turn is combined with other components to form the finished product.

Supervisors: individuals who assign jobs to and supervise production workers.

Surface coating: a finishing process that adds a layer of a different material to the surface of a base material.

Surface measure: a method used to measure many materials. Two typical measures are linear measure and area. Pipe is measured in linear units, while land is measured by area (such as square miles or acres).

Surface model: a computer model in which an object is defined only by its surface.

Surface texture: the way a material's surface looks or feels: smooth, rough, shiny, dull, etc.

Suspension: a liquid where particles of a solid material are mixed with but not dissolved in the liquid. Slip, which is used to cast ceramic products, is a suspension.

Synergism: a state in which the whole effect is greater than the sum of the individual effects.

Synthetic polymers: polymers designed and produced by humans from natural organic materials.

System goals: the meeting of human needs and wants as they relate to this specific system.

Systems drawings: engineering drawings that are used to show electrical, pneumatic (air), and hydraulic (fluid) systems. They show the location of parts in the system and connections.

T

Taconite: a low-grade iron ore. Taconite must be preprocessed at the mine.

Tape rule: a common measuring tool that has markings on a flexible tape of steel, cloth, or plastic.

Taper turning: machining done by the cutting action of a single point tool to produce a uniformly changing diameter along the length of a rotating workpiece.

Task setting: the process of setting job requirements.

Technical data sheet: information sheets that describe the characteristics of a product to a designer or engineer.

Technological developments: advancements in science and technology that give designers new product ideas. Companies must never stop gathering information about new developments.

Technological systems: physical systems consisting of five essential, interrelated elements: inputs, processes, outputs, feedback, and goals.

Technology: the study of how humans develop and use items to change or control the world around them.

Tempering: a thermal conditioning process that removes internal stresses in materials.

Temporary fasteners: devices designed to allow easy removal for disassembly of a product or component. Cotter pins and wing nuts are examples.

Tensile strength: the ability to resist forces that would pull the material apart.

Termination notice: a form used to tell employees that their services are no longer needed.

Territory: an area that is assigned to a salesperson. A territory may cover a city, several counties, an entire state, or several states.

Test market: a geographic area, considered to be "typical," that companies use to test the appeal products and the effectiveness of advertising programs.

Thermal conditioning: a process that changes the internal structure of a material through controlled heating and cooling.

Thermal conductivity: a material's ability to conduct heat.

Thermal emission: ability of a material to give, off, or *radiate,* heat.

Thermal expansion: the amount that a material expands when heat is applied and contracts when heat is removed.

Thermal processes: processes that use heat to change a material, such as steelmaking changes iron into steel.

Thermal properties: characteristics of a material that determine how it will react to temperature and to heat and cold at different levels.

Thermal resistance: a material's ability to resist melting.

Thermal shock resistance: the ability of a material to resist shock caused by sudden temperature changes.

Thermoforming: a forming process that holds and heats sheet material in a frame. The hot material is lowered over a mold. The air in the mold is drawn out. Atmospheric pressure forces the plastic into the cavity.

Thermoplastic adhesives: adhesives usually resins suspended in water (solvent). They form a bond when the solvent evaporates or is absorbed into the material.

Thermoplastics: synthetic polymers that can be heated and formed repeatedly.

Thermosets: synthetic polymers that resist heat once they are molded.

Thermosetting adhesives: powdered or liquid polymers that cure by either chemical action or the application of heat.

Thickness: the smallest dimension of a part.

Threaded fasteners: devices that use the friction between two threaded parts, or between threads and the material itself, to hold pieces together.

Time: the human and machine time needed to process, produce, and sell products and structures.

Time clock: a device that automatically prints an employee's starting and ending times on a card.

Tolerance: the amount a dimension can vary and still be acceptable.

Tool: a device that enhances hand work.

Tooling: devices such as jigs, fixtures, patterns, and templates that help workers make products better and faster. Tooling is designed for three purposes: to increase speed, accuracy, and safety.

Top down system: one in which a program is developed by managers and handed down to the workers to be carried out.

Top management: the persons who make decisions about the day-to-day operation of a company.

Torsional strength: the twisting force needed to cause the material to shear and separate.

Torsion strength: the ability to resist twisting forces.

Trade credit: the delay allowed by suppliers (usually 30 to 90 days) between the time material is purchased and the time it must be paid for.

Trademark: a name, a symbol, or combination of name and symbol used to identify a product.

Trade name: the legal name of a company; the name that identifies a company to its customers.

Training: the process of preparing workers for jobs.

Transformation processes: processes that change resources (inputs) into outputs, such as products.

Transformation technology: the appropriate use of tools, machines and systems to convert materials into products.

Transportation: converting energy into power to move people and goods from one location to another.

Transportation technology: the use of technological systems to move people and goods from one place to another.

U

Underground mining: a mining method that uses digging tunnels to reach the material. There are three major underground mining methods: shaft mining, drift mining, and slope mining.

Unfair labor practices: tactics that interfere with union-organizing activities.

Union: a legal organization that represents workers' interests.

Union contract: agreement between a company and workers represented by a union.

Union organizer: representative of the national union sent in to help start a local union.

Union security: the level of labor agreement a company has, affecting whether workers must belong to a union.

Union shop: all workers must become union members after a specified period of time.

Union steward: a union official who represents workers in a single department or plant area.

Units: single, complete items used for counting purposes. For example, plywood is sold by the sheet.

Unlimited liability: exists when the owner of a business is responsible for all debts of the company (if it is a proprietorship or a partnership). The owner must pay the debts with his or her own money.

Usufacturing: making products for personal use

V

Van der Waals forces: a secondary form of bonding important in determining how a plastic will behave: if the bond is strong, the plastic is thermoset; if weak, a thermoplastic.

Variable path systems: methods of material handling that are not restricted to following fixed routes from point to point.

Variable sequence robot: a robot that can be programmed to do one task, then easily reprogrammed to do a different one.

Varnish: a clear coating material made from a mixture of oil, resin, solvent, and drier.

Vendor: a supplier of materials to a company.

Veneer: a thin sheet of wood that is cut from a log. The log is unwound much like paper from a roll.

Vertical drilling: the most common type of drilling; the result is a hole that runs straight up and down.

Vestibule training: formal classroom and technical training used before a worker is assigned to jobs of a more complex nature.

Vice presidents: people who are in charge of a major part of the company such as sales, marketing, engineering, manufacturing, or personnel.

Video display terminal: a television-like screen upon which the computer displays information.

Volume: the three-dimensional space that a thing or substance occupies. Volume is described in cubic units.

Voluntary dissolution: this occurs when the owners of a business close the company because they want to do so.

W

Wage control: the process of measuring performance against job requirements.

Wage rates: a type of labor cost, the amount of money the worker is paid.

Warm forming: the forming of material at temperature between room temperature and the recrystallization point.

Waste: unusable surplus material resulting from a manufacturing process.

Weight: how heavy an object or substance is, compared to a standard. Weight is used to measure materials that are hard to break into units, such as sand or clay. It is easier to weigh an amount of clay than to sell a certain volume of clay.

Width: the second largest dimension of a part.

Wildcat strike: an illegal strike called during the life of a contract.

Wire fasteners: mechanical fasteners, formed from wire, that make use of friction between the fastener and the material to hold pieces together.

Wireframe model: a series of lines, arcs, and circles that together form the outline of an object.

Wood screws: fastening devices that depend on the friction of the threads against the material.

Workers: the people who do the actual production.

Work force: the employees of a company.

Work hardening: a structural change in metal caused by such processes as hammering or rolling.

Working capital: the money at a company's disposal from sales income, stockholder investments, and loans.

Work orders: forms used to direct machine operators.

Workstations: locations on an assembly line or other operations where workers perform tasks.

World-class product: products of high quality that work well, are easy to operate, and are relatively inexpensive.

Y

Yarding: the method used to move fallen trees to a central point for transport to a sawmill.

Yield point: beyond this point, additional stress will permanently deform a material.

Index

A

B

D

G

H

I

N

O

P